MR. CAPONE

Also by Robert J. Schoenberg

Geneen

The Art of Being a Boss

MR. CAPONE

ANTHONY BERARDI

Robert J. Schoenberg

Quill
William Morrow
New York

It is the policy of William Morrow and Company, Inc., and its imprints and affiliates, recognizing the importance of preserving what has been written, to print the books we publish on acid-free paper, and we exert our best efforts to that end.

Mr. Capone/by Robert J. Schoenberg

Library of Congress Cataloging-in-Publication Data has been applied for.
Library of Congress Catalog Card Number: 92-5228
ISBN 0-688-12838-6

Printed in the United States of America

2 3 4 5 6 7 8 9 10

BOOK DESIGN BY NICOLA MAZZELLA

To
Anita Kane and Dorace Schwartz
and
to the memory of
Sam Kane
(*1904–1981*)
and
Phil Schwartz
(*1910–1975*)

Preface

In the 1948 film *State of the Union*, when the character played by Spencer Tracy repented of having sold out his principles, he said, "I thought I could hijack the Republican nomination. . . . I became an Al Capone of politics." Capone had been dead for over a year, and before that had been either in prison or vegetating in Florida for over sixteen years. But director Frank Capra assumed his audience needed nothing further to know exactly what Tracy meant.

When Imelda Marcos returned to the Philippines in November 1991, her lawyer scoffed at the government's charges of failure to file income tax returns. "They can't get her on the more serious charges," he said, "so they're trying to get her the way the U.S. government got Al Capone." That prosecution had occurred almost exactly sixty years before. But the lawyer knew that even in the Philippines everyone would know what he meant without further explanation. So did a *New York Times* op-ed writer in June 1989, when he argued that Vice-President Quayle's asking El Salvador strongman Roberto D'Aubuisson not to "embarrass" the country's new president, Alfredo Cristiani, was "like asking Al Capone to pay his taxes." So did a letter-writer to the *Los Angeles Times* in October 1991, who, to express dismay at news that a convicted murderer had gotten hold of the jury's names and addresses, penned an impassioned "Shades of Al Capone."

So do an astounding number of other letter-writers, news writers and novelists, who still evoke the name of a man who did nothing much to attract attention outside Chicago before 1926 or after 1931. A Chicago criminal lawyer, Julius Lucius Echeles, has taken to clipping such instances, commenting, "Amazing how often his name is used to spice up a story.'' Al Capone is more than a person; he has become an allusion.

Al Capone was, beyond question, the world's best-known gangster, and one of the best-known Americans. His name is still recognized everywhere without any explanation needed about who he was or what he stands for—his value, of course, as an allusion. The city fathers of Cicero, Illinois, still occasionally ponder the wisdom of changing their town's name to ''Hawthorne,'' its citizens weary of the snide looks and cracks they receive wherever they go because Capone made Cicero his headquarters.

Yet apart from the lurid events that have figured in a number of movies—like the St. Valentine's Day massacre—Capone's story is actually so little known that moviemakers have shown no compunction about ludicrous inventions and liberties with the facts, confident that audiences will not know enough simply to laugh at them. (The powers of Cicero, for instance, are evidently unaware of Capone's intimate connection with the name ''Hawthorne.'')

With Capone, the truth is not only stranger, but a good deal more fascinating than the fiction that has surrounded him. His story needs no embellishment.

It does need serious treatment. This book is an attempt to demonstrate *why* he did what he did, how things looked to him, what led to his decisions. It's pointless to spend time deploring his criminality, inaccurate to make him out a grotesque. It's both more interesting and more instructive to see why even the lurid things happened. He was a businessman of crime and had lucid, rational and discoverable reasons for his actions. That applies with equal force to most of those with whom he dealt on both sides of the law. This is a story of human beings (with a few arguable exceptions) whose actions, even when vicious or corrupt, stemmed from discernible human motives.

''There are three ways to make a gangster or hoodlum interesting,'' wrote film critic John Simon. He can be so brutally and unrestrainedly criminal that he compels our attention; or he can be a master criminal of such cunning and competence he mocks all punishment until some simple, unexpected human flaw betrays him; or he can be, for all his criminality, so human he inspires thoughts of ''There, but for the grace of God . . .''

Again and again, Al Capone exemplified, in turn, all three.

Acknowledgments

First, profound thanks to the management and staff of the Chicago *Sun-Times*—and most especially to Terri M. Golembiewski, head librarian, and Judith Anne Halper, chief reference librarian, for their kindness and patience in sharing with me their remarkable expertise. The *Sun-Times* library contains clippings from all Chicago newspapers during the period this book covers; without access to them my task would have been immensely harder.

Thanks to those who took the time and trouble to talk with me. I want to express particular gratitude to three who gave me an extraordinary amount of their time and effort, with repeated interviews, correspondence and phone consultations.

William Balsamo, of Brooklyn, is a researcher, historian and author. His specialty is crime, particularly in the New York area. Although not himself involved with crime (before retirement, he worked as a longshoreman on the Brooklyn docks, as his father did), his great-uncle was Battista Balsamo, a Mafia don, and Bill grew up knowing some who are "connected." After the police and reporters leave, Brooklyn residents—because Bill is one of them—willingly tell him what *really* happened, like his exclusive interview with the man who gave Capone his scar. He unstintingly shared his knowledge with me and I relied heavily on it and on his counsel.

Mark LeVell, of Chicago, who fixes computers for the phone company, has made a lifelong study of Chicago's criminal past. He painstakingly delves into the most minute points, filling many notebooks with double-checked data. He is a treasure of information and detail, and he is the source for documents like the

St. Valentine's Day massacre police reports I rely on in Chapters 19 and 20. Mark has been most generous with his time and knowledge.

Michael Y. Graham, of Libertyville, Illinois, has researched a number of important TV and movie projects about Capone and the era. He is also an entrepreneur who will soon open a show that celebrates Chicago in the twenties. His information about the era is encyclopedic; his collection of photographs and artifacts is astounding. He very kindly gave me the benefit of them all.

Those of the following who are not identified here are identified in the text or notes; my thanks to them for talking with me. Santa Russo Baldwin, Detective Sergeant William Baldwin, Commander Nelson S. Barreto (Public and Internal Information, Chicago PD), Anthony C. Berardi, Dennis Bingham (Public and Internal Information, Chicago PD), Roy Bosson (writer-historian, Hot Springs, Arkansas), Howard Browne, Harry Busch, Esq. (an assistant state's attorney at the end of Capone's era), Joseph Davis, MD (chief medical examiner, Dade County), Reverend John E. Delendick (pastor, St. Michael the Archangel, Brooklyn), A. A. Dornfeld, Julius Lucius Echeles, Esq., Jerry Gladden (chief investigator, Chicago Crime Commission), Commander William Hanhardt (retired as chief of detectives, Chicago PD), Detective Tom Hoolahan (Miami Beach PD), John Ingraham, Norman Kassoff, Detective Sergeant Ron Koivu (Miami PD), Edwin P. McNichols, Senior Judge Abraham Lincoln Marovitz, George Meyer, Henry T. Morrison, Michael J. Mullins, Reverend James P. Murphy, Jesse George Murray, Jerome F. Nachtman, DDS, Arthur G. Ristig, Joseph A. Refke, Charles Trilling, Howard L. "Pat" Purdue, Hy Saxe (Miami resident around 1940), Superintendent Emil J. Schullo (Cicero PD), David J. Shipman, Esq. (Chicago law student during end of Capone era), Milt Sosin (Miami reporter in Capone's era), Walter S. Spirko, and Mack Staley.

Thanks also to the few who talked with me on the condition that they remain anonymous.

As the notes to this book show, seven people gave me the benefit of their special professional knowledge. My thanks to Lydia Bayne, MD (San Francisco General Hospital), Richard C. Froede, MD (medical examiner, U.S. Armed Forces), Terence F. MacCarthy, Esq. (executive director, Federal Defender Program, Chicago), Prentice H. Marshall (senior judge, U.S. District Court, Chicago), Thomas R. Mulroy, Jr., Esq. (Jenner & Block, Chicago), Robert Rolfs, MD (Centers for Disease Control, Atlanta), and Roger P. Simon, MD (San Francisco General Hospital). If the text errs in the technical areas on which they advised me, mistakes are the result of my imperfect understanding.

Special thanks to Michael Kertez (assignment editor, WGN-TV, Chicago), who gave me many leads, introductions and a number of highly useful documents. Also thanks for the leads they gave me to Jeffrey Kane, Esq., and Steve Sanders (news anchor, WGN-TV) in Chicago, Burton L. Wellenbach, MD, in Philadelphia, and Sgt. David Rivero (Miami PD).

Thanks to the following organizations and people for their help in locating information and documents.

Archdiocese of Chicago, Archives and Records Center, especially Patrick Cunningham (associate manager);
Board of Education of the City of New York, especially Etta Grodinski (director of PR) and Shirley Poch (school secretary, P.S. 133);
Brooklyn Historical Society, especially Clara Lammers (associate curator);
Brooklyn Public Library, especially Elizabeth White (curator, Brooklyn Collection) and staff of the microfilm room, which houses back copies of the defunct Brooklyn *Eagle*;
Cadillac Motor Car Division, GM, especially Vincent Muniga (manager, product publicity);
Chicago Crime Commission
Chicago Historical Society, especially Archie Motley (curator of archives);
Chicago Police Pension Board, especially Kay Hylton;
Chicago Public Library, especially staff of the newspaper microfilm room;
Circuit Court of Cook County, Criminal Division, especially Frank J. Baley (director), Carolyn Barry and Maria Blazquez;
Clerk of the Circuit and County Courts, Eleventh Judicial Circuit, Dade County, Florida, especially Gordon W. Winslow (archivist-historian);
Cox Newspapers, especially office of the Palm Beach *Post*, which maintains an index and clippings from the defunct Miami *Daily News*;
Historic Newspaper Archives, Rahway, New Jersey, especially Hy Gordon (owner of this service that provides customers with the front pages of major newspapers on the day of their birth, etc.; Mr. Gordon kindly let me go through his collection of bound volumes of the *Chicago Tribune*); thanks also to his kind and helpful staff: Lemuel Blackwell, Alex Brown, Willie F. Davis, Patricia Elliott, José A. Espinal, Tyrone Franklin, Alex G. Heim, Hazel Ince, Tara M. Ladagon, Rajnikant Patel, Charles Peterson, Edward J. Phipps, Brendan Rhodes, Michael A. Robinson, Sonia Sanchez and Andrea E. Zambrano;
Historical Association of Southern Florida, especially Rebecca A. Smith;
Historical Society of Pennsylvania, especially the manuscripts staff;
Kings County Court House, Records Section, especially Ralph Mancaruso and Frank Siclari;
Long Beach Public Library, especially staff of the microfilm room;
Los Angeles Public Library, especially staff of the Central Library sociology section and periodicals microfilm room;
Miami *Herald*, especially Nora Paul, also Joyce Tullo and Edward Sorin;
National Archives, Great Lakes Region, especially Susan H. Karren;
National Archives, Washington, D.C.;
New York Department of Health, Bureau of Vital Records, especially Caroline Durant (chief clerk);
New York PD Museum, especially Officer Dominick Palermo;
New York Public Library;
Philadelphia *Inquirer*, especially M. J. Crowley (head librarian);

Philadelphia Public Library, especially staff of the microfilm room, which houses the defunct Philadelphia *Public Ledger*;
Santa Monica Public Library, especially staff of the reference desk and periodicals desk;
Sentinel-Record, Hot Springs, Arkansas;
Temple University Library, especially George Brightbill (curator, urban archives, which houses the library clips of the defunct Philadelphia *Bulletin* and *Record*);
UCLA Library, especially staff of the University Research Library and its microfilm room;
Union Memorial Hospital, Baltimore, especially Carol Ristau;
U.S. Department of Justice:
Federal Bureau of Investigation, FOI reading room,
Federal Bureau of Prisons, especially John W. Roberts (archivist) and Anne Diestel.

The author of one of the books listed as a major source in the bibliography deserves special thanks. If I do not cite John Kobler's 1971 *Capone* more often in the notes, it's because as policy I rely on the earliest accounts available, contemporary ones for choice, and cite only those. But Mr. Kobler was the first to pull many disparate early sources together and the first to weave a chronologically coherent account of Al Capone's career. In a number of instances, I am able to cite original sources only because of his work in uncovering them. Even where I quarrel with his interpretation, it's not without awareness of my debt.

Thanks to three friends: Bud Gobler, proprietor of Book Buddy, Torrance, California, who was always able to locate hard-to-find books for me; to Tom Krebs, who was always ready to help me with computer problems; and to Jack Langguth for his invaluable comment, criticism and suggestions.

Thanks to those whose hospitality made it possible, both financially and emotionally, to spend so much time on the road. In Chicago, Anita Kane; in Philadelphia, Burt and Shirley Wellenbach; in New York, Nick and Janet Wedge (Ossining); in Florida, Dorace Schwartz (Miami) and Marty and Joan Conroy (Captiva); in La Quinta, Lois Ebeling Marcum; and for six months in Los Angeles, Ted and Marj Schoenberg.

To my immediate family, love and thanks as always for their love and support, literal and figurative: Shirley and Burt Wellenbach, Art and Jane Schoenberg, Ted and Marj Schoenberg.

Thanks to Tom Sheridan, *Sun-Times*, Sandra Spikes, *Tribune*, and Eileen Flanigan, Chicago Historical Society, for help with pictures.

Finally, thanks and admiration to my editor, Frank Mount, for a fine eye and a firm hand. Also Eivind Allan Boe for the most thoughtful and painstaking copy editing I've ever seen. And, as always, to Don Congdon, without whom, not.

Contents

All I ever did was to sell beer and whiskey to our best people. All I ever did was to supply a demand that was pretty popular.

Why, the very guys that make my trade good are the ones that yell the loudest about me. Some of the leading judges use the stuff.

When I sell liquor, it's called bootlegging. When my patrons serve it on silver trays on Lake Shore Drive, it's called hospitality.

—Al Capone

CHAPTER 1

A Twig Grows in Brooklyn—and Is Bent

GABRIEL CAPONE PICKED a rotten time to bring his young family to America. At age twenty-eight, with a twenty-three-year-old wife, the former Theresa Raiola, their year-old son and another on the way, Gabriel left Castellammare di Stabia, his native village sixteen miles down the bay from Naples. They landed in New York just in time for the Panic of 1893, which would wrack the country's economy for years. Gabriel wisely chose Brooklyn as home in preference to the even greater squalor and density of Mulberry Bend, Manhattan's Lower East Side Italian colony.

Not that the depression spared Brooklyn. Unemployment would soon idle one quarter of the borough's workforce, making it no time for the unskilled. Yet most Italians who arrived in America then lacked skills that could land them decent urban jobs. The Industrial Revolution had largely bypassed their part of Italy; nearly 97 percent of them had been peasants.

Why did so many flock to the cities? Why didn't they look for farm jobs or continue west to homestead what remained of the frontier? First, they had emigrated to escape a rural life they could conceive of only as brutal and dehumanizing; they came to better themselves, not suffer more of the same. The second reason bore more directly on Gabriel Capone's experience in America.

However hard, his lot was easier than that of most Italian immigrants because he *did* possess an urban trade. He was a barber, which implied considerable skill at a time when many still visited barbers to be bled and have teeth yanked. Even so, Gabriel could not practice his trade right off the boat, because like his fellows, he had no money. The average Italian immigrant family in the nineties had just $17 when they landed, enough to sustain them at best for ten to twelve days.

That meant most could not have searched for farm work or traveled to it even had they wanted. They took what they could get—which they could seldom get on their own. Most spoke no English. They typically became virtual chattel, recruited by one of the *padroni*, entrepreneurial countrymen who would sell the newcomers' services in work gangs to perform the most backbreaking labor at the lowest pay. One Italian later recalled bitterly his daily ten hours with pick and shovel for only a dollar, with a Saturday-night kickback of one day's pay to the foreman, followed by the present of a chicken at each Monday's shape-up if he hoped to work that week. That arrangement was extreme. More usual was the hod carrier who pulled down a good $1.50 for his ten hours—fifteen cents for each hour lugging bricks up ladders.

For Gabriel, lack of capital meant he could not open his own shop; and with haircuts and shaves a nickel each, no one could support a growing family barbering for someone else. Those nickels represented the other side of depressed wages: prices had to match, which usually meant immigrants could afford only the dregs. Four dollars a month rented a two-room apartment with bare walls, no gas or electricity, water carted in from a pump in the yard, a communal privy out back. A *really* poor family might cook on a kerosene stove, which doubled as their only source of heat. The better-off fed chestnut-sized coals into a potbellied iron stove. No one considered heating both rooms, not with coal at thirty-five cents for a hundred-pound bag. "In winter," says someone who lived like that, "our place was just a little hotter than outside." His mother refrigerated food by storing it in the bedroom. Some weren't that well off. One investigator found five families—twenty people—sharing a single room, twelve feet by twelve feet, with two beds, no partitions, screens, tables or chairs.

The Capones lived better than most. Though Gabriel could not ply his trade at first, he avoided the drudgery and extreme low pay of manual labor because he boasted another skill that went with his trade: he could read and write. In Italy, as well as in America, the illiterate expected their barber to read them any letters that came their way. Gabriel's learning earned him a job in a grocery store until he could gather enough of a stake to open his own barber shop, a storefront in the tenement at 69 Park Avenue.

Children came. The Capones christened theirs with Italian names, though except for one, all of them grew up known by American equivalents. Vincenzo, born in Italy the year after the Capones' 1891 marriage, was called Jimmy in America. Raffalo, born not long after they landed in 1893, was Ralph. Salvatore, always known as Frank, was born in 1895.

Theresa had her fourth son on January 17, 1899, a mild Tuesday.

Exactly three weeks later, on February 7, the infant's godmother, Sophia Milo, bundled him off to the church where Theresa prayed with devout regularity, the cramped, in-a-basement St. Michael the Archangel, on Lawrence Street at the corner of Tillary in downtown Brooklyn. The Reverend Fr. Joseph Garofalo baptized him Alphonsus, Latin for Alfonso. Later, many would claim that his family name was *really* "Caponi," the implication being that he had Americanized it. In fact, his last name, at baptism, was spelled with an *e*. And since the Church required only one sponsor, he would grow with the guidance of *no* godfather.

When little Al came along, the Capones lived at 95 Navy Street, five blocks from St. Michael's, bordering the New York Naval Shipyard—better known as the Brooklyn Navy Yard. Predictably, the neighborhood was rough, Sands Street especially. Conveniently dead-ending on Navy Street at the Navy Yard gate, Sands offered all the diversions suggested by the phrase "drunken sailors." In uniforms customized at Max Cohn's, sailors could swagger into the saloons that lined Sands, then stagger out to the attendant pickup dance halls, brothels, hot-bed rooming houses, pawn shops and tattoo parlors. Sands was a street, as one of its historians put it, of "cheap liquor and even cheaper women." The occasional penitent might patronize the Naval YMCA. The few other savory establishments that graced Sands underlined what made this a strange neighborhood for a family named Capone. McIntire's candy store stood on Sands, as did McLean's livery stable, Seeney's harness shop and Martin Connally's saloon, for Sands marked the southern boundary of Irishtown. Italian families had colonized the Navy Yard district, but the Capones inhabited the borderland. At the turn of the century even tiny St. Michael's could comfortably accommodate the number of Italian communicants thereabouts.

Not long after little Al's birth, Gabriel took his brood away from the blatant vice of Sands Street to one of three apartments over his barber shop at 69 Park, removing the family even further from their countrymen. The Capone household included two boarders, a help with the rent: Michael Martino, also a barber, perhaps Gabriel's shop assistant; and a greenhorn musician, just off the boat, Andreo Callabrese. The other families at 69 Park were Irish, the McBrides and the Ratigans. Although Irish names predominated on the block and surrounding streets, other nationalities also flourished: Swedes and Germans, three Chinese at 79 Park, and, great rarity, a native-born family of at least the second generation, the Swifts at 55 Park. A few other Italians lived near, like the three-family enclave around the corner on North Portland Avenue; but little Al spent his first six or seven years surrounded by "foreigners," escaping the insularity felt by most immigrants of all nations who usually clustered in homogeneous ghettos. That exposure would strongly influence a vital factor in his development as the businessman of crime.

Like the Capones, most Italian immigrants to America between 1881 and 1911 came from the impoverished and unremittingly downtrodden south of Italy or from Sicily. That region had seldom known anything but raids, pillage and tyrannical misrule, usually by literal foreigners, never by anyone except those they considered outsiders. Greeks, Carthaginians, Romans, Arabs, Normans, the Spanish, the French—all in turn had ravaged Sicily and Italy's Boot. Smoldering resentment against foreigners and mistrust of all authority became tradition. For those crushed people, Italian history was neither an academic subject nor a recital of national glories. Most immigrants first heard the names Michelangelo and Leonardo in America. When they saw a statue of Columbus, it appeared to them another monument to another *padrone*. For them history remained a living, cautionary tale of outrage and injustice. "Once upon a time," an immigrant in Brooklyn used to intone as a bedtime story for his sleepy nephews, "there were in Sicily many *latifondie* [estates]. Then after that *figlio puttana* from France who call himself Napoleon invade Sicily, there was great trouble. . . ."

Just before that, in the late 1700s, Italy resembled a butcher's chart of meat cuts: the leg belonged to Spanish Bourbons, both northern shoulders to Austria, the rump of Tuscany to Spanish Hapsburgs, the midsection to the pope. Italians ruled themselves only in the northwest Piedmont. When brief unification came, it naturally started with foreign invasion. The *figlio puttana* twice swept down the peninsula, welding all but the Spanish Bourbons' reactionary Kingdom of Naples into an "Italian Republic"—with himself emperor, his family and generals viceroys. Even this interlude of mock nationhood ended after Waterloo; the Congress of Vienna returned Italy to political chaos. The country remained pastoral and poor, owned by everyone except the Italian people.

That finally changed after 1848, in the Europe-wide ferment of popular revolts. But for the benighted south nationhood meant little. Now they found themselves oppressed by the more prosperous north. Lashed by hunger, the peasants sporadically erupted in revolt. The government was centered in Rome, peopled largely by better educated, more successful northerners. Naturally, it had to squash insurrection; but that meant half of Italy's army, northerners mostly, garrisoned the Boot, another "foreign" army of occupation.

Given such a history, mistrust of outsiders became a southern Italian heritage—directed not just toward other nationalities, plus obvious oppressors, like Romans and Tuscans, but among themselves. Neapolitans learned to fear and despise Calabrians, who loathed Puglians, who scorned Basilicatese, and back again, a round-robin of detestation. All suspected Sicilians, who trusted *no one*.

For little Al, such attitudes, inevitably absorbed from his Neapolitan parents, were offset by spending his most formative years in the midst of the "others." That may account for his later lack (with, as we'll see, one startling exception) of *any* demonstrated national, regional or religious prejudice, a notable factor in his success.

On May 25, 1906, Gabriel forswore allegiance to the king of Italy, and in a fine, firm hand signed his naturalization certificate, using the Italian spelling of his first name, "Gabriele." Native-born little Al already possessed citizenship.

Theresa bore two more sons, Amadeo Ermino, in 1901, later called John and nicknamed Mimi (presumably from Ermino), and a few years later, Umberto, whom everyone called just Albert from the start, though more formally Albert John. The year 1907 found them a family of eight, however, not nine, because Jimmy had run off two years before, lost to the family ken until years later, when he would suddenly reappear, a lawman from Nebraska.

The homogenized nature of Park Avenue made no professional difference to Gabriel; almost half the barbers in greater New York were Italians, a people thought to possess some special gift for snipping hair, patronized by all. But he and Theresa—more likely she, whose English never progressed far—hankered to live among their own. In 1907, when they could afford to, they moved about a mile and a half south to a much better neighborhood, called South Brooklyn, just over the line from Park Slope, next door to the tough Irish section of Red Hook, yet the very heart of Brooklyn's Little Italy. They moved first to 21 Garfield Place, later into the upstairs of a frame duplex at 38 Garfield Place.

Little Al had started school at John Jay, P.S. 7, at 141 York, near the Navy Yard. After the move, he transferred to William A. Butler, P.S. 133, at 355 Butler Street, seven blocks away from Garfield. Until he reached sixth grade, he maintained a B average, testament to his natural brightness, since he devoted much time and energy to his favorite extracurricular activity, playing hooky. In the sixth grade it caught up with him. One term he attended class only thirty-three days of the required ninety, falling so far behind in arithmetic and grammar he had to repeat the grade. He never did. A red-haze temper that would occasionally overmaster him all his life exploded one day, and he hit a teacher who was lecturing him on some schoolroom misdemeanor. Sent to the principal, he got a whipping, and quit school in chagrin. He was fourteen at the time, and about ready to quit anyway. That was practically a family tradition for Capone sons: older brother Frank had lasted barely through half of the sixth grade at Butler. Only the last son, Matthew Nicholas, always called just Matthew, born 1908, made it to high school.

In a nearby pool hall Al played with his father, becoming renowned for his ability with a cue. He may have pitched sandlot baseball well enough to cherish dreams of turning pro, except for the same problem he had in school, lack of control. Big for his age, he combined athletic agility with heft and size—as shown by his later prowess as a barroom bouncer *and* ballroom dancer. He would grow to five feet ten and a half inches in a day when Jim Jeffries could punch his way to the world heavyweight boxing title weighing 175.

Between racks and pitches Al made a stab at an assortment of honest jobs: clerk in a candy store, pin boy in a bowling alley. For a while he earned $23 a week working in an ammunition factory; he also worked as cutter in a book bindery, following his older brother Ralph, who had worked in the print shop of a newspaper. Apprenticeship for Al Capone's life's work, though, came on the streets.

Immigrant Brooklyn existed in their moldering tenements; they *lived* on the streets, the Italians more than any others. No one else's streets in Brooklyn, wrote one social historian, "were so given over to unrestrained life." The immigrants re-created faithfully as possible the outdoor piazza living they had known back home. "Everything possible was done out-of-doors and in the companionship of neighbors and friends."

The streets Capone traveled as a boy were ruled by gangs, or more precisely *kid* gangs, because the word by itself can easily mislead. Members of these kid gangs could not be called gangsters; by today's standards they would barely qualify as delinquents. Excepting petty pilferage and occasional lunch-money extortion, few engaged in activity anyone would consider downright criminal. One gang's notion of fun was to storm down a street, bowling over pushcarts and milk cans, scattering bread bins while helping themselves to trifles of merchandise, breaking windows and street bulbs, pulling old men's beards. Another group specialized in bonfires heaped with palings ripped from fences and the wood of unattended wagons, the pep rally usually dispersed by cops and firemen. One of the fiercer Jewish gangs in the Williamsburg section regularly smashed the windows of a Christian mission set in their midst; milder rivals contented themselves with periodic snakedances through the mission to disrupt meetings, chanting, "We'll all stand up for Jesus, we'll all stand up for Jesus" until they exited on the punch line, "For the love of Christ, siddown!"

Gangs with longer traditions might possess a "clubhouse," a room in an abandoned building or a storefront, sparsely furnished, where members could play cards, shoot craps, smoke American Beauty cigarettes—sold four for a penny—or just hang out. None of these pleasures touched the real reason for membership. Slum kids had to belong to a gang for protection and survival—certainly psychic, perhaps physical. Their vocation was fighting. One Irish kid later recalled finding himself in a fistfight at least two or three times a week from age twelve. No one picked on him in the classic sense of bullies tormenting a scrawny kid. On the contrary, he was strong for his age, like Capone. His *existence* presented a challenge.

One-on-one fistfights were the least of it. They merely settled intramural discipline, disagreements and pecking orders. Real clashes came between gangs for violations of turf or to gain prestige. All saw themselves as beset, surrounded by enemies, and all were right. A Jewish gang member in Brooklyn later recalled that "the Irish could be said to fight almost for the fun of it; while the Jews always fought in self-defense." That strong Irish kid remembered a different Brooklyn. Anytime he and his pals crossed the line into a Jewish neighborhood, even in innocent passage on a trolley, trouble threatened. "The Jewish boys did not like anyone in their neighborhood who did not look like themselves," he wrote years later, "and would drive us away."

Italian kids faced that extra hazard of their parents' traditional enmities. Had Capone, the Neapolitan, ventured to eastern Flushing Avenue, not far from his old home, he might have become "Scarface" several years earlier; Sicilians reigned there. Like the other kids, he had to join a kid gang for safety. What distinguished him from most was that he joined as though answering a call.

Al joined the South Brooklyn Rippers, a junior gang that inducted kids as young as eleven. Hanging out part time on Manhattan's Lower East Side, where brother Ralph now operated and would soon come fleetingly to police attention with a crooked auto sale, Al also joined the Forty Thieves Juniors, a more significant alliance. This was an elite offshoot of the junior branch of the notorious Five Points, an adult gang. Membership in the Manhattan gang showcased his tough scrappiness under the noses of those who could best appreciate and employ it if, like some, he chose to join a serious adult gang at age fifteen or sixteen.

No reliably authentic details about Al's activities in his late boyhood and early teens have survived,* but he evidently did not stand out or apart. Not many years later, a former Brooklyn kid-gang member remembered him "as something of a nonentity, affable, soft of speech and even mediocre. . . ." So, "when this Brooklyn boy made good in the world, the surprise was general among his old friends and acquaintances." Of course he chose the path that led from kid gangs to the real thing. As the same gang member put it, "The greater number . . . passed on to become good citizens." They would drop out of gangs to get jobs or go on in school. Plainly each kid's choice had to be at least partly a conscious decision. But circumstance could force the issue, as could association and opportunity, as with Al's membership in the Forty Thieves Juniors. Two influences—the first overarching and long-run, the other immediate—tipped the scale for Al.

From the start, John Torrio was the thinking man's criminal. Born 1882 in Orsara di Puglia, almost sixty miles east of Naples, Torrio came to New York in his mother's arms at age two, his father dead in an accident before migration. Maria Torrio and John lived two years with her brother while she worked as a seamstress.

Then she married Salvatore Caputo, and Torrio got his start. Exactly what kind of start depended on which story Torrio told to which audience for which purpose. Once, when a judge was about to pass sentence, Torrio claimed that stepfather Caputo ran a blind pig behind windows that made 86 James Street look like any other Mulberry Bend grocery store—selling moonshine and beer, unlicensed and untaxed, therefore cheaper. Little Johnny toiled there as a porter, starting at age seven, able to snatch only thirteen months of formal education. Plainly he had missed all the advantages that might have kept him straight. On a different occasion, to convince U.S. Immigration of his worthiness to remain a citizen, he limned himself as a Horatio Alger lad pluckily pursuing high school education at night after exhausting days as delivery boy for his stepfather, this time a real grocer. Torrio always knew what people wanted to hear.

In 1901, age nineteen, he promoted boxing. New York then permitted only amateur bouts, a fatuous attempt to curtail fixing. Canny impresarios like Torrio,

*Close students of Capone literature will wonder why I don't cite the material in *My Years With Capone*, which purports to be an interview with Jack Woodford, a mild pornographer from the 1930s and '40s. The book includes a wealth of supposed confidences by Capone about exactly this period. It is a double hoax. The interview was not with Woodford but with a Chicago lawyer, Luis Kutner; and evidence suggests that Kutner never knew Capone. Certainly most of the statements of fact about himself that I could check proved false. I also ignore a couple of other published anecdotes about this period in another book—otherwise fine and honest—which don't make sense, even internally. Interested readers will find a complete discussion in the Notes.

who campaigned under the name J. T. McCarthy, could rig odds and dictate profitable dives with amateurs as readily as with pros. Prospering, Torrio bought a bar on James Street at the corner of Water. He soon expanded, leasing a rooming house down the block which he filled with whores, a natural business fit offering the kind of service his clientele wanted, as did the nearby store he converted into a pool hall. From such a customer base he easily recruited likely looking layabouts for the capstone of his small conglomerate: the James Street Boys, his own gang, whose jobs he planned and whose loot he fenced. He was always scrupulous about the split. John Torrio never looked to cheat a soul. He proclaimed what would become his lifelong motto, "There's plenty for everyone."

Few gang leaders shared that principle, but then Torrio differed in many ways. At the minimum, most matched their henchmen for brawn. Not that many in the gangs towered; police measured the average arrestee at only five feet three inches, weighing under 135. But hard labor from boyhood had steeled most of them amazingly. Torrio, however, dominated his swollen-muscled thugs with a large brain abetted by a colossal nerve and will, all seated in a puny, flaccid body, with chipmunk cheeks, a little potbelly, and dainty soft hands and feet. He knew how to turn his very puniness into an asset of control, the shrimp being somehow all the more terrible when he can set a hulking subordinate cringing because the hulk knows other hulks stand ready to enforce the shrimp's word. Equally important, Torrio knew when to make alliances, and with whom, and when to cut ties that had started to bind.

Monk Eastman typified the era's gang leader, shading over the average at five feet five inches and 150 pounds, a walking slab of muscle who prided himself on never blackening a woman's eye ("putting a shanty on her glimmer") without first removing his brass knuckles; and never, no matter how sorely provoked, using his club on her.

Eastman's gang unceasingly fought territorial wars with Paul Kelly's Five Points gang, historic successor to the Whyos (named probably for their battle cry). Gangs had ranged over lower Manhattan for nearly a hundred years: first the antebellum Forty Thieves (ancestor only in name of Al Capone's kid gang), Shirt Tails (they wore them as a uniform outside their trousers), Plug Uglies (protected their heads with stiff plug hats), Dead Rabbits (*rabbit* was slang for "rowdy," *dead* for "very," as in "dead game"), Chichesters, Roach Guards, Black Eagles. At the turn of our century two survivors terrorized the area. The Five Points gang took its name from the heart of its territory, the confluence of five streets (now merged into three: Baxter, Worth and Park Row) notorious then as a Coney Island of vice for the poor, now site of several courthouses. Of ninety-nine places of amusement on the Bowery in 1898, the police counted only fourteen "respectable." Without dispute, the Five Pointers preyed on everything between Broadway and the Bowery, Fourteenth Street and City Hall Park. Eastman's unchallenged territory lay from Bowery to the East River, Fourteenth Street to Monroe. The clash came over who owned whoring, gambling, booze and extortion rights in the strip between Pell Street and Bowery. Two years of mayhem and murder had settled nothing.

John Torrio allied his James Streeters with the Five Points gang—which, about 1,500 strong, figured to win eventually. More important, their leader was to Torrio what Torrio would become to Al Capone: model and mentor, almost idol.

Paul Kelly's real name was Paolo Antonini Vaccarelli. He pegged in somewhere between Torrio and Eastman as the image of a turn-of-the-century New York gang leader. He stood below average in height, but swung the muscular bulk of a onetime boxer. Yet he had educated himself, read extensively, liked good music, dressed with conservative good taste, spoke softly and well, even a smattering of French, Italian and Spanish—an accomplishment that then marked (except in immigrants, of course) the cultured gentleman. Kelly admired Torrio's mind, its acuity reflected in the slickness of the James Street Boys' coups; but he found that mind unformed, untutored. He adopted Torrio as protégé, offered him hints on how to develop himself. Torrio sponged it up. Soon he too dressed in chaste dark suits and stolid derbies instead of checkered trousers and yegg's caps. Soon Torrio too hummed opera motifs.

Warfare continued until an overly public shoot-out left three dead and twenty wounded—including three bystanders who hadn't dived for cover quickly enough. Tammany, fearing political backlash, dictated a truce. It held, but only uneasily, broken by frequent border skirmishes. By the early 1900s, Torrio recognized what had become a deteriorating situation. Lower Manhattan no longer recommended itself to one who simply wanted to conduct the peaceful *business* of crime. Torrio sold his James Street holdings, gave the Boys his blessing, bade Kelly sad good-bye, and relocated to Brooklyn, where he could expect respectable Italians to accept what he handed them with quiet dignity and in peace, whether as customers or victims.

That was part of the Italian immigrants' tradition and of their American experience.

Until about 1875, the United States viewed Italians with sympathetic admiration, even affection. When Giuseppe Mazzini and his warrior disciple, Giuseppe Garibaldi, led Italy in its long struggle for independence and unification, they seized the imagination of Americans. The two Italians seemed obvious analogues of our own founding fathers. In the city of Washington a crowd celebrated the Italian people's "glorious struggle . . . to free themselves from the despotism of a foreign power." When the nominally republican French armies rushed in to prop up the old regimes, a Fourth of July resolution in Philadelphia's Independence Square voted France's Louis-Philippe "the Iscariot of liberty, the Benedict Arnold of the Old World." Even the Know-Nothings, America's first organized movement in reaction to immigration, exempted Italians from their bigotry. The repression Italians had suffered and their freedom-loving bravery somehow absolved them of being foreign and Roman Catholic in Know-Nothing minds.

It helped immensely that, at first, most Italians stayed in Naples and Palermo and Rome to suffer their martyrdom. The 1850 census counted only 3,045 Italians throughout the United States. Not until 1870 did Italian immigration to the United States exceed two thousand a year. Furthermore, those who came posed no threat, even to Know-Nothings. They were mostly artisans and professionals from north and central Italy. As one writer put it in 1881, with a retrospective sigh, "Higher walks of American life . . . long included many talented and charming Italians"—the likes of Lorenzo Da Ponte, once Mozart's lyricist, who arrived in 1805, and of Garibaldi himself, who fled to U.S. exile in 1850.

Increasingly horrid conditions impelled southern Italians to emigrate. Corsair raids that lasted until almost 1800 had forced the people inland to defensible hill towns, forsaking the fertile coast and lowlands, which devolved into marshes and malarial swamps, while the refugees scratched flinty hillsides. Centuries of overcultivation and deforestation had blasted even that subsistence-farming land. Now political upheavals further deranged the economy.

At first, the unskilled, job-seeking competition chose the warmer, more familiarly Latin South America. Many worked there in the local summer, returning to Italy for spring planting. Long after that pattern changed, southern Italians called both American continents *laggiu*, "down there." By the 1870s, South America could not absorb all those who had to escape starvation in the South of Italy, where no one tasted meat save perhaps a scrap at Christmas and Easter; where families could afford a glass of wine only for the men, who dragged themselves back from the fields each day, desperate for some restorative after the day's sweat. *Spaghetti* became a treat. Well into our century, wrote an Italian newsman, "millions of Italians still lived a life of prehistoric squalor."

America's post–Civil War expansion demanded new pools of unskilled labor. Though only 12,354 Italians landed in 1880, that was more than the total number even of Italian *descent* in the United States eleven years before. Soon came a deluge: 100,135 in 1900, over 200,000 in 1903, a peak of 285,731 in 1907.

These were not Mazzinis, Garibaldis, Da Pontes; they weren't even John Martinis, Custer's trumpeter—providentially sent away with a message for reinforcements before the Little Bighorn. "They were," observed one commentator, "the most disadvantaged and humble white people that other Americans had ever seen." At the turn of the century 22.9 percent of all immigrants to the United States could neither read nor write. Northern Italians cut that rate in half, 11.4 percent. Among southern Italians illiteracy soared to 57.3 percent. They also tended to look the most different of all immigrants—shorter, darker.

The differences, the disadvantages, and especially the competition for jobs in such numbers occasioned extremes of prejudice.

In the 1850s New Orleans had voted a resolution praising Italians for their "struggle of right and justice against brute force and tyranny in their most odious forms . . ." On the 14th of March, 1891, after Italians had moved among them, a New Orleans mob broke into the jail, shot nine Italian fishermen in their cells, and hanged two. The eleven had been arrested on the most tenuous evidence that some *might* have murdered the city's police superintendent, though six of them had already been acquitted, three had faced hung juries and two had not yet been tried. Other vigilance committees in such communities as Tampa, Denver and Johnson City, Illinois, lynched Italians in 1893, '95, '96, '99, 1901, '06, '10, '14 and '15.

Yet despite the prejudice and stunted opportunities, the Italians boasted one of the lowest rates of pauperism of any immigrant group; only a tiny fraction ever applied for relief. Those who could not dig, picked rags. Some paid the Irish who operated garbage scows to let them "trim" the loads, salvaging rags, bones and other such salable treasures. "That's what most people have missed," says a former Chicago chief of detectives, himself Italian. "When you look at Capone or any Italian gangsters back then, the key is the aggressiveness of all the Italians. They would do anything to get ahead." An Italian newspaperman, contemplating all they had to put up with, wrote that "the Italian immigrant who did not become a criminal, or go mad, was a saint." Actually, the overwhelming majority were none of the above. In 1924 Dr. Antonio Stella, protesting America's newly restrictive immigration policy, pointed out that the "average criminality of the foreign born Italian . . . is less than that of most of the other races and only slightly more than that of the white Native American population." The arrest rate averaged 158.1 out of 100,000 Italians, while the figure for German-born was 218.9, for English 488.3, and Irish 1,540.1; even the Swiss outdid the Italians in crime, at 167.4.

Still, in southern Italian tradition crime remained a reasonably honorable alternative way of getting ahead. An 1863 Italian parliamentary report on conditions in the south had admitted that "the existence of a bandit has many attractions there for the poor laborer . . ." who anyway harbored "an absolute lack of confidence in the law and the exercise of justice. . . ." In America, an understanding Yankee writer pointed out that "insufferable tyranny" had "made every outlaw dear to the hearts of the oppressed people. . . . Even if he robbed them, they felt that he was the lesser of two evils, and sheltered him from the authorities."

The hostile reception they met in America did nothing to decrease the bunker outlook of southern Italians or incline them to side with authority against even the depredations of those who were at least their *own*.

John Torrio quickly reestablished himself in Brooklyn. He may have patronized Gabriel's shop occasionally. In any case, young Al could hardly have avoided absorbing the lesson of someone who had attained money and power without the drudgery that weighed on most others. The John Torrio Association, probably his headquarters, occupied the second floor over a restaurant where

Fourth Avenue, the main north-south arterial through South Brooklyn, met Union, the neighborhood's principal market street. The corner sat on Al's way to P.S. 133, gilt letters brave on the association's windows, constant reminder that some crime pays handsomely.

In 1909, still before ten-year-old Al could fall directly into the orbit of twenty-seven-year-old Torrio, the chubby little man received a call to Chicago. As we'll see, that move would occasion the biggest single influence in Al's life. More immediately, Al had caught the eye of someone who could exert the most portentous influence possible at that stage. It was Frankie Yale, six years older, who had ushered Al into the Forty Thieves Juniors. When Al entered his mid-teens, Yale welcomed him into the adult gang—and gave him a job.

The family baptized their children and buried their dead under the name of Ioele, pronounced *yo-ay-lee*. Frank—Francesco, universally called Frankie—was born in 1893 in Calabria, near the toe of Italy's boot, and brought to New York at age nine. A member of Five Points Juniors kid gang, as soon as he was old enough he joined the Five Pointers. Their days of glory had passed, but the Five Points remained a serious gang in the modern, adult sense. Joining was a career decision, on the order of entering some corporate training program.

At about the time Torrio left for Chicago, Frank Ioele Americanized (and disguised) his name to Uale. Though of only medium height, with a tendency toward plumpness, he was chunky and tough, with a square face, broad, fleshy nose and lips, and lobeless ears that hugged his head. Like Torrio, he had small, delicate hands, unusual for his body size, the more so because he brawled happily and with authority. At age seventeen, he and a friend, who wrestled under the name of Booby Nelson, cleaned out Kister's Coney Island pool hall, hurling billiard balls and swinging cues.

Kister's location was meaningful. Also like Torrio, Uale spied more opportunity in Brooklyn than in lower Manhattan. Two years later police picked him up on a gun-toting charge, which—like the time he stole some $300 worth of sheep and goat skins—came to little. Extortion paid better. After some unimaginative juvenile racketeering, Uale organized the ice trade. He sold the icemen protection and monopoly territories, which he rigorously enforced.

With the money he made from this scheme he opened a bar in Coney Island. At about the same time, 1917, at age twenty-four, he married Maria Delapia, with whom he would have two daughters, Rosa and Isabella. He and Maria lived in Brooklyn in a house owned by her parents. At the same time, he started using the name Yale in place of Uale despite Maria's disapproval. The most successful Coney Island nightspot then was the College Inn. Yale, who tried parting his thick black hair in the middle for a while, in emulation of the prevalent collegiate style, hoped to hitchhike on that success, calling his place the Harvard Inn. That made the new spelling of his name a regular thighslapper, good for business. It also further disguised him, further shielding his family.

A dive, though a large one, the Harvard Inn occupied a one-story building on Bowery, a short but seedy alley running between Surf and what would soon become the Boardwalk. An orchestra in the back played for drinkers and diners who could grope each other as they shuffled about the forty-by-twenty-foot dance floor. Drinks flowed from a twenty-foot-long bar.

Frankie Yale put Al Capone behind that bar.

CHAPTER 2

Early Maturity

BY THE TIME Al Capone turned sixteen—maturity in that hothouse of dangers, pressures and responsibilities—he battled both as a member of the Five Points gang in Manhattan and for Frankie Yale in Brooklyn. The latter connection was the more important because while the Five Points gang's importance and scope continued to dwindle, Yale's blossomed. "Capone learned all there was to know about extortion and slugging and the rest on the banks of the Gowanus Canal," says William Balsamo, Brooklyn native and historian of New York crime. Yale was fashioning a complex of enterprises beyond the Harvard Inn: a mortuary; racehorses; prizefighters; another nightspot, the Sunrise Cafe, around the corner from his home; a line of cigars—all based finally on his main line, strong-arm terror.

Though his troops might pull the occasional free-lance robbery or stickup, Yale was not a thief in the street sense, and Capone would later take pride in not being one himself. Instead, Yale specialized in varieties of extortion; shylocking, lending workmen money at 20 percent each week; exacting tribute from the bookmakers, pimps and gamblers in his territory; offering shopkeepers "protection" for a regular fee; organizing vertical associations—as with the ice sellers—from whose members he collected "dues," in return fixing prices that guaranteed artificially high markups, with any who dared undercut prices instantly punished.

Muscle in one form or another powered most of Yale's dealings. True, he offered those breaking the vice laws real protection from arrest and prosecution; and what he offered honest merchants included real protection from unlicensed predators—*other* extortionists—and from the more unreasonable exactions of such petty authorities as beat cops and fire and health inspectors. Nevertheless, the main protection he sold was from what he himself would otherwise order done. Bookies, pimps and shopkeepers alike kicked in tribute, or else; borrowers paid his juice, and on time, or else. So Yale needed a stable of strongarms who could not only break arms and heads but would kill. As with most gangs at the time, the availability of such muscle made Yale, its proprietor, courted by politicians who needed election help, and who, in turn, provided the political clout inherent in the protection Yale sold.

This was classic racketeering. An FBI study identified the practice as age-old, but traced the word to the late 1800s when gangs like the Five Points gave dances called "rackets." The gangs unloaded blocks of tickets to merchants who had no intention of attending but feared the consequences of refusal.

Even Yale's seemingly legitimate enterprises ranked as rackets, since muscle animated them also. The cigars, for instance, whose boxes bore his likeness, commanded more shelf space than their quality merited because merchants dared not say no. Once, when he needed someone to run his new Sunrise Cafe, Yale simply appropriated the popular manager of another club, never consulting contrary wishes of the man or his former boss. At the same time Yale recognized the need to treat the buying public as much as possible like customers, not victims. His sales pitch could not widely be "Use my undertaking service or get buried yourself" or "Dance at the Harvard Inn or I'll have your legs broken." As we'll see, the need for greater subtlety in pressuring the public gave Capone his chance to gain Yale's favor and to reveal the first glimmer of the person Capone would become.

Capone joined Yale at the pregnant moment, just as Yale started his play for the big time. Yale had his eye on possibilities throughout Brooklyn, especially on its five-mile stretch of waterfront with its sixty-odd piers, dense with pluckable merchandise and men.

Frustratingly, the Irish were already there plucking away. Dennis Meehan led the White Hand gang. They called themselves that to spite Yale's outfit, which newspapers called a "Black Hand gang" because that's what they automatically called all Italian criminals. Actually, by then Black Hand extortion—which we'll examine in the next chapter—figured only modestly in Yale's operation. What's more, few gangs had formal names outside newspaper columns. A gangster who had to identify himself to others would say, "I'm with Frankie Yale," "I'm with Torrio"; eventually, "I'm with Capone."

Those with Frankie Yale found those with Dinny Meehan hard to displace. Irish immigration had antedated Italian by decades; the Irish had long since organized the docks, cowing the shipping companies and the longshoremen. The Italians could supplant Irish authority only pier by pier, shylocking a few stevedores here, muscling a shipper there, each time having to convince the prey that they, no longer the Irish, were the ones to fear, then defending gains against Irish counterattack. While the fighting never escalated to pitched battles, gangsters did die, and need for muscle never abated. As a shylock and extortion collector, and as a territorial brawler, Capone shone—a tough, supple, hefty hooligan.

An added dimension impressed Yale. His people did not spend full time shaking down dockers or potting at or grappling with White Handers; they helped man Yale's enterprises. Although muscle is unneeded for rolling cigars and inappropriate at funerals, it's treasured in a raucous night spot.

Coney Island, on Brooklyn's south shore, had started as an Atlantic Ocean bathing and amusement resort of some pretension when the principal transportation to it consisted of one's own tallyho. The railways, the trolleys and finally the subway ended that. "Never again," mourned one Tory historian, "would the world of fashion come to Coney except in small groups and those, indeed, slumming." Elegant restaurants had served clams and champagne to swells. Clams gave way to Charles Feltman and his discovery that a warmed Frankfurter sausage in a roll appealed to the masses (the name *hot dog* followed much later). Champagne? By 1900, on any warm Sunday, New York City police made half their drunk arrests at Coney Island, and champagne was not to blame. Prostitutes—"blisters" to the Coney locals—teemed. Some were streetwalkers, others inmates at famous houses like The Gut, Madame Korn's, Lillian Granger's Albatross and Mother Weyman's, where Princess Zaza titillated clients with a fish.

Even in such company the Harvard Inn ranked as a dive. Exactly because of that, Capone's job as bartender and bouncer demanded a certain finesse. In a dive, bounceable behavior tonight seldom makes the customer unwelcome tomorrow. The trick was to bounce without alienating, and only after considered efforts to calm, placate and subdue had failed. Ideally, the bounced would recognize themselves as out of line. Capone combined the mass to bounce authoritatively and the intelligence to do it with tact. Yale liked what he saw. Capone became his pupil, favored by invitations after a hard night at the Harvard Inn to sleep over at Yale's house. That happened often enough so that much later Yale's daughters would show visitors "Al's room."

Capone liked what he saw, too. Yale had six years on him, with the swagger of a young man, already boss yet still a comer. Inevitably Capone would take Yale for a model as well as teacher.

Yale enforced discipline with unending obscenities, unrelenting and indiscriminate brutality. When displeased with his brother Angelo, ten years his junior, Yale beat his brother into a hospital case. Concurrently—and this trait Capone would not have to unlearn—Yale played the gracious don, a river of small benefactions to all who honored and respected him. Thieves robbed a poor delicatessen dealer; he found the cash on his counter next morning, replaced by Yale. A fish peddler lost his pushcart; Yale pressed $200 on him saying, "Get a horse, you're too old to walk." When two free-lance hoods tried to shake down Frank Crespi, popular and colorful hat check operator at a neighboring restaurant, Yale personally beat the interlopers senseless.

* * *

Dispensing largess lay in the future for Capone; Yale's example of how brutality and coarseness built business made a more immediate impression. At eighteen, for all that his brain might grasp the need for on-the-job finesse, some buried beast in Capone threatened to burst out any moment. It happened one night with literally scarring consequences when his eighteen-year-old glands spoke louder than his brain.

They called Frank Galluccio ''Galluch.'' Briefly a merchant seaman, a sometime barber's assistant, he discovered his true calling as a nothing hood, eventually a spearcarrier in thc Genovese crime family. One summer night in 1917 he strutted into the Harvard Inn, his date, Maria Tanzio, on his arm, his kid sister, Lena, tagging along. The sight of Lena set Capone's glands talking. He persisted in trying to make up to her, every time his rounds took him past her table trying to chat with her. Lena snubbed him. Her brother, half drunk, did not know Capone, but assumed from his familiarity that Lena did. At last, his kid sister's growing anger penetrated Galluccio's alcoholic fog.

''You know that guy?'' he asked.

''I never saw him before. He's got a lotta nerve. He won't give up, Frank. He can't take a hint. But I don't like him; he is embarrassing me. Maybe you could ask him to please stop—in a nice way.''

Capone headed their way again, and Galluccio was ready to take him aside like a gent:—*Hey, mister, please do me favor, okay? She's my kid sister, you know . . .*

Sweet reason never got a chance. Beforc Galluccio could speak, Capone hove to, leaned over Lena, and ''whispered''—loud enough to startle the party at the next table, heads swiveled in amazement—''You got a nice ass, honey, and I mean it as a compliment. Believe me.''

Her brother sprang to his feet. The insult was bad enough; the fact that strangers had plainly heard made it insupportable. ''I won't take that shit from nobody,'' Galluccio shouted. ''Apologize to my sister now, you hear?''

At a moment, Capone's brain reasserted itself, perhaps kicked in by ''sister.'' Family meant everything, and its evocation put this customer unarguably in the right. With his most ingratiating, placating smile Capone turned to Galluccio, arms spread wide, palms up and open—*Hey, whatsa matter, pal, a little misunderstanding, a joke, no offense . . .*

''This is no fucking joke, mister,'' cried Galluccio.

Capone stopped smiling but kept coming. Some guys you can't reason with . . .

Galluccio stood five-foot-six, weighed under 150. Caponc looked like a mountain avalanching toward him. The Galluch knew he could be badly hurt unless he struck first and quickly; but his punch would never do the job. He clawed his knife from his pocket and lunged, as the streets had taught him, for his attacker's throat. Take into account Capone's athlete and brawler reflexes, plus Galluccio's drink-induced muzziness, and the little fellow aimed pretty well. One slash started two inches in front of mid-ear, curved four inches down the left cheek to just below the corner of Capone's mouth; the other two each measured two and a half inches, one on the left jaw, the other on the neck under the left ear.

Galluccio grabbed his sister and date and beat it out the door. Someone rushed Capone to the Coney Island Hospital, where doctors took about thirty stitches in his face.

Soon Galluccio heard that some cut-up bruiser who said he was with Frankie Yale had been asking around for him. Galluccio appealed to Joseph Masseria, overlord of all New York, for justice. "Joe the Boss" decreed a sit-down at the Harvard Inn, where their betters agreed that Capone had indeed been wrong and would not be allowed to seek vengeance, while Galluccio must apologize for his disproportionate reaction—which he readily did, contrite at sight of how he had disfigured Capone.

Capone also recognized the justice of the settlement and the dishonor of his scars. He later put out the story that they had come to him in service with the Lost Battalion of World War I. In fact he had never been called up in the draft. Until he left New York, whenever he encountered Galluccio he studiedly smiled his no hard feelings. He once told Galluccio he realized it had been wrong to insult his sister, especially so publicly. In his days of greatness, on visits to New York, Capone would hire Galluccio as a supernumerary bodyguard for $100 a week.

That incident in no way diminished Frankie Yale's appreciation of his star pupil. Nor did three rousts by the law. Indeed, Capone was probably on some mission for Yale when police nabbed him on a disorderly conduct charge in Olean; what else would a Brooklyn boy be doing among the apple-knockers of upstate New York? As for Capone's two early murders, they demonstrated a useful willingness to kill. The one about which details have survived displays Capone's hot head *and* cool reason. An acquaintance beat a neighborhood crap game for $1,500 or so. Downstairs, in the hallway, Capone poked a gun in the winner's belly and took the money.

—You oughta be ashamed doing this to me. I know you good!

What could Capone do? That clearly implied retaliation, either personal or, breaking the code, through the police. Capone shot the fool.

—The kid was wrong (he explained to Yale). *He never should have said that, it was the wrong thing to say. It was his own fault he got shot.*

Yale understood. Although questioned by police, Capone was in the clear. No one knew or had seen a thing. He was getting dangerous, just the way Yale liked him.

In 1918 Capone met an Irish girl at a dance and fell in love. She was pretty, slim and tall with a round piquant face and large eyes framed by a helmet of blond hair. Baptized "Mary," she would be known all her life as Mae, daughter of Michael Coughlin, a construction worker, and the former Bridget Gorman. Born April 4, 1897, Mae had almost two years on Capone, perhaps an embarrassment to them because each fudged a year of age on their marriage registration. Capone appears in the church records as ''Albert,'' maybe a mistake, or maybe a typical bit of criminal obfuscation.

Italian-Irish matches were not common in those days. But Italian men did willingly marry young, whereas Irish men notoriously wanted to wait until much later in life, a point in favor of Italian men with Irish women. For his part, Capone harbored no prejudices. Mae's parents may have resisted the match—as much in distaste for Capone's vocation and associates as his origins. But on December 4, 1918, Mae gave birth. Eighteen days later, Mae's sister, Kathleen, and James DeVico, an otherwise obscure friend of Capone's, became godparents to Albert Francis Capone. Since Capone would soon marry Mae willingly (it's hard to imagine how an honest construction worker could enforce a shotgun marriage on a killer), why didn't he do it before the birth of Sonny, as the boy would always be called? The answer is lost. But Capone did hurry the process, seeking dispensation to skip the publishing of banns. December 30, 1918, eight days after he baptized Sonny, the Reverend Fr. James J. Delaney married Mae and Capone in the Coughlins' family church, St. Mary Star of the Sea, with Mae's sister Anna as maid of honor, DeVico ''Albert's'' best man.

Before marriage could mellow him, Capone's temper erupted again and changed everything. One day, on his collection rounds, he stopped in a dockside saloon for a drink. The stranger who ambled in was with Dinny Meehan's White Hand gang. Neither Arthur Finnegan nor Capone had achieved prominence (Finnegan never would); they did not know each other. Finnegan's hobby, an extension of his vocation, was abusing Italians. This day in this saloon he picked the wrong one. Capone used nothing but those great ham-like fists, yet beat Finnegan so badly he thought he had left him for dead, an opinion shared by doctors for a considerable while. Finnegan languished weeks in the hospital.

Police presented no problem, since as usual no one had seen or heard a thing or recognized a soul. The White Handers were another matter, especially Meehan's chief lieutenant, William Lovett. He evoked fear from any sensible person. At 5 feet 6 or 7 inches, a frail-looking 150 pounds, brown-haired, with cupid-bow mouth, doe eyes, faun ears, he presented an elfin appearance. After the Meuse offensive in the Great War, the 77th Regiment had decorated Lovett with the Distinguished Service Cross for headlong bravery. *Friends* trod softly in his presence. He shot one of them for pulling a cat's tail; he could not abide the sight of small animals in distress. Gang bosses routinely traveled with a retinue of bodyguards; when Lovett succeeded to leadership after Meehan's murder in 1920, he still walked alone. He habitually packed a .45, and everyone knew he would as soon shoot you as look at you. This was a dangerous man.

True, Artie Finnegan occupied a deservedly low rung in the gang. But someone had put a White Hander in the hospital barely clinging to life, and Bill

Lovett would not stand it, all the more because that someone had almost surely been a greasy ginzo, one of Yale's. Question was, which one? Like Finnegan, no one in the saloon could put a name to the beefy youngster.

Soon Yale got word that Lovett had set himself the task of identifying and tracking down Finnegan's mauler. How long could it take with the scars to help identification? Everyone knew what would happen; a highly efficient nemesis now stalked Capone.

Had Capone remained single and childless, would he have waited for Lovett to come? Whether yes or no, Yale insisted on a different solution. A call to Johnny Torrio set everything. Of course Torrio could find a spot for the kid, always room for another reliable man. A year or two at the most, time enough for Lovett to cool down and stop looking . . .

—Al, you can come right back to Brooklyn. Always a place for you in Brooklyn.

Now, though, in late 1919, Yale sent Capone to Chicago.

CHAPTER 3

How a Town Toddles

CHICAGO AND AL CAPONE made a perfect marriage. Each received joyfully what the other had to offer. While each learned refinements and new techniques in their years together, neither could complain of being further corrupted by the other. It takes two to toddle, and those two were meant for each other—Capone and Chicago.

Chickagou or Checagou. That's what the Potawatomi Indians called both the river that emptied into the great lake and the skunk cabbage that choked the bogs the river drained. The word also meant "bad smell." In the late 1670s French trader Pierre Moreau built a crude shelter on the site and established two Chicago traditions. He sold the Indians firewater, an illegal act in New France, and he did it under protection of his friend, the governor. That made Moreau Chicago's first bootlegger and Count Frontenac the area's first corrupt government official.

In 1779 a mulatto from Hispaniola, Jean Baptist Point du Sable, built a cabin where the north and south branches of the river joined, the first permanent settler. In 1804 Chicago's first businessman and civic booster arrived. Silversmith and trader John Kinzie bought Sable's holding and began the energetic business hustle that soon attracted other settlers, a more beneficent Chicago tradition.

Although the community had grown enough by 1825 so that its citizens appointed a constable, Chicago recorded no arrests until 1833, the year it received a town charter; a vagrant and a thief occupied a makeshift jail. The following year Chicago tried its first murder case—and saw its first botched criminal prosecution. An Irishman who had unarguably killed his wife got off because of confusing instructions from the judge. On March 4, 1837, Chicago incorporated as a city. Three years later it first executed a criminal, a rapist-murderer who maintained his innocence but proved himself a stand-up guy, saying that even if he knew whodunit he would not squeal, start of another tradition.

The lack of crime depended on definition. During the 1830s gambling flourished, technically against the law, but tolerated except for brief jailings after preachers inveighed insistently enough. Soon more gambling halls flourished in Chicago than anywhere else north of New Orleans and west of Pittsburgh.

The same held true for prostitution. Banned since 1835, so many brothels ran so openly on Wells Street that in 1870 the city council changed the street's name to Fifth Avenue lest brothels besmirch the memory of Captain Billy Wells, a noted killer of Indians, martyred by them in an early uprising. (Chicago restored the name after the turn of the century.) No one seriously contemplated getting rid of the brothels instead of the street's name, a preoccupation with appearance over substance that would also become hallowed. Unsurprisingly, an itinerant temperance booster soon declared that he had "never seen a town which seemed so like the universal grog-shop as did Chicago." Equally unsurprisingly, a horde of parasites descended. Newspapers bemoaned the precipitous rise in crime.

That rise engendered another significant tradition. Until 1855 a "city watch"—steadily derided for its inefficiency—had policed Chicago. The city needed more. Chicago's anti-immigration Know-Nothings, who dominated the city council despite a 60 percent foreign-born population, created an eighty-man police force, every member native born and a Know-Nothing party regular. "The political complexion of Chicago's first police department," a longtime head of the Chicago Crime Commission later wrote, "was to become a permanent tradition—a tradition which has habitually rendered law enforcement in the city relatively ineffective."

Other factors helped, principally open and pervasive corruption. In the late 1800s, when gambler Cap Hyman wed Annie Stafford, a madame, Deputy Superintendent of Police Jack Nelson attended as guest of honor. Ties like that made the police, in the words of one social historian, "notorious among criminals and decent folk alike for venality and inefficiency." In turn, the obvious consequences of such a reputation helped win for Chicago "worldwide renown as 'the wickedest city in the United States,' a distinction which it has never relinquished."

Not all factors that subverted police effectiveness were discreditable. The tradition of John Kinsie's hustle also survived. Chicago displaced New York as leader in meat packing and grain shipping, became terminus for twenty-one rail lines, and developed as a new manufacturing center. During and after the Civil War growth became exponential. Chicago was gateway to the West, a boomtown magnet for legions wanting to get in on the easy money. Such an atmosphere would have taxed even an efficient, trained, honest and adequately staffed police force.

The actual force never stood a chance. The public would not tolerate too much policing of their vices. From a civics viewpoint there is nothing inherently vicious about gambling. Its presence in no way automatically corrupts police—as horse races and lotteries in many states, dog races in some of them, and bingo in a surprising number of churches all attest. Neither does prostitution, as witness Nevada. But when a community *declares* them vicious, and only a minority means it, corruption follows. In Chicago, the people repeatedly voiced a majority *pro*-vice opinion at the polls. Just before the Civil War, when Mayor John Wentworth stamped down too hard, the voters turned him out. Next time in office he tiptoed, realizing that no matter what sentiments they said "amen" to on Sunday, a voting majority of Chicago's people cherished their anything-goes tradition. Again, in 1873, when Mayor Joseph Medill tried reform, gambling boss Michael Cassius McDonald called a mass meeting of saloonkeepers, gamblers and thieves to enunciate an open-town platform on which their candidate ran and won.

Such incidents set the tone for Capone's era. When the people do not want enforcement of laws, when the police bosses and politicians consort with those who supply the many with what the few insist they should not have, when a policeman achieves not advancement but ridicule and demotion for doing what the book calls duty but the public calls officiousness and common sense calls folly, then the policeman who refuses bribes for not doing what almost no one wants done is a fanatic or a fool.

Reasonably responsible officials hoped to placate if not please everyone by segregating vice in tolerated pockets, the solution tried in seventy-seven American cities. With characteristic energy, Chicago did it better, its brothel and gambling-joint districts bigger and badder, congregated mostly in the First Ward, which then extended south from the river to Twelfth Street, including what is now the Loop. Each of the city's thirty-five wards then elected two aldermen to the city council. From 1893, when thirty-three-year-old John J. Coughlin first won election, until Capone displaced them, Coughlin and his co-alderman, thirty-five-year-old Michael Kenna, treated First Ward vice like a fiefdom, swapping protection for tribute.

Coughlin was known as "Bathhouse John" for his early career as a rubber or masseur—before bathhouses, or his calling, implied sexuality. He was built like a wrestler, but remained always sweet-tempered, a gentle giant and handsome gladhander who affected remarkable costumes. At one ball, his billiard-table-green tailcoat, lavender trousers and cravat, mauve vest, pale pink kid gloves and luminous yellow pumps moved a colleague to murmur, invoking a chic North Shore suburb, "Bathhouse, you look like an Evanston lawn kissed by dew." He inspired amused affection even among those who deplored his grafting and exactions, an engaging character.

The most colorful thing about Bathhouse's partner, Michael Kenna, was his nickname, "Hinky Dink," meaning "tiny," as in *Petey Dink* the day's popular comic strip about a runt, and the word, "dinky." Barely five-foot-one and spare, austere and taciturn, Kenna owned two money-coining saloons though he neither drank nor smoked nor jollied patrons. He always delivered whatever votes the machine needed, though no voter's back ever felt his slap. Instead, Kenna fed down-and-outers at the free lunch of his bar, eight thousand a week during one spell of especially hard times; he also let them flop in the upstairs portions not dedicated to gambling. H. G. Wells, after a visit, allowed that Kenna was probably "the only decent influence" in the lives of his regular clientele. Today, a Loop department store proudly calls a bar in its basement after Hinky Dink.

Coughlin and Kenna ran the corrupt First Ward exactly the way the corrupt majority of its denizens wanted it run—ultimately preparing it for Capone, and explaining the pair's astonishing political durability.

In 1894, John P. Hopkins took the mayor's seat with visibly honorable intentions—and presided over the start of a time when Chicago routinized corruption as never before. Politicians, police, prosecutors, defense bar, judiciary, citizenry and lawbreakers became so bound by ties of graft, bribery and intimidation that honesty appeared eccentric. By the time Capone hit town, one of the few eccentric aldermen would say, "Chicago is unique. It is the only completely corrupt city in America." In such a place, a Capone became not just logical but inevitable.

By now aldermen openly sold their votes to award streetcar, railway and utility franchises, not bothering to deny bribery. When Charles T. Yerkes sought franchises peculiarly disadvantageous to the city, the amounts he offered—up to $50,000 for key votes—stirred common admiration. Routine bribes ran on a schedule: $100 for a saloon license, $500 to restore a license that had been revoked for cause.

Gambling and prostitution could count on absolute immunity from any lasting interference no matter how loudly reformers howled. One group presented Mayor Hopkins with addresses of gambling joints, demanding he have police close them. Hopkins first waffled, then admitted he permitted "a few" to run because "it is surprising how many reputable businessmen want gambling to continue. I have had representatives of prominent wholesale houses tell me that they have great difficulty in entertaining their country customers because they cannot take them around to gambling houses." As an analytic Chicagoan once remarked, "No YMCA ever growed a big town."

True reformers—not just prigs or one-issue tub-thumpers like prohibitionists—could never fathom why The People did not stand with them, aghast at open corruption; why they so seldom turned the rascals out; and why, the few times they did, next election they usually voted back the same lot or worse. The reformers never understood that corruption remained a silk-stocking issue, a thorn only to those who did not want anything from their rulers but honesty and chaste governance. Most voters wanted specific help to ameliorate their blighted lives.

A contemporary historian observed that street politics meant "an unending and ceaseless devotion to getting a job for Tom, taking care of Dick's sick mother, and getting Harry out of the clutches of an over-savage or vindictive public prosecutor.'' If the prosecutor in fact proceeded with neither ferocity nor animus but only evenhanded justice against the plainly guilty, the fix occasioned even greater gratitude. A longtime Chicago judge observed that a reformer in trouble generally displayed as much distaste for justice as did the most recidivist criminal. As Al Capone once put it, ''Nobody's on the legit. . . . Your brother or your father gets in a jam. What do you do? Do you sit back and let him go over the road, without trying to help him? You'd be a yellow dog if you did. Nobody's really on the legit when it comes down to cases.'' Maybe not *nobody*; but not enough to carry most precincts.

It would have taken an unnatural voter not to reciprocate favors at the polls, anticipating further ones. It would have taken the suicidal to vote instead for secular saints whose principles, if elected, barred them from municipal payroll padding, favoritism, subversion of justice, and the other dulcet arts of ward politicking, and whose platforms offered only ''good government.'' Of course, honest reformers also spoke vehemently to the issues of minimum-wage-and-hour laws, rigorous housing codes, and like measures that would eventually shatter the poor's dependence on their ward boss's intervention. But until such reforms arrived, along with relative prosperity, talk of even modestly clean big-city government was so much wind.

Not that no honest people worked in Chicago government. Indeed, the proportion was probably the same as in any machine-dominated big city. But for political survival the most implacable idealist had to make political accommodations. During Capone's era, that need would lurk behind many otherwise mystifying alliances. Someone who participated in politics then says that honest politicians in those days sometimes ''had to make compromises with evil, in their judgment, in order to create a greater good.'' The corrupt and the criminal controlled too many situations. ''You couldn't ignore them—and win. Not to cooperate with them now and then might cost the election, where you might be electing a very decent man. Really, very decent people had to acquiesce in support of that kind. It's a tough thing for the public to understand.'' The system, as it had developed, *mandated* corruption.

Capone once said, ''I can't change conditions. I just meet them without backing up.'' The man who ultimately made that possible for Capone in Chicago was James Colosimo. Brought to the United States at about age ten, he lived most of his life in the First Ward. As a boy he shined shoes, hawked newspapers, and lugged drinking water to rail layers. He grew to a strapping fellow, broad shouldered, tall, dark and handsome, with a heroic moustache and jet hair, a popular, jolly extrovert. He positively broadcast ambition. By eighteen, he had mastered the art of picking pockets, and had turned out half a dozen adoring women to whore for him. He tried Black Hand extortion for a year or two in his mid-twenties. Then something went wrong; probably the law came close enough to frighten him. Whatever the exact cause, in 1897, at twenty-six, Big Jim Colosimo joined the "white wings," Chicago's street sweepers, mostly Italian, who wore white uniforms. Wags said they followed the horses.

Colosimo soon organized his fellow sweepers into a social and athletic club, whose bloc vote he placed at the disposal of Aldermen Coughlin and Kenna. They made him a precinct captain, and Colosimo again flourished. He opened a pool hall. He became a collector for the protection the aldermen sold brothels and gambling joints ($100 a month, per floor, and $25 a day, respectively; saloons that wanted to stay open after hours kicked in $50 a week; all had to insure through Coughlin's brokerage, and buy from four provisioners who kicked back to the aldermen). Everyone liked Big Jim, even the madames from whom he collected, because his visits unfailingly cheered their inmates with his playful yet courteous badinage.

One madame loved him. Victoria Moresco owned two large brothels on his route. She had six years on him, was plain, and running to fat. She stood no chance against this devil's blandishments even if she suspected her properties had attracted him more than her frowzy charms. They married in 1902. Colosimo promptly affixed a sign to one of the brothels, declaring it "The Victoria." His wife called that the finest wedding present any woman could want. Colosimo continued his scramble upward.

About the time they launched Colosimo, Coughlin and Kenna saw the basis of their own power and affluence dribble away. As early as the 1880s—pressured by Chicago's commercial mandarins—police had chased the gamblers, streetwalkers and brothels from the growing business district in what would become the Loop. In 1893, more shifted south voluntarily to be nearer the Columbia Exposition, Chicago's first World's Fair. Now, at the close of the century, Mayor Carter Harrison, Jr., ordered police to clear the Loop and its environs. Most joints moved to the area bounded by Eighteenth and Twenty-second streets, Clark Street and Wabash Avenue. It had always harbored brothels, and now it became infamous around the world as "the Levee," which, before the migration, had been the nickname of one egregiously sordid stretch of south State Street, tribute to what was traditionally the raunchiest part of southern river towns. Trouble was, the new Levee sat entirely in the *Second* Ward. Since Coughlin and Kenna could not bring it back to the First Ward, they plotted to take the First Ward once again to the Levee, redrawing ward boundaries. Mayor Harrison owed the two aldermen and would normally have obliged. Unfortunately, another strong supporter occupied one of the Second Ward's aldermanic seats, and the mayor dared not offend him. Nonetheless, hope existed. Charles F. Gunther, Democratic incumbent in the normally Republican Second Ward, had to run for reelection in 1900. His opponent would help make Capone an inevitability for Chicago. Friends had teased William Hale Thompson, Jr., into running. A millionaire, he was trying to win a $50 bet.

* * *

In Big Bill Thompson, Chicago found the exemplar of its political tradition. The future mayor learned about it at age fourteen. Billy and his friends, all sons of rich men, delighted in clattering their ponies on cowboy-and-Indian pursuits over the Chicago River's rackety bridges. In April of 1881, the river at flood, bridges dangerously infirm, bridge tenders warned the kids to slow down. They careered on. At the far end one bridge tender snagged Billy's bridle. Almost full grown, the boy grappled with the runty bridge tender until police arrived to drag Billy to jail. His father sprang him at 2 A.M., and in many places that would have ended matters. In Chicago, later that morning, the city's chief of police and the commander of the station where Billy had been locked up stood in the mayor's office and apologized, trying to wheedle Mr. Thompson from his original demand that every policeman who had touched his son be sacked. Billy saw the effect of clout in Chicago.

Clout sat easy on Thompson shoulders. The family had been in America six generations, since 1700, when Robert Thompson immigrated to New Hampshire from England and established a family fortune in land. Big Bill's great-great-grandfather led the Revolution's forces in New Hampshire; his grandfather fought the English as a naval captain in the War of 1812. His father served as fleet paymaster to Admiral Farragut in the Civil War, during which he married Medorah Eastham Gale, daughter of a prominent Chicago family—her father among those who had obtained the town's original charter. Before the war, William Thompson, Sr., had moved to Boston, where, as a commission broker, he added to his inherited fortune. By 1867, when his first son, William junior, was born, the older Thompson had tired of Boston's counting-house narrowness. Next year the family moved to bustling, booming Chicago.

Big Bill's first biographer called him a "perennial boy." He yearned to quit school, trek west and be a cowboy. After the bridge scrape his father agreed. If he saved enough from his job in a grocery to pay his own way, he could go, provided he returned each winter for further schooling. Before he turned fifteen Bill had hit the trail for Wyoming. For six years he returned to Chicago each winter, finishing high school, attending the Metropolitan Business College, and spending the rest of each year out West. By age twenty-one he had cleared $30,000 in cattle deals. Impressed, his father bought an interest in a Nebraska ranch for Bill to manage. In 1891, William senior died, and Thompson returned permanently to Chicago at his mother's tearful urging. He would carry on the family's hugely successful real estate business.

Actually, William's former bookkeeper managed the business, leaving the perennial boy free to substitute sports for cowpunching. About six feet tall, hardened on the range, Thompson excelled. He joined the Chicago Athletic Club, captained its water polo team to a national championship, then led the football team to another national championship as captain *and* coach. When it came out on the eve of the game that one club member had regularly bought dinner for six players, a gross violation of amateurism, Thompson told the directors of the

Chicago Athletic Association, "There's only one thing to do. We want the championship, but we've got to expel those men." When the club won anyway, Chicago hailed him as purest of the pure.

Nights, however, sometimes discovered him in the Levee with Gene Pike, boyhood companion in the bridge-storming. They shared bachelor digs in the Hotel Metropole. Pike also inveigled Thompson into politics. Pike, who had become a Second Ward alderman, badgered Thompson to run when the Republicans needed a candidate to unseat the Democrat Gunther in 1900. Having stood unsuccessfully the previous spring for the athletic association presidency, Thompson refused. "I don't ever want to run in any other election of any kind," he said. "It's all right when you win, but it's not so good when you lose." Pike persisted, and one day at the club another member slapped a $50 bill on the card table, offering to bet Pike that their friend was too scared to run. Ah, a *dare*! The perennial boy's meaty hand covered the money; he'd take the bet himself.

Aldermen Coughlin and Kenna rejoiced at the news. Because of the Second Ward's peculiar demographic hash, Thompson might be the candidate to beat Gunther, allowing Mayor Harrison to put the Levee back under their control. The rich lived on the east of the Second Ward, toward Lake Michigan. The poor lived west of Michigan Avenue. The so-called Black Belt started on the ward's south, Chicago's black ghetto.

Thompson's dimness has always been a shibboleth for writers. Yet he always campaigned brilliantly. Second Ward silk-stocking voters normally treated aldermanic elections with apathy, viewing the city council and Levee with equal disdain. Although Republicans, many had mildly supported Gunther for two terms. He was a rich reformer. For that reason, the ward's nominally machine-Democrat poor and its Levee hustlers had never warmed to him. Thompson figured he could count on silk-stocking support; he was one of them. He concentrated on wooing the Levee. He pressed flesh in every one of the ward's 270 saloons, spending $175 a day setting up drinks. Even shrewder, he worked the Black Belt. In those days blacks still voted for the party of Lincoln. The trick would be to get them to the polls in great enough numbers. Thompson went among them and convinced them he cared—which he genuinely did; the plight of an oppressed people naturally engaged a perennial boy's sense of fair play.

Thompson won, 2,516 to Gunther's 2,113.

Flattered and cozened by Coughlin and Kenna, Thompson voted for redistricting, gerrymandering himself out of another term. He did pay off the Black Belt, pushing through an appropriation to build there the city's first public playground. This move would bring later rewards.

In 1902 he won a seat as a county commissioner with the ticket's second-highest plurality. Then he bided his time.

Meanwhile, Big Jim Colosimo had moved past prosperity to the start of riches. With Victoria's two parlor brothels as a base, he added a saloon in 1905 and ever more $1 and $2 crib brothels. He also developed a lucrative wholesale trade in prostitutes. Averaging between twenty-three and twenty-four in age, they lasted in parlor houses seldom more than five years, after which they hit first the cribs, then the streets. Demand for what the pimps (or "cadets") called new "stock" stayed constant. Colosimo combined with Maurice Van Bever, a pimp who affected elegance, maintaining a carriage driven by a top-hatted coachman in gold-buttoned livery. They organized what everyone called "a white slave ring."

The term sensationalized and in a perverse way softened and glamorized the miserable truth. Newspapers in the first decade of the 1900s loved to linger over detailed reports of fifteen-year-old farm girls on visits to the city plied with drugged lemonade by plausible strangers who had been talking marriage. The girls came to in a brothel where they were repeatedly raped, a process called "breaking in." They were beaten and kept more or less naked while being prostituted until, if lucky, they managed to smuggle a message out to kin, the favorite method of the tabloids a scribbled note tossed from an unguarded window.

A Chicago sociologist's close study concluded that such lurid cases were rare. That they *ever* occurred only gives added testimony to the whoremasters' depravity, because drugs, beatings and rape were superfluous recruiting tools. Most prostitutes at the time turned to the life in desperation. In 1913 an Illinois State Senate investigating committee took testimony from William F. Bowler, a Chicago police sergeant with thirteen years' experience, the last four in the Levee:

> "Now, what would you say, sergeant, is the most direct cause of a girl's downfall?"
>
> "Poverty."
>
> "Unquestionably poverty?"
>
> "Why, certainly. You can take a dozen contributory causes, you can say it is fondness of dress, hasty marriages, divorces, Greek restaurants, Greek cafes, and so forth, but if you sift it right down, it is poverty."

One woman newspaper reporter moved to a $2-a-week room, landed a department store sales job—8:15 to 6:35, Monday through Saturday, furtively leaning on counters to relieve her aching feet. She tried to live on her $6.50-a-week wage. Fifteen-cent breakfasts, store canteen lunches that ran 14 to 17 cents, and 20-cent restaurant dinners all left her hungry an hour later. On Sunday she reckoned her expenses for the week. Even with 15 cents in tips, they left her 11 cents short. "To balance my accounts," she wrote, "I would have to do without my 20-cent dinner on Sunday; but crackers and milk are a frequent diet with the poorly paid woman. I had hoped to buy a paper, but I could not afford to do so."

Who needed knockout drops? The investigating committee summed up their findings after 805 pages of testimony and exhibits: "Poverty is the principal cause . . . of prostitution. To avoid this conclusion is to quibble."

Taking advantage, Colosimo and Van Bever had established one of the first rings with ties to pimps in other cities: New York, St. Louis and Milwaukee. They interchanged whores, a smart move that assured a supply of new faces who could be sold to brothels for as much as $400.

The money rolled in, and Colosimo's success inevitably excited the greed of Black Hand marauders, whose tactics were standard with all victims. A note would arrive in the mail of a successful Italian. It usually bore a drawing of a hand in outline, filled in with black ink, perhaps accompanied by a skull and crossbones, crosses, a knife. It would demand however much the extortionist calculated to be just this side of victim desperation, with threats against the victim, family, property, or all. The language might be flowery with Old World politeness:

> Most gentle Mr. Silvani: Hoping that the present will not impress you too much, you will be so good as to send me $2000 if your life is dear to you. So I beg you warmly to put them on your door within four days. But if not, I swear this week's time not even the dust of your family will exist. With regards, believe me to be your friend.

More often it was direct and utilitarian, covering all outs and contingencies that might occur to victims:

> You got some cash. I need $1,000.00. You place the $100.00 bills in an envelope and place it underneath a board at the northeast corner of Sixty-ninth and Euclid at eleven o'clock tonight. If you place the money there you will live. If you don't you die. If you report this to the police, I'll kill you when I get out. They may save you the money, but they won't save your life.

Another Black Hand letter ended, "Even if you leave the city we will think to revenge ourselves with a severe vengeance and precaution." Enough of the brave or foolish ignored letters and had their homes or businesses bombed, or ended up murdered, to freight all Black Hand letters with terrible credence. In cities across the United States, between forty and sixty independent gangs calling themselves Black Handers murdered about four hundred examples in thirty years—until federal authorities used postal laws to intervene and crack down.

Most victims had amassed money honestly, with hard work and thrift. They lacked the ability to fight back, or scrupled to do so with sufficient ruthlessness. Colosimo had no compunctions about killing Black Handers who risked targeting him. But their lack of organization, their anonymity and ubiquity defeated sensible attempts to retaliate. He could not kill whom he did not know and could not locate. He did what he could, setting traps, personally shooting at one extortionist, sending his men after another with buckshot. In 1909 he called for help.

Colosimo's wife, Victoria, had a cousin in Brooklyn, John Torrio, who had achieved a fearsome reputation. Newspapers later called him "Little John" and "Terrible John," both apt nicknames, though no one except reporters used them. His businesslike colorlessness resisted any except, occasionally, "J.T." On arrival he dispatched Colosimo's immediate Black Hand problem with his usual efficiency.

Three strangers had demanded $25,000, upping it to $50,000 when Colosimo ignored them; they may even have waylaid him, their threats punctuated by guns shoved in Colosimo's gut. Torrio took over negotiations, pretended to accede, a money drop arranged for midnight at a railroad underpass. Two Colosimo gunmen, Joie D'Andrea and Mack Fitzpatrick, crouched in the carriage with Torrio. Three Black Handers materialized from the shadows. Torrio's gunman opened fire, killing two, leaving the third mortally wounded. He lasted long enough to ask police at the county hospital to summon his "friend" Big Jim to his bedside. When Colosimo arrived the Black Hander croaked his dying curse in Italian. "You traitor!" A coroner's jury found that the three interlopers from Pittsburgh had been murdered by the usual "persons unknown." The drama of three killed at one time immediately chilled the enthusiasm of other free-lancers for this particular target.

Understandably, Colosimo wanted Torrio to stay, and Torrio was willing. Manhattan remained in turmoil; the Irish grip on the Brooklyn docks had not yet loosened, so that borough offered less scope than his talents warranted. At age twenty-seven Torrio started a new career as manager of Colosimo's number two Levee brothel, the Saratoga on Twenty-second Street.

From manager of the Saratoga he graduated to overall supervision of Colosimo's operations. Now chief lieutenant, he received a split of the take. With his patron's permission he opened his own places, a saloon and a brothel not far from the Victoria, importing as manager of the brothel a cousin, Rocco Venillo, who liked being known as Roxy Vanilla, and claimed to have been a Montana gunman.

Only one cloud lowered. In 1912 Mayor Harrison's hand had been forced by a maverick state's attorney backed by vociferous reformers. The mayor had established a special fifteen-man Morals Division led by an Illinois National Guard officer, Major Metellus Lucullus Cicero Funkhouser. A fanatic who censored Chicago's movies with the same gusto he showed in monitoring its morals, Funkhouser also happened to be incorruptible. Although nominally "second deputy police commissioner," his mandate let him operate apart from police control. In six months his squad, under his uncompromising eye, had arrested about 350 Levee whores, pimps, owners and gamblers, and had shuttered some of the worst dives.

Of course, the Levee persisted despite Funkhouser, who received only grudging, minimal support from the courts and regular police. He was actively opposed and subverted by the district's police commander, a Coughlin and Kenna employee, Captain Michael J. Ryan, who tipped off joints before Morals Division raids. But even such an imperfect crackdown hastened a move Torrio had been contemplating anyway. Automobile registration in America had about doubled between 1910 and 1912, and would nearly double again to 2,309,666 by 1915. Torrio realized that roadhouses, already dear to the upscale trade, could appeal to the scads of potential working-stiff customers in the steel and mill towns that surrounded Chicago, some just across the Indiana line, like Gary and Hammond.

Torrio found the perfect spot for his pilot operation in Burnham, a scant fifteen-mile crow's flight southeast of the Levee, about one square mile, home to roughly one thousand, a stroll from the Indiana border. He would encounter no police problems. John Patton became known to newspaper readers as "the Boy Mayor," though his actual title was village president; he took the job before he reached age twenty because no one else wanted it. He became one of Torrio's top executives for vice, later playing the same role for Capone. Soon Torrio and Colosimo, who were now partners, had three combination nightclub-brothel-gambling joints there, first the Burnham Inn, then the Speedway and the Arrowhead—the latter managed by the Boy Mayor, with Burnham's chief of police tending bar while three village trustees waited table.

The concept of suburban vice, where the whole government and constabulary could be *owned*—safe from the vagaries of big-city politics and reform movements—would spread with significant results when Capone came along. But it demonstrated its brilliance and utility in providing refuge for vice much sooner.

By mid-1914 the Levee hummed with tension. A suspected Morals Division informer had been murdered, a detective investigating the incident knifed; the owner of the dive where the murder took place had his license lifted and, embittered, had been drunkenly blabbing about a plot to kill Funkhouser and his top aides.

In that steamy context, about nine o'clock the night of July 16, the squad raided the Turf, a brothel on Twenty-second Street, leaving two of the squad's least-experienced officers with those arrested to wait for a paddy wagon while the rest rushed off on another raid. After the rookies handed their prisoners into the wagon, they set off on foot to join the others. Almost at once they found themselves dogged by a growing crowd of Levee hooligans, hurling threats and curses. When someone hurled a brick, the officers stopped and drew their guns as a third man joined them. He was another of Funkhouser's sometime informers.

Two regular detective sergeants had happened past the crowd, unconcerned. But when they heard a shout, "Watch it, they've got guns!" Stanley J. Birns and his partner rushed back to discover two strangers in civilian clothes brandishing weapons. Naturally, the two detectives drew their own guns.

No one could later establish who started it. Behind the crowd rolled Torrio in his new, unmistakably fire-engine-red roadster, with him cousin Roxy Vanilla and Mack Fitzpatrick, the gunman who had helped dispatch Colosimo's Black

Hand tormentors. One of the rookies claimed that a man in a light gray suit and straw hat had fired first; a witness swore that the man had leapt from the red car, pistol leveled, but others disputed his testimony. Roxy Vanilla matched the description and he had received a bullet in his foot. On the other hand, an anonymous "Officer 666" wrote a newspaper piece that tagged Mack Fitzpatrick as the gunman. Crooked Captain Ryan smugly charged the fault lay with Funkhouser's "green cops" who "got scared and let loose."

No matter who fired first, Sergeant Birns ended up sprawled on the pavement, dead, his partner hit in the thigh, one novice hit seriously in the hip, the informer in the groin, the other novice nicked in the leg.

Torrio holed up for a while; Colosimo, to his amazement, was hauled off to jail for a few hours, his only time behind bars. No one ever came to trial except the murderer of the suspected informer, and he beat the rap on appeal.

Even so, the attendant scandal slammed shut the Levee's wide-open days. The mayor banished the aldermen's man, Captain Ryan, from the district, replaced by Max Nootbaar, an honest cop. Ike Bloom—once a Levee leader, still a prominent dance hall owner—came around to make the usual "arrangement" with the new captain. Nootbaar kicked Bloom downstairs, then ostentatiously removed from the squad-room wall a framed picture of Bloom that had long occupied a place of honor. Harrison also revoked many licenses, including, for a time, those of Colosimo and Torrio, who rejoiced all the more in their Burnham operation. One saloon owner, husband of a Levee madame, paid off his people, locked his doors and announced that for the first time "it looked as if reform might stick."

He did not have the last word. That belonged to Big Bill Thompson.

The 1912 battle between Theodore Roosevelt and William Howard Taft had handed Democrat Woodrow Wilson the presidency and left Illinois Republicans in hostile factions. Fred Lundin, a Swedish immigrant with authentic American back-room political genius, saw in such regular party chaos his chance to snatch the 1915 Republican mayoral nomination for a candidate of his own. He chose Thompson. He started two years before the primary, fashioning an organization, classically, precinct by precinct.

Brought to America at age twelve, not long after the Chicago fire of 1871, Lundin slogged through the usual poor-boy jobs until at age twenty he set himself up as a medicine-show huckster with a wagon, a couple of blacks strumming banjos to gather a crowd, and a fantastic costume (black, wasp-waisted frock coat, string tie, broad-brimmed felt hat, round amber spectacles) that accentuated his buck-toothed, shock-haired, offbeat looks. He developed a slick pitch for his home-brewed Juniper Ade: ". . . that wholesome, delicious, incomparable, refreshing, foaming but nonalcoholic beverage. . . . Recommended by all doctors. . . . It's cheap! It's tasty! Add a teaspoon to a gallon of water! Add anything else you like! It's good! Step up, folks, step right up!" He did well, entered politics, served in the state senate and one term in 1908 as a U. S. representative.

But his forte was behind-the-scenes maneuvering, preferring to call himself "just a poor Swede" and "insignificant me."

While Lundin organized and Thompson speechified about an "unbossed candidate," the regulars dithered. They did not announce a candidate until almost a month before the February primary. Harry Olson, chief judge of the municipal court, barely deigned to campaign, assured that upstarts could never derail the machine. Thompson courted the newly enfranchised women and befriended blacks as in 1900. With World War I raging in Europe, he denounced British atrocity propaganda to the city's 600,000 Germans and Austrians. Just before the primary, his campaign issued leaflets charging that Olson's Catholic wife would prevail on him to wreck the public-school system.

Thompson won the nomination by 3,591 votes.

In April he faced Cook County Clerk Robert M. Sweitzer, who had thumped Carter Harrison for the Democratic nomination. During the campaign Sweitzer complained that Thompson told different tales to different groups. As one newspaper backing Sweitzer put it, "Thompson tries to be all things to all manner of people. In the Polish wards he's anti-German. In the German wards it's '*Unser Wilhelm für Bürgermeister*.' . . . In the Irish wards he damns the King of England." Exactly. He spoke directly to his several audiences' distinguishable interests. He told women, "I'll appoint a mother to the board of education! Who knows better than a mother what is good for children?" In silk-stocking wards he promised, "I'll clean up the city and drive out the crooks!" while in blue-collar wards he crooned in his whiskey baritone, "I see no harm in a friendly little drink in a friendly corner saloon," and in the Black Belt, "When I'm mayor the police will have something better to do than break up a friendly crap game." More to the point, he promised blacks, "I'll give your people jobs." To everyone he promised, "We'll have new wide streets, new buildings, new factories, new pay rolls. We'll have full prosperity and a full dinner pail!"

Throughout his career, through eclipse and disgrace, Thompson would remain in rhetoric *and sometimes action* the populist candidate. As we'll see, that's the key to understanding Chicago's otherwise gaga politics of the 1920s. Sweitzer, personally, was honest enough, and madly popular, "the man with a million friends." But everyone knew him to be the creature of Democratic boss Roger Sullivan, who figured in the natural-gas rate scandals of the nineties. "Boys," Thompson told reporters, "this campaign's real issue is the gas trust! It gets its profits by gouging unfair prices from the little fellow. . . . [T]he question in this election is: 'Do the people want to turn over the government to Roger Sullivan and the utilities?' Ask a man with a family. . . . Ask any mother who must take the food from the mouths of her children to feed the greedy coffers of the gas trust. Take a look at your own bills. . . ." A week before the vote he filed a showboat petition with the utilities commission to slash the gas rates 30 percent. Meanwhile, Lundin kept dealing in the back rooms. Democrat Carter Harrison hated Sullivan so much he countenanced a bolt to the Republican candidate by some of his top people, including his secretary, Charles C. Fitzmorris, who would be rewarded by appointment as Thompson's secretary and, later, chief of police.

On election eve Sweitzer addressed a group of political leaders, while Thompson more cannily romanced voters in the black Second Ward. The "white hope," Jess Willard, had beaten Jack Johnson, a black, for the heavyweight boxing title that afternoon. "Only a good cowboy like Willard," Thompson consoled his audience, "could beat a good man like Johnson. Tomorrow the cowboy will be on your side. Remember this: Bill Thompson is going to win for you at the polls. . . ." They knew he *was* for them. He had shown it before, would show it again. Next day, Thompson won, 390,691 to 251,502—almost exactly Lundin's prediction—the largest margin ever run up by a Republican in Chicago.

When Thompson took office, he soon transferred honest Captain Nootbaar to the sticks, and fired morals inspector Francis D. Hanna for penning a report that brothels again ran unmolested by regular police. As for the Morals Division, he sniped continually at Major Funkhouser, throwing up legal barricades to his work, finally inducing the city council to cut off appropriations.

Even so, Torrio and Colosimo knew the lush days were over; vice in the Levee would have to become less splashy. To Colosimo, it didn't matter. In 1910 he had opened Colosimo's Café at 2126 South Wabash, the southeast corner of the Levee's heart. Except for gambling upstairs, Colosimo's stayed straight and posh, all gilt and velvet and furbelows. Big Jim employed a world-class chef, laid in a fine cellar, booked top entertainers and orchestras. He saw Colosimo's become the nightlife cynosure for Chicago's café society, an oasis of lush security in the wicked Levee for everyone from Gold Coast sports to visiting firemen. Big Jim soon wanted nothing but to wander his café, the adored, gracious host. Then, too, he raked in at least $50,000 a month personal profit—thanks mostly to his partner.

Torrio tended to the humdrum of their empire. He replicated the Burnham operation in other villages surrounding Chicago, west and southwest. Even where local leaders were as eagerly corrupt as John Patton, Torrio thoughtfully visited neighbors at each projected site and made friends: paying off mortgages, fixing leaky roofs, whatever householders wanted so long as they agreed not to cause trouble.

Back in the Levee he acquired, as headquarters, a four-story red-brick building down the block from Colosimo's at 2222 South Wabash, the Four Deuces. Its layout reflected the Levee's new reticence. The first floor housed a saloon, no more raunchy than most. A staircase ascended from the rear of the taproom, screened off. Another climbed the side of the building. On the second floor Torrio established headquarters and bookkeeping offices. On the third floor, behind a steel door, was gambling, on the top floor, a brothel.

One worry lingered: that someone might replace the complaisant Thompson as mayor in the 1919 election.

"He is an industrious and remorseless spoilsman," wrote *The New York Times*. "One of the worst mayors in the city's history," claimed the Springfield *Republican*. "Poor Old Chicago," the Kansas City *Star* headlined a story that called Thompson "an impossible mayor." When he won renomination, the country boggled at the notion of Thompson's reelection. Was Chicago demented? What did they see in him?

"Thompson is," wrote John Bright, a not overfriendly contemporary, "simply the kind of fellow most Chicagoans like and respect"—big, breezy, confident, "immensely good natured," towering over them "like a successful uncle." After "gaining their respect he obtains their love by talking in their language, vulgar, slangy and alive. . . . As a result he is one of the best showmen in America, and one of the best shows." He provided more than theater. Some politicians, one journalist has put it, "are unconscious interpreters of their times and have an intuitive and prophetic understanding of what their countrymen long for. You must ignore their explanations, which are often ridiculous. . . . They speak as if they are giving words to the crowd's own dumb sentiments."

Thompson sensed what large numbers of his constituents wanted because he wanted it, too, at least partly through his perennial-boy fervor that enshrined the rhetoric of loyalty and fair play. While he spoke, he became a true believer. That meant he remained untrammeled by considerations of reality, which made him seem guileless. No matter that he owned some $1,670,000 of utility securities; he could bash the gas trust with unmistakable sincerity because it *wasn't right* that they should gouge his friends who elected him. Equally, it didn't matter to those friends that he never seemed to *do* much to curb the trusts—so long as he boisterously demanded lower rates.

It helped Thompson immeasurably that no palatable alternative challenged him in 1919. Democratic boss Roger Sullivan again ran Robert Sweitzer, heading a slate less venal than those in power only in proportion to how much farther they stood from the trough. Their only issue was "corruption." One social worker planned to vote for Thompson because the opposition "had not a single constructive or progressive plank in their platforms." Meanwhile Thompson kept insisting that stories of his corruption were "lies, all lies!" fabrications of "the crooked, lying newspapers of Chicago." He said, "I'm for home rule, I'm for reduced gas rates, I'm for the five-cent streetcar fare. I'm for the people and against the selfish interests!" He continued to champion blacks, giving them jobs on the public payroll and in his administration, and naming Louis B. Anderson, Black Belt alderman, his city council floor leader. He cut some blacks in on the spoils and attacked any opponent who voiced bigotry.

One magazine writer summed it up: "The truth is, the machine has been allowed to become the exponent of progressive ideas, of democratic reforms."

"Everybody to his taste, as the old woman said when she kissed the cow," commented the Lincoln, Nebraska, *State Journal*. "A plurality of Chicago voters wanted more of Mayor Thompson. . . . That is their business and their funeral."

On April Fool's Day of 1919 Big Bill Thompson had won his second term as mayor of Chicago, this time by only a little over 21,000 votes.

Later that year Frankie Yale arranged for Al Capone to lie low in Chicago until Bill Lovett tired of trying to track him down.

On arrival, Capone did not step into a grand position. He may have started as a bouncer at the Four Deuces at $35 a week. Or maybe he was a capper: A reporter claimed to have spotted Capone in front of the Deuces "a dozen times, coat collar turned up on winter nights, hands deep in his pockets as he fell in step with a passer-by and mumbled: 'Got some nice looking girls inside.' " In any case, he did not remain a flunky very long. Early in 1920, back in Brooklyn, Yale had wearied of his tit-for-tat war with the Irish. Risking public reaction, he conceived a coup that might permanently tip the balance. For it he needed some especially reliable, unrecognizable out-of-town killers. He called Capone, aware that his former bouncer had already gained enough of Torrio's confidence to broach the subject and make the arrangements.

Barely twenty-one, Capone's intelligence and imagination already shone through his unsophistication. He was fortunate to fall under Torrio's tutelage, just as Torrio had been to encounter the emollient influence of Paul Kelly. For his part, Torrio quickly appreciated the young ruffian's qualities. Capone displayed dimensions far beyond those of the run of hooligans Torrio generally had to use. For instance the Four Deuces building incorporated an unused storefront which boasted its own address, 2220 South Wabash. In line with the Levee's new attempt at surface seemliness, Capone outfitted the storefront with a showcase containing leather whatnots. He hauled in an old piano, three tables, a planter, a rocking chair, some rugs and an aquarium. A shelf of books included a family Bible. He printed up cards:

ALPHONSE CAPONE
Second Hand Furniture Dealer
2220 South Wabash Avenue

and obtained a telephone listing, trading up slightly, as "A. Capone, antique dealer." While this did not represent impenetrable cover, it did give him the color of legitimate business. When asked, he'd laugh that he sold "any old thing a man might want to lay on."

John Torrio had no sense of humor, but the imagination and foresight appealed to him. Capone was someone he could use, and as for doing Yale a favor and arranging for killers, the answer was yes. Torrio might want the favor returned, and soon. Business boomed and the new wrinkle of national Prohibition promised absolute bonanza. But to cash in, Torrio would have to do something about Big Jim Colosimo.

CHAPTER 4

Opportunities

AL CAPONE COULD not have arrived in Chicago at a more yeasty moment than now, when John Torrio would most need him and most prize his talents. Big Jim Colosimo meant nothing to Capone. From the first, Capone reserved his loyalty and admiration exclusively for Torrio, and Torrio's views automatically became Capone's. Although to Colosimo their business looked unimprovably marvelous, Torrio could discern the problem lurking in such complacence. Colosimo would neither grasp the unparalleled opportunity Torrio saw nor let Torrio do it for him.

True, Colosimo had cause for complacence. He had outgrown Coughlin and Kenna. As proprietor of the First Ward's Italian votes—the Italians rapidly displacing the Irish as majority—he commanded his own political base so firmly that he could assert his own claim to political and police consideration. Not that any rupture embittered Colosimo's relations with the aldermen. If anything, Bathhouse John Coughlin welcomed the disengagement. The violence that enforced collections and maintained order for Colosimo and Torrio had always made that gentle giant nervous. The new arrangement suited him better. Coughlin and Kenna still sold booze to most First Ward joints and still retained political control, with Colosimo delivering the votes as usual. But now Colosimo alone controlled the area's vice.

When Mayor Thompson's candidates swept the 1920 elections, he exulted, "The roof's off!" and every development in his administration showed what he meant. His appointees and allies were, with one exception, unequivocally corrupt. Having broken with outgoing Illinois Governor Frank O. Lowden, whose administration *The New York Times* characterized as "a model of integrity, effectiveness, economy," Thompson scuttled Lowden's bid for the Republican presidential nomination, helping give the nation Warren G. Harding. Then, instead of Lowden's reasonable choice as successor, Thompson backed Len Small, banker-farmer from Kankakee and a favorite of Fred Lundin's. While governor, Small would be tried for a fancy form of embezzlement, and would escape conviction only through jury intimidation and bribery. He would also establish a brisk business in pardons for well-connected criminals.

Thompson's one major associate not unequivocally corrupt was Robert E. Crowe, state's attorney (the Illinois term for district attorney) from 1920 to 1928. Born in 1879 in Peoria, graduate of Yale Law School, the Irish Crowe married into the powerful Italian Cuneo family; his father-in-law had established the city's oldest wholesale produce company. Stocky, of medium height, Crowe had a pugnacious, outthrust jaw that matched his imperious manner.

Nearly everything about him was *equivocal*. He had posted a fine record as an assistant state's attorney and assistant corporation counsel for Chicago. He then served four years as a circuit court judge, youngest on the state bench, at thirty-eight being named chief justice. He was a fine lawyer. As state's attorney he did not hesitate to try cases himself—unusual in so politically sensitive a position, where a loss could cost dearly at the polls. He personally faced Clarence Darrow in the Leopold-Loeb trial.

On the other hand, under his direction not a single top gangster was convicted of any serious crime. He recorded not a single conviction for gangland killings. Most questionable of all, in at least two cases of great interest to Capone, as we'll see, Crowe's very *policy* subverted the possibility of conviction. Crowe also regularly used gang slugging help at the polls (though that made him far from unique). As for political principle, though he despised Thompson, he had thrown in with the mayor for his own 1920 race.

John Garrity represented the more usual Thompson minion. He was the second police chief in a row that public clamor forced the mayor to dismiss. Garrity was caught ineptly covering up corruption and was probably in cahoots with Mike Heitler, a swinish pimp (though this scheme involved liquor). Thompson replaced Garrity with Charles Fitzmorris, who at thirty-six became Chicago's youngest-ever police chief. His previous experience had been as a newspaperman, followed by service as secretary to Carter Harrison, then to Thompson when he sold out before the 1915 election. The mayor gave the new chief the customary noisy public mandate to "Clear out the crooks." Fitzmorris's raids resulted in maximum fuss and minimal ruffling of any feathers that mattered. Shocked to discover gambling in Chicago, he hit some card rooms and knocked over Izzy Lazarus's three roulette wheels and broke up a few crap games. He also rousted hundreds of layabouts and minor civic annoyances, some of them with police records. Soon, even such desultory raids declined.

Curiously, Torrio welcomed the raids while they lasted. Apart from Colosimo's apathy and opposition, only Chicago's current crime wave posed a plausible threat to Torrio's plans. Since the end of the war, street crime had grown alarmingly. Chicagoans murdered over 300 of each other in 1919, with about 250 robberies during a single week in November, a record. At the start of the year, prominent citizens had formed the Chicago Crime Commission in the hope they could do what the police notoriously did not. Although a wrenching postwar recession had hit the country, the Chicago Crime Commission tartly observed that crime in Chicago "is not due to poverty or hard times. Crime is a business here." That kind of "business" adversely impacted Torrio's kind, since it inspired citizen indignation that could trigger unfixable crackdowns against *all* forms of illegality. A real crackdown could ruin Torrio's plans for an illegal service that promised to dwarf all others.

"No tendency is quite so strong in human nature," wrote William Howard Taft, "as the desire to lay down rules of conduct for other people." The majority of Americans probably never wanted Prohibition, but that turned out not to matter. Members of the Anti-Saloon League, founded 1893 in Oberlin, Ohio, sincerely believed everyone would be better without alcohol. They "looked forward," one historian has observed, "to a world free . . . from want and crime and sin, a sort of millennial Kansas. . . ." Their campaign, which quickly enveloped the nation, combined such animating idealism with the most brutal, brass-knuckles politics. The league terrorized Congress. In the words of a popular song of the era, "What Have They Got on You, Mr. Congressman?":

> We've heard just how those drys,
> Keep cases on you Congress guys.
> They say a careful record's kept,
> Of cash you took and where you've slept.

One "wrong" vote and reports would wing back home, broadcast by the league's fifty thousand field workers.

America's April 6, 1917, entry into the war sanctified the dry cause as patriotism even for the doubting majority. The liquor trade gobbled up enough grain for eleven million loaves of bread a day, the league insisted; drunk workers could not turn out war matériel any more than drunk soldiers could shoot straight.

Caving in to these pressures, Congress passed a resolution calling for a prohibition amendment to the Constitution and sent it to the state legislatures for ratification in December 1917. On January 16, 1919, Nebraska provided the necessary three-fourths majority by becoming the thirty-sixth state to approve the resolution. The Eighteenth Amendment would become law in one year.

Meanwhile, the league rammed "wartime" Prohibition through Congress. Until the Eighteenth Amendment kicked in at midnight on Friday, January 16, 1920, the interim law forbade anyone in the United States to make, sell or transport (without a permit) any beverage containing more than one half of 1 percent alcohol. The ban—both permanent and interim—cunningly did not include owning, drinking or buying liquor. The league had been careful not to offend voters or members of Congress, many notoriously wet in habits no matter how dry they voted.

In one of the earliest popular songs of the era, Sophie Tucker suggested that citizens would obey the law, explaining,

> For if kisses are intoxicating as they say,
> Prohibition, you have lost your sting.

But America would soon change its tune. By the time of the Ziegfeld Follies of 1919, Irving Berlin complained, "You Cannot Make Your Shimmy Shake on Tea."

For John Torrio, Prohibition was an answered prayer. He always strove to turn crime into a regular business; now the fools had obliged him by making a regular business criminal. That meant he could enter it fearing little or no competition from those already practiced in the business, while its new illegality meant opportunities for enormous price hikes. A schooner of beer that used to sell for a nickel or a shot of booze that went for a dime might now cost twice that or even ten times as much. With limited competition to meet only modestly diminished demand, suppliers could almost pick their number. They also had the pick of personnel, since bootlegging plainly offered better pay, with less risk and lower penalties, than any other branch of crime.

One consideration gave Torrio pause. Although the Eighteenth Amendment mentioned "concurrent powers" of the states and federal government to enforce it, the National Prohibition Act was a federal law (and was better known as the Volstead Act after the congressman who introduced it, Andrew J. Volstead of Minnesota). Enforcement would be by federal agents, under the authority of the Treasury Department. Torrio could not forget the time their usual exchange program had sent a prostitute to Bridgeport, Connecticut, and she had talked, naming Colosimo. This threatened a Mann Act conviction with a stiff sentence. She had to be murdered; the federal agents involved had proven persistent and incorruptible. How much of a hazard would *these* feds be? How seriously enforced this new law?

The answer came quickly. The Anti-Saloon League, drafters of the Volstead Act, had anticipated no great enforcement problems. The nation shared their disgust with saloons, which had become sinks of vice and the seat of corrupt politics. But the league confused that anti-saloon sentiment with the implacable hatred they felt toward all alcohol. They assumed the public shared that, too, so imagined that outraged Americans, particularly women, would inform on lawbreakers. The league felt free to let the enforcement mechanism become an object of patronage. Prohibition agents did not at first come under civil service; when they finally had to take the exam, only two fifths could pass after two tries. They received less pay than garbage collectors ($1,200 to $2,000 a year at first,

$2,300 in 1930), an invitation to corruption. One in twelve ended up dismissed for cause—a failure rate one dry senator defended as being no worse than Jesus had experienced with disciples.

One disgusted official called the agents ''ward heelers and sycophants named by the politicians.'' But had they all been miracles of honesty and efficiency, not enough of them existed to do the job. They numbered only about 1,500 to begin with, and their ranks never exceeded 2,300 field agents nationwide. What's more, *these* feds heartened Torrio from the start by exhibiting an almost Chicagoan capacity for corruption. Under the wartime Prohibition law, the superintendent of the Illinois revenue district was indicted for accepting bribes, along with two assistants. Treasury's own estimates put bribe payments in Chicago at $200,000 for the period of the wartime measure.

Equally heartening were the paltry penalties for being caught. A first offense called for a maximum $1,000 fine and a year in jail. Even federal judges fined few the maximum, sent fewer still to jail. From the first, nearly everyone saw the law as a joke. For instance, the government could seize and sell any vehicle used to transport liquor. A small furor erupted over the fate of the vehicle that Chicago bank president Charles N. Thomas used to transport a hip flask of booze to a cabaret. Would the United States of America *really* auction off his trousers?

Torrio hesitated not a moment over the prospect of energetic enforcement by Chicago authorities. Chicago's citizens would not stand for it. Before, they had voted 391,260 to 144,032 against closing even the admittedly awful saloons. Now they treated Prohibition with open disdain. With the rest of the nation, but with particular gusto, Chicago sang a song that explained why ''All the Boys Love Mary'': she might be ''the homeliest girl in town,'' but the lookers

> . . . ain't got what Mary's got—
> A daddy with a cellar full of you know what,
> And all coming through the rye.

As for those Boys,

> It's the smart little feller,
> Who stocked up his cellar,
> That's getting the beautiful girls.

Besides, decades of corruption had built to this historic moment, as evidenced by the program for the William Hale Thompson Business Men's and Women's Club ball. A reform group noted ads from ''seventy-five saloons, disorderly cabarets, hotels and restaurants, many of which are known to be violators of the law.'' ''Prohibition enforcement in Chicago,'' complained the federal district attorney, Charles Clyne, ''is a joke.'' Police Chief Fitzmorris admitted that about half the force was involved in bootlegging—not just as solicitors and recipients of bribes, but actively pushing booze!

Ordinary people learned from their betters, like George Babbitt wowed by the casual elegance with which a rich man served him a drink. Citizens goggled at stories of how Prohibition agents brought confiscated booze to President Harding's Ohio Gang hangout, the notorious House on K Street, and at tales of members of Congress staggering drunk on the floor of House and Senate. Visitors to the United States would ask, "When does Prohibition start?"

Torrio could see many examples of how *not* to capitalize on Prohibition. A truckload of whiskey was hijacked on West Lake Street in the early morning of January 16, 1920, with Prohibition not due until one second past midnight. Six men, probably led by Herschel Miller, West Side leader of a mostly Jewish gang, drove a truck into a railroad yard. Masked, armed with revolvers, they tied and gagged two watchmen, locked six trainmen in a lean-to, and made off with $100,000 worth of medicinal whiskey from two sealed freight cars. Across town, another gang looted four barrels of alcohol from a warehouse.

Torrio could see that the boys were thinking small, selling only in their own neighborhoods. Moreover, basing a booze and beer business on stolen supplies was three times foolish. First, the theft bore intrinsic danger; its victims—especially other bootleggers—tended to be equally violence-prone and armed. Second, the theft itself constituted a criminal offense carrying far greater penalties than the sale of the loot. Finally, such operations were hopelessly retail, with quantities limited to what other people owned or made. Torrio thought wholesale.

He could profit from a current example. Terence J. Druggan and Frank Lake were boyhood buddies, in their mid-twenties when Prohibition started. They were born and raised in "The Valley," a flat wasteland of squat brick tenements southwest of the Loop. After adolescent pilferage together, Druggan had toiled on his father's garbage truck while Lake worked first as a railroad switchman then as a city fireman. During wartime Prohibition, Lake, guided by an older fireman, started selling five-gallon cans of bootleg alcohol to local saloons. He prospered so dramatically that Druggan joined him. Soon they established a connection opportune enough to impress and instruct even Torrio.

The Stenson family had long owned breweries. Legal "near beer" contained less than 0.5 percent alcohol. To make it, brewers made beer as usual, then "bled off" the alcohol to a permitted level. That meant the breweries legally swam in real beer, which obviously could be diverted by those willing to break the law. The Stensons very early displayed downright eagerness. Before the first month of wartime Prohibition ended, the Stenson Brewing Company was in court for a violation of the law. Three of the four Stenson brothers were fined for violations, including the youngest and most active, Joseph Stenson. Before long he would cut a deal with Druggan and Lake to supply them with beer. Eventually the pair would buy into five Stenson breweries, yet from the first they unimaginatively stuck to sales in their Valley neighborhood, branching out only with purchase of the Little Bohemia, a North Side cabaret.

Legend grew that Torrio had bought an abandoned brewery intending to outfit it as one of his barracks-like brothels, but that somehow the light dawned. The connection of Druggan and Lake with the Stensons provided all the outside illumination a mind like Torrio's needed. Indeed, a far greater vision seized him: an amalgamation of organizations that would serve the entire thirsty city, plus the existing suburban operations and many more to come. Only two things stood in Torrio's way. Love and fear had demoralized Big Jim Colosimo.

Love came in the delectable shape of a singer, Dale Winter. Pictures show her tall and slightly moonfaced, but wholesomely, winsomely pretty, deep-bosomed yet slim, a transitional figure from the upholstered-hourglass Lillian Russell era (Russell would be buried in Pittsburgh on June 8, 1922) to that of the boyish flapper.

About age twenty-five, Winter arrived in Chicago in 1917 for a vaudeville job that turned out not to exist. Broke, she had to find work. Arthur Fabri, a violinist, then led the five-piece band at Colosimo's. He ran into Winter making her rounds in the Loop, and recognized the lovely woman, having played on the same bill with her some three years earlier in Grand Rapids. Back then they had become close enough friends (if not lovers) so that now Fabri bought her some requisitely chic clothes and arranged an audition with Big Jim Colosimo, who delighted in his role as the café's impresario. She came, she sang, she conquered.

At first Winter supplemented her salary from Colosimo by soloing at the starchy South Park Avenue Methodist Church. When the congregants discovered her secular calling, even the pastor's sermon on the text, "He among you that is without sin, let him first cast a stone at her" cut no ice from their hearts. Winter resigned. No matter. Colosimo soon moved from boss to protector, hushing the crowd for her performance, fending off groping drunks. She wore white gowns with a red rose at the bosom, setting off her raven hair and golden glow, accenting her apparent innocence while showing off her figure. Her looks, her daintiness, her voice, her manner, what one reporter called her "undeniable charm," left Colosimo moonstruck.

He had concluded some liaisons with fashionable patronesses who thought it thrilling to slum sexually with the big, handsome, reputedly dangerous but visibly amiable owner of the city's premier café–society joint. As a result, the once popular josher of whores and madames had developed a taste for "refinement" in women. But those were other men's women; Dale Winter might be his own. He arranged for Enrico Caruso and Tito Ruffo to hear and judge her voice. When the Chicago Opera Company's maestro, Cleofonte Campagnini, pronounced it educable, Colosimo promptly enrolled Winter in the Chicago Music College.

About this time, Colosimo and his wife, Victoria Moresco, began living apart. Later, Winter would insist, "I wouldn't go with him until we could be married and I told him so. I loved though—I loved him with all my heart!" His love for her was palpable. Astoundingly, given Colosimo's past and character, her account of their premarital abstinence may have been true. Colosimo's chauffeur, Woolfson, claimed never to have witnessed even a good-night kiss when his boss escorted Winter to her far South Side apartment after the café closed each night; he walked her to the door and chastely parted from her there, leaving time for no intimacies. Woolfson noted Colosimo's buoyancy afterward. The great bear behaved like a kid whistling his way home after a date with the prom queen.

Love visited Big Jim just as Prohibition brought its opportunities to Chicago. He had increasingly surrendered operation of his brothels, gambling joints and dives to Torrio, bemused by his status as café society's raffish darling. Dale Winter's refinement suggested possibilities of a sanitized, respectable life. What a consort she would make for a newly genteel Colosimo! But those hopes, fueled by love, led to fear.

A new life and new Colosimo seemed improbable were he to become king of bootleggers, as Torrio wanted. Neither his position nor income as vice lord was especially visible. Patrons might dally with elegant courtesans they met at the café, but even the most concupiscent among his society patrons were unlikely to frequent Torrio's two-buck cribs. From the outset, however, bootlegging attracted massive press notice, and none could predict that bootleggers would ever be held socially acceptable.

Worse, bootlegging was dangerous. Colosimo worried more about the federal angle than Torrio. What's more, Torrio's envisioned empire would inevitably excite the same sort of Black Hand attacks Colosimo's brothel prosperity had. In fact, he apparently suffered further approaches in the fall of 1919, this time with the added agony of threats against Winter. Then there were all those hijackings. So whether hounded by feds, victimized by extortionists or targeted by rival bootleggers, Colosimo dreaded the consequences of expansion. For what? Bathhouse John Coughlin had often retailed the advice he had received from his own mentor, Senator William Mason: "Never take anything big. Stick to the little stuff. It's safer."

To be sure, Colosimo was already into bootlegging—in a safe little way. Torrio had prevailed on him to back Jack Guzik, a roly-poly, bright, repellent sometime pimp. Guzik had used $25,000 of Colosimo's money to buy into a small brewery that now supplied much of the beer for their suburban dives. Furthermore, Torrio had bought the Malt-Maid Brewery in the spring of 1919, before even wartime Prohibition. But those remained inconspicuous operations; the combine Torrio contemplated would encompass all Chicago.

If Colosimo allowed Torrio to proceed while himself remaining detached from operations, that presented its own dangers. It meant virtually surrendering to Torrio complete power, all of which would be needed for protection in this new, highly exposed enterprise. Again, what for? Why risk any of it, jail, a bullet, or dispossession, with life suddenly so sweet?

Colosimo would not lend his prestige to the action, and without the top man, Torrio could not swing the weight to put together his combine. National Prohibition dawned and Torrio had to witness Druggan and Lake and others doing—however stupidly, haphazardly, nigglingly or sporadically—what he wanted to do in his methodical, wholesale way.

Whatever hope Torrio might have cherished of changing Colosimo's mind went glimmering when the big man confided his intention of divorcing Victoria and marrying Dale. The highly improbable story has survived that Torrio replied, "It's your funeral." Torrio would not have said anything so impolitic to someone still unchallengeably the boss; moreover, he would not have *felt* that way about love and marriage with a good woman.

In 1912, Torrio had married someone with the qualities Colosimo found in Winter: Anna Jacobs, then twenty-two years old to Torrio's thirty, an attractive if hardly stunning redhead. None of the Levee crowd could figure out where or how Torrio had met Anna, a woman from a small town near Lexington, Kentucky, visibly and audibly not of the Levee world. A lawyer who dined as the Torrios' guest avowed that Anna Torrio was "a gracious, intelligent person of good breeding."

As soon as Torrio married, he moved from his hotel in the heart of the Levee to an apartment at Nineteenth and Archer, at the district's northern fringe, taking his bride away from the raucousness. He did not drink or smoke or sample his other stock in trade. He was mild and temperate in speech: a reporter on the "Torrio beat" never heard him swear. Every evening, generally by six o'clock, he flew straight home, where he and his wife spent the kind of quiet slippered evening they both liked best. Anna once called Torrio "the best and dearest of husbands," her marriage "one long, unclouded honeymoon."

Colosimo divorced Victoria on March 20, 1920. He gave her $50,000, charged desertion, and won an uncontested decree. Victoria later charged that his clout had let him contrive the action without her ever being notified, but Victoria did not languish inconsolable. Two weeks later she married Antonio Villano, twenty years younger than herself; he dabbled in tombstone cutting and petty crime.

Colosimo and Winter eloped to French Lick, Indiana, Chicago's quickie marriage spot. He hired a touring circus to entertain the reception guests (a bear cub bit one of them), then the couple retired to West Baden, Indiana, for the honeymoon before returning to housekeeping in his mansion at 3156 South Vernon, about a mile south of the café.

Meanwhile, Torrio prepared. He consulted his rising young lieutenant, Capone, on the choice of the right man to do the job, and then left him to make the arrangements. A lot might turn on the choice made. This would be no routine "crime-wave" murder; Torrio knew the police would have to exert themselves on this one, so he and Capone could afford no clues and no trail. No matter how carefully plotted and mounted, the action might develop snags. If everything was not exactly right, the killer would need the judgment to postpone. Such cool heads were not easily found.

While Colosimo honeymooned, Capone put in a call to Frankie Yale. It was the sort of favor that would not be refused, and indeed would be returned less than a year later.

The setup came about a week after Colosimo brought his bride back to Vernon Avenue. Torrio called with the good news that Jim O'Leary, in whose mother's barn the Great Chicago Fire had supposedly started, had two truckloads of whiskey for Colosimo—the good, prewar stuff. O'Leary would meet Colosimo at the café at four o'clock Tuesday afternoon, May 11, 1920, to arrange delivery.

Dale had planned to go shopping with her mother, who lived with them. Colosimo promised to send their chauffeur back with the car after Woolfson dropped him off. Colosimo seemed agitated during the ride, muttering in Italian, which the chauffeur did not understand.

Inside the café, Colosimo made his way to a small office in back, where bookkeeper Frank Camilla sat working on accounts. They shared a daily ritual.

"Hello, Frank," Colosimo would ask with a laugh, "what's going on?"

"Nothing," was Camilla's unvarying reply.

Colosimo asked if anyone was waiting for him or had called. No. They talked business for a bit, and Colosimo called his lawyer. He soon left the office, chatted a few moments with some other employees in a back room, and then went to the front of the café as if to leave.

Two doors led out to Wabash, separated by about fifty feet along the café's front. The one Colosimo chose had swinging doors out to the sidewalk; across a small foyer, a glass-paneled door opened into the café. On the other side sat a good-sized vestibule, a door opening into the cloakroom.

A minute or so after Colosimo went to the front of the café—hardly more than fifteen minutes after his arrival—those in the back heard two noises, close together and indistinct, just loud enough to require investigation. No one thought immediately of gunshots; the noises sounded more like tire blowouts, and might have come from the front or the back. Chef Antonio Caesarino checked the back alley. Frank Camilla left his bookkeeping to look out front, using the southernmost door to the street. The spring lock closed behind him, so when he found nothing remarkable outside he had to return by way of the northern, main entrance. Pushing through the swinging doors he immediately spied Colosimo through the glass panels, facedown on the tile floor of the vestibule, blood pooled at his head, his body sprawled so close to the door that Camilla could barely squeeze through.

Torrio had been right; the police went all out. Before long those milling about the café included Chicago's police chief, his first deputy and the chief of detectives. The position of the body and the angle of fire told police that Colosimo had been turning when hit. They saw two possible scenarios: either Colosimo had gone to the vestibule to watch for his caller through the glass panels when the killer, hiding in the cloakroom, came up behind him, the sound causing Colosimo to turn at the last moment; or the killer waited for him up front, they talked and Colosimo had abruptly turned away from him—a surly habit he employed to end conversations that displeased him—when the killer shot. One bullet had pocked the plaster wall of the cashier's booth twenty feet away; the other got Colosimo behind the right ear. His own pearl-handled .38 was still tucked away in his hip pocket.

The police found two intriguing, mysterious notes in the café. One, found at a table, had been scribbled on an order form and bore the phone number of a company, National Rubber Products, with the unhelpful words, "So long vampire," and "So long Lefty," along with "Saturday evening" and the name of Samuel Lavine, whom no one knew. The other, found on a phone booth shelf in the vestibule, was in Colosimo's handwriting. It gave innocuous instructions to someone, never identified, named "Swan": "Don't keep over thirteen men. If you've got more, ask someone to lay off," and ended, "P.S. Anything you make over $50 belongs to me."

Word of the sensational killing flashed around Chicago. Dale, still shopping downtown with her mother, heard the news and was prostrated. "Jim is gone," she wailed two days later, "and everything is gone with him. . . . There will be no happiness for me now."

The police nominated many initial suspects, but even at first hung onto only Victoria's brothers, Louis and José, soon freeing them, too. Both had plausible alibis. Victoria was visiting her new in-laws in Los Angeles, along with her husband. Arthur Fabri, the violinist who had introduced his old friend to his boss, and arguably had his nose put out of joint by their romance, drew only brief suspicion, as did an escaped lunatic who had sworn vengeance for Colosimo's testimony that helped commit him after a murder. Other theories postulated Black Hand revenge for nonpayment—or the reverse, that a Colosimo shakedown victim had rebelled. Or maybe it was a losing gambler driven to desperation, or part of a labor-racketeering war, or a gang rival, maybe bootleg-connected. In short, for all their effort, the police were stumped.

They naturally questioned Torrio and Capone, already known as Torrio's lieutenant. Torrio had been loud in public grief at the killing. "Me kill Jim?" he cried. "Jim and me were like brothers." Both had established alibis, part of the point in importing Yale.

As for Yale, the police came as near to nailing him as could be expected. Since two of the more improbable theories of the crime involved people in New York, the police asked their New York colleagues to check out the only partial identification they had. A porter at the café, Joseph Gabrela, had almost certainly glimpsed the killer, who had been waiting briefly up front before Colosimo had arrived. Gabrela described a man, stocky, no more than five-foot-eight with a fat face, dark complexion, and wearing a black derby. The description fit Yale closely enough to elicit a mug shot from New York. "That's the man!" Gabrela said. Chicago detectives hustled him to New York for a lineup. But eye-to-eye with Yale, the porter wisely said, "The fellow I saw ain't here."

No one doubted that Yale had pulled the trigger or that he acted for Torrio. But the police lacked proof, and the murder remained officially unsolved.

The Church had its own certainties. Colosimo's notorious life as "a public sinner," his lack of attendance in church, his constant failure to make Easter duties, along with his divorce and remarriage, all convinced George Mundelein, Chicago's archbishop, and later a cardinal, to forbid the body's presence in a Catholic church or cemetery. That upset plans for a service at Sancta Maria Incoronata and interment at Mount Carmel. The archbishop counseled "simple rites" for whatever funeral family and friends could arrange.

They contrived the grotesque. A crowd had gathered outside Colosimo's mansion long before the service. Inside were crammed fifty-three pallbearers, working and honorary. Three municipal judges jostled almost enough aldermen for a city council quorum. A congressman also attended, along with an assistant state's attorney, two state legislators, two ward committeemen, luminaries of the Chicago Opera Company, a scattering of Chicago's high-life restaurateurs, and virtually all of its most honored gamblers, dive operators, pimps and whoremasters, including John Torrio, plus the likes of Ike Bloom and Sol Van Praag, who had combined gambling operations with service in the legislature. The guest list was, commented one newspaper, "a strange commentary on our system of law and justice."

After a prayer by the Reverend Pasquale R. De Carlo, pastor of the Italian Presbyterian Church, Alderman John Coughlin sank to his knees by the $7,500 silver and mahogany casket to lead mourners in a Rosary and prayer for the dead. While the Apollo Quartet sang suitable selections, Dale Colosimo shakily descended from upstairs on the arm of Colosimo's attorney, Rocco de Stefano. Pallbearers carried the body out to strains of a band's "Nearer, My God, to Thee."

Dale and de Stefano followed the hearse in a shrouded car. Aldermen Coughlin and Kenna led one thousand First Ward stalwarts, soon swelled to five thousand in a mile-long cortege that weaved through the old Levee to pause ten minutes outside Colosimo's beloved café, where two brass bands played a dirge. Onlookers blocked the surrounding streets and jammed fire escapes seeking a better view.

At the secular Oakwood Cemetery, de Stefano started the graveside eulogy. When emotion choked him, Bathhouse John Coughlin finished up. John Torrio mourned. He may well have been unaffectedly sad that the man who gave him his start in Chicago had to die. Al Capone had gone unshaven between the death and the funeral, an old-country tradition. Colosimo's funeral inaugurated a competition among Chicago gangsters for lavishness—what might be called Obsequies Chic.

No will surfaced, and Colosimo died unwed according to Illinois law, which demanded a year between divorce and remarriage. Neither Victoria nor Dale would get the estate—which to everyone's astonishment turned out to be relatively trifling. Everyone expected $250,000 to $500,000 *in cash*, plus some list of Colosimo's interests. Instead, the office safe and bank safe deposit boxes disgorged only $28,000 in government bonds, a $2,000 diamond lavaliere, small change and stock in a hotel. All told, the estate amounted to $81,000. Colosimo's father, Luigi, already held title to the mansion; the probate court awarded him the estate. He gave Dale $6,000 in bonds and diamonds, $12,000 to Victoria. Dale drifted back into show business, marrying an actor-producer in 1924; Victoria divorced her young husband a year after the murder and disappeared from Chicago.

For Torrio, and even more for the impressionable and still student Capone, Colosimo's life held a significance summed up by two postmortem statements. "There wasn't a piker's hair in Big Jim's head," said Ike Bloom. "Whatever game he played, he shot straight. He wasn't greedy. There could be dozens of others getting theirs. . . . He brought the society swells and the millionaires into the redlight district. It helped everybody, and a lot of places kept alive on Colosimo's overflow. Big Jim never bilked a pal or turned down a good guy and he always kept his mouth shut." A newspaper observed that his name, still a talisman of power, had grown "less and less sinister as the years rolled by."

He had demonstrated the wisdom of making the provisioning of customers with the forbidden a rational business, not a deadly competition, its purveyors valued (and potentially popular) entrepreneurs, not dreaded hoodlums. He had died only because he tried to evade the logic of his life.

CHAPTER 5

Torrio Ascendant

TORRIO ANTICIPATED A bootleg "business" in the classic nineteenth-century sense of trusts in joyous restraint of trade. He saw that there would be enough for everyone with trivial risk. A Prohibition official soon called Chicago "the wettest city in the United States." All it would take, Torrio realized, was organization. He brought to the task a trait born of his heritage, a passion for organization, especially strong among those southern Italians who acquired it during generations of working arid hardscrabble patches.

Italian journalist Luigi Barzini maintained that fear lay "behind the Italians' peculiar passion for geometrical patterns . . . and symmetry in general . . . mainly the fear of the uncontrollable and unpredictable hazards of life and nature. . . ." Barzini saw it in everything from the elaborate pyramids of Italians' fruit and vegetable stands to the complicated patterns of their gardens, the "rigid symmetry" of their city streets, and their nearly oriental admiration for bureaucracy. "The word for this," Barzini wrote, "is *sistemazione*. To *sistemare* all things is considered to be the foremost, perhaps the unique mission of man on earth. . . . *[S]istemare* means to defeat nature. . . . '*Ti sistemo io*' is a much abused threat. It means, 'I will curb your rebellious instincts.' . . . Industrialists often dream of being able to *sistemare* competition, by establishing strong cartels and ironbound agreements. . . ."

In time, Al Capone would outstrip his mentor in the trait; a contemporary observer noted Capone's positive "genius for organization." To Torrio's credit, he saw the potential in Capone and encouraged it. Both shared the problem of having to make disciplined "businessmen" of thugs.

Torrio's favor quickly convinced Capone that his future lay in Chicago, not back in Brooklyn—where, anyway, vengeful Bill Lovett now ruled the White Hand gang, following Dinny Meehan's murder, arranged by Frankie Yale on March 31, 1920, a little more than a month before Colosimo's murder. Torrio sponsored Capone's little Sonny for confirmation, thereafter presenting the boy a $5,000 bond on each birthday, a gesture well calculated to capture Capone's heart, well known for intense loyalty and often soppy sentimentality.

Capone's commitment to Torrio also had a rational base. Ike Bloom claimed that Colosimo "had what a lot of us haven't got—class." Capone could recognize Torrio's brand of class as different in kind from Colosimo's or Yale's. Torrio was profoundly more worthy of Capone's admiration because his was a *useful* class, essential for the *sistemazione* of gangs into an efficient bootlegging cartel.

Take Torrio's insistence that suburban neighbors of his brothels be cajoled and bribed into complaisance, *not* intimidated. Capone learned, as one of his earliest observers put it, "the value of a bland smile and ready handshake." Torrio's slogan became his own signature remark, "We don't want any trouble." Torrio insisted Capone attend night school to improve the way he spoke. Capone seemingly *willed* himself to contain that volcanic temper, which nevertheless would erupt on random provocation. The transformation of Brooklyn thug into Chicago businessman of crime was slow going, but it progressed steadily.

Much later, Capone would claim for tax purposes that Torrio paid him only $75 a week during this period. More disinterested sources pegged his income at about $25,000 a year, representing a quarter share of the brothel profits, with half the bootlegging action promised. He needed it because his responsibilities expanded geometrically.

Back in Brooklyn, Capone's father had been nursing a weak heart for quite a while. The evening of November 14, 1920, Gabriel Capone, then fifty-five, collapsed in a pool hall near the family's flat on Garfield Place and died of myocarditis before a doctor could arrive. His son would eventually have him reinterred in Chicago's Mount Olivet, a tombstone shared with Al's brother Frank, whose killing we'll examine later.

It delighted Capone to provide, one way or another, for the whole family, a mark of success that made up for having had to flee Brooklyn ignominiously not so long before. It required considerable provision. In addition to mother and brothers, the Capones now included Rosalia (called Rose), then about ten, and eight-year-old Mafalda (whose name, after the Italian princess, resisted Americanization). Soon they all joined Capone in Chicago, along with three cousins, Charles, Rocco and Joseph Fischetti.

Older brother Ralph had come first, lured by Capone's ability to get him a job. Ralph had gone reasonably straight since his 1916 fraudulent car sale: he'd held jobs as a telegram messenger, in a bindery, as a streetcar conductor, a salesman, and a longshoreman. During World War I the Marines had washed him out of boot camp because of flat feet. He'd returned to Brooklyn, where he'd tended bar in a speakeasy, twice arrested and fined—though few considered that criminal behavior. In Chicago he started as a barkeep, soon succeeding Al as manager of the Four Deuces. He shared an apartment near the Deuces with Al and Mae. The Fischettis—who also worked for Torrio—had their own flat in the same building.

For the rest of the family, Capone bought a fifteen-room, brick-faced duplex at 7244 South Prairie. The neighborhood offered solid, petit-bourgeois comfort, and the house sat flatteringly close to the new home of John Torrio, an apartment at 7011 South Clyde.

Little pleases Chicagoans more than Carl Sandburg's description of The City of Broad Shoulders, the pragmatic *working* city, the city that works. In the twenties, its friends, admirers and apologists endlessly extolled its explosive growth, its industrial might and wealth. Many loved it. One Nebraska farm boy hitchhiked in to watch Babe Ruth play the White Sox, alighting at the corner of Michigan and Roosevelt Road and walking north. "By the time I reached the near North Side," says Howard Browne, "I said, 'I'm not leaving this damn town.' I fell in love, the way you fall in love with a woman." A ninety-six-year-old retired policeman agrees. "It's a wonderful man's city," says Michael J. Mullins. "If you can't get a job there you can't get a job nowhere."

The portrait limned by the city's lovers did, however, show another face. About the time Al Capone hit town, one visitor found Chicago a "valley of factories," a mean hodgepodge of breweries, saloons, stores, telegraph lines, railroad tracks and "hideous small houses." He called it "The Ugly City." Chicago had already experienced what the rest of the country would later bemoan as urban blight and suburban flight. The Gold Coast still sheltered along the lake behind the barrier of Michigan Avenue, but the solid citizenry had long since removed to North Shore suburbs or way south. For instance, of the thirty-four families that shared the block with the Capone home on Prairie in 1929, Capone's was the only Italian family. The others included three police officers—an undersalaried yet highly lucrative calling in those days—a clothing manufacturer, a druggist, a grocer, and a retired Presbyterian minister. That left the city proper to successive waves of immigrants, living in squalor and poverty until they could engineer their own flight.

Chicago's potpourri ethnic mix contributed as mightily to its bootlegging as to its industrial and cultural strength. The 1920 census counted 11.81 percent of Chicago's 2,701,705 population as "Polish," 10.55 percent "German," 5.40 percent "Irish," 4.80 percent "Italian," and so on, down to 0.29 percent "Belgian" and 0.11 percent "Indian, Chinese, etc." To a good many of these, the forbidding of a traditional, almost culturally mandated stein of beer, shot of slivovitz, glass of Chianti—*that* was the crime, not their provision. Even without special urgings of culture or tradition, many of the 23.80 percent termed "American" shared a strong disaffection from temperance.

* * *

Although Chicago glistened as the right city at the right time for what Torrio proposed, the most prominent of the other gang leaders seemed numb to the lessons of what even their casual, unorganized bootlegging promised. They insisted on "crime as usual."

Consider Terry Druggan and Frankie Lake. They certainly should have known better; for them, the usual kinds of crime *hadn't* paid. Bootlegging rescued them from lives of honest toil, and made them so rich that by late 1923 the IRS would chivvy them about taxes. Nevertheless, they continued throughout the first years of Prohibition as they had back in their juvenile "Valley Gang" days. In 1921 and 1922, police arrested Druggan for involvement in a measly $11,000 jewel robbery, arrested Lake for shooting former policeman Timothy Mulvihill in the partners' club, Little Bohemia, and arrested both for hijackings that netted them as little as one truckload of beer—at a time when they had or were acquiring an interest in five breweries! That they never stood trial for any of those offenses is beside the point.

Equally purposeless, though of more historical significance, was their July 1921 murder of Steven Wisniewski. He was nothing, the operator of a greasy spoon, branched into bootlegging. But one time he anticipated Druggan and Lake, looting an alcohol warehouse before they could strike. Always choleric, Druggan scolded him for his presumption. Wisniewski, a powerful brute whom newspapers called "Big Steve" and "the blond giant," replied by knocking Druggan senseless. Three weeks later, Druggan and Lake caught up with a drink-befuddled Big Steve in a Valley saloon. With help of two others, they wrestled him into a car. His body was found on the roadside near Libertyville, about twenty-five miles northwest. Wisniewski's was the first clearly identified Prohibition gang killing, and the first instance of someone being taken for a ride, that is, transported alive to a more tactful spot for unobserved killing.

Herschie Miller, the West Side gang chief, and Samuel Morton were another case. "Nails" Morton (born Markowitz) had won a Croix de Guerre and battlefield commission in France when he led his squad over the top, although already twice wounded. A high-stakes gambler, he also dabbled in bootlegging with Miller.

On August 23, 1920, a little after one A.M.—then the legal closing time in Chicago—Miller and Morton left Ike Bloom's dance hall and went to The Pekin Inn, which upstairs housed the Beaux Arts, a black and tan cabaret that ran full blast after the more exposed Pekin closed. About three A.M., Detective Sergeant James A. Mulcahy ambled in, leaving his fellow sergeant, William E. Hennessy, still outside talking with a former Prohibition agent. Miller, Morton and Mulcahy moved to an inner alcove, a wine room, evidently to discuss a past liquor deal. An argument erupted over the split and quickly degenerated into a fistfight between Morton and Mulcahy. Morton knocked the detective down, turning his mouth into a bloody mess. Witnesses saw Morton standing over the fallen policeman, Miller holding him back. Mulcahy's partner, Hennessy, hearing the fracas, rushed upstairs and into the room. The prostrate Mulcahy pulled a revolver and hollered ''stick it on 'em'' to his partner, who pulled his own gun.

Nails Morton wisely backed out of the alcove, but tripped over a chair and sprawled on the floor. Mulcahy leapt up, followed and kicked Morton in the head. Miller drew his own gun. He and Hennessy fired. Hennessy died with four bullets in him, his own single shot shattering a mirror. Miller killed Mulcahy with one more shot.

Nothing much resulted. The Pekin Inn was closed, eventually to house a police station and courtroom. Miller and Morton claimed self-defense, won acquittals for each killing.

What kind of ''business'' was this? Two years later, Capone would give his own display of boozy hotheadedness, but at least it was page-five stuff, not a front-page, headline cop-killing. Torrio was teaching Capone to keep centered on the *business* of supplying the public's illicit desires, not wandering off into jewel robberies and warehouse raids and random killings.

Dion O'Banion presented the most egregious example of wrongheadedness. Early a thief and jackroller (mugger of drunks), he graduated to journeyman safecracker ''in a small way,'' as one newspaper put it; he once managed to blow out the walls of a building, yet leave the safe intact. He had scarcely amassed a fortune at it; five jobs throughout 1918 had yielded a total of only $9,919. Then came Prohibition.

O'Banion may have been the era's first hijacker. On the bitterly cold morning of December 30, 1919—before the start of national Prohibition—Deany (as his many friends called him) found a sidewalk in the Loop blocked by a flatbed truck, its driver oblivious to passersby. The truck stood in an alley next to the Bismarck Hotel, loaded with a delivery of Grommes and Ullrich whiskey. On impulse, O'Banion mounted the running board, got a headlock on the driver through the window and knocked him senseless with his fist. He dumped the driver out and drove off. ''I had no idea what to do with the stuff,'' he said later. ''I drove around for fifteen minutes and then headed for Nails Morton's garage out on Maxwell Street.'' (Besides gambling and bootlegging, Morton sidelined in processing stolen cars.) Twenty minutes of phoning found saloon buyers for bargain whiskey; they sold the truck to a Peoria brewery.

Easy money, modest risk. But the small change earned and hazardous working conditions compared badly with profits from wholesale manufacture and sale of beer and booze. Incredibly, O'Banion *understood* that—he bought into breweries and soon became one of Chicago's richest bootleggers—and *still* continued his mindless hijacking and warehouse raids. Witness the indictment for his raid of March 14, 1921, on the Price Flavoring Extract Company, where he scored twenty barrels of alcohol, worth $5,500; or his indictment for a hijacking on April 12, 1922, which netted 225 cases of Susquemac Distilling Company liquor worth $6,075. Forget that he always managed to buy or intimidate his way out. The risk plainly outweighed any possible gain, yet the madcap kept it up almost to the end. He even persisted as late as mid-1922 in blowing safes—having to pay about $30,000 for one ''not guilty'' verdict after he was caught red-handed!

In the midst of such crackbrained goings-on, Torrio's problem is transparent as is the reason he so took to Capone, who was eagerly malleable and already half-persuaded by observation of Frankie Yale's enterprises that the proper business of crime was business. Torrio had a hard enough time channeling his own thugs, let alone independent bravos. He strove for *sistemazione*; wild colonial boys like O'Banion craved *action*. All Torrio could do, with Capone's help in the field, was pursue his plan and use the allure of its success as empiric argument.

Aldermen Coughlin and Kenna were naturally amenable. They wanted turmoil no more than Torrio did. Torrio lacked Big Jim's charisma. He did not enjoy the same personal sway over the ward's Italian votes; but that sway was finally rooted in money, clout and patronage, so Torrio's position as leader of vice made up for deficiencies in congeniality.

Torrio moved immediately to consolidate and expand his power base. He put Capone in charge of lining up those pool halls and cigar stores not already handbooks. Where persuasion proved necessary, Capone displayed technique precisely to Torrio's taste. With his masterful size and manner he seldom had to take the velvet glove off the iron fist. As one observer put it, ''With no conscious effort he emanated menace while saying please.'' In reserve he had the ''Look.'' Like most gangsters he had been taught, young, and had assiduously practiced ever since, the knack of staring a victim out of countenance. It meant emulating the puff adder, swelling the neck muscles and widening the eyes in a fixed glare that said, ''I'm not fooling now; give in before I have to whack you.'' Fledgling gangsters might practice the Look before mirrors to invest it with maximum menace. Capone had it pat.

While Torrio used Capone to advantage in town, he concurrently intensified his move into the suburbs, places like Stickney, Forest View, Posen, Burr Oaks, Blue Island, Steger, Chicago Heights and other towns and villages south and west of the city. Suburban brothels did marvelous business, some running three 8-hour shifts, with as many as sixty whores. Torrio liked the control possible in the sticks. Although brothels still dotted the city, they now had to operate small-scale, almost furtively. Some thirty whores made the Four Deuces relatively Gargantuan. At that, its inmates had to stay three discreet flights up. More typical were the Cort, around the corner on Twenty-third Street, which masqueraded as a hotel and harbored only eight whores, and the Covart on West Madison, which could support only three by day, a night shift of no more than eight, sometimes only six.

Moreover, urban brothels had become cheerless places. At the Four Deuces, for example, dispirited customers sat on rude benches set against the bare walls of an otherwise unfurnished, harshly lit reception room waiting their turn with the more comely whores while an obese henna-haired madame chanted from her rocking chair, "Come on boys, come on boys, pick a girl," to wheedle more efficient utilization of the assets.

By contrast, the suburban places could operate openly and grow ever bigger once Torrio's saccharine wooing had subverted neighbors and local authorities. That meant concentration, which allowed Torrio better control, easier administration and protection. Openness also meant that combination cabaret-brothels like the Arrowhead in Burnham could foster a party atmosphere. Some suburban dives featured so much festivity that management bolted down benches and tables to spare their being smashed as weapons during inevitable drunken brawls.

Occasional breakage was a small price to pay. The conviviality of suburban operation hugely expanded the market for beer, always 90 percent of alcohol sales in Chicago. Torrio could supply his own beer after his deal with the Stensons—all to his further profit.

Politics also spurred Torrio's suburban flight and added weight to his gangland arguments for combination. By mid-1921 Mayor Thompson found himself embattled, his administration and machine disintegrating. He was on the outs with Fred Lundin, who resented not being consulted in Thompson's appointment of Charles Fitzmorris as police chief. In turn, Thompson resented Lundin's monopoly of patronage and his squandering the machine's total campaign fund in a futile effort to reelect a particularly noisome county judge. The mayor didn't care, either, for the continuing public babble about how Lundin had made him and how Lundin really pulled the strings. And he *hated* the embarrassment of a scandal over school board graft with actual indictments of many, including Lundin and his nephew, Virtus Rohm.

Despite the tattered condition of his machine, for the June 1921 judicial election Thompson advanced a slate whose only visible recommendation was loyalty to himself. The bar association, reformers, the public, even Hearst's *American*—the mayor's only major newspaper support—all voiced outrage. A bipartisan ticket trounced Thompson's. Press comment ran to talk of "crushing defeat" and hopeful warnings that the whole country would watch Chicago's 1923 mayoral elections to see "whether Chicago really meant what it said in 1921." Thompson appeared washed up.

While Torrio could not foresee how relatively troublesome Thompson's successor would turn out, he could scarcely conceive another administration as *good* for his business. Positioning operations beyond the reach of possible city reform made increasing sense.

So did rescuing the governor. Len Small's gubernatorial pardons had made him in effect John Torrio's recruitment and personnel officer; his complaisance helped ensure tranquil operation in rural areas that the state policed. Now the governor was in trouble, indicted in July 1921 for corruption during the war when he had been state treasurer. He had placed state deposits with the Grant Park Bank, an institution extant only in the imaginations of the two Curtis brothers, their only depositor the state. They paid interest, duly turned over to the state—but barely 2 percent, though the war had multiplied interest rates. In all, prosecutors claimed, the scheme had netted at least $819,690.31 in four years.

Torrio sent intimidators and bribers to Waukegan, where Small's trial would be held. The bought and browbeaten jury acquitted Small on two ballots. The governor would later show himself suitably grateful.

Another sort of politics in the city council elections of February 1921 showed all gangs why Torrio was right about their needing a united front.

Anthony D'Andrea—perhaps a defrocked priest, certainly a convicted counterfeiter—decided to challenge Alderman John Powers in the Nineteenth Ward. D'Andrea had failed in earlier races, but had meanwhile become president of the Chicago chapter of the Unione Siciliana. This organization mixed fraternal and benevolent activities with the deeper purpose of providing national cover for gangster operations. With that base, D'Andrea exuded serious strength, and John Powers sought to placate him. The alderman offered D'Andrea the ward committeeman's post. The state supreme court blocked that ploy, due to D'Andrea's criminal record, but D'Andrea wanted it all, anyway, and declared war.

In late September 1920, a bomb destroyed the porch of Powers's house. Then D'Andrea announced his candidacy for Powers's aldermanic seat in the February 22, 1921, election. "Why shouldn't the Italians have one of their own race to represent them in the city council?" he asked. "The Nineteenth Ward is composed largely of Italians. This vote is with me solid. And I want to tell you something else—the Irish are with me."

Wrong on both counts. On bread-and-butter issues like jobs, and even on sharing political clout, Johnny Powers had been ecumenical, winning many unshakably loyal Italian supporters, who jokingly called him "Gianni Pauli." Several bilateral bombings, shootings and sluggings later, Powers won by 435 votes that survived a fraud challenge. Now came payback time.

Paul A. Labriola, a municipal court bailiff and Powers precinct captain, had spurned all appeals to switch sides. When he told Salvatore Ammatuna (always called "Samoots") and Frank Gambino, two hoods allied with the Genna brothers, D'Andrea's main muscle, that he owed the Italian nothing and the Irishman a lot and would stick with his benefactor, they threatened him. The threats were renewed after the election. "I hear they're out to get me," Labriola told a policeman friend.

On March 8, two weeks after the election, three gunmen waited at the corner east of Labriola's home at 735 West Congress Street, and two more waited at the corner of Halstead, to the west. Just after nine o'clock that morning Labriola left for work, heading west. The two gunmen at Halstead walked toward him. He seemed to recognize one of them, nodded, and was about to speak when they opened fire. The three behind him ran forward and also opened fire as Labriola fell. One stood straddling his body for a final shot. Nine bullets were in him. Labriola's wife heard the shots. She rushed out her door, saw her husband's body, and collapsed. He was thirty-nine; they had two children. The gunmen included Ammatuna and Gambino, plus John Gaudino and Angelo Genna, youngest, toughest and meanest of the brothers. No witness ever identified the fifth gunman.

At one o'clock that afternoon they hit again. Newspapers already carried the story of the Labriola killing. Four men bought a paper from a newsboy at the corner of Garibaldi Place and Taylor Street, read the story and walked down Garibaldi, commenting on the killing. Halfway between Taylor and Polk they came to the cigar store of Harry Raimondi, who styled himself "Raymond" except in Little Italy. Raimondi and D'Andrea had been friends, but Raimondi remained faithful to Powers in the showdown.

Two of the gunmen checked out the shop. After they left, the other two entered and purchased cigars. When Raimondi reached for change, out came pistols. One shot got Raimondi in the left temple, the other two in the chest.

D'Andrea called the killings "regrettable," but complained, "I knew nothing of it and I can't see why my name should be dragged into it." Powers offered $2,500 reward for each killing's solution and quoted his colleague, Alderman Bowler, "It is worse than the middle ages."

The slaughter was indeed gothic, and seemed to continue about as long as the age, with more killings on both sides. D'Andrea may not have borne direct responsibility. At any rate, he tried to cry quits. He disclaimed any further interest in politics or the Nineteenth Ward. Too late. In mid-April, Abraham Wolfson found two notes in his mailbox. He lived in the first-floor apartment of the building D'Andrea owned. D'Andrea lived on the third floor with his wife and three daughters. The first note said cryptically, "He killed others. We are going to do the same." It was signed "Revenge." The second gave instructions:

> To the Tenant, First Floor, 902 South Ashland avenue: You are going to move in fifteen days. We are going to blow up the building and kill the whole D'Andrea family. He has killed others. We do the same. Move and save your lives.
>
> Revenge

The police thought the notes were a joke; D'Andrea thought they were Black Hand, and not worth worrying about. Wolfson, a real estate dealer, thought it was time to move. He vacated the ground floor before May 1, 1921.

About two A.M. on May 11, after a night of cards in Amato's restaurant, D'Andrea's bodyguard and driver dropped him off outside his home and drove away. D'Andrea mounted the nine steps from the sidewalk to the building's porch. A shotgun exploded from the empty apartment's bay window to his left. Thirteen slugs hit him. He crumpled, but before he died he managed to pull his own gun and fire into the now empty window. The killers had entered through a rear basement window. One had gotten his hand in coal dust and left smudges on the wall. As they fled out the back to a waiting car, the killers also left a new sawed-off shotgun and a size 7 hat with a $20 bill in the lining and a note that read "For Flowers."

D'Andrea's bodyguard, Joseph Laspisa, was killed just over a month later, suspected of having betrayed his boss. Joseph Sinacola, Laspisa's close friend, had sworn to avenge the murder. He was gunned down on July 6, 1921, practically on his own doorstep. The killings, back and forth, continued through the summer and fall; about thirty died in all. Despite a number of eyewitness identifications, Chicago convicted no one.

Immediately after it happened, a colleague declared that D'Andrea's murder would surely end the battles. "His followers may avenge his death, but I doubt it," he said. "They know now that they have lost the powerful political influence of D'Andrea and they are not likely to be so bold. . . ." Although he was wrong, his reasoning was sound. The upshot of D'Andrea's challenge to Powers enforced a great lesson on all Chicago's gangsters at the pregnant time for Torrio. It demonstrated the ascendancy of political clout throughout the city. No more fearsome group of killers existed than the Genna brothers and their hirelings, yet even when they had D'Andrea's clout behind them they could not cow people who knew *they* had a single powerful alderman. *Without* D'Andrea, they should have realized they stood no chance at all. For all the killing, the Gennas could not prevail against equally ruthless and determined forces backed by a politician's clout.

Clout meant everything. In the unsettled conditions attending Mayor Thompson's decline, who knew what fixes still held, what sources could still deliver protection? Chicago gangland *had* to combine.

Torrio proposed the right kind of combination. To the small-time operators he offered employment and protection. To the large independents he offered cooperation, assured supplies and protected markets. Members would respect each other's territories. No one outside the combine could move in; no speakeasys could buy from anyone but the designated gang-overlord of that territory. All would collectively enforce those rules. Although the gangs could buy beer and booze from any source they wished, or like Druggan and Lake make their own, Torrio stood ready to supply anything they needed at a price quite fair considering the markup thirsty patrons had shown they would pay.

Torrio would also supply the more precious commodity of experience and contacts-in-place for the routine corruption bootlegging required—a dramatically more complex and ongoing proposition than the ad hoc bribery felons were used to arranging for themselves each time they were caught. Bootlegging had to be fairly open, as Torrio's brothels were, subject to unstrenuous police detection. He already knew how to keep the continuous fix in.

Torrio did not demand or insist, he cajoled and reasoned. And prevailed. Chicago entered the second year of national Prohibition with at least the skeleton of a criminal combination in place.

CHAPTER 6

Capone Rising

BOOM TIMES CAME to Chicago crime. As early as 1921, before Torrio really got rolling, an Amazon of beer gushed from at least eight full-scale breweries. Volume grew so large so fast that on January 6, 1921, three of Torrio's men incorporated the gang's own trucking firm, World Motor Service Company. Incorporation helped confound vehicle seizure laws: ''respectable'' owners of bootlegging trucks swore they had no idea what the vehicle was being used for and got their property back.

By 1922, local bootleggers supplied a vast and increasing demand. A Chicago referendum on modification of the Eighteenth Amendment to allow beer and wine passed five to one. It carried no legal weight, but citizens also voted with their throats.

One of Capone's duties was to buy trucks for Torrio's operation—which those in it called ''the outfit,'' especially after Capone took over. That was *not* a name of consequence or significance, like crime reporters' and novelists' lurid ''The Syndicate.'' Members used it casually when talking among themselves about their group—''I joined the outfit two years ago''—as they would say ''I'm with Capone,'' when talking about their affiliation to an outsider. Many in the outfit were involved in transporting beer; for most of them, being part a gang meant little more than being a truck driver: that was the entry-level job for most, with promotion to muscle and racketeering work. But Torrio relied on few to buy new and used trucks. One was Capone. By mid-1922 Capone already ranked as Torrio's number two, a fact long obscured by several misunderstandings.

In the early morning of August 30, 1922, Capone, roistering with three men and a woman, careened his car east along Randolph Street and crashed into a taxi parked at the corner of Wabash. Capone's companions wisely beat it; he rushed over to the cab spoiling for trouble, his alcohol-sodden temper exploding. Revolver in one hand, the star (badge) of a special deputy sheriff in the other, Capone ranted at the taxi driver, Fred Krause, still slumped over the wheel, so badly injured he needed a hospital. Capone threatened to shoot the man. The motorman of a passing streetcar, Patrick Bargall, stopped and said he'd seen the whole thing. Why was Capone waving that gun? *He'd* been to blame for the accident, not that fellow bleeding there. Capone now threatened to shoot Bargall. Fortunately, the police clanged up with an ambulance. They rushed Krause to the hospital and hauled Capone to the police station, where they locked him up, pending bail, to answer three charges next day in police court: assault with an automobile, driving under the influence and carrying a concealed weapon.

So far no misunderstanding. Capone was twenty-three; Torrio could not remake him overnight, and nothing would ever quite curb his temper or his recurrent taste for debauch. The misunderstandings started in 1930, when Capone's first biographer quoted only the first news report of the incident, commenting that it demonstrated Capone's "comparative obscurity." He noted that only one paper used the story, and on an inside page, further suggesting nonentity by giving the name as "Alfred Caponi."

The story Fred Pasley referred to was by the City News Bureau, a cooperative owned by all the major Chicago papers, a sort of one-city Associated Press, staffed by young and green reporters. The City News story had Caponi "living at the notorious Four Deuces," and ended with his arrest at the scene.

The *Tribune* changed the story before printing it. The newspaper's more knowledgeable editors knew Capone was "alleged owner of the Four Deuces," and also knew what had happened at the police station. He had, the *Tribune* story concluded,

> threatened the arresting officer with immediate discharge, and promised his "pull" would make things unhealthy for his prosecutors. "I'll fix this thing so easy you won't know how it's done," he is alleged to have boasted.

Police and reporters hear vainglory like that a dozen times a day, almost all of it vaporing. Capone spoke only the truth. He was soon bailed out, and the case was never even called in court, the record expunged. He already commanded that kind of clout, scarcely the mark of a nobody.

One policeman never doubted Capone's status even that early. Charles Trilling had joined the force the year before. About the same period as the accident, he was on traffic duty at the corner of Twenty-second and Wabash. Capone always put himself out to make friends with policemen. He took to the young, bright Trilling. "Hi, Red," he would boom on his way to the Four Deuces. "Hi, Al." One day Capone said, "Hey, kid"—his habitual vocative; he was one year older than Trilling—"any time you want to make sergeant, just let me know." Trilling pondered before passing up the offer. Trouble was, he knew he'd live thereafter under the big man's thumb. But everything about Capone—how he acted, the deference he received, what Trilling had heard—all persuaded him that Capone could make good.

Also in 1922, the IRS knew him as someone earning so much beyond the wages of a salaried dogsbody that agent Edward P. Waters sought out "Al Brown" in a Cicero dive to talk about taxes. Like most criminals, Capone had adopted an alias to frustrate easy identification and to spare his family disgrace. In time, calling him "Mr. Brown" became a mark of respect.

The news story's mistake about Capone's first name meant little. Papers were much less exacting about such matters then. The spelling "Caponi" meant less. The *Tribune* of Colonel Robert Rutherford McCormick did not need anybody to tell it how *anything* should be spelled. An element of McCormick's autocratic tetchiness was spelling reform; the *Tribune* long stuck to the likes of "clew" for what detectives seek and "boro" for what Manhattan is one of five. The paper continued to spell it "Caponi" as late as December 31, 1928, long after Capone had achieved *world* notoriety, and long after the *Journal*, the only other Chicago paper that had regularly ever spelled it that way, had switched. It had nothing to do with obscurity.

As shown on the endpaper maps, the shoreline of Lake Michigan curves south-southeast from Waukegan, running straight north-south opposite downtown Chicago—about North Avenue to Cermak Road (then called Twenty-second Street, heart of the old Levee). The north branch of the Chicago River parallels the lake's shoreline. Six miles from the shore at the north suburb of Evanston, it is barely over a mile away where it meets the south branch and runs due east to the lake. The south branch heads due south, then swings sharply southwest just before Cermak. The shoreline and the two branches of the river form a misshapen H, with the river's main branch the crossbar.

Madison Street, four blocks below the crossbar, divides street numbers north and south. State Street divides numbers east and west. Since only three blocks separate State from Michigan Avenue—the last major boulevard before Lake Shore Drive—downtown Chicago has almost no "east" addresses.

Street addresses aside, Chicago's North Side started north of the river-crossbar. In the Torrio era, "West Side" meant west of the river-upright, effectively that part west and southwest of the crossbar. The South Side did not start immediately south of the river, because the Loop was a special case, referring to the loop made by the elevated train tracks that bounded the area of Chicago's biggest stores and brightest lights. State, for instance, did not become "that great street" until it hit the Loop. Down near Roosevelt Road (Twelfth) it was (and is) site of police headquarters and the east border of the railroad tracks. North of the river State boasted only slightly greater distinction. The South Side, once it started, reached down to the Stockyards area well below Twenty-second Street.

Before Torrio's plan, territory meant little, because most gangs practiced traditional crimes of thievery and mayhem against targets of opportunity wherever found. They were territorial only in where they hung out, though with preference for home-ground marauding. Torrio's plan codified those areas, ratifying old stomping grounds as new territories for exclusive bootlegging rights.

Most of the North Side belonged to Deany O'Banion. Like him, his lieutenants were jacked-up safecrackers and strongarms, and O'Banion presided over the most collegial Chicago gang. After his death the rumor mill ran briefly to talk of gang leadership residing in a sextumvirate of his top associates, not just one boss. Also, more than with any other gang—even Capone's at the height of his fame and popularity—O'Banion bound his people to him by personal affection. A fearsome enemy and casual killer, he *personally* dispatched twenty-five victims according to one police chief, sixty-three by another reckoning. But friends found him a *good* friend, and loved him. When not killing, he was almost a caricature smiling Irishman, open, radiant, unfailingly jolly. ''Oh, no,'' he'd cry when someone started to defame a mortal enemy, ''swell fellow!'' Then he might order the swell fellow's death or do the job himself on whim. Once his driver noticed gore on the seats after O'Banion had returned from a lone drive; ''Clean it up,'' O'Banion said, offhanded and cheerful, with no explanation. Contemplated murder seemed no reason for grumpiness in what one psychologist called a mind of ''sunny brutality.''

It was no master mind, and perhaps O'Banion knew it, which accounted for his sometime willingness to take counsel. His brightest subordinate was Earl Wojciechowski, born in Poland, brought to the United States at age three. For reasons no contemporary recorded, he adopted the name Hymie Weiss, an oddly Jewish-sounding choice for a convinced Catholic who wore a crucifix and often used a rosary as worry beads. Lean, mean, with sharp features dominated by intensely staring eyes and a lava temper, Weiss nevertheless had greater vision and cooler judgment than his boss. Police thought him the one who had steered the gang into bootlegging.

Vincent Drucci (his real name Di Ambrosio), the only Italian in a leadership position with O'Banion's North Siders, had started small, pilfering telephone coin boxes as a kid. He stayed rather juvenile; they called him ''Schemer'' for his harebrained robbery and kidnapping plans. Like Weiss, he had been with O'Banion since adolescence. (At the start of Prohibition O'Banion was only twenty-eight, most of his gang in their early twenties.)

Louis Alterie, born Leland Varain in California, son of a French immigrant, was the gang's flake. He had boxed as a pro, a palooka, had a ranch in Colorado and liked to be called ''Two-Gun'' (.38s tucked in shoulder holsters, with a spare in his car); one reporter called him a ''wild Chicago gunman and wooly Colorado ranchman.'' He started in Chicago as companion to Terry Druggan in a mad jewel robbery, formed and ran the Theatrical Janitors' Union, ran a small gambling operation and hooked up with Herschie Miller and Nails Morton before gravitating to O'Banion, whom he idolized. Once, to delay a murder prosecution until witnesses could be discouraged or made to disappear, he feigned insanity, and was entirely convincing. Eventually the gang would exile him for his flakiness.

George Moran was the gang's most stolid big shot, despite periodic rages so abrupt and unmanageable newspapers called him "Bugs." He preferred the alias "Morrissey," but in fact was of Polish extraction—his real name either unknown or forgotten by authorities and newspapers of the time. Moran was slow moving and slow thinking.

Two more rounded out the top echelon. Newspapers considered Daniel J. McCarthy so elegant they called him "Dapper Dan." Originally a labor racketeer, business agent for the Journeyman Plumbers' Union, he once shot a policeman who had come to arrest him for desertion from the army in 1918. He became O'Banion's social companion and partner in one of his early breweries, Cragin Products. The other heavyweight, Nails Morton, was less gang member than friend and occasional partner in O'Banion's escapades. He joined O'Banion in buying an interest in a florist shop, O'Banion's chief delight, which became North Side gang headquarters. Having acquired some trappings of the gentleman along with his battlefield commission, Morton tutored O'Banion in certain social graces, introducing him to dinner dress, for instance. O'Banion, in turn, made dinner jackets a fashion statement among Chicago gangsters.

Little Sicily—southwest corner of O'Banion's territory—formed a small enclave that O'Banion could exploit only partially and warily. Conversely, O'Banion's bootlegging rights bulged a little south of the river, down to Madison, into Torrio-Capone territory.

O'Banion's gang, under its successive leaders, presented Torrio and Capone with their only challenge for domination of the city. The gang's strength and esprit derived partly from the depth and forcefulness of leadership, something untrue of most other Chicago gangs. For instance, when Torrio advanced his plan, the leader of the South Side O'Donnells was in jail and the rest of the gang counted for so little they could be ignored—with results we'll examine later. The other factor that gave the North Siders such importance was the size and wealth of their realm.

About one and a quarter miles south of O'Banion, the same distance north of the Four Deuces, and just across the south branch of the river, the six Genna brothers owned Little Italy, the old Nineteenth Ward, battleground of Powers and D'Andrea, renamed the Twenty-fifth after a 1920 redistricting. Samuel, James, Peter, Antonio, Michael and Angelo Genna had been obscure before Prohibition. Contemporary reporters were even unsure whether they had come to the United States (from Marsala, Sicily) as children or adults. Whichever, they had established themselves firmly in the neighborhood. Sam was a Black Hand extortionist, with tough Angelo and Mike as enforcers; Peter and James ran saloons. They acquired a pool hall and a fruit store, importing olives and cheese. Tony had some intellectual and artistic pretensions. He was a patron of the opera and a student of architecture who in time would put forward plans for low-cost housing. His brothers were meaty men; Tony stayed relatively slim and elegant. Called "Tony the Aristocrat" and "Tony the Gentleman," he once had a pedicure in preparation for an operation. He did most of what thinking had to be done for the family's operations.

The lessons of the Powers-D'Andrea battle necessarily hit the Gennas first and hardest, and they were most eager to throw in with Torrio. That left them free to capitalize on an idea which came from the man who would become Angelo's brother-in-law, lawyer Henry Spignola. The Gennas organized alcohol cooking as a cottage industry throughout Little Italy, conglomerating so many small home stills that they soon found themselves masters of a giant distilling enterprise, a money fountain.

Their territory was not big. Centered at Taylor Street and Halstead Avenue, it sprawled along Taylor from the river to about Western Avenue (2400 west) with Congress Avenue (500 south) and Sixteenth Street roughly its north and south borders. On a map it looks like a morsel about to be devoured by jaws composed of two other gangs, the upper being O'Banion's.

The lower mandible belonged to Druggan and Lake's gang, at first roughly coterminous with the Valley. Eventually their territory grew a northern extension that reached up to Madison and snaked west, past Cicero Avenue to about Laramie. The extension nestled against the territory of the West Side O'Donnells (distinct from the South Side O'Donnells) to form the jaw socket. The West Side O'Donnells' territory stretched from just west of Halstead to Austin, then spread south, surrounding the west end of Druggan and Lake territory, past Roosevelt Road, into parts of the town of Cicero, a presence that would give Torrio an opportunity, give Capone trouble and finally give Chicago fits.

Only the oldest of the three West Side O'Donnell brothers, William, really counted. For unknown reasons, he was always called "Klondike." Thick-bodied, with red face and black hair, he was older than the general run of gangsters. When Klondike went to jail in the later twenties, the gang no longer figured as a player. His much younger brother, Myles, slim, fair and sickly, lacked the grit to take charge; the middle brother, Bernard, was lucky to have a job. After Klondike, the most vibrant member of the all-Irish gang was James J. Doherty, who might have kept things going in Klondike's absence had he not been killed in the 1926 episode that would exercise the city and astound much of the nation.

About two and a half miles due south of the Gennas, starting about Forty-third Street, lay a kidney-shaped territory, trending southwest to around Kedzie and Sixty-third to embrace the Stockyards. This belonged to Ralph Sheldon's gang, a continuation of Ragen's Colts, an old-time Irish club devoted to sport, politics and social activities, all flavored by frequent mayhem. Started as the baseball team of the Morgan Athletic Club, the organization disbanded and reconstituted itself as Ragen's Athletic and Benevolent Society to avoid a damage suit by the Santa Fe Railroad after members playfully devastated a number of coaches en route to their annual picnic. The ball club became Ragen's Colts, a catchier name that soon identified the entire group.

Frank Ragen was the club's leader, a pol who used his followers' muscle to reach the post of city commissioner. In his youth, Chicago mayor Richard J. Daley had been a Colt—and probably a headbreaker, too; they combined superpatriotism with racism and joyful violence. Roughly a quarter of their two thousand members had enlisted in World War I (membership numbers notwithstanding, their watchword was "hit me, and you hit one thousand"), and Colts worked particular mischief during the five-day race riot of July 1919 that left some forty-three dead.

Ralph Sheldon recruited his gang from these ranks, which also provided a fine fierce pool of talent for Torrio and later Capone. Sheldon was a slight, consumptive young man, only eighteen when national Prohibition came, but already a veteran malefactor. He first appeared in court at sixteen for highway robbery, a charge he beat (although his two companions were convicted), impelling the judge to rebuke the jury for a "miscarriage of justice." Although the huge majority of his gang were comparably young, its most prominent member after Sheldon was Daniel Stanton, somewhat older (born 1896) and much sturdier than Sheldon. A teenage Stockyards hand, driver and muscle for a labor racketeer, slugger for the Checker Cab company, Stanton had been cited for valor in France with the Illinois 131st Infantry.

Sheldon, Stanton and the rest had a reflexive aversion to Italians, their Irish Catholic sensibility especially shy of Torrio's brothels. On the other hand, they knew the value of political connections. Equally convincing, they had no other ready source of beer, and they recognized the profits possible at Torrio's wholesale supply price of $35 a barrel, including political protection. Sheldon signed up.

West of Sheldon, the district called "Back of the Yards," belonged to Joseph Saltis, a shambling, broad-shouldered, six-foot and two-hundred-pound, wrinkle-jowled, pasty-faced mess. Saltis would have been a joke save for his murderousness: anyone believed to have bludgeoned to death an elderly woman for refusing to sell beer in her soda parlor finally resists being thought a figure of fun. Yet everything else about him seemed hapless; not even those who feared him respected him. Before Prohibition, he kept a saloon. Despite his heft, one of the Ragen's Colts used to love dropping by, friends claimed, to "pull Polack Joe's shirttail out."

Born in 1894 in Hungary, brought to Chicago at age eleven, Saltis had the ineradicable nickname of "Polack Joe." He spelled his name with an *o*—*Soltis*—but police, prosecutors, the newspapers, everybody spelled it *Saltis*. Finally, when seeing the police about an arrest warrant, he signed his name with an *o*, and his lawyer, thinking to keep the record straight, said, "The name is S-A-L-T-I-S." The detective chief goggled at the prisoner: "Don't you know how to spell your own name?"

"Spell it," said the Hungarian Polack wearily, "any way you want."

Torrio had little trouble selling him on the plan. In time, Saltis would operate at least three of his own breweries, owning one outright; but at the start of Prohibition he had tried to run beer in from Wausau, Wisconsin, too uncertain and too long a supply route when he had some 200,000 Slavic thirsts to slake southwest of the Stockyards.

Torrio also indirectly supplied Saltis with the ferocity he lacked, except with old ladies. Frank McErlane, despite a habitual glower, looked to one reporter like "a butter and egg man," a portly five-foot-eight and 190 pounds, with blue eyes, a rosary ever-present in his pocket. But his face habitually glowed a choleric red, and when drunk (also habitual) his eyes would glaze over, at which sign his closest friends edged for the door. The Illinois Crime Survey tagged McErlane "the most brutal gunman who ever pulled a trigger in Chicago." In 1922, when he faced another jail sentence, a friend, George Karl, asked Torrio to intercede. Torrio gave Karl the $12,000 needed to fix the matter. McErlane hitched up with Saltis.

A few other minor, neighborhood gangs existed, delighted to be left in possession of their minor holdings and equally delighted to secure assured supplies and potent protection. Those of Claude Maddox, especially, and of Martin Guilfoyle, both on the North Side, but west of the river, would become important in Capone's later struggles with O'Banion's successors.

By the start of 1923, Torrio had achieved an understanding with all but one gang. The South Side O'Donnells looked as though they no longer existed. Until 1920 the four brothers, led by the eldest, Edward J. O'Donnell, then thirty and always called "Spike," had slugged, robbed and burgled in the far South Side, usually below Sixty-third. But Spike, who had won two murder-trial acquittals after he'd shot some half-dozen victims, had been convicted of the daylight, $12,000 Stockyards Bank and Trust job. Prohibition dawned with Spike doing time in Joliet. Brothers Walter, Thomas and Steven were lost on their own. They mooched around the Four Deuces, thankful for whatever odd jobs Torrio tossed them. Torrio, pardonably, assumed they were no issue. Ironically, that miscalculation would occasion proof that his plan really did work.

CHAPTER 7

A Spike in the Plan

By The Start of 1923 almost no one in the Thompson administration was speaking to anyone else. A year before, three members of the school board had told State's Attorney Crowe details of a fishy land deal. On the outs with Thompson, Crowe investigated, and the board's counsel drew a one-to-five-year sentence. Then Crowe looked at Lundin's role in the scandal.

After a grand jury indicted Lundin in January 1923, Thompson withdrew as a candidate for reelection. The indictment had merely capped a mountain of damning revelations about his administration. He later sulked, "My friends have crucified me."

A jury would acquit Lundin and the others next summer. The defense team, which starred Clarence Darrow, proved too adept. (Despite Darrow's image of high-mindedness today, he was a criminal lawyer who quite properly would represent nearly anyone who could afford his fee; he would soon file an appearance for Frank McErlane, Chicago's "most brutal gunman.") Defense counsel easily made the witnesses sweat, since most had rooted in the trough along with the accused. Thompson interrupted a Hawaiian vacation to testify for Lundin, the finale of their partnership.

To capitalize on the Republican disarray, George E. Brennan, successor to Roger Sullivan as Democratic machine boss, needed a candidate who would not, as usual, alienate the good-government wing. He hit upon William E. Dever.

Dever, then sixty, had spent ten years as an alderman and twelve as a judge. Not a whisper of scandal touched him in either post. Transplanted from Woburn, Massachusetts, where his Irish father had settled, Dever had come to Chicago in 1887. A skilled tanner, he put himself through night law school. He was one of five Democrats a bipartisan commission pronounced acceptable as a mayoral candidate in the April 1923 election.

Dever's good odor with reformers notwithstanding, Brennan knew him to be fiercely ambitious and willing to play machine politics. As price for the machine's support, Dever agreed to let Brennan handle patronage, renouncing the lever that might have pried Chicago corruption apart. It's astounding that he was able to cause Torrio and Capone as much bother as he did.

He won election by over 100,000 votes.

Dever's biographer called him "a dripping Wet who enforced Prohibition." He would tell a meeting of beer-guzzling Germans, "I have never pretended to be, and am not now, a Prohibitionist." But he *would* have law and order. He *would* "enforce the law to the limit."

He chose as police chief Captain Morgan A. Collins, a onetime medical student and saloon owner whom reformers reckoned an honest, capable policeman. True, when he was a captain, an investigator had spotted sixteen bookies operating in his precinct; but Collins correctly said that he'd need at least three hundred more men watching the front and back of every place in order to shutter permanently his depraved Chicago Avenue district.

Within a month, Collins's men were raiding places citywide with, wrote a reporter, "unabated enthusiasm," arresting five hundred in one sweep, 450 in another. Within six months his men had closed over four thousand blatant saloons and some five hundred "soda parlors"—notoriously fronts for selling beer. Collins told his captains he wanted more action. "If I don't get it," he later said to the press, "some of them will be looking for jobs." Chicago had heard that kind of rhetoric before. But two weeks later license revocations reached 1,400.

More chilling to Torrio, Capone and the rest of the boys, Collins turned down handsome opportunities to be reasonable. The Four Deuces had become the outfit's headquarters. Collins raided and padlocked it. That created no overwhelming problem; they had subsidiary headquarters in the Pershing Hotel at Sixty-fourth and Cottage Grove. But the trend was unsettling. Torrio offered Collins $1,000 a day to lay off, soon upping it to $100,000 a month. Another outfit offered $5 a barrel on 250 barrels a day just for permission to *move* the beer. Collins said no to everyone, even shutting down two hundred bookie joints owned by Mont Tennes, an oldtime gambling chief who had always bought *total* immunity.

Before long, commentators could write of "drying up Chicago." That never happened; but even the possibility gave further impetus to Torrio's plan. The way the combine worked in response to Spike O'Donnell's challenge provided the clincher.

O'Donnell once said, "I can whip this bird Capone with bare fists any time he wants to step out in the open and fight like a man." It would have been an epic street brawl. Spike was a fighter, long and lean, leathery tough. Yet the lines that creased his long, sharp-featured face, and that gave him the air of a particularly melancholy basset hound, were laugh lines. He had a nice wit. Later, with his best men murdered and his own hide frequently peppered, he said, "Life with me is just one bullet after another. I've been shot at and missed so often, I've a notion to hire out as a professional target."

After friends had bought him a pardon from Governor Small, about the time Dever became mayor, Spike O'Donnell called his brothers around him and recruited a dozen or so drivers, strongarms and "salesmen." They were generally parolees—most notably Jerry O'Conner, another Joliet alumnus. The new gang started well. Their first source of supply was hijacked loads of beer, often Torrio's. Then O'Donnell trucked in beer from a Joliet brewery, and it was all the real stuff, none of it near beer needled with alcohol, the cheap slop often palmed off on customers. Moreover, his beer retailed at $45 a barrel against Torrio's $50. He started with saloons in his own Kerry Patch neighborhood on the South Side, then rapidly edged into Saltis's Back of the Yards territory, the New City police district.

Ironically, what made O'Donnell's challenge possible was Dever's crack-down policy. Brothels, gambling joints, breweries and speakeasies, with their relatively large and elaborate plants and need for reasonably fixed addresses, cannot exist "ten minutes," as one observer put it, "without the knowledge of the district police captain"—which meant "protection." Before Dever, city hall had dispensed protection through ward bosses like Coughlin, Kenna and Powers, who designated the protected and collected from them. A police captain who allowed anything but show raids on the favored courted transfer or early retirement.

Mayor Dever and Chief Collins changed that, ending centralized corruption. Although captains were now instructed to shutter everyone permanently, implicit in that power was the freedom to ignore instructions. No one could *prove* they knew about a speakeasy they permitted to operate. Under this equal opportunity corruption, Spike O'Donnell could arrange his own protection. Captain Thomas C. Wolfe, district commander, would make the rounds with Edward Nealon, one of his sergeants, urging O'Donnell's beer on saloons in place of the Torrio beer that Saltis sold.

Competition remained peaceful at first. Torrio retaliated only by cutting his price $10, to $40 a barrel, a reduction O'Donnell could not meet. So the great Beer War of 1923 started as nothing more sinister than a classic exercise in cartel tactics to smother competition. There matters stood, unresolved through the spring and into the summer.

Meanwhile, three unrelated events bore heavily—though not at the time obviously—on Capone's accession to power.

The first seemed simply a bizarre accident with slapstick sequel. Deany O'Banion and Nails Morton had become great pals. Morton, at twenty-nine (two years younger than O'Banion), was a bachelor who lived in the Congress Hotel. O'Banion had married two years before.

On Sunday, May 13, 1923, jodhpured, spurred and booted, they went horseback riding with Mrs. O'Banion and a commission broker named Peter Mundane. Morton evidently fancied himself a fine horseman. At the Lincoln Riding Academy, 300 North Clark Street, he chose a horse later described as "particularly nervous and mettlesome." Out in the street, the brute reared, took the bit and galloped south on Clark. Morton, rather portly, stood in the stirrups for better purchase on the reins. The stirrup strap broke and Morton pitched over, head first, still holding the reins. The horse, further terrified, lashed out, kicking Morton in the head. A hotel doorman and a policeman who were first to reach Morton found him unconscious; he died without waking up.

Grief so smote Louis Alterie that he took the horse from the stable to a field northwest of the city and executed it. He then called the stable. A reporter later claimed, with family-newspaper demureness, that Alterie had said, "We taught that ——— ——— horse of yours a lesson. If you want the saddle go and get it."

O'Banion had lost more than a buddy and sometime partner. Morton had exerted a moderating force. Who else could talk restraint to him? Hymie Weiss was, in his own way, more impulsive than O'Banion. Moran really was "Bugs," Drucci hyperkinetic and vicious, Alterie a flake. Had Morton lived, he might have argued O'Banion out of the actions that precipitated the final blowup with Torrio and Capone.

A fractious horse would help make Capone "The Big Fellow" in American crime.

The second portentous event started in 1921. Harry and Alma Guzik, who had managed a Torrio brothel and who apparently shared a taste for authentic white slavery, hired a sixteen-year-old farm girl who had advertised for a city job as a maid. After five months of forced prostitution, the girl prevailed on a patron to bear a message to her father. He came, backed by ten friends, and carted her away.

The evidence, incontrovertible, convicted the Guziks—after many continuances and an attempt to buy off the girl's father. The Guziks appealed to judicial review. More effectively, they appealed to Torrio, who prized Guzik as an experienced, energetic and profitable brothel keeper as well as a reliable payoff man for police and politicians. Besides, how better could Torrio demonstrate his clout to colleagues—at a time when his authority and the feasibility of his plan faced O'Donnell's challenge? How could *anyone* hope to get such an antipathetic and incorrigible felon off after so unappetizing a crime?

Torrio cashed in chits the outfit had earned the year before with Governor Small's acquittal. On June 20, 1923, while their case was still under appeal, and before they had served a day, Harry and Alma Guzik received a pardon despite the din of outrage that reverberated through the state. *This* was clout. And because it was the outfit acting, it imparted an impression of institutional power transferable to Capone, smoothing his way at the crucial moment.

* * *

A week after the Guziks' pardon, Torrio could have used one himself, although the third event appeared trivial at the moment.

Brice F. Armstrong *looked* even more ineffectual than most Prohibition agents: short, stocky, twitchy with nervousness, hair thinning, eyes owlish behind enormous horn-rims. But on duty he packed two guns and a pocketful of nickels—to call newspapers. He further cemented press relations by letting reporters walk off with a bottle or two of evidence after he had posed for pictures. Partly for Armstrong's bumptiousness, partly for his unbecoming effectiveness, authorities often banished him to the hinterlands. During one of these exiles, in early 1923, he took under observation Puro Products in West Hammond, southeast of Chicago.

The brewery had gone bankrupt in 1915. Joseph Stenson, Torrio's uptown brewery connection, had bought Puro in October 1920, assigning ownership to Torrio two years later. On February 23 and March 9 of 1923 Armstrong watched Puro deliver beer to Chicago speakeasies. On June 27, 1923, he raided.

The government indicted Torrio and two others, one a former U.S. deputy marshal who owned Puro stock. Torrio entered a not guilty plea to buy time, not because he intended to battle the charge or for fear of consequences; first convictions drew only a fine with the brewery padlocked for a year. But Torrio could not afford the record of a conviction at the moment.

His mother and stepfather lived back in Brooklyn. When she voiced a desire to return to Italy for her last years, Torrio planned a family trip to get the old couple settled. His stepfather, Salvatore Caputo, had to tell him the truth: Torrio would have trouble getting a passport. In order to avoid a long waiting period, Caputo had told some lies on his citizenship application. The perjury discovered, Immigration had revoked his citizenship—and with it, his stepson's derivative citizenship. Torrio would have to apply on his own. He filed a petition on July 18, 1923, three weeks after the West Hammond raid. He had to swear that he had never been convicted of a crime—which, with the trial not scheduled until October, was true. Of course he also had to swear he had never been *charged* with a crime, but that exposed him to only the most exiguous risk. He listed his residence as his parents' Brooklyn address, which made him virtually unconnectable to a trial scheduled three months and a thousand miles away.

Come fall, Torrio shuttled between Chicago and Brooklyn. A federal judge ordered Puro Products padlocked on October 10, with the trial to start October 19. On November 1, his citizenship having been granted, Torrio changed his plea to guilty. He returned to Brooklyn for his oath of allegiance on November 11. On December 17, 1923, a judge fined him $2,500 for West Hammond. A second conviction, though, would carry mandatory jail time.

Back on Chicago's South Side, Spike O'Donnell could answer the combine's price-cutting only one way. He and his brothers, with Jerry O'Conner and his other hired muscle, would invade a saloon, pistols blatant in their belts, and tell owners they had a new source of supply. They met sales resistance by breaking steins, furniture, windows and heads.

Torrio never scrupled to order heads broken when necessary, but muscle wasn't generally needed. Most bar owners were either content enough or smart enough to stick with Saltis-delivered Torrio beer—as long as protection included immunity from the O'Donnells as well as the authorities.

Torrio's master plan would work. Ralph Sheldon, with his collection of Ragen's Colts and his territory right next door, shared Spike O'Donnell's contempt for Saltis, but Torrio's dispensation made Sheldon and Saltis allies. They would jointly battle the intruder.

Early in the evening of September 7, 1923, Walter, Steve and Tom O'Donnell, along with Jerry O'Conner and two other O'Donnell sluggers, George Bucher and George Meeghan, started a round of hard-sell calls. Some days before, Jacob Geis—a Saltis customer (and tough customer)—had not only refused to switch suppliers but had bounced the O'Donnell rep out of his West Fifty-first Street saloon. Now, that drizzly Friday night, Geis and his barkeep, Nicholas Gorysko, were serving about a half-dozen patrons when the O'Donnell party walked in and gave him a chance to change his mind. When he refused they dragged Geis across the bar and pistol-whipped him hard enough to fracture his skull; only a hardy constitution kept him alive. When the barkeep tried to intervene they pounded him unconscious. Then they moved on to make similar sales calls on five other holdouts.

One of these, Frank Kveton, his saloon a block and a half west of Geis's, called the New City police station to complain. Perhaps he more sensibly also called Saltis or Sheldon or Torrio; perhaps one of the police on the combine's payroll passed on the message. Somehow word got through. At the end of the evening's work, Spike O'Donnell joined his troops for a drink and a bite at Joseph Klepka's saloon, nearby at Fifty-third and Lincoln, a regular O'Donnell hangout.

Four gunmen surprised Spike and the boys at the bar. Ralph Sheldon led them, at his side Daniel McFall, a Ragen's Colt and part-time deputy sheriff with the county police. McFall, .38 in hand, shouted, "Stick up your hands or I'll blow you to hell." Walter O'Donnell challenged them, "Now give us a square deal. Come outside and fight it out." In answer, a shot hit the wall just above Walter's head. McFall grabbed Jerry O'Conner, covering him with his .38 while the O'Donnells, Bucher and Meeghan dashed for the back and side doors. A shot at Steve O'Donnell's heels sped them along. Just then two men came in the front. One was Joe Saltis's murderous partner, Frank McErlane, wearing a raincoat and toting a sawed-off, double-barreled shotgun. At a nod from McFall, McErlane stepped back outside and McFall herded O'Conner out. A shotgun blast tore away half of O'Conner's head.

The operation bore Torrio's unmistakable stamp. He wanted a lesson and warning, not massacre headlines. O'Conner was the O'Donnells' most effective persuader. He was also a nobody, a paroled "lifer," his death apt to cause less stir than would that of any O'Donnell, while hurting the gang more than would the death of the lesser brothers—yet less apt to infuriate Spike past the point of heeding the warning. Spike himself remained inviolable. Not until later, when Torrio concluded it was unavoidable, would he (then Capone) order the assassination of a rival gang leader. That sort of thing would unfailingly excite a public reaction that could lead to *real* crackdowns. Besides, who knew where the precedent of killing gang leaders would end? As it was, Torrio took to carrying a pistol, prudently obtaining *two* concealed weapons permits.

Ten days later, on September 17, 1923, when Spike had not heeded the warning, the combine hit again, harder.

Laflin Street runs north and south, two blocks east of Ashland. At 5500 south it crosses Garfield Boulevard, which boasts a strip of greenery down its middle, dividing the traffic lanes. As usual, near dinner time, Georgie Meeghan was driving himself and "Spot" Bucher home in his Ford flivver after another uninhibited day of pushing O'Donnell beer. They were headed south on Laflin, the street on which they both lived. Meeghan stopped for cross traffic in the westbound lane at Garfield. As he started across, a green sedan waiting in the westbound lane of Garfield swung away from the curb and made the left turn onto Laflin as Meeghan stopped again in the divider zone for eastbound Garfield traffic.

In the green sedan sat Frank McErlane and Danny McFall, armed. Thomas Hoban drove. The gunmen began firing as soon as they pulled abreast of the flivver, a pistol from the front seat, a shotgun from the rear, both gunmen snatching up loaded guns when empty, riddling both victims, killing them instantly. They hit Meeghan, for instance, in the head, chest, thorax, abdomen *and* limbs.

Mayor Dever blistered Chief Collins and Chief of Detectives Michael Hughes. He suspended the district's commander, Captain Wolfe (even before he learned of Wolfe's beer-hustling), and replaced him with tough Lieutenant William H. Shoemaker. The mayor insisted that until the killers were caught, "every officer of the law and every enforcing agency should lay aside other duties and join in the common cause—a restoration of law and order." He promised that "the police will follow this case to a finish as they do all others." Exactly. Chief Collins spoke more realistically. "It's plainly a beer runners' feud," he said, "and I'm glad no honest citizen was killed."

Police questioned Capone about all the killings and released him when the O'Donnells refused identification. After Torrio's arranged surrender, and some grilling, only his possession of a pistol remained an issue, easily resolved because of those two permits.

Gun permits for known criminals bore a significance beyond what they said about Chicago law. One had been signed by Joseph Mischka, justice of the peace in Cicero, a community on Chicago's western border. Roosevelt Road marked Cicero's northern border, lined with raunchy, tough saloons. The city limits extended from there, at 1200 south, about three and a third miles to Pershing

Road, 3900 south. Cicero started some eight miles west of State Street, its eastern border about 4600, and ran west to just beyond Austin, 5600, a rectangle set on end. The main drag was Twenty-second Street (now Cermak Road).

Its multitudinous saloons and ad lib gun permits notwithstanding, Cicero was a very solid community, first-and second-generation middle-European, principally from Czechoslovakia's Bohemia, 68 percent living in homes they owned, many working at the giant Western Electric plant on the east edge of town. With a population of seventy thousand, Cicero ranked as fifth-largest city in Illinois, probably first in literacy. Although part of Cook County, it had its own government and police; the writ of Chicago's mayor did not run in Cicero.

Torrio's move to the suburbs found its logical culmination in Cicero. Of course he'd long had a presence there, as witness the gun permits (Capone had one too); and Cicero was where the IRS man had sought out Capone the year before. But the town stayed relatively closed to outsiders. A local, Edward D. Vogel, dominated gambling, especially Cicero's ubiquitous slot machines. Vogel raked off 60 percent of the net, but the saloonkeepers and merchants who gave the slots space counted on their 40 percent to pay the rent, the utilities and maybe a barkeep's or clerk's wages.

Torrio's move into Cicero would not break the combine treaty. Klondike O'Donnell and his brothers monopolized the Roosevelt Road business, but most of Cicero was unaffiliated. Undoubtedly the West Side O'Donnells liked Torrio better at a distance, but they were not the problem. Vogel, with his middle-European primness, disapproved of an Italian pimp's presence. There were few Italians in Cicero, no whorehouses at all, and Vogel intended to keep it that way. Backed by Edward Tancl—an ex-boxer, a saloonkeeper and a potent vote-broker—Vogel controlled Cicero's otherwise independent government and police.

In October of 1923, during a lull in the South Side Beer War, Torrio sent Capone to Cicero. Capone's Four Deuces madame, the rocking-chair redhead, rented a rooming house on Roosevelt Road where Capone installed a half-dozen whores, complete with a capper outside to pull in trade. Cicero police closed them down within hours. Capone next opened shop at Fifty-second and Ogden Avenue, an arterial cutting diagonally through town. Cicero police promptly wrecked the place and arrested everyone in sight.

Vogel had forgotten where he was. As part of Cook County, Cicero fell under the jurisdiction of the county sheriff, as well as its own police. That sheriff was Peter B. Hoffman, a man firmly in Torrio's pocket. At Torrio's behest the sheriff mounted a raid on Cicero slot machines, as was his right and long-neglected duty. He impounded every slot he could find, and since they had always operated openly he knew where to look. Suddenly Eddie Vogel faced loss of his main income. The lesson was clear: he might tie up some local protection, but the man with enough state clout to get a Harry Guzik pardoned also swung enough county clout to make the sheriff forget whatever "arrangement" the locals thought they had with him. If Torrio and Capone couldn't operate in Cicero, no one could.

Klondike O'Donnell joined Vogel in asking for a peace conference. Torrio did not make them crawl. He would have Hoffman return the slots at once. He ratified the O'Donnell hold on Roosevelt Road and other parts of Cicero where they had already penetrated. All Torrio wanted was the right to sell beer in the rest of town and to operate gambling centered on Twenty-second Street. He would not sully Cicero with brothels; Capone had opened those two only to make a point. Torrio could set up all the whores he wanted across the town line in Stickney and Forest View to the south, Berwyn to the west.

Soon Eddie Vogel would be known as "an associate of Capone in gambling," and Klondike O'Donnell's brother, Myles, would help murder a troublesome holdout who dared reject Torrio's beer.

Cicero settled, attention returned to the other, South Side O'Donnells. Spike was as irrepressible a troublemaker as ever, scorning his losses. The combine had eliminated three of his top muscle salesman; now it would try cutting into his source of supply.

About one-thirty A.M., on December 1, 1923, a Saturday, two O'Donnell trucks with seventy barrels of beer between them left the Joliet brewery for Chicago, some twenty-three miles northeast, traveling on Sag Road. Thomas Morris Keane, always known as "Morrie," drove the first truck, William "Shorty" Egan beside him; Martin Brandl and Joseph Belice manned the second truck. The convoy passed the village of Lemont, about a third of the way, when two sedans overhauled them and forced them to the side with a shotgun blast.

Shorty Egan later gave an account of the action. Two of the hijackers—Frank McErlane and William Channell, a parolee who had killed a woman during a holdup—motioned Egan and Keane out of their truck, others handling the second truck. Just then another car flashed by, freezing the action in its headlights. Edward Triebel and his family were returning to Chicago from a Thanksgiving trip. The hijackers peppered his car with shotgun blasts, but Triebel sped on, unharmed. McErlane and Channell bound their captives, manhandled them into the back of one sedan, and tore off down the road with them, Channell driving, McErlane riding shotgun.

Channell stopped when he spotted a truck and sedan by the road. But when he and McErlane got out they were greeted by a volley of shots, rocking the car. Shorty Egan could not imagine who the shooters were. Other hijackers? Police? In any case, his captors dove back into the car and zipped away.

Not long afterward, tall, skinny Willie Channell said to McErlane, "Where you gonna get rid of these guys?"

McErlane laughed. "I'll take care of that in a minute," he said, fussing with the double-barreled shotgun across his lap. He turned toward the back and pointed his gun at Keane. Without a word he fired. The blast took Keane in his left side, a spasm twisting him around. McErlane gave him the other barrel in his now exposed right side.

Calmly, still silent, he reloaded, turned and fired both barrels at Keane again. After another quick reload, he glared at Egan and said, "I guess you might as well get yours, too." He shot him in the side.

"It hurt like hell," Egan said later, "so when I seen him loading up again, I twist around so it won't hit me in the same place. This time he got me in the leg. Then he gimme the other barrel right in the puss."

Egan slid off the seat, a heap on the car floor. He heard McErlane firing again at Keane, then came the anguish of a final shot into his own undamaged side.

McErlane, fat but agile, scrambled over the front seat into the back. He opened the door and shoved Morrie Keane's body out with a piston kick of his leg. Then he dragged Egan over to the door. "We was going about fifty from the sound," Egan remembered. McErlane shoved and Egan ended in a half-frozen puddle in a ditch.

Unbelievably, Shorty Egan survived, though half his face had been torn away. When he came to, he dragged himself to the nearest light, the Palos Park Golf Club, deserted except for a caretaker, who called for help. In the hospital, still in shock, he "partially identified" Channell's mug shot, then retracted when he'd had time to reflect. A garage attendant who worked on the sedan, pocked with bullet holes and bloodstained, also identified both Channell and McErlane at first; Channell had driven into the garage at dawn. After weeks in custody as a material witness, like Egan pondering the consequences, the attendant also recanted.

The state's attorney arrested McErlane, held him for a while in the Hotel Sherman, then released him. Finally indicted, he walked when State's Attorney Crowe entered a nolle prosequi for want of witnesses.

The other two drivers, Brandl and Belice, ended up in police hands, Brandl telling a quickly hushed-up tale of turning the beer over to six highway policemen (under the ineffable Sheriff Peter Hoffman), who saw to its delivery. Hoffman promised to look into the matter.

A grand jury finally indicted Danny McFall—after much pressure on Crowe—for killing Jerry O'Conner. But McFall had packed a .38 that night, and a shotgun had killed O'Conner. The case fell apart. McFall, McErlane and Hoban were indicted for killing Meeghan and Bucher (that's when Darrow filed his appearance for McErlane), but the case was nol-prossed after the O'Conner dismissal.

Though Spike O'Donnell would make a pest of himself for nearly a year more, his gang was in retreat.

With all more or less quiet on all fronts, Johnny Torrio took his parents to Italy, where he bought his mother a villa, staffed by fifteen servants, making her the envy of all her girlhood friends. He also thoughtfully secreted some $1,000,000 among Italian and Swiss banks against a day of need. He made it a long, leisurely trip. There was no need to hurry back; Torrio had left Al Capone in charge.

CHAPTER 8

Capone in Charge, in Cicero, and in Trouble

WHILE TORRIO LAZED in Italy during the winter and early spring of 1924, Capone consolidated their gains, studiously avoiding conflict with the other gangs. Specifically, he and Deany O'Banion never tangled. As O'Banion grew wilder and wilder, Capone visibly matured. Though his temper could still overmaster him, the occasions grew more rare.

For instance, gangsters routinely tried to evade news photographers, hoping to thwart easy identification. In the early days, in public, "Al Brown" reflexively covered his face when cameras appeared. Less publicly he might kick photographers senseless. One photographer vividly remembers an incident at a police station sometime in 1922. The police had run Capone in for something minor. "Capone and a couple of other guys," says Anthony C. Berardi, "were slapping photographers around like you couldn't believe." The police, who periodically had to arrest their meal tickets because of newspaper pressure, simply watched, amused.

Not much later, Capone had attained such prominence that identification stopped being an issue. Berardi's editor talked public relations sense to Capone. "Why do you want to be mean? Why don't you be sociable, be a gentleman? You're a public figure and you ought to act like it. We shouldn't have pictures of you covered up."

It registered. Capone became by far the most accessible of gangsters, a favorite with newsmen—which of course added to his eventual fame. (Significantly, much later in Florida, where he was anxious *not* to be a public figure, he reverted to covering up at the sight of cameras.)

The same maturation showed in Capone's consolidation and expansion of the suburban empire he and Torrio had built. He remained single-mindedly ferocious in attaining the control he needed, but his way of brutalizing dissidents changed. Take the difference between what happened in Forest View and Cicero.

World War I veterans founded Forest View as a memorial to their fallen comrades and as a nice place to bring up families away from the city hurly-burly. Chicago lawyer Joseph W. Nosek, while visiting a client there, got the notion to turn this cluster of farms about a mile south and two miles west of Cicero into a village. American Legion pals joined Nosek in the project. He became police magistrate; his brother became village president. As police chief they chose one William Dillon, who claimed to be a vet, too, but may have been another convict pardoned by Governor Small.

Soon Chief Dillon told Nosek that two brothers named Capone proposed building a hotel and club in Forest View. Nosek was delighted. ''I didn't know just who the Capones were,'' he later explained. ''It looked like a good chance to improve our village.'' When he saw what kind of thugs came to town, he inferred the sort of operation they had in mind, and insisted that Dillon run the whole bunch out. Next day, Ralph Capone threatened to throw him into the village sewage ditch. Nosek assumed the squat, glowering man was joking, a poor bit of character reading; someone who knew Ralph remembers him as ''a son of a bitch, crude, mean—crazy!''

At four A.M. next morning two armed heavies invaded Nosek's house and dragged him to the village hall, where seven more armed men waited. They said they were going to kill him; he believed them. They whacked him on the head with gun butts, and when he sank to the floor, streaming blood, they kicked him. ''I'm not ashamed to admit,'' said Nosek, ''that I got down on my knees and prayed that they let me keep my life.''

The price exacted for not killing him was his instant departure from the village he had founded. Then Capone's men beat and chased out of town another twenty or so other ''troublemakers.'' In the next village election a more accommodating slate won, and Capone built the largest of his suburban brothels, The Maple Inn, generally known as ''the Stockade'' for its size and bleak decor. Forest View soon became known in the papers as ''Caponeville'' (''Caponiville'' to the *Tribune*).

Capone had buffaloed Forest View with terror, making no effort to reconcile the troublesome to his tenure. He was hated there.

Cicero would be different. Capone acted as barbarously as need be to establish unquestioned control. But then he corrupted officials and tried always to co-opt influential rebels, rather than terrorize, and he took pains to ingratiate himself with the citizenry. In Cicero, Capone ended up a hero to many.

Early in 1924 he established a headquarters in the Hawthorne Hotel at 4823 Twenty-second Street, two blocks west of Cicero's eastern border, one west of the main north-south arterial, Cicero Avenue. The outfit occupied at least one entire floor of the three-story brown-brick hotel. Capone had steel shutters installed at the windows. Visitors entered by a twenty-five-foot-long passageway

The Capones' home at 21 Garfield
WILLIAM BALSAMO

Frankie Yale, Capone's first boss
WILLIAM BALSAMO

New York police picture (c. 1917–1918) of the teenage Capone, with acne, before he left for Chicago
MICHAEL GRAHAM

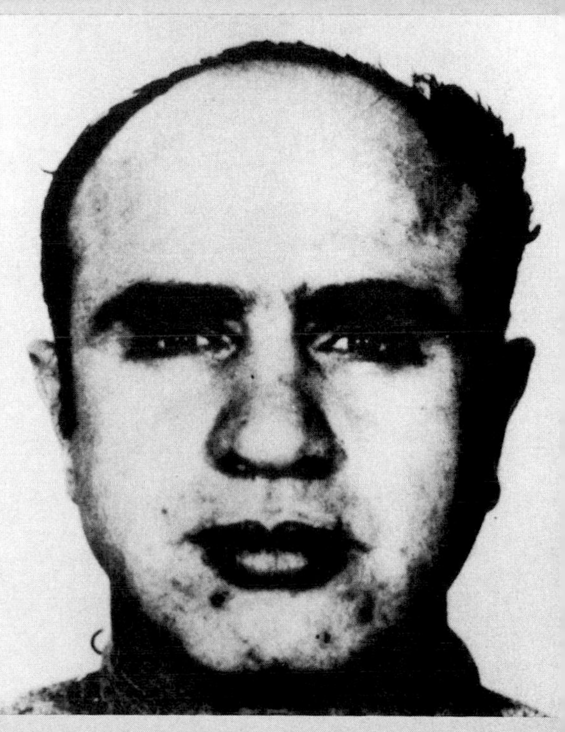

Dale Winter with Jim Colosimo, crime overlord when Capone reached Chicago
ANTHONY BERARDI

John Torrio, Capone's mentor, in jail after his near murder in 1925
ANTHONY BERARDI

Thompson campaign poster in 1927
CHICAGO SUN-TIMES

Big Bill Thompson, when he first became Chicago's mayor in 1915
CHICAGO SUN-TIMES

William Dever, Chicago's reform mayor, 1923–1927

Alderman Michael "Hinky Dink" Kenna, 1903
CHICAGO SUN-TIMES

Alderman "Bathhouse" John Coughlin celebrates repeal, 1933.
CHICAGO SUN-TIMES

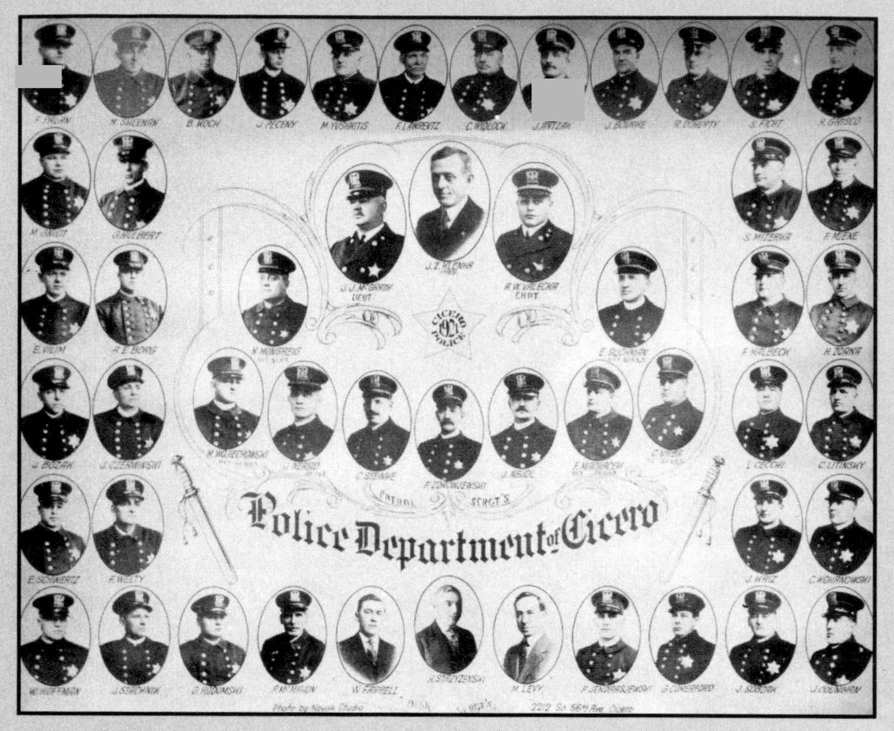

The entire Cicero police force, 1921, before Capone moved in—and knocked town president Klenha (center, top) *down the City Hall stairs*
EMIL SCHULLO, CICERO PD

"Spike" O'Donnell fought the Torrio-Capone combine.

ANTHONY BERARDI

Dion and Viola O'Banion
MICHAEL GRAHAM

CHICAGO TRIBUNE

Schofield's flower shop, North Side gang HQ, after O'Banion's 1924 murder; also the scene of Hymie Weiss's 1926 murder

Crowd at O'Banion's lavish funeral

ANTHONY BERARDI

CAPONE FAMILY

Alphonse Capone
Complainant.

Theresa Capone

Gabriele Capone

The signatures of Capone, his mother and father, in 1930, 1937, and 1906, respectively

Al, giving The Look

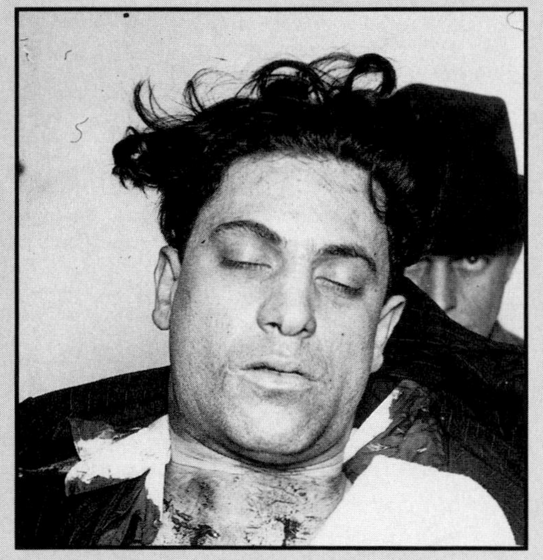

Frank, killed by police, 1924
ANTHONY BERARDI

Ralph, with beer for reporters at the gate of Al's Palm Island, Florida, estate as Al lay dying, 1947

Matt

John, known as "Mimi"

CHICAGO SUN-TIMES

Jimmy, long-lost oldest brother, as he appeared when a lawman during Prohibition
CHICAGO SUN-TIMES

Rocco Fischetti, Charley's younger brother
CHICAGO SUN-TIMES

CHICAGO SUN-TIMES

Mae, almost never photographed (shot with telephoto lens), on her way to see Al in Alcatraz

Mafalda at her 1930 wedding
CHICAGO TRIBUNE

Charley Fischetti, Capone cousin
CHICAGO TRIBUNE

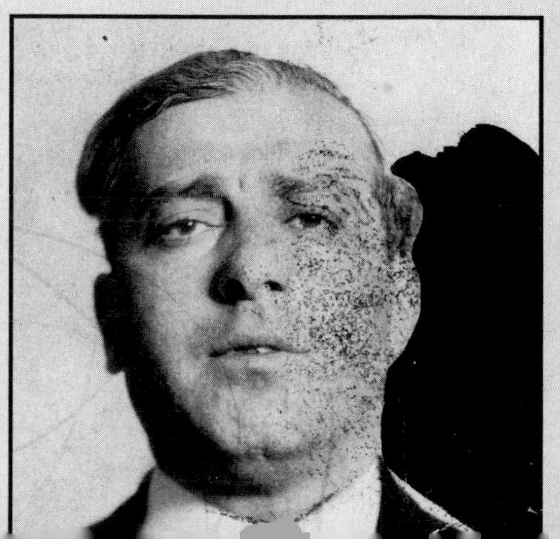

CHICAGO SUN-TIMES

State's attorney Bob E. Crowe

Joseph L. Howard, killed by Capone, personally, in 1924

DAVID SCHOENBERG/DICK DUFFIN

Twenty-second Street, heart of Capone's Cicero

MICHAEL GRAHAM

under constant observation by the guards Capone posted in the lobby, whose chairs, front desk and cigar-stand all angled on the passageway.

Capone soon inaugurated his gambling operation, its flagship the Hawthorne Smoke Shop, next door in the Anton Hotel. Other joints would open, including the Subway, across Twenty-second and east a block, and the trendily named Radio, which shifted among a half-dozen locations, all within a couple of blocks, each used after the occasional show raids. One would close in the morning, the next open that afternoon.

Capone and Torrio did not try to control all the gambling joints. For instance, the Ship—a more permanent operation around the corner at 2131 South Cicero Avenue—belonged to four others, although soon Capone and Torrio joined them, cutting O'Banion in also. The biggest games in town, maybe in the country, were at Lauterbach's, where $100,000 might be spread on a crap table or roulette layout at one time; it remained independently owned.

In 1924, according to the government's rock-ribbed estimate, the outfit's gambling take ran to over $300,000. In addition, Capone and Torrio extracted a 25 to 50 percent slice of the other owners' action for protection, with an agent posted in each joint to oversee the split. This was not exactly extortion; it was service. Capone and Torrio could guarantee protection because they owned the Cicero town government.

Capone had not thought up this masterstroke, though he seized on the idea, organized the operation and implemented it with vigor. Edward G. Kovalinka, a Cicero native son, had started as a soda jerk, affable and political. After he had risen to precinct captain and then ward boss, Governor Small had anointed him Republican committeeman.

Chicago's 1923 election worried Kovalinka. Cicero's administration had long been corrupt in a cheerfully bipartisan way, with Republican Joseph Z. Klenha president of the village board since 1917. But following Dever's example, Cicero Democrats had begun rumbling about reform. They proposed a separate ticket for the April 1, 1924, election. With the county election machinery in Democratic hands and with Election Commissioner Anthony Czarnecki striking 3,063 Republican names off the rolls and replacing cooperative election watchers, clerks and judges with his own choices, Kovalinka rightly figured the party would need help. Ed Vogel sent him to Louis La Cava, one of the original partners in the Ship. La Cava introduced him to Capone, who instantly grasped the possibilities. *Of course* he would help reelect the Klenha slate!

Capone coordinated his attack with the O'Donnells. He also borrowed some muscle from O'Banion, now a partner with Torrio and Capone in a brewery, the Mid-City. What one newspaper termed "outlawry unsurpassed in any previous Cook County political contest" began the Monday night before the election, when gunmen invaded the office of William K. Pflaum, Democratic candidate for clerk, beat him and shot up the room.

Next day at least twelve seven-passenger touring cars full of gunmen patrolled the streets. Thugs, guns drawn, allowed voters to cast only ballots marked the right way. When Cicero policeman Anton Bican interfered outside one polling place, the gunmen put him in the hospital. Opposition workers and citizens who fought back or had been targeted as probable centers of resistance were kidnapped and held until the polls closed.

Citizen complaints finally reached County Judge Edmund K. Jarecki. Appalled—not least because he was a Democrat—Jarecki conspired with Mayor Dever to circumvent the law that prohibited Chicago police from duty outside city limits. He deputized them as special agents of the county court.

The force of seventy uniformed patrolmen, nine patrol cars and five squad cars of detectives did not hit the streets of Cicero until late afternoon, and then could not make those streets instantly safe for democracy or Democrats. A little over an hour before the polls closed, about ten gunmen stormed one precinct and ran seventy-five citizens out of the polling place, taking time to search the policeman on duty. Klenha took that precinct by almost a quarter of his total plurality. He won reelection 7,878 to his rival's 6,993. The rest of Kovalinka's slate also won, and by strikingly similar margins.

Jarecki's special deputies had accomplished one thing. They killed Capone's next older brother, Frank.

Toward dusk, the detective squad commanded by Sergeant William Cusack spotted Frank Capone, his cousin, Charley Fischetti, and a short squat gunman they did not recognize, near the poll at Cicero and Twenty-second. They piled out of their squad car across the street from the three and started toward them, guns drawn.

According to several perhaps not unbiased witnesses, the three gunmen never fired. According to the police, Frank Capone shot first; at the inquest they produced the gun—with three rounds missing—that they swore Frank used. They were probably telling the truth. In those days detective bureau squad cars were the same style of seven-passenger touring cars gangsters favored, and naturally were unmarked. Just as naturally, detectives wore plainclothes. The confusion possibly engendered by that lack of identification would figure in a later cop killing that had a sensational aftermath. It certainly explains the otherwise inexplicable gunplay. Frank and his fellows, unaccustomed to any police opposition that day, might well have mistaken their armed assailants for a rival gang or vigilantes and shot first.

If Frank did fire his piece, he missed. Sergeants Phillip J. McGlynn and Lyle Grogan returned fire, McGlynn getting Frank through the heart. Charley Fischetti ran into an open field, then surrendered to Cusack and the other two members of the squad who chased him. The third gunman ran south, firing with a gun in each hand, and escaped into the gathering dark. A myth grew that this was Al Capone. But the third gunman was later named as one David Hedlin; the police had wounded him, too.

As usual, Capone stopped shaving until Frank's funeral, which was an impressive display of Obsequies Chic, with over $20,000 worth of floral tributes—arranged by O'Banion, of course, gangland's official florist. Flowers arrived at the Capone house on South Prairie in a profusion that overflowed first to the porch, then the front lawn. The coffin was silver plated and satin lined, the procession lengthy. Capone ordered Cicero saloons closed for two hours during the funeral as a mark of respect—the driest period that town had ever known.

So far Capone's actions may seem no less rebarbative in Cicero than in Forest View, but the way he enforced the control he had established signaled a real change. Joseph Klenha soon got above himself. He forgot that he had struck a bargain, and with whom. Klenha blathered about cleaning out the gangsters; worse, he ignored Capone's orders. Capone went to the Cicero City Hall and had Klenha summoned outside. Klenha came, accompanied by a policeman. Capone ignored the escort and started slapping Cicero's chief executive silly, knocking him down the City Hall stairs. The police escort prudently toddled away. A little later, when the town council seemed bent on introducing hostile legislation, Capone sent his men in. They broke up the meeting, dragged a councilman out and bludgeoned him.

Capone's capacity for unmanageable rage had not disappeared, but these were *not* incidents of it. He was thinking about consequences. As he explained to a friendly newsman, he realized he couldn't permit such liberties from people he had bought. He'd have to kill someone if defiance continued. Better to act less drastically now. "This way they learn their lesson," he said, "and nobody really gets hurt."

From then on, he paid Klenha public respect and cut him in on the action so generously that a few years later the village president qualified (along with Kovalinka, plus Cicero's police chief and other town fathers) for a place on a federal indictment. Nor did Capone again have to discipline the newly tractable town council. In return, he stayed tender toward their sensibilities. He kept the whorehouses out and never allowed gambling to spread into residential Cicero. As for *street* crime, Capone made Cicero what Klenha claimed: a place he would stack "against any city in the country for orderliness and observance of the law" (if you didn't count the Eighteenth Amendment). The fifth-largest city in Illinois got along with three shifts of seventeen policemen each, yet recorded only about three holdups and eight burglaries a month—committed by those brave, foolhardy or desperate enough to ignore what Capone would order if the outfit caught them.

Following Torrio's motto, Capone always tried to avert trouble. Anytime it developed anyway, and others in the outfit countered with reflexive brutality instead of first exhausting the possibilities of quiet bribery mixed with credible threats, Capone fumed at their stupidity—as in the case of Robert St. John.

In 1922 St. John had started a newspaper, the Cicero *Tribune*, and from the first attacked Capone and the corrupt administration. He exposed paving-contract kickbacks. He endlessly wrote stories about a brothel Capone had opened just south of the town line, near the Hawthorne Race Track in Stickney, stories calculated to inflame Cicero burghers.

Capone and Klenha tried freezing St. John out, giving the rival Cicero *Life* all the town's public-notice ads, pressuring merchants not to advertise on penalty of increased tax assessments and a sudden sea of "No Parking" signs in front of their shops. At the same time, they repeatedly had an intermediary ask how much St. John and his two partners would take for the paper.

Finally, after a particularly acerbic story, Ralph Capone sent a messenger with word of his annoyance. St. John sent back defiance. Ralph exploded. Early in the morning, two days later, as St. John crossed the street to his office, a big gangster car screeched to a halt. Mean crude crazy Ralph and three thugs sprang out. He supervised while they beat St. John with gun butts and the old kid-gang weapon of a cake of soap in a sock. St. John curled up on the expensively paved street, his arms guarding his head. One blow penetrated and struck him unconscious. The two uniformed policemen he had noticed before the action started did nothing to intervene; Cicero policemen knew better.

Upon discharge from the hospital, St. John stopped to pay his bill. The cashier said it had been paid by a husky, richly dressed stranger in a blue suit, sporting a diamond stickpin.

St. John hobbled to the police station. He asked his friend, Chief of Police Theodore Svaboda, to swear out arrest warrants for Ralph Capone, along with one thug he had recognized and two "John Does." His friend tried to talk him out of such a futile and probably dangerous gesture. "You know Al," St. John remembered Svaboda saying. "You know how he is with friends and with his family. If you could get convictions on all those charges you've mentioned, you could send Ralph to the jug for the rest of his life. But do you think for one minute that Al would let that happen?"

St. John persisted, and the police chief told him to return next morning for the warrants. When he did, his friend sent St. John to wait in an upstairs office. Very soon the door opened and in strode Capone.

"He smiled as he came across the room," St. John wrote, almost thirty years later, "holding out his hand." Capone wanted to settle the fuss, to ingratiate himself. "I'm an all-right guy, St. John," the newsman remembered him saying, "whatever they say. Sure I got a racket. So's everybody." But most of those with rackets hurt people, and that wasn't Capone's way. "I don't hurt nobody, especially not you newspaper boys." They gave him free what he could not buy. "Take the Ship. That's a big operation. Lots of overhead. But how can I advertise? I'd like to buy a page in the Chicago *Tribune* every day. Come to the Ship. Best gambling joint in the country. But I can't advertise. So you guys write stories, exposures or whatever you call them, and they get right on the front page and I get my advertising for free." Besides, Capone realized newsmen had to live, too; exposés were *their* racket. "Why should I get sore?"

He lit a cigarette and brushed imaginary lint from the razor crease in his blue suit's trousers, diamond stickpin glinting.

"Now this little trouble you and Ralph had," Capone continued. "Too bad. Never should have happened. Sure the boys were sore at you." But they were dumb. "I tell them, 'Let the kid alone.' "

Usually that would have ended it. But Ralph and the boys had been out drinking all night—more stupidity! "I tell 'em, 'We sell the stuff, we don't drink it.' . . . So they were boozed up and they forgot what I told them and they made a mistake and now I gotta straighten it out. Always I gotta fix up their mistakes. Christ, how I hate people that ain't smart!"

Capone whipped out a horse-choker of a roll, mostly hundreds, and began peeling them off. This would take care of St. John's lost time, this for his damaged clothes, more for incidentals. . . . St. John lost count after seven hundreds fluttered off the roll.

"I was," St. John wrote,

> juvenile enough to like dramatic gestures, so without saying a word I turned abruptly, opened the door, went out, and slammed the door shut as hard as I could. My only regret years later was that I had been unable to see the facial reaction of Mr. A. Capone, antique dealer, after I left.

He likely would have seen a study in resigned disgust. Capone had been trying for amicable accommodation because cooperation meant less trouble. In fact, he could already command what he wanted. He held a controlling interest in St. John's paper. The editor and an associate each had 49 percent of the stock, the other 2 percent owned by a friend to conform with state law that required three corporate directors. During St. John's hospital stay, the associate, whom St. John had accurately mistrusted, readily sold to Capone. The friend surrendered his 2 percent to a go-between on receipt of a forged note from St. John asking him to do so.

Torrio reappeared from Italy to find that he and Capone effectively owned Cicero. A bonanza. Besides their own saloons and gambling joints, there were many local-owned saloons to supply. Scattered about Cicero's approximately six square miles, at least 120, perhaps as many as 165, saloons ran full blast.

Capone made a fortune. Many people estimated with varying reliability how much the various enterprises netted, based on records seized in raids—$5,000 a week average at the Stockade, and so on. But that was the outfit's profit. The government pegged Capone's 1924 share just from gambling and just in Cicero at $123,101.89, the total laughably low, but documentable. That year Capone bought himself a customized $12,500 McFarland—when $1,335 bought a four-door Maxwell and a new Chevy coupe cost $680. Capone traded in his Lincoln, paying the $4,500 difference in cash. Torrio had made Capone a fifty-fifty partner, mightily pleased at the growth of his protégé. The outfit's men obviously liked, admired and would obey him.

He was in no sense a Boy Scout leader. In newsrooms around town they said that for internal light discipline the brawny young ex-bouncer liked to lift the errant by their ears, rabbit-style, hurl them on the floor and kick their wisely unresisting bodies a bit. On the other hand, he treated them fairly, and with understanding, often with generosity. George Meyer, now in his eighties, still remembers the first time he drove Capone to the racetrack. "How much money have you got, kid?" Capone asked from the back seat as they arrived. Only $80 or so. Capone wadded up four or five $100 bills and tossed them over. "Be back here after the sixth," he said as he strolled off.

Equally important, Capone reposed inspiriting trust in subordinates. The go-between with Cicero newspaperman Robert St. John had been a skinny gnome, Louis Cowan, who stood a scant five feet high. He tended a Cicero newsstand. Capone saw potential, and made him the outfit's bail bondsman, putting in his name apartment buildings worth $500,000 to use as security. He also made Cowan publisher of the Cicero *Tribune* after maneuvering control away from St. John. Cowan worshipped Capone; but he was only an extreme case.

"The underworld is like the upper," says Jesse George Murray, who as a cub reporter met Capone during the gangster's waning days. "It's held together by jobs and opportunity—*and* loyalty. That's all U.S. Steel can offer. They knew they could rely on Capone."

He proved it with Jack Guzik. Few people could abide Harry's equally repulsive younger brother. "I used to hate that man with a passion," says George Meyer, who sometimes drove for Guzik, too. "Everything he ate for a week you could see on his vest. And the B.O.!" Looks, personality and daintiness aside, Capone found Jack Guzik capable and useful. In time the fat little ex-pimp would become his business manager.

However repellent, Guzik would have excited sympathy from the most fastidious early the evening of May 8, 1924, when he tottered into the Four Deuces, face blood-smeared. Who did this? asked Capone.

Joseph L. Howard was small-time even in his three suspected murders. Later, some guessed that Guzik had refused Howard a loan, others that Guzik had been making up to a girlfriend of the handsome, twenty-eight-year-old roughneck. Whatever the cause, toward evening on that May Thursday, Howard started slapping the flaccid butterball around, Guzik whining for mercy until the blood flowed and Howard let the little wreck wail his way back to the Four Deuces.

A few minutes before six-thirty, Capone barged into the saloon of Henry Jacobs, a half block down Wabash from the Four Deuces, and marched up to Howard, lounging at the cigar counter.

"Hello, Al," Howard said, never imagining that anyone might object to his pummeling a slug like Guzik.

Capone grabbed Howard's coat by the shoulder. What did he *mean* by hitting his friend? Howard snarled something along lines of "G'wan back to your whores, you Dago pimp."

With those words he transmuted what otherwise would have been merely an exemplary beating into his own murder. Capone yanked Howard to him, off balance, while he clawed a pistol out of his pocket, shoved the muzzle against Howard's cheek and fired, then pumped five more rounds into the fool.

No one in ''Heinie'' Jacobs's saloon had seen or heard anything that could help the police. Still, Capone had to disappear for a month while his friends made *sure*. Late in the afternoon, June 11, 1924, Capone ambled into the Cottage Grove Avenue station house. ''I hear the police were looking for me,'' he told the precinct commander, ''and I was curious to know what it is for.'' They took him to see Robert Crowe's whiz-kid assistant, twenty-four-year-old William H. McSwiggin. Capone didn't know a thing about Howard; in fact he was out of town that day. The case stayed unsolved.

It was probably the most strategically important killing of Capone's career, coming eight months before he assumed leadership of the outfit. Now, the scruffiest gang member could tell himself, If Al would go the limit for that *pig* Guzik, what wouldn't he do for *me*?

CHAPTER 9

The Wild Colonial Boy . . .

IRONICALLY, IT WAS Deany O'Banion who said, "There's thirty million dollars' worth of beer sold in Chicago every month and a million dollars a month is spread among police, politicians, and federal agents to keep it flowing. Nobody in his right mind will turn his back on a share of a million dollars a month."

Then what about the mental health of those who blithely jeopardized their share of the *thirty* million, not to mention their lives—especially since it was in hopes only of appropriating an insignificantly larger share even though they could not *throw away* money fast enough to dent their fortunes. But then it was never really the money any more than with our modern raiders and junk bond kings who threw away legal billions trying to snatch a pittance more. Ego can efface even self-interest.

How else explain Deany O'Banion?

He shared Bill Thompson's perennial-boy likability, which evoked the same indissoluble loyalty from followers. Despite a boyhood accident that left his right leg four inches shorter than the left, giving him an odd rolling gait, despite later fleshiness, despite small delicate hands with slender fingers on which he prided himself, O'Banion kept the toughness of his days as newspaper circulation slugger and jackroller. Broad shoulders and powerful arms, short neck and round face made him look shorter than his middle height. So did the way he often tilted his head to one side when talking. He never *acted* tough. His habit of calling even enemies "swell fellow" mirrored an ingrained cheeriness and courtesy. He chronically beamed at the world; it amounted to a fixed grin, belied only by unblinkingly cold blue eyes. He was an indefatigable handshaker and backslapper, though never at the same time: at least one hand stayed free to go for one of three gun pockets tailored into his clothes.

Like Anna Torrio, O'Banion's wife declared their home life an idyll and her husband's domestic behavior blameless, with no suggestion of his professional wildness. They had met at a Christmas dance—Viola Kaniff back from her Iowa school—and married on February 5, 1921, when she was about eighteen, he twenty-nine. Although O'Banion owned a house on the far North Side at 6081 Ridge, the newlyweds settled into a large apartment closer in at 3600 North Pine Grove Avenue. He never smoked, drank only moderately. Dinner time usually found him at home, where he *stayed*—often beguiling the evening by singing to the accompaniment of the $14,000 player piano on which he doted. "He was not a man to run around nights, only to take me to a show," Viola told a reporter after his death. "And never one of those men with women calling him up. He was home loving, wanting his friends about him, and never leaving without telling me where he was going."

He got his kicks from work—again, with a puzzling gap between his sensible shopkeeper's manner in the flower store and his mad extravagances outside.

O'Banion loved flowers, arranged them with flair. When he and Nails Morton took an interest in Schofield's, they had *bought* in, not muscled in, and Bill Schofield grieved at both their deaths, especially O'Banion's, calling him "The swellest fellow that ever lived." Although they had another shop far north on Devon, O'Banion headquartered at the main one, 738 North State Street, opposite Holy Name Cathedral, a location surrounded by the arty bohemian section of Chicago. He turned almost unctuous with customers, accepting any stranger's check as though the customer were doing *him* a signal favor. Schofield's became gangland's official florist for funerals, the core of a marvelous business, since orders under several hundred dollars constituted an insult, orders into the thousands *de rigueur* if the dead gangster enjoyed the least éclat.

Out of the shop, O'Banion's high jinks were often so bizarre they impelled one reporter to ruminate on the "childishly irresponsible gangsters" Torrio and Capone had to deal with. O'Banion's impulsiveness could be benign and amusing. One blazing summer's day, he noticed two police sergeants stationed across from his Malt-Maid Brewery, plainly staking him out. "It's too hot out here," he hailed them, "come on in and have a glass of beer and talk it over." Some of his impulses could be praiseworthy. Once, at his own arraignment for a murder, he saw a bedraggled girl from Missouri, only twenty, hauled in for riding a railroad boxcar. On the spot he started a fund to buy clothing for her.

Equally, his whims could be plain loony. One of his earlier safecrackings was "solved" (fruitlessly as usual) partly because at three-thirty A.M. a watchman saw O'Banion and his confederates "perched on a refuse box alongside the curbing, lustily singing a popular melody," a performance that more circumspect safecrackers might expect to attract attention and be remembered.

O'Banion could just as easily turn impetuously murderous. A police wiretap of the flower shop intercepted a call to him from some of his men who had been stopped by two policemen while ferrying a truckload of Torrio beer. They had only $250 among them, and the police demanded $300 more to release the beer. ''Three hundred dollars?'' O'Banion roared. ''Three hundred dollars to them bums? Why, say, I can get 'em knocked off for half that much.'' Alarmed, the wiretappers sent a backup to the hijacking site. Meanwhile, the beer runner checked and called O'Banion back. ''Say, Deany, I just been talking to Johnny, and he says to let them cops have the three hundred. He says he don't want no trouble.''

Charles Dion O'Banion was born on July 8, 1892, in Aurora, about twenty-five miles west of Chicago. His father, Charles H. O'Banion, had been a farmer and then a house painter and plasterer. In the mid-nineties Charles moved the family to Maroa, a village about 150 miles south of Chicago. Neighbors remembered him there as a barber. Much later Charles claimed to be in the oil business, upset at talk that his son had to provide for him.

Schoolmates in Maroa called Dion ''Brinigan'' and thought him ''full of the devil and always having fun''—though it often had a hectic quality and could betray a mean streak, like the time he kicked the lunch box out of a classmate's hand, gusting with laughter when its contents spilled in the dirt. He was a ''real daredevil, who had to do everything best.'' He broke his arm because he *had* to walk on stilts higher than anyone else in the class dared try.

His mother, Emma, died of tuberculosis in 1898 at age thirty. Much later, O'Banion would erect a simple, handsome stone at her grave with ''Mother'' engraved on top, her name and dates on its face. Three years after her death, when Brinigan was nine, Charles took him and his younger sister, Ruth, to Chicago. Older brother Floyd had long since run off to join the navy.

O'Banion must have cherished memories of life in Maroa. In his prosperity he urged a local banker to let him know whenever folks from the village were in a Chicago hospital. He wanted to send daily flowers. He would have picked up the bill if anyone asked; he once sent a crippled child to the Mayo Clinic, and when treatment failed he continued support until the child died. His benefactions were extensive, but typically idiosyncratic and impulsive. He would strew largess in cash, clothes and food personally to those who engaged his sympathy, but scorned all organized charities. ''My money goes straight to those who need it,'' he once said.

Certainly life in Maroa would have seemed retrospective paradise compared with what he found in Chicago. The family took an apartment at Chestnut and Wells, in the North Side section called ''Little Hell.'' His father later insisted that Dion had served four years as an altar boy at Holy Name, acolyte to Father William D. O'Brien. The folks in Maroa doubted that: none of the family had ever attended church back there. Maroa was right. At best, Dion could have served at the altar only one year, and that unlikely. The Reverend D. L. McDonald did not baptize him until January 29, 1905, when Dion was twelve, and the boy did not receive his first communion until June 11, 1905, less than a month before his thirteenth birthday. He stayed at Holy Name only one more year, leaving after the sixth grade, and only students could be altar boys.

In any case, religion did not take. After leaving school he hawked newspapers and eked out those pennies by joining the Market Street gang of pilferers and election sluggers. At sixteen he became a singing waiter in a dive called McGovern's, improving his earnings this time by jackrolling drunks. A year later, in 1909, he did his first of two short terms in the Bridewell—Chicago's house of correction—for burglary. In 1911 he was in for assault with a blackjack. Together they totaled less than twelve months, his only jail time.

He became a slugger for the *Tribune* under circulation manager Max Annenberg. In 1917 O'Banion met Charles Reiser, the man who tutored him, however imperfectly, in safecracking. Those relatively slim pickings meant O'Banion also continued newspaper slugging, though by 1919 he had shifted to the Hearst side under Moses L. Annenberg, estranged brother of Max and father of Walter, who became Ronald Reagan's ambassador to Britain, and was philanthropist heir to many publications, including *TV Guide* and the *Racing Form*.

Prohibition rescued O'Banion from all this scrabbling, made him rich, and changed his spur-of-the-moment nature not one whit.

He started off 1924, characteristically, by removing 1,750 cases of whiskey from the Sibley government-bonded warehouse with forged withdrawal certificates. Police Lieutenant Michael Grady and his five-man squad helped him transport the loot. O'Banion capped that by palming off on an accomplice, as his share, a $41,000 "certified" check he had bought for $100 from the same forger.

The very next night—January 20—at the Sunday opening of a new comedy, *Give and Take*, the audience bundled out of the La Salle Theater into subzero weather. Among them came Davy Miller, a prize-fight referee as well as partner with his brother—West Side gang leader Herschie—in the family's gambling, extortion and bootlegging enterprises. Davy had attended the show with his youngest brother, Max, and Max's wife, his own wife at home nursing a cold.

O'Banion spotted the Millers outside the lobby. He may have been in the audience, too, or passing by; he had not been stalking them. At O'Banion's side stood Hymie Weiss and an enigmatic creature named Julius Schwartz, whom everyone called "Yankee." A ham-and-egg boxer and fight-game hanger-on, Schwartz had fled to Chicago from New York after mild implication in baseball's 1919 World Series fix. He had gravitated to bootlegging and boxing, bringing him in tolerably friendly commerce with the Millers, especially Davy. But a few weeks before, Davy had accused Schwartz of having cheated a friend of Davy's in a bootleg deal. Davy refused to speak to him anymore, publicly cutting him.

O'Banion had always enjoyed amicable relations with the Millers, pals of his own great dead pal, Nails Morton. Had Nails been alive it surely never would have come to this. Schwartz had insinuated himself with O'Banion, complaining of his treatment, perhaps also trying to convince his new patron that Davy had been bad-mouthing O'Banion.

The Millers exchanged friendly greetings with O'Banion and Weiss, but pointedly snubbed Schwartz. O'Banion took that as an affront to himself as well, and trouble started.

"O'Banion came up to me," Davy later told Herschie, "and pulled me aside."

O'Banion asked why Davy was down on Yankee.

"He isn't any good—I don't want anything to do with him," Davy said, starting toward Max, who had hailed a cab.

O'Banion stopped him. "He's as good as you are, you ———," (as the newspaper account decorously put it).

"That got my dander up," Davy told his brother. He had snapped at O'Banion, "I can lick all three of you, but this isn't the place." O'Banion shoved him. Davy knocked away the offending arm, and O'Banion whipped a pistol from one of those three special pockets. He fired point-blank into Davy's belly.

At the sound, Max whirled and started for them. O'Banion fired again, the bullet glancing off Max's belt buckle, knocking him down. In the pandemonium O'Banion and his two companions melted into the crowd.

At the hospital, when police asked who'd shot him, Davy had at first murmured, in approved gangland fashion, "Never mind, I'll take care of that myself." But when doctors prepared him for surgery and allowed that he stood only a fighting chance, he called Herschie and talked. Herschie scandalized all right-thinking hoods, and doubtless stupefied the attendant police, by passing the story on.

Davy beat the odds, recovered, and refused to press charges.

Herschie's version may have been nonsense. Very quickly, reporters became convinced the quarrel arose because O'Banion thought the Millers had cheated him of $60,000 in a booze deal. O'Banion explicitly denied that rumor. He even denied he had started the ruckus, asking, "If I had wanted to bump this Miller off, do you think I'd have done it in the middle of the Loop?" He answered himself, "I'd have fixed him out in some dark alley, on the South Side." He claimed that Miller had drawn a gun on him first—but could hazard no motive. In another version, Yankee Schwartz had pretended to have caught some slighting remark by Davy that night:—*Hey, did you hear what he just called you?*

The only explanation O'Banion ever gave that made sense in fact contained the key to the entire affair. "I'm sorry it happened," he sighed. "It was just a piece of hotheaded foolishness."

That was the chilling feature for Capone, and for Torrio when he returned from Italy in the spring: the plausibility of *any* version. No one doubted that O'Banion might gun down someone before a thousand witnesses for slighting an acquaintance *or* a $60,000 grievance *or* a passing slur on himself. And this was the man they had to do business with!

Two days later, at eleven in the morning, with a dragnet out for him, O'Banion was in the midst of hijacking 251 cases of whiskey, valued at about $30,000, when Lieutenant William O'Conner and his detective squad happened along. Next day, at arraignment, O'Banion and Weiss offered to post their respective $40,000 and $35,000 bails in cash.

O'Banion's behavior continued to deteriorate. In February he killed a man for, at best, muddled reasons. Yankee Schwartz had introduced around town one John Duffy, also known as Dougherty, a blowhard drunk from Philadelphia. Duffy tried to connect with the Millers, who rejected him. Even O'Banion found him too squirrelly—especially when Duffy said the Millers had offered him $10,000 for O'Banion's murder.

Duffy lived with Maybelle Exley, an Ohio farm girl whom he had met in a Louisville brothel. On February 20, 1924, they were having a drinking bout with a young sidekick, William Engelke. Duffy and Exley started to scrap. It ended with a shot by Duffy, Exley dead. Suddenly sobered, Duffy and Engelke needed help, and wound up at O'Banion's.

Sense of mischief in full cry, O'Banion promised to take care of Duffy, arranging to meet him later outside the Four Deuces. Engelke tagged along. He watched the two meet and drive away. That naturally implicated Capone—after police found Duffy's body in a snowdrift on the road to Joliet—when Engelke told police where he had last seen the victim.

Capone surrendered to the police, accompanied by a lawyer. "I'm a respectable businessman," he said with indignation born of a rare sense of total innocence. "I do not own nor have any connection with 'The Four Deuces.' I own a furniture store adjoining the place and for no reason at all somebody is always trying to drag me into something." He had never even met Duffy or Engelke and had not spoken to O'Banion for three weeks.

Engelke, who considered himself a dead man after police beat the story out of him, eagerly snatched at a chance to recant when confronted with O'Banion. "If that's O'Banion," he said, "he ain't the fellow I thought he was. I never met O'Banion, but I understood he was the man I saw climbing into a sedan with Duffy the night after Duffy killed Maybelle Exley. This ain't that guy, though."

Why did O'Banion kill Duffy? Maybe he figured that Duffy, desperate, might try to kill him and collect from the Millers. Or maybe it was something Duffy said. With O'Banion no one could tell. Whatever the reason, the murder's pointlessness and that gratuitous nastiness of dragging in the Four Deuces simply added to Capone's growing sense that someday he would have to do something about O'Banion.

Now was not the time. The conquest of Cicero entirely occupied Capone. Besides, only Torrio's personal authority could resolve the *real* problem because it involved Torrio's first and nominally closest allies, the Gennas. Actually, it was they, not O'Banion, who kept violating terms of the basic combine agreement. As a result, O'Banion and the Gennas were ready to murder each other.

* * *

An "alley cat breed" one observer called the Gennas. They had dominated Little Italy before Prohibition. Prohibition made their fortune. They procured a federal license to deal in industrial alcohol, then they illegally redistilled the denatured alcohol to render it potable. They leased a three-story warehouse at 1022 Taylor, which doubled as headquarters, and watched demand instantly swamp their supply. Henry Spignola—a relatively crime-free lawyer, garage owner and local politician whose sister would marry tough young Angelo Genna in 1925—conceived the solution. Financed by Torrio, the Gennas installed crude stills in hundreds, eventually thousands of tenement flats, houses, and spare rooms throughout Little Italy. Before long, says Howard Browne, who at age nineteen plied a butter and egg route in the neighborhood, "You walked down Taylor Street and you could damn near get drunk on the fumes."

"The average family," a reporter explained, "was paid $15 a day on the salary rolls of the Genna combination. It was a soft snap for the Sicilians. The man of the house had little to do save smoke his pipe, keep the still stoked, and scratch his back."

Not quite. Alky cookers had to spread mash with sugar and yeast, then wait for fermentation before cooking, then tend the still with some diligence, on pain of carelessness inducing an explosion. Spignola's idea of decentralizing such a time-intensive chore constituted brilliance. The Gennas supplied everything each family needed: equipment, mash, yeast, sugar. They showed still tenders how to tap into gas and water lines, not only to save money but because such a large metered use of those utilities could pinpoint stills, and any cop who found one he didn't already know about demanded a raise from the bootlegger.

At peak, the Gennas paid off five captains, about four hundred uniformed police in the Maxwell Street station (a handy four blocks from the warehouse), a number of squads from the central detective bureau and representatives from the state's attorney's office. So many police trooped in and out of the warehouse that neighbors took to calling it "The Police Station." The brothers received a monthly duty roster of badge numbers from the station house so outsiders couldn't mooch in. When some shipments were stopped by police outside the precinct, the Gennas complained and thereafter had a police escort for their convoys outside the district. Individual policemen might get as little as $15 a month—or as much as $125, which coincidentally equaled the usual *year's* merit bonus a policeman could earn for outstanding efficiency or valor. Captains drew $500 a month.

With alky cooking, the Gennas' capacity soon outstripped demand in their own territory and from their extraterritorial customers. They began to edge east and north of the river, selling to O'Banion's customers. What they sold was rotgut, while much of O'Banion's booze had pretensions to palatability, but the Gennas overcame this disadvantage by undercutting his price, $3 a gallon against O'Banion's low-end $6 to $9.

O'Banion naturally would not stand for it. Beer remained his principal business, and the Gennas did not handle beer; their slop could woo away only the dregs of O'Banion's trade. But *nobody* took anything from him. Respecting Torrio's stature, O'Banion first complained to him. Torrio managed to restrain the feral Gennas just enough so that no one declared war, despite many border excursions and guerrilla skirmishes. Capone also kept the alley cats from O'Banion's throat while Torrio vacationed.

Torrio and Capone had no reason to let the Gennas' greed undo the combine. They found the brothers almost as savagely unpredictable and troublesome as they found O'Banion. Within a year, Capone would have problems with the Gennas unsolvable short of murder. For now, though, they had help from Mike Merlo.

Born 1880, brought from Sicily at age nine, Merlo had succeeded Anthony D'Andrea at the head of Chicago's Unione Siciliana branch. As Torrio was the thinking man's gangster, Merlo was the social worker's Unione chief (one of his sons in fact became a playground director, keeping many boys from turning criminal). Merlo's vast prestige among Chicago Italians derived from his palpable concern for their welfare. He did exploit them for political power, and he sanctioned their exploitation by the Gennas. But he also strove to improve their lot, and he insisted on peaceable solutions. As one magazine writer had it, "He absolutely rejected violent death as a weapon." He refused to permit the Gennas any open assault on O'Banion, and throughout Little Italy Merlo's word was fiat.

The standoff continued through early spring of 1924. Torrio returned about the time of the Cicero election, approved vigorously of what Capone had accomplished, and took over peacekeeping chores. Even after O'Banion hijacked a truckload of Genna booze, Torrio (and Merlo) kept the brothers from retaliation, though Torrio must have wondered how long the situation could continue to degrade. Suddenly O'Banion came to him with what seemed a providential out for everyone.

O'Banion had grown tired of it all. Those Gennas were *not* swell fellows, they were crazy. Who knew what could happen? And for what? O'Banion had his. He was thirty-two. He loved Colorado, where he delighted in visits to Louis Alterie's ranch. He'd like his own spread there. If Torrio would buy out his interest in the Sieben Brewery for, say, a half million, he would clear out.

O'Banion received payment and arranged for Torrio to be there very early the morning of Monday, May 19, 1924, when they planned a particularly large shipment. Torrio could take the occasion to assure Banion's boys that he would keep them supplied with the beer they would continue to distribute throughout the North Side.

The Sieben occupied 1464–75 North Larrabee Street, not far from where Ogden dead-ends at North, a big, unconcealable plant. Prohibition agents had visited the summer before, on August 29, 1923, when the Sieben operated under a permit to make near beer. They discovered the genuine article in racking-room barrels. The brewmaster and ostensible owner, George Frank, said he was sorry he had been caught. Although the government revoked the near beer permit, production and payoffs continued unabated.

Police Chief Collins got wind of the Sieben's continued operation and resolved on a secret, lightning raid. Unfortunately, he had to let some colleagues in on it. One of them tipped the date and hour of the raid to O'Banion, who hurriedly cut his deal with the unsuspecting Torrio.

Collins and the new district commander, Captain Michael Zimmer, led several squads at five A.M. They arrested armed hoodlums patrolling the area in cars. Inside the brewery gate they found four trucks loaded with 144 barrels and 20 half barrels of beer, another truck being loaded from a supply of 300 barrels standing by. A total of 128,500 gallons of beer sat roundabout the Sieben Brewery that morning. Collins personally ripped the badges off the chests of two patrolmen who had been stationed at the brewery to enforce the permit revocation, jailing them and a third who was off duty.

In all, the raiders bagged thirteen trucks, six sedans, a passel of revolvers and shotguns, and thirty-one bootleggers, including Torrio and O'Banion. They also recovered a little black book belonging to George Riley Jacobs, a relative of Anna Jacobs Torrio. It listed the names of six police sergeants stationed in the district near Torrio's Manhattan Brewery, with payoff amounts: $40 for day duty, $30 for the dogwatch, midnight to eight A.M., $25 for those on evenings. It also revealed that Jacobs had wired $3,000 to Torrio in Milan as late as February 7.

Capone missed the whole thing, in hiding until all witnesses to the Joe Howard murder, eleven days before, could be squared or intimidated.

Collins turned the case over to federal authorities, not to Crowe, a calculated swipe at the state's attorney. Collins blandly explained that the feds were geared up to cooperate.

At the police station, Torrio, "silent and sullen," gave his usual alias, "Frank Langley." O'Banion, on the other hand, appeared remarkably chipper. He sauntered out of the holding tank on the heels of two detectives, whistling merrily, nabbed only at the outer door of the marshal's office. He slipped a janitor $20 to fetch breakfast for the crowd. He had no worries. This was his first Prohibition offense, which meant only that laughable fine.

For Torrio it would be different. After arraignment, he peeled $7,500 off his wad for his own bail and a further $5,000 for a mysterious "James Casey"—who turned out to be a Democratic pol named Daniel J. O'Conner, publicity manager for William McAdoo, vociferously *dry* candidate for the Democratic 1924 presidential nomination. Torrio hoped to whisk O'Conner away before police or reporters could penetrate his alias. He left the others, including his own men, to wait for the arrival of regular mob bondsmen Ike Roderick and Billy Skidmore, little Louis Cowan not yet on the job.

Torrio's sullenness made sense. He could fight it; he would indirectly, if futilely, offer Assistant U.S. District Attorney William F. Waugh $50,000 to drop prosecution; he could delay it for eight months with continuances. But his

West Hammond conviction meant that this second time, Johnny Torrio must go to jail.

Three days later the body of Bobby Franks was found where Leopold and Loeb had left it, a story that made Chicago forget the front-page news of the Sieben raid.

CHAPTER 10

. . . Out

NOT ONLY WOULD O'Banion's perfidy land Torrio in jail, the wretch refused to return the money Torrio had paid for a brewery now unfixably padlocked! Even so, Torrio might have been willing to swallow the incident if that would keep peace and keep the combine intact. The jail sentence figured to be under a year, and padlocking lasted only a year, which meant Torrio now owned outright a very valuable property. But O'Banion gave him no option.

A story of remarkable silliness circulated to account for the final rupture and its consequences. Hymie Weiss supposedly counseled caution during the dogfight with the Gennas, and O'Banion snapped, "Oh, to hell with them Sicilians." When his words reached the Gennas, Torrio and Capone, the story went, that *did* it.

A few, like O'Banion himself, might kill a man for any daft reason. But these were not exquisites whose feelings were easily bruised, especially not by secondhand insults, not even directed to their faces. Besides, this was an era of easy racial, ethnic and religious slur. Not that the Gennas would have considered "Sicilians" a mortal insult. As for Torrio and Capone, they weren't Sicilian, and if anything, they harbored stronger feelings than O'Banion did. According to a former subordinate, Capone would say, "You can't trust those damn Sicilians," explaining that they "really aren't Italian and they're no damn good." If so, it marked his only known prejudice; and in fact he worked closely and well with many Sicilians, trusting them implicitly. But it certainly described his feelings about the Gennas, whose greed had detonated fighting in the first place, and whose ambition would soon become insupportable. If asked, Capone and Torrio would have said "amen" to anyone who called for the Gennas' damnation.

It was not what O'Banion said, but what he did, that got him killed. "Dion was all right," Capone later summed up,

> and he was getting along, to begin with, better than he had any right to expect. But like everyone else, his head got away from his hat. Weiss figured in that. Johnny Torrio had taught O'Banion all he knew and then O'Banion grabbed some of the best guys we had and decided to be the boss of the booze racket in Chicago. What a chance!
>
> O'Banion had a swell route to make it tough for us and he did. His job had been to "smooth" the coppers and we gave him a lot of authority with the booze and beer buyers.
>
> When he broke away, for a while it wasn't so good. He knew the ropes and got running us ragged. He was spoiling it for everybody. Where we had been paying a copper a couple of hundred dollars, he'd slip them a thousand. He spoiled them. Well, we couldn't do anything about it. It was his funeral.

Three things kept Torrio from immediate attack after the Sieben double cross. First, his field general, Capone, would not emerge from hiding in the Howard murder until June. Next, Mike Merlo's hand remained fixed against killing O'Banion. Finally, for all the sniping, O'Banion had not yet escalated to all-out war, distracted as he was through the spring and half the summer of 1924 fighting the legal consequences of his earlier antics.

Although State's Attorney Crowe had nol-prossed all three charges, a federal grand jury had indicted O'Banion for that hijacking two days after the Miller shooting, and for his looting of the Sibley warehouse. A judge set trial on the first charge for July, the second in November.

The Sibley trial would never take place, because O'Banion was dead by then. But the July proceedings typified gangster trials of the era, and explain the lack of convictions. First off, the court had a terrible time finding jurors. "Say," said O'Banion as the voir dire dragged on, "it's gonna take them longer to get a jury than it'll take the Democrats to pick a candidate for President" (which took 103 ballots!). In an earlier case, one prospective juror, begging off, declared he'd vote innocent even though convinced of the defendant's guilt for fear of being slugged or shot.

Beyond that sensible fear, the court had to unearth twelve veniremen willing to swear that they had no knowledge of notorious gangsters who haunted the front pages and, even less likely, no prejudice against Prohibition. One of O'Banion's lawyers, Michael J. Ahern, questioned a juror who had allowed that, yes, he thought all laws should be enforced:

> Ahern: You aren't prejudiced one way or another are you?
>
> Juror: Why, about seven years ago I contributed some money toward an anti-saloon organization.

Ahern: Perhaps you've changed your mind in the last five years.

Juror: Yes, sir, I have changed my mind.

When testimony started, driver Charles Levin suffered what O'Banion once termed "Chicago amnesia." He couldn't identify his attackers. The prosecutor showed him his statement right after the hijacking. He must have been confused; actually his mind was a blank from the moment that sedan crowded him to the curb. Then how about his later, similar statement to the federal grand jury? Another dazed spell. He couldn't say whether those were the fellows or not.

Ten of the jurors heard police testimony and mustered laudable courage. But the panel declared itself hung after thirty-six ballots, ten to two for conviction. One of the majority, at midnight, after eight hours of haranguing the holdouts, had knelt to pray God that the two would at least *listen* to the others' arguments. Case dismissed. According to later reports, the two shared $50,000.

The judge held Levin for a grand jury probe of perjury charges, whereupon the hijacked driver admitted he had indeed been intimidated. "They warned me," he said. "I was told to keep my mouth shut or get it closed for good. And there is reason to believe something like that might happen." He offered to purge himself by a return to the stand, but the government could only shilly-shally about trying O'Banion on witness-intimidation charges. Levin would probably crumple again.

After the trial, in late summer, O'Banion kept himself out of mischief and out of Torrio's sights, taking a long vacation in Colorado. He returned only in time for the November 1924 elections, back to help reelect Bob Crowe.

The day before the election, O'Banion became a dead man without knowing it. For his help in Cicero the previous April, Torrio had awarded him points in the Ship. On November 3, 1924, O'Banion dropped by for his weekly split. With Torrio busy somewhere else, Capone sat in charge, surrounded by Frank Maritote (alias Diamond), Frank Rio (Capone's chief bodyguard), and Frank Nitti, Capone's eventual successor. Capone mentioned that heavy losses by young Angelo Genna had swelled the week's take. Angelo had dropped a bundle of cash, plus the amount of a sizable marker. Capone suggested they tear up the marker as professional courtesy. O'Banion's response was to limp for the nearest telephone. He called Angelo, abusively demanding that he pay the marker within the week.

That did it. When Torrio heard of it, he sent O'Banion his percentage, then crumpled the marker. But the Gennas, especially savage Angelo, could not overlook this insult—a personal affront, not just a matter of business. No one could have restrained the Gennas now, not Torrio, not even Mike Merlo.

Torrio no longer wanted to. He no longer saw a prayer that O'Banion would continue as even an irregularly peaceable member of the combine. Killing him might set off a blood vendetta; but how much worse could that be than the clearly deteriorating reality? O'Banion's likely successor, Weiss, notably smarter, might see through his own hotheadedness to the business sense of forgiving and forgetting. The murder of a gang chief would make unwelcome headlines and force a show of police repression; but that, however costly, would be temporary, whereas the turmoil O'Banion caused seemed endless. In any case, Torrio's patience had expired.

As for Mike Merlo, he was dying of cancer. Doctors did not expect him to live out the week and his cranky prejudice against murder would die with him. In fact, he succumbed on Saturday, November 8, 1924—the last head of Chicago's Unione Siciliana to die in bed for over five years.

Capone again called Frankie Yale. They needed him now for the same reasons they had needed him with the Colosimo murder: unknown to the victim, he'd also be harder to identify by potential witnesses. Most of all, he'd stay cool to adjust plans, even call off the kill, if circumstances warranted caution.

Merlo's funeral, which Capone and City Sealer Carmen Vacco had orchestrated for Thursday the thirteenth, was to be an extravaganza. Honorary pallbearers would include Mayor Dever, State's Attorney Crowe, Police Chief Collins and Cook County board president Anton J. Cermak. Three thousand of the distraught would mill outside Merlo's home on Diversey Parkway and follow in the rain to St. Clement's for the funeral mass, where a priest would say in eulogy, not without reason, "His gentle kindness made him loved by all his people"; ten thousand would choke Mount Carmel Cemetery. An open car would lead the cortege, in it sitting a life-size wax effigy of Merlo, its suit woven of blue flowers. Eighty-eight cars of Unione members would follow, then the rest of the mourners behind the hearse. The 266-car cortege would stretch a mile.

Many of the cars would be laden with flowers, about $100,000 worth, and a good many of those flowers would come from Schofield's. Funerals invoked truce. Torrio ordered a $10,000 arrangement and Capone, becomingly subordinate, one for $8,000. For the occasion, O'Banion would accommodate even the Gennas without undue wariness. Jim Genna arrived at Schofield's on Sunday with Carmen Vacco to pick up a $750 wreath—and to case the place. That night, after O'Banion had left, having spent a long day filling orders, Yale called. He placed a $2,000 order with Bill Schofield. Could he pick it up tomorrow near noon?

Monday morning, November 10, 1924, O'Banion got to the shop around ten-thirty, ready for a big day. The shop was twenty-five feet wide, the showroom running about twice that deep, a jumble of plants, ferns and flowers obscuring the walls. In the back, a floor-to-ceiling showcase, five feet deep, filled much of the shop's width, flaunting a glory of American Beauty roses. The side of the showcase and the side wall formed a narrow passage that continued to a back workroom and the upstairs offices, the workroom screened by a wicker gate, usually propped open.

Shortly before eleven-thirty, O'Banion stood in the workroom trimming stems from a mass of chrysanthemums. The shop's manager, the bookkeeper and a delivery driver were busy in the rear offices. ''Bill,'' said O'Banion to the black porter, William Crutchfield, ''you better clean up those petals,'' nodding to the detritus of flower arranging in the showroom. Crutchfield had just finished sweeping when the street door opened. He had never seen the three men who entered, the one in the middle taller, clean-shaven and well-dressed. Later, Crutchfield said he ''might have been a Jew or a Greek,'' lighter in complexion than the two who flanked him. They were ''short, stocky, and rather rough looking.''

O'Banion rolled in from the back, clipping shears in his left hand, his right extended for his habitual handshake. ''Hello, boys,'' he boomed as Crutchfield headed for the rear, ''are you from Mike Merlo's?''

Frankie Yale, standing between John Scalise and Albert Anselmi, the Genna's most pitiless gunmen, said ''Yes,'' extending his own hand. Crutchfield glanced back as he passed the wicker gate to see the men shaking hands.

Yale kept hold, perhaps grabbing O'Banion's other arm as well, while Scalise and Anselmi whipped out .38s.

Seconds later, Crutchfield heard a fusillade of five shots, tightly bunched. Two shots hit O'Banion in the right chest. Two more tore into his throat, and the fifth shattered his right jaw. The force hurled O'Banion back and he crashed to the floor against the showcase. A beat or two later, Crutchfield heard the sixth shot as one of the killers bent over the supine O'Banion to shove the muzzle of his .38 against the victim's skull for a last, make-sure shot, the blast staining the left cheek with powder burns.

Schofield's three other employees lit out for the back alley. Crutchfield ran up front in time to see the three gunmen disappear out the door.

Outside, witnesses saw the killers sprint around the corner to Superior, a few doors away from the shop, and pile into a dark blue, nickel-trimmed Jewett that waited with the motor running, its driver (who turned out to be Mike Genna) at the wheel.

Getaway cars with motors running were no news. But as this one sped off west, six other cars pulled from the curb, crosswise on State and Superior, blocking traffic, frustrating any possible pursuit until the Jewett turned south on Dearborn. Then at the toot of a horn, the six swung back into their lanes and vanished. ''That,'' *The New York Times* marveled, ''so far as can be remembered, was quite new.'' In the ploy's brilliant simplicity, ''the existence of a master mind was revealed.''

Bill Shoemaker would be in charge of the investigation; he was the honest cop Dever had chosen to replace Spike O'Donnell's beer-selling police captain. His first guess was that O'Banion had known his killers because of the handshake the porter saw. ''A handshaker, yes,'' said Shoemaker, ''but not with strangers. . . . He knew them—at least by sight—and he did not suspect them.'' With strangers, O'Banion normally stood straddle-legged and wary, arms akimbo, thumbs back, fingers curled a second away from one of those gun pockets.

Shoemaker was wrong. There is no reason to believe O'Banion had ever met Yale. And *no one* who recognized Scalise and Anselmi would relax. The Gennas had imported them from Sicily. In 1920, when Frankie Yale had needed two unhesitating killers to invade an Irish dance hall in Brooklyn, and slaughter the gathered White Handers, he had called Capone. The ones Capone borrowed for him were Scalise and Anselmi. But they were not yet prominent enough to be recognized by most enemies any more than by witnesses—another bonus of covering up around photographers.

Besides, why would O'Banion have asked, "Are you from Mike Merlo's?," had he recognized them? Yale's flower order, following Torrio's, Capone's and Jim Genna's, constituted more of a master stroke than the six cars. It disarmed O'Banion's suspicion and had him *expecting* to see "strangers" there to pick up another large order, secure on his own ground in his privileged position as florist to the boys.

Captain Shoemaker rounded up the usual suspects. Capone led the list, followed closely by Torrio. Neither knew a thing; Torrio said he and O'Banion had been pals—look at that $10,000 order! Shoemaker made a pro forma effort to question O'Banion's stricken followers. When he asked Hymie Weiss, "If you knew anything about this murder would you tell me?" Weiss set the tone of the inquiry: "Well, to be frank, I guess I wouldn't."

Only Louis Alterie made a statement, naturally the flakey, wrong thing. "If I knew who killed Dion," he said, "I would shoot it out with the gang of killers before the sun rose. . . . If I can make an appointment with the killers of O'Banion I will shoot it out with them." If they got him, "I'll go with a smile because I'll know that two or three of them will go with me."

That bravado overloaded Mayor Dever's circuits. Here was what the papers persisted in calling the "king of the underworld" lying in state (some forty thousand would see him) in a funeral home owned by an assistant state's attorney (John Sbarbaro, later a municipal judge). Now this hood had the effrontery to issue public challenges for a duel. The mayor exhorted the police to "shoot to kill" any who resisted arrest or questioning and generally to run all gangsters into jail or out of town.

Cardinal Mundelein as usual forbade a Funeral Mass and burial in consecrated ground. A spokesman explained that "a person who refuses the ministrations of the Church in life need not expect to have the ministrations of the Church in death."

That interdiction could not tarnish the Obsequies Chic of the send-off O'Banion's North Siders arranged on November 14, the day after Mike Merlo's extravaganza: the mountain of flowers (including a basket of roses whose card read "Al Brown"); the tastefully matte, $10,000 silvery bronze coffin, rushed by special freight car from Philadelphia, resting on a marble plinth inscribed "Suffer the little children to come unto me"; the two-mile cortege to the unconsecrated section of Mount Carmel Cemetery, the streets so jammed that scalpers could demand a dollar for vantage points on surrounding buildings. The boys did Deany and themselves proud.

Five months later, Viola O'Banion had her husband's remains quietly moved to consecrated ground, near a mausoleum sheltering the bones of three high churchmen—inducing a scandalized police captain to say, "Look at him now! Eighty feet from a bishop!"

As expected, the police investigation got nowhere. Chief Collins said, "The killers won't be caught unless we're lucky," and even then he doubted they could make a case. Their best witness, the Schofield porter, declared he'd never be able to identify the killers even if he saw them again. The police kept trying. "We have run down a thousand and one clues," sulked Chief of Detectives Hughes, "only to find them all worthless." No one would talk. "The more we run around the more we are convinced we are wasting our time."

Almost. They had picked up Frankie Yale in the sweep after O'Banion's funeral, caught with a pistol in his pocket, then released at night court, his identity ringing no official bells. The following Tuesday, November 18, Chief Collins got a tip that Yale, prime suspect in the Colosimo murder four years before, was in town and involved. The police grabbed him at the La Salle Street station minutes before his train left for New York.

Yale was indignant. He had a permit for that gun, signed by a New York judge. He'd come to Chicago for Merlo's funeral, and stayed to see old friends. He was lunching with one of them, Genna henchman Samoots Ammatuna, when O'Banion was killed. A waiter at the Palmer House would confirm it. Sure enough, Nick Delassandro could remember even what they'd had to eat and drink.

Yale caught the next train home.

It's strange that the people who would soon start the combine unraveling now provided a final demonstration of how well it worked.

Eddie Tancl, born in the Valley of Bohemian extraction, had been a prizefighter, a lightweight with a brutal punch that had once battered an opponent to death. He became a successful saloonkeeper in Cicero. The Hawthorne Park Café at Forty-eighth and Ogden Avenues brought him so much local power he fancied he could defy the combine. Although he scorned Torrio and Capone, at first he accepted their beer from the O'Donnells. But when Dever's pressure cut supplies of real beer in 1923, the O'Donnells necessarily delivered needled beer, and Tancl refused the slop, declaring he'd henceforth buy beer wherever he chose.

The combine told him to get back in line or get out of town. Tancl laughed. Two hoods, one O'Donnell's, one Sheldon's, were discovered touching a match to the fuse on sixty-five sticks of dynamite outside Tancl's place. Phillip Corrigan's bullet-riddled body was found several weeks later; Joseph Brooks went into hiding.

Once O'Banion was out of the way, it came time to deal with a petty annoyance who might inspire others to opposition. After a Saturday-night drinking bout at Tancl's place on November 22, 1924, Klondike O'Donnell's kid brother, Myles, and their best gunman, Jim Doherty, returned Sunday morning about six o'clock and started drinking again. At eleven that morning they started a rhubarb over a $5.50 check. Myles hit waiter Martin Simet with his gun as Tancl lumbered into it. When the shooting on both sides ended, Myles and Doherty fled, wounded, leaving bartender Leo Klimas dying and Eddie Tancl dead.

Next spring, after stories had been changed and witnesses had disappeared, a jury took only nine minutes to find Myles O'Donnell and Jim Doherty not guilty; the judge had warned he would set aside any other verdict.

The coroner's jury impaneled to examine O'Banion's murder declared itself unable to name his killers. They complained that they had been confronted "with much conflicting testimony" and that "some of the witnesses were evidently withholding the truth, while other witnesses refused to make any statements whatsoever and some material witnesses that were asked for could not be located."

O'Banion's survivors were less uncertain.

CHAPTER 11

Torrio Out

AS PART OF their huffing and puffing over O'Banion's murder, police finally nailed shut the Four Deuces in December 1924, though by now it served only as a business office. Capone established another, two blocks east at 2146 South Michigan. Behind a name plate, "A. Brown, MD," visitors discovered what looked like any prosperous doctor's waiting room, complete with old magazines. In the back, shelves of bottled booze samples lined the walls to facilitate customers' ordering and Jack Guzik supervised the outfit's bookkeeping amid stacked records of major customers, details about supply, smuggled liquor, alky cooking, distilleries and breweries, balance sheets for brothels and gambling, schedules for payoffs to politicians and police.

Torrio may have been out of town when Capone set up the new office. A newspaper noted that he had not been seen from shortly after O'Banion's funeral until the second week in January 1925. Rumor had Torrio and his wife trailed by some of O'Banion's fiercer mourners to places like Hot Springs, New Orleans, St. Petersburg, Palm Beach, Havana and the Bahamas. On such scant evidence, later writers romanticized it into a hegira of hairbreadth escapes, the gunmen missing their quarry "sometimes by only hours."

Maybe Torrio and Anna did take such a trip, but his later behavior makes the chase scenes improbable. He never took the least security precautions, never even had a bodyguard up to the moment O'Banion's avengers caught Torrio at his own doorstep. Plainly he did not think himself a target—a miscalculation that could hardly have survived being harried about the Gulf, Florida and the Caribbean. Somehow he fancied himself immune from attack even after someone tried to kill Capone.

Early the morning of January 12, 1925, a sedan—side curtains pulled down, license plate obscured—cut Capone's car off, forcing it to the curb at Fifty-fifth and State. Shots rattled from the sedan's front and rear windows, raking Capone's car with pistol and shotgun slugs. "They let it have everything but the kitchen stove," commented an impressed police sergeant when he viewed the riddled hulk.

Capone was in no danger because he was not in the car. Amazingly, two of the three men who were in it suffered not a scratch. Capone's driver, Sylvester Barton, turned away when the sedan curbed him, and caught one slug in the back before he slumped below the line of fire. The two in the back seat—one of them Capone's cousin Charley Fischetti—had flung themselves safely to the floorboard before the shooting started.

Although the story later circulated that Capone had just stepped into a restaurant, that seems unlikely. It was too early for him to have been up and about, especially inspecting his restaurant properties. Contemporary news accounts certainly made no mention of his presence.

Even though the assault misfired, it shook Capone enough so that he immediately ordered a Cadillac chassis on which he had built a steel-armored body with bulletproof glass. Where the usual Cadillac sedan might weigh two tons and cost $7,000, his rolling fortress weighed seven and cost at least $20,000. It was the first of a series of armored cars he would own.

Torrio pleaded guilty to the Sieben Brewery charges on January 17, 1925. As he had hoped, federal judge Adam C. Cliffe gave him a meaningless $5,000 fine and only nine months in the jail of Du Page County, just west of Cook County. Federal prisoners spent such negligible terms in county lockups, and Torrio surely knew what most of Chicago would discover only later in the year: Terry Druggan and Frank Lake, recently jugged for running their Standard Brewery despite a federal padlock, spent about as much time outside the Cook County jail as they did inside. It was run by Torrio's faithful old retainer, Sheriff Peter Hoffman. Torrio could anticipate a similarly understanding management in Du Page County.

Two of the policemen supposedly guarding the Sieben got three months each; Nick Juffra, one of the outfit's heavyweights, got six months. Edward O'Donnell—a politician, not "Spike"—owned one share of stock in the Sieben, so drew eight months because as a former mayor of East Chicago he had no business in such company. Judge Cliffe fined six truck drivers $500 each and let twenty-two other defendants off. He gave Torrio a ten-day stay to settle his affairs before reporting for execution of sentence.

Torrio celebrated his forty-third birthday on January 20, three days after Capone turned twenty-six.

On Saturday, January 24, 1925, Torrio had business downtown, including a visit to Michael Kenna. Reorganization had reduced each ward to a single alderman, but Hinky Dink still ruled as First Ward committeeman. John Coughlin remained alderman, his fellow city fathers voting Bathhouse the spiffiest dresser in City Hall.

Torrio, Anna and her brother, Tom Jacobs, were driven downtown at about eleven o'clock that morning, an unusual treat. Torrio's own car was in the garage for some work, and he did not employ a regular driver. Jack Guzik, on business in New Orleans, had turned over to the Torrios his Lincoln and his chauffeur, Robert Barton, a small, dark-haired man always called ''Bobby,'' brother of Capone's driver, Sylvester. While Torrio worked, Anna busily shopped. They met later that afternoon for the drive back to their South Shore apartment at 7011 South Clyde, packages piled high. Tom Jacobs was not with them.

A little past four o'clock that afternoon, a mild day for January in Chicago, Walter Hildebrandt could look forward to the pleasures of Saturday night, since he was near the end of his deliveries for the Oriental Laundry Company. He had one on Clyde.

Peter Veesaert, seventeen, son of the janitor at 6954 South Clyde, diagonally across Seventieth Street from the Torrios' apartment at 7011, loitered at the top of his building's basement stairs.

Mrs. James Putnam lived at 7016, directly across from the Torrio's building. She had just looked out the window at the dimming afternoon when an unusually luxurious sight on this narrow, residential, apartment-lined and determinedly middle-class street caught her eye. A big black glistening Lincoln—with a chauffeur in brown livery!—pulled up in front of 7011. Before the chauffeur could leap out and hasten around to open the curb-side rear door, a man in a gray fedora and dark blue overcoat stepped out—*why, it was Mr. Langley!* He helped his wife out, her arms stuffed with packages. Like others on the block, Mrs. Putnam knew them only as affably nodding neighbors, and by Torrio's usual alias, Frank Langley. Mrs. Putnam understood that he was some sort of broker in La Salle Street finance.

Parked just around the corner on Seventieth, with a line of sight on the Torrios' apartment, a gray Cadillac had been waiting for about an hour, motor idling.

Anna Torrio, the gray fox collar of her moleskin coat raised despite a balmy forty-degree temperature, did not wait for her husband but scurried ahead, up the short walk to the door of the apartment's foyer. She turned, arms full, to push open the door with her back. She saw Bobby Barton holding a large bundle, her husband backing out of the Lincoln, his arms chockablock with packages. She also saw the men who had emerged from the Cadillac.

Vincent Drucci stayed at the wheel while Hymie Weiss, with a shotgun, and George Moran, with a .45, leapt out and charged the Lincoln. They spread out, still in the street, Weiss at the tail of the Lincoln, Moran at the nose, and made their first mistake. The two opened fire through the heavy car body instead of continuing around for a clear shot.

One slug ripped Barton's leg, just below the knee. Torrio, unharmed, dropped the packages and dashed as fast as his fat little legs would take him toward the apartment. A half-dozen steps put him in position for the clear shot the gunmen had lacked before. Moran's .45 hit Torrio in the right arm, spinning him around. Torrio fumbled for his own gun while still stumbling backwards, but now Weiss's second barrel caught him, one buckshot slug smashing Torrio's jaw, tearing into the neck, three more in his chest and belly, five wounds in all.

Torrio collapsed. Moran dashed forward, ready for the sure kill, muzzle to the head. But that early, futile firing had emptied his automatic as it had Weiss's shotgun. Moran scrambled to ram another clip into his .45 for the final shot. Torrio sprawled, helpless; Anna stood transfixed in horror.

At that moment, Walter Hildebrandt turned his van onto Clyde.

Drucci honked an urgent signal. To Moran, any coup de grace must have looked superfluous. Ragged, bleeding holes gaped in Torrio's torso. His lower face and neck were raw meat. Moran stopped reloading, turned and followed Weiss back to the Cadillac. When they jumped in, the big car, its curtains drawn, took off along Seventieth west to Stony Island. To the credit of his grit, if not sense, Walter Hildebrandt took in the scene and chugged after them, noting that the Cadillac lacked license plates. Luckily for him, when they turned north on Stony Island the sedan soon outdistanced his laundry van.

Bobby Barton limped back into the Lincoln and drove off.

Anna Torrio ran to her husband and dragged him by the shoulders to shelter in the vestibule.

Across the street, Mrs. Putnam put in a frenzied call to the nearby Woodlawn police station, which dispatched two officers. An ambulance rushed Torrio to the Jackson Park Hospital.

Thomas J. Conley had retired from the police as a desk sergeant, but when he spotted Barton's bullet-torn Lincoln speeding along, he trailed it through many strange, disoriented turnings until the chauffeur pulled up at a drugstore only a block south and three blocks east of the Torrio's apartment. He hobbled in to phone Capone at the Hawthorne in Cicero.

Capone burst into the hospital, distraught. "Did they get Johnny?" he exclaimed, and when told the situation, kept muttering tearfully, "The gang did it! The gang did it!" When pressed for specific nominations, he told Assistant State's Attorney (and gang undertaker) John Sbarbaro "I'll tell you more when he gets well."

Even if he meant it, that did not seem like much of a promise. The wounds alone figured to finish Torrio, and even if he survived those, Torrio himself suggested what he considered a supplementary peril. Gunmen had imported from Sicily a devout assurance that bullets boiled in onion-water, then rubbed with garlic, would promote necrosis, failing an outright kill. The fetish had spread throughout gangland: garlic mixed with gunpowder, they all believed, generated gangrene. Today's forensic medicine brands that nonsense: garlic does not promote or generate infection any more than any other foreign matter would. Even if it did, the heat of the slug's flight through the gun barrel would burn off outright poison, let alone any onion and garlic decoctions. But in the mid-twenties, Torrio's attending physicians, Drs. Omens and Byrne, entertained the possibility. Torrio begged them to cauterize his wounds, treatment gangland thought sovereign against gangrene.

Soon, Capone's gangster discipline reasserted itself, and he stopped babbling. Now he had no idea who had done it—unless it was the unknowns who had shot up his own car twelve days before. He added, "There's nothing to the theory that friends of Dean O'Banion shot Johnny Torrio. They were the best of friends."

From the first, Torrio refused to name his assailants. "Sure, I know who they were," he told a reporter at his bedside. "That's all I'm going to say right now." Or ever. His wife, when asked, asked back, with some reason, "What's the use of telling the police?"

Young Peter Veesaert told what he could. When shown mug shots, he unhesitatingly picked out George Moran, identifying him particularly by his pronounced chin dimple. Later he twice picked him out of lineups.

"No," said Anna Torrio, when the police hauled Moran around to the hospital, "that's not one of them." Like her husband, she had announced that she would never identify the assailants. The police dragged Moran to Torrio's bedside nonetheless, along with Weiss, Drucci and so many others that the doctors protested. Such continual disturbance might impede Torrio's recovery. Torrio said, "No use bringing anyone here, I won't rap them" ("rap" was then slang for "identify," a "bum rap" originally meaning a spiteful false identification).

Stymied, police arrested Capone, questioned him overnight and released him. While they were at it, about two A.M., three carloads of men pulled up to the hospital. One of the men marched in and told Dorothy Beck, night superintendent of nurses, that he wanted to see his pal John Torrio. She starchily informed him that these were *not* visiting hours; besides, Mrs. Torrio had forbidden visits by any but family and *closest* friends. The man insisted—until Beck added that two police guards were stationed in Torrio's room. The next day, when Capone heard about the attempt, he moved in. "And while I'm there," he said, "nobody will bother him." Torrio occupied the middle room of a three-room suite, Anna Torrio in the room on the right, Capone in the other. Two uniformed policemen guarded his door, two more at the stairwell, others sprinkled around the corridors, while still others watched the entrances from outside.

Doctors despaired for the first few days. Torrio's fever raged and his pulse galloped because the neck wound had indeed become infected. Yet by the fifth day doctors reported him "gaining" and by the next day, "recovering." On February 9, just three weeks after the shooting, Capone smuggled Torrio down a back fire escape and away in a closed sedan.

Torrio headed for jail. His lawyer had arranged a transfer of sentence from the Du Page jail in tiny, rural Wheaton to the Lake County jail in urban Waukegan, about thirty miles north, where the still-shaky prisoner could receive better medical treatment. Although the court had granted a further stay to February 27, Torrio thought it prudent to seek the safety of bars and walls right away.

A municipal judge released Moran on $5,000 bail despite howls from police and prosecutors. But with the Torrios' adamant refusal to rap him or anyone else, and with an arranged alibi offsetting Peter Veesaert's sole identification (neither the neighbor nor the laundry driver had seen any faces, and Moran's was the only one Veesaert had glimpsed), the state's attorney did not even bother to indict.

Although it looked like a dungeon—thick walls crowned by a huge dome—the Lake County jail offered Torrio a downright homey sanctuary. The sheriff, Edwin Ahlstrom, proved little sterner than Cook County's Peter Hoffman. He let Torrio furnish his inordinately roomy cell—which boasted the unusual amenity of running water—with a brass bed, carpet, easy chairs, bookcase, dresser, record player and, a rarity in jail, a radio. He let Torrio screen the barred windows with heavy bulletproof mesh, the insides masked by blackout curtains so not even a shadow would provide any sniper a target when the room was lit at night. He let Torrio hire deputy sheriffs as personal guards. He let Anna join Torrio for cozy lunches in the shrieval home, where Torrio also took most breakfasts and dinners, and let the couple enjoy fine afternoons in rocking chairs on the porch. Certainly he let Torrio alone when Torrio wanted conferences with colleagues from the outside.

It may have been sometime in March when Torrio summoned Capone and his lawyers.

Torrio wanted out. Even as early as March it was evident that the combine faced disintegration. All of Chicago gangland—not just O'Banion's successors—seemed intent on eviscerating it. As the mayhem worsened, Torrio could see that any "peaceful" business cooperation would have to be reasserted and enforced with the gun, not diplomacy. He was too old for what he could see coming.

Torrio turned everything over to Capone. When his jail term was up, he would leave Chicago, perhaps to live in Italy. Capone would remit part of the profits—one rumor put it at 25 percent for ten years—and Torrio would be available for consultations.

Various analysts later assured the world that Torrio "could dish it out, but couldn't take it." Certainly the trauma of his shooting and Anna's anguish shook him. But he had never before displayed cowardice. More likely, it was what the man who never wanted any trouble saw happening and what it suggested was on the way. Capone was welcome to it.

CHAPTER 12

Capone Versus Them Sicilians

WHAT TORRIO COULD see coming was epitomized by Chicago's gang killings. They totaled over seven hundred during Prohibition. Even in 1922, the combine's year of maximum peace, one observer counted thirty-seven. That jumped to fifty-seven during Spike O'Donnell's 1923 incursions. With the surface harmony of 1924, the total dropped to sixteen, but soared to forty-six in 1925, the pace steadily rising—up to seventy-six the next year.

Although Torrio's own orders had done much to swell the totals, that was not the point. Killings disrupted business no matter who initiated them or for however exigent and substantive a reason. Eliminating a rival might expand or protect one's territory, but no one made a penny from gang murders. Quite the contrary, when the prominence of victims or volume of killings occasioned crackdowns—no matter how temporary—murders could cost a bundle.

Torrio could see that the trend was up. More dismaying, he could see that their nominally strongest *allies* must soon be eliminated.

Success had not exactly changed the Gennas, just made them more so. Even though they lorded it over Little Italy, the coming of Prohibition had found them small fry, James and Angelo first arrested in 1920 for trifling offenses, Angelo for possession of one gallon of booze. But they exhibited even in small things a characteristic meanness. Early in 1921, when Mike Genna was arrested for transporting a stolen auto, he skipped, risking the home that two poor illiterate benefactors had pledged as bail.

Angelo, already surliest of the Gennas, became arrogant when in a single year he beat two murder charges: one for killing Paul Labriola during the D'Andrea fight and one for killing Paul Notti. He escaped Notti's deathbed statement accusing him because the defense showed that the victim had received an opiate for his pain; the judge threw out the case on grounds that Notti's mind might have been befuddled. That was in June 1922.

In August, two friends asked for Angelo's help. They had taken fifteen-year-old Genevieve Court to Milwaukee where they had "mistreated her," as she was prepared to put it delicately in a federal Mann Act prosecution. The day before the trial, Angelo stopped her on the street to explain that he would kill her and her mother if she testified against his pals. After the prosecution stalled, she told of his intimidation. Next November, Angelo went to Leavenworth for a year and a day.

That only fleetingly ruffled his self-esteem; it had taken the feds to put him away. And with the Gennas' alky-cooking success, the whole family soon shared Angelo's arrogance. Though still hated and feared, they spread enough money throughout Little Italy to buy deference. They quickly loomed as powerful as their political patron, ward boss Joseph Esposito, known as "Diamond Joe."

By late 1924, with the entire police district on their payroll, the Gennas felt strong enough to challenge Torrio and Capone. They seized the Unione Siciliana presidency, frustrating Capone, who had wanted his friend Anthony Lombardo to succeed Mike Merlo. A sagacious, calm commission broker and cheese merchant—with bootlegging and gangster ties—Lombardo commanded general respect. He would have seen to it that no one could use the Unione for leverage against Torrio and Capone. The Gennas capitalized on Capone's distraction following O'Banion's murder to maneuver brother Angelo quietly in as president. From that moment, the interests of the two gangs sharply diverged.

Just before Torrio was shot, Angelo Genna married Lucille Spignola, eighteen, pretty, vibrant, and from a prominent family—further demonstration of the Gennas' surging power. Lucille's brother Henry, who had thought up alky cooking, was a lawyer, a graduate of McKinley High and John Marshall Law, and was intimate with stars of the Chicago Opera. The match did not delight the Spignolas, but Lucille really loved Angelo. And no one said no to the Gennas anymore.

Invitations took the form of newspaper ads: "Come One, Come All." Some three thousand guests packed the reception after a glittering church ceremony. The wedding cake weighed literally a ton, stood twelve feet high, was surmounted by a toy bride and groom on a frosting balcony, and had "Home Sweet Home" embossed in icing. In fact, the couple temporarily set up house in a Belmont Hotel suite on Sheridan Road, facing Big Bill Thompson's home, an old-money enclave slightly northwest of the Gold Coast.

A happy marriage did not mellow Angelo any more than success muted his brothers' greed, which meant that Capone had a problem. The North Siders had made no dramatic moves after shooting Torrio, but that situation remained tinder. How could Capone fight the conflagration, when it erupted, with "allies" like the Gennas at his back? He would first have to defang the brothers, yet avoid the debilitations of open warfare.

One element of his strategy became apparent the night of April 10, 1925—but so subtly and casually it did not alert the Gennas.

The previous Monday, Mayor Dever had orchestrated a raid on Dr. Brown's "luxuriously furnished suite" at 2146 South Michigan. Plans stayed so secret that Capone's spies got him no word, and the raid's leader, Detective Sergeant Edward Birmingham, was either honest or closely scrutinized enough to resist an offer of $5,000 from Jack Guzik to come back later.

The raiders had seized the outfit's complete records, but the police brass and Bill McSwiggin, the hot young assistant state's attorney in charge, decided "it would be illegal" to make public any customers' names or business details. On Thursday, municipal judge Howard Hayes obligingly impounded the records when gang lawyers asserted they had been unconstitutionally taken without a warrant. Federal D.A. Edwin A. Olson immediately informed Judge Hayes that the government wanted to examine everything and specified that it would seize the records on a federal warrant if necessary.

The next day, without notifying the government or the state's attorney, Judge Hayes held a quiet hearing and on the spot turned everything over to Capone's lawyer. Olson exploded, publicly calling Hayes's action *at best* "peculiar." The state's attorney's office declined to join in condemnation of the judge.

That night, a detective squad spotted Capone in a car on Roosevelt Road, no unusual sight, but since Mayor Dever and Chief Collins were in a lather, the detectives flagged down Capone for speeding and arrested him and his companions when they found two revolvers in the car. One of those companions was John Scalise, half of the Gennas' star murder team. Capone had enough legitimate business with Scalise so that their being together did not excite Genna suspicions. In retrospect, we can guess that Capone was already wooing the Gennas' best killers.

On the morning of May 25, 1925, Angelo Genna set off in his $6,000 roadster, $11,000 in his pocket to pay for a house he and Lucille wanted in Oak Park, suburban hometown of Ernest Hemingway. Angelo had driven only a short way on Ogden when a closed sedan with four men in it tried to overtake him. Shotguns barked and Angelo hit the accelerator, jerking from a belt holster one of the two guns he carried, firing out the window as he wheeled down Ogden at better than sixty miles per hour.

The more powerful sedan caught up, and Angelo tried to maneuver. He began a screeching turn into a road intersecting Ogden at an acute angle. Losing control, he crashed into a lamppost near Hudson Avenue. He sat stunned, gun empty on the seat beside him as the sedan rolled beside the roadster. Shotguns blasted again, slugs tearing into Angelo, one shattering his spine. Bystanders found him, semiconscious, clawing at his other still-holstered gun.

Angelo lived long enough to shrug when a detective, at the hospital, asked him who had done it and long enough for Lucille to get there and call him "sweetheart." But he died before brother Sam arrived, so never breathed a word about who had shot him.

Without saying why, the police announced they were satisfied that three of the four men glimpsed in the sedan were Hymie Weiss, George Moran, and Vincent Drucci; they *thought* the driver to be Frank Gusenberg, one of three brothers in the gang. The circumstances of the next Genna murder, though, suggest that the police simply *assumed* that North Siders had killed Angelo to avenge O'Banion. In fact, the killers almost surely worked for Capone.

The Gennas evidently believed the police. First they laid Angelo away with a brave show, pointedly buying a coffin that cost $1,000 more than O'Banion's. The main feature of the traffic-choking Obsequies Chic cortege to the unsanctified part of Mount Carmel was Angelo's riddled, crepe-hung roadster in tow, a display and turnout that the *Tribune* termed "an interesting commentary on our city." Next, the Gennas went gunning for North Siders.

Early Saturday morning, June 13, 1925, eighteen days after Angelo's death, five of the Genna gang, including brother Mike, as well as Scalise and Anselmi, lurked in an alley near Sangamon and Congress, near the northern edge of Little Italy. They had somehow set the North Siders up for an ambush. When George Moran and Vincent Drucci cruised slowly by, east on Congress, the Genna gunmen sprayed their Hupmobile with shotgun slugs. The two returned fired and escaped. Drucci was grazed and the car shot up so badly they abandoned it. Moran hightailed it home, where he later told police the car had been stolen.

Of course, the attackers didn't know that their targets were temporarily out of action. An hour later, about nine-thirty A.M., Mike Genna, Scalise, Anselmi and an unidentified driver were heading south on Western Avenue. By chance, at the corner of Forty-seventh, about four miles southwest of the ambush site, a detective squad car going north on Western passed them. The squad commander, Michael J. Conway, recognized Mike Genna and ordered his driver, Harold F. Olson, to *follow that car*. Maybe they could surprise the Gennas in bootlegging or some allied villainy.

It was probably the U-turn of the big, unmarked detective sedan that caught the attention of the gangsters. The Genna driver immediately stepped on the gas.

A moment later, pedestrians saw two cars hurtling south at over seventy miles per hour in a classic chase, dodging traffic and fishtailing on the asphalt and streetcar tracks still slick from heavy rains earlier that morning. Although it later became a point of dispute, the detectives may have been sounding their gong, more often used in those days than sirens.

At the corner of Sixtieth a truck nosed into the intersection. The Genna driver mashed the brake to avoid collision. The car skidded around to a halt, hood jutting into the roadway, rear over the curb, with the right rear fender crumpled against a lamppost. The police car also slewed around, stopping at right angle to the Genna car, so when both sides leapt out, the gangsters had the cover of their car while the police stood exposed.

Witnesses of varying reliability differed as to who shot first. It was almost certainly the gangsters, if only because their car sustained but one hit while the detectives' car became a sieve with over seventy shotgun slugs buried in its body. Driver Harold Olson was first out of the police car, foot still on the running board when a shotgun blast killed him. Officer Charles B. Walsh dropped next, mortally wounded. Conway and the fourth officer, William Sweeney, still crouched in the car, had their guns out and had started to return fire. Another blast smashed into Conway's chest and he went down, severely wounded, though he would survive.

Sweeney ducked out of the car and continued shooting, emptying his service revolver without scoring any hits—except perhaps that one on the gangsters' car.

The Genna driver had scuppered off almost before the action started. Now the other three fled west through a vacant lot, where one of them hurled aside an empty shotgun.

Sweeney snatched up two of his partners' guns and started after the gangsters. He had almost caught up with them by the time they reached the far side of the lot. While Scalise and Anselmi disappeared between two houses, Mike Genna whirled around, pointed his shotgun full at Sweeney's chest, pulled the trigger . . . and heard the hammer thonk on a spent cartridge. Sweeney fired his pistol. He hit Mike—a thigh wound eight inches above the left knee. The gangster clumped down the passageway between the two houses, Sweeney after him, blazing away. Frantic for refuge, Mike rounded the corner of one house, saw a basement window, smashed the glass with his useless shotgun and dove in headfirst, disappearing as Sweeney also rounded the corner.

While the policeman took stock, two more panted up—off-duty officers, one who had seen the action from a passing trolley, the other from his nearby home. Together, they kicked in the basement door. Mike lay on the floor, propping himself up, in his hand a blue-steel Spanish .38. He jerked a single shot, way wide, then slumped back. The officers rushed up and now saw the blood gushing from Mike's only wound: that single shot had severed an artery.

Before he died, Mike Genna, age thirty, managed one feeble kick at the face of an ambulance attendant who had bent to help him. He growled, "Take that, you son of a bitch!"

Meanwhile, Scalise and Anselmi looked like favorites to escape. They had run several blocks away when they loped into a store, panting, to buy caps to replace those they'd lost in the getaway, uncovered heads being an unusual sight in those days. The storekeeper didn't like their looks or condition and turned them away. They two evidently panicked, because they now doubled back to Western at a dead run. Policemen in a passing squad car had heard the shooting and noticed two men, conspicuously disheveled, hatless and blowing hard, boarding a streetcar. They overtook the trolley and hauled the now gunless gunmen off. Of course word flashed around that fellow officers had been killed. Scalise and Anselmi showed up at headquarters badly beaten, having "resisted arrest"—the only *real* punishment they would receive.

Later, an Italian identified only as "prominent" told police that, in fact,

Mike Genna was a dead man that day no matter what happened. Capone had completed his wooing of Scalise and Anselmi and they were speeding Mike south toward Chicago's outskirts—and discreet death—when the detective squad had spotted them. (After the next killing put all the Genna murders in a new light, and Capone came under suspicion, one newspaper analyst realized that the shotguns used to kill Angelo were much more a South Side Italian weapon than one favored by the North Siders; Weiss's use of a shotgun on Torrio was exceptional.)

Next to go was Anthony Genna. On July 8, 1925, Tony the Gent got an early call from someone he knew well enough to meet at ten-thirty that morning near the grocery store of Charles and Vito Cutaia on Grand Avenue near Curtis. When they met, the caller grasped Tony's hand in a firm shake, just like O'Banion's last. Two gunmen came from behind and pumped five shots into Tony's back.

He hung on in the hospital until four that afternoon. Told he was dying—and at the urging of his brother Sam and his mistress Gladys Bagwell—Tony three times gasped a name that non-Latin police ears at first misunderstood, sparking a futile search for some gunman named "Cavallero."

Soon everyone realized Tony had identified *Il Cavaliere*, sobriquet of Joseph Nerone, alias Anthony Spano, called The Cavalier for his lean aristocratic bearing and manner. Mathematics teacher turned criminal in Sicily, he had fled to Chicago around 1921. He imagined that his financial acumen was central to the Gennas' success, and had demanded a percentage. The brothers refused, keeping Nerone on a miserly salary, perhaps as little as $50 a week. The disgruntled Cavalier left to join a bootlegging cousin in Chicago Heights, a suburb south of the city.

Nerone's involvement in setting up Tony strongly advances the supposition that Capone ordered all the Genna deaths. The connection was clear. Nerone and his cousin prospered in Chicago Heights at Capone's sufferance; it had long been a Torrio-Capone suburb. And when Nerone was himself murdered, over a year later, it would be by a member of a family trying to muscle in on Capone, not by one who had anything to do with the Gennas. Informants claimed that Tony had been the next target only because Jim, the chief Genna, was in Sicily at the time. Either way, the death roster made perfect sense: first disarm retaliation by eliminating Angelo and Mike, the clan's fighters, then strike at the head, Jim, or at Tony, the brains.

The plan worked. The three Gennas had died within forty-two days of each other, losses that broke the family's power. Sam and Pete got ready to skip to Marsala, Sicily. Jim stayed there, later serving two years for his part in a Sicilian jewel robbery. When he did return to Chicago, about five years later, he stayed out of trouble and stuck to importing cheese and olive oil.

Eliminating the Gennas did not entirely settle Capone's problem with remnants of their gang. Salvatore Ammatuna's first name turned into the nickname

Samuzzo, then Samoots. A contemporary said that Samoots "wore silk gloves on his soul." Once a professional musician, a violinist of some accomplishment, and gangland's nearest sartorial rival to Bathhouse John Coughlin, he was still no tulip. When a laundry returned one of his silk shirts marred by an iron's scorch mark, he whipped out a gun and aimed at the delivery man, then mastered his rage enough merely to shoot the man's horse. While still a working musician, he was arrested with three others for the attempted shotgun murder of his union's business agent.

With prosperity, he swore he would never carry a gun again. He claimed they were unnecessary to his business—running Citro's, a cafe he owned with John Scalise as a silent partner. He also functioned as payoff man for the Gennas.

At their downfall, daring as ever, Ammatuna succumbed to visions of regrouping the organization. Control of Little Italy could come best through control of the Unione Siciliana. In an imaginative piece of ecumenical thuggery, Ammatuna recruited a saloonkeeper, Edward Zion, and a free-lance bootlegger, Abraham Goldstein (nicknamed "Bummy"), both experienced West Side gunmen. They invaded Unione headquarters, where Ammatuna announced himself successor to Angelo Genna.

Capone had not broken the Gennas for Ammatuna's benefit. He had not arranged to vacate the Unione presidency only to watch someone besides his own candidate, Tony Lombardo, take over. Ammatuna had stolen a march, but Capone would have something to say about it.

In the fall of 1925 Scalise and Anselmi faced trial for killing Officer Olson, the Walsh killing to be tried later. In June, Samoots Ammatuna had raised a $100,000 defense fund which hired the crack criminal defense team of Mike Ahern and Thomas D. Nash, plus Patrick J. O'Donnell, a white-haired old-school jury-rouser who thoughtfully provided the court clerk and some jury tipstaffs with bottles of booze.

The prosecution, led in person by State's Attorney Crowe, seemed to have an easy job. The defense could not dispute the participation of Scalise and Anselmi in the shooting. They did maintain it was not murder—at worst, manslaughter, at best self-defense—a theory that enjoyed the unusual advantage of being true. As a reporter wrote three months before the trial, even the police now recognized that the pair "didn't for a moment suspect that they were battling policemen; they thought they were continuing a gun fight with two members of the Dean O'Banion gang. . . ." It was those unmarked, gangster-style detective cars. Frank Capone had made the same mistake and so had *innocent* citizens. Not too long afterward, the police tacitly admitted the problem by flying two planeloads of drivers to Detroit to ferry back distinctive yellow Cadillacs for the detectives.

Defense counsel Ahern stated his case too broadly. "If a police officer detains you, even for a moment, against your will, and you kill him," he said, "you are not guilty of murder, but only of manslaughter. If the police officer uses force of arms, you may kill him in self defense and emerge from the law unscathed.'' Bob Crowe and the press leaped on that, pretending it was an invitation to slaughter police. Ahern had maladroitly failed to make it clear that he was talking about police who acted without a warrant or probable cause to suspect that the people they detained had committed a crime. Even that overstated the case (false arrest grants no license to kill), but the principle Ahern articulated was essentially correct.

Yet when the jury asked if they could return a manslaughter verdict, the judge said no. At one point the jury hung at eight for guilty, three for manslaughter, one innocent. Finally, at three-thirty A.M., November 12, 1925, they compromised on a manslaughter verdict despite the judge, specifying a fourteen-year prison sentence.

The trial for killing Officer Walsh started the following February 10. Venireman Orval W. Payne begged off, saying, ''I would have to carry a gun the rest of my life if I serve and found the two guilty. . . .'' This time Crowe left prosecution to his first assistant, George E. Gorman, and to Bill McSwiggin. It followed the same course, but Gorman made a most peculiar summation. ''There can be no middle ground,'' he told the jury. ''Either this is a hanging case or a case of justifiable homicide.'' Rot. Given the first jury's verdict, that amounted to a prosecution plea for acquittal. The second jury obliged: after only three ballots they found the killers not guilty of killing Officer Walsh.

The two killers solemnly shook hands with their deliverers and with the defense-fund contributors who clustered about. After the first trial, Officer Olson's deaf-mute mother had signed, ''The verdict is a blow to justice.'' Now, Walsh's widow glared at the scene of congratulations and said, ''My husband and his friend were killed by these men who now have a crowd waiting to shake their hands. I give up.'' She was spared the scene down in the cells where Scalise and Anselmi waited to be taken back to Joliet to continue their manslaughter sentence. They capered in joy, embraced and shouted their glee.

On December 23, 1926, the Illinois Supreme Court granted them retrial for Olson's killing, holding that if they committed murder, fourteen years was a travesty, if manslaughter, excessive. The following month, the two got out of prison on $25,000 bail. On June 9, 1927, four days short of the killing's second anniversary, the new trial started. Two weeks later the jury found that Scalise and Anselmi had merely defended themselves against, as the defense said, ''unwarranted police aggression.'' Not guilty. They walked away free—with not quite two years to live before Capone killed them with his own hands.

Meanwhile, Capone's problem with Samoots Ammatuna ended at seven-thirty the night of November 10, 1925, one year after O'Banion's murder.

Ammatuna had popped into Isidore Paul's barbershop on Roosevelt Road before taking fiancée Rose Pecorara and another couple to hear *Aïda*. He had just emerged from the chair when two men, one short and one tall, entered. The short one fired first, four shots, one of which hit Ammatuna in the neck. The other also fired four times but missed. They bolted from the shop to escape in a waiting car. Ammatuna died at age twenty-six on November 13.

Three days later, while returning from Ammatuna's funeral, Eddie Zion was shot and killed. After another three days, someone stole a shotgun from a police car and used it effectively on Bummy Goldstein, Ammatuna's other Unione "campaign manager."

Although the state never filed charges, a later writer claimed that gangland talk had fingered as Ammatuna's killers Jim Doherty (Klondike O'Donnell's best triggerman) and Vincent Drucci (O'Banion's faithful Italian). Almost impossibly unlikely. Rumor said that eleven days before, Ammatuna had hijacked from Klondike's gang a load of liquor that the O'Donnells had just hijacked themselves. But witnesses in the barbershop had agreed that *both* gunmen were swarthy, which could have described only Drucci, and only barely: he did not have an especially dark complexion. Anyway, why would either the O'Donnells or former O'Banions have wanted to punish Zion and Goldstein, who were not implicated in the hijacking?

Capone, on the other hand, certainly might have wanted them killed, along with Ammatuna, to forestall reprisal and to emphasize his determination that this time Tony Lombardo must become Unione president. Only Capone gained from the three murders: they left his man in charge of the Unione, with Capone himself free to regather the fragments of the Gennas' fabulously profitable alky-cooking setup.

Most conclusively, Ammatuna's nephew later confided to crime historian Bill Balsamo what the family had known all along: Capone had indeed ordered Ammatuna's murder.

Not that all the killings could put Capone back on top, or do anything to reestablish the combine.

In October 1925, after the Gennas' murders but before Ammatuna's, Torrio's jail term expired. Three carloads of gunmen, led by Capone, spirited him out of Chicago to catch an eastbound train in Gary, Indiana. Had he stuck around, nothing he saw would have changed his mind about leaving.

CHAPTER 13

Disintegration

THE TWO KINDS of turmoil that continued throughout 1925 were so severe that they drastically reduced the outfit's profits.

The lesser damage came from legal and citizen action. Just before Torrio had been shot, the Juvenile Protective Association prodded a judge into ordering Sheriff Peter Hoffman to padlock the Harlem Inn, Capone's principal Stickney brothel. Under Hoffman's benevolent eye it could of course surreptitiously reopen, but closings like that disrupted the cash flow. Legal action also quickened a smoldering vigilante spirit, emboldening the West Suburban Ministers' and Citizens' Association to raid the Harlem Inn, closing it again. By May, they felt feisty enough to attack even in Cicero.

Early Saturday afternoon, May 16, 1925, Capone still dozed in his Hawthorne Hotel suite. Unlike Torrio, he was a night person who kept a personal, lickerish, carouser's eye on booze and brothel operations, sampling both. When woken with news of trouble across Twenty-second Street, at number 4818, one of the outfit's many alternate gambling sites, Capone pulled trousers and collarless shirt over his pajamas and padded over, still muzzy and unshaven.

Chester Bragg, a robust insurance salesman from neighboring Berwyn, guarded the door. Capone shoved it part way open.

"What the hell do you think this is, a party?" Bragg growled, pushing back.

"Well, it ought to be my party," Capone grunted as he pushed harder still. "I'm the owner of this place."

"Come on in, Al," said Bragg. "We've been waiting for you."

Capone bounded upstairs to the main gambling hall, trailed by David Morgan, like Bragg from Berwyn, a $40-a-week investigator for the association. Capone strode the length of the hall, past the craps and pool tables, roulette wheels, chuck-a-luck layouts and racing tote boards (this was the day of the Kentucky Derby). As he passed a knot of raiders he snarled, "This is the last raid you'll ever pull," and continued to the back of the hall, partitioned to make an office.

The Reverend Henry C. Hoover, Berwyn Congregational minister, had led the raid. The pince-nez on his long thin nose and his weighty manner only partly offset the youth visible in his thin but rubicund face. He had been standing with three officials: the police chief of La Grange (a hamlet southwest of Berwyn), the La Grange judge who had signed their warrant for the raid, and their police escort, a highway police lieutenant on duty with Crowe's office. The group followed Capone into the back room in time to see him stuffing money from the till into his pockets.

"Who is this man?" Hoover asked the police lieutenant.

Capone looked up from his task. "I'm Al Brown, if that's good enough for you."

"Oh," said Hoover, "I thought it was someone like that, someone more powerful than the President of the United States."

"Why are you fellows always picking on me?" asked Capone. Nothing personal, Hoover explained, just principle.

Capone had his bookkeeper, Leslie Shumway, take charge of the money. Capone's cousin, Charley Fischetti, helped Shumway convey it across the street.

Capone followed them. He shaved, then returned wearing collar *and* tie. Suitably clad, the man who above all wanted to avoid trouble sidled up to Hoover and tried soft soap. "Reverend," he said, "can't you and I get together—come to some understanding?"

"What do you mean?"

"If you will let up on me in Cicero, I'll withdraw from Stickney."

"Mr. Capone, the only understanding you and I can have is that you must obey the law or get out of the Western Suburbs."

End of discussion. One of the raiders asked their attendant judge to issue a warrant for the arrest of the man who had announced himself owner of this illegal enterprise. While the judge drew up the papers, Capone disappeared. And when the raiders left it was through a gauntlet of gangsters mixed among the onlookers. The gangsters attacked with blackjacks and brass knuckles, breaking Chester Bragg's nose, beating David Morgan.

As usual after a raid, Capone's gambling operation resumed later the same afternoon. Hoover, Bragg and Morgan all later reported harassment, Morgan saying he had been shot and left for dead. That ended their raiding careers. Morgan gave up investigating and reverted to his trade as a mechanic.

An accommodating Cicero judge dismissed charges that resulted from Hoover's raid when they came to trial. Still, that kind of action hurt business.

* * *

The more serious damage throughout 1925 came from gang warfare all over town. Nobody tended to business; they spent the year shooting each other. In April 1925, the number of murders set a new city record; by October, Cook County averaged over one murder a day. Not all were gang killings, but it was they that swelled the totals enough to set records. The combine had come apart almost overnight, rent by the same mix of greed and ego that had seduced O'Banion.

Disintegration fed disintegration until they were *all* at each other's throats, leaving Capone with nice judgments to make about whom to woo, whom to whack and even what killings to ignore until he could strengthen his hand. The formerly solid South Side became as bloody a killing ground as the North Side and Little Italy.

Spike O'Donnell again presented a target, but now with an unsettling difference. Capone's most usual admonition to errant colleagues was "You keep your nose clean, you understand?" Rocked by the killings of 1923 and early 1924, Spike had kept his nose pretty clean. Now he tried a comeback, and when he imported New Jersey gunman Henry C. Hassmiller and again tried to muscle in during the winter and spring of 1925, no combine stood against him as it had before. The Saltis-McErlane and Sheldon gangs were now gunning for each other.

Everything was happening at once. In June, while Capone started to deal with the Gennas, Sheldon's men killed Hassmiller and Walter O'Donnell. On September 25, when Capone had turned his attention to Ammatuna, McErlane made one of what would be ten attempts to kill Spike O'Donnell; next day, Spike tried to kill Walter Stevens, Ralph Sheldon's mentor. On October 4, Frank McErlane shot up the Ragen's Colts clubhouse, killing two of Sheldon's people. In mid-October (when Johnny Torrio cleared out of Chicago), McErlane turned his gun again on Spike O'Donnell, missing him but wounding Spike's brother Tom. A month later, two bullet wounds put Joe Saltis in the hospital. Ten days after that, gunmen attacked McKeone's saloon, a Saltis-McErlane hangout, three men dead in the ensuing firefight. The following week, December 2, 1925, fourteen bullets riddled Ralph Sheldon's car; two days later, two shots pinked Danny Stanton, Sheldon's eventual successor.

Except when Torrio and Capone did the planning, gang leaders tended to end up wounded rather than killed—if hit at all—because shots at them had to be on the fly. Gang leaders were seldom out without an entourage, so they could not be quietly snatched and finished at close range. A double killing three days before Christmas, 1925, underscores the difference. Two Ragen's Colts, both in their mid-twenties, left a saloon together near six o'clock that evening. Joey Brooks, a saloonkeeper married just one week, had tried to dynamite Eddie Tancl a year before, while the combine still functioned. Edward Harmening was a county policeman. About two hours after leaving the bar their bodies were found in the back seat of Harmening's car, abandoned in the drifted snow of Marquette Park in the city's far southwest. Crossfire from the front seat got them, six bullets in Brooks, two in Harmening. They had been waylaid, forced into the car and driven to the deserted park—police thought by McErlane in retaliation for those killed in McKeone's bar the previous month.

This was the preferred style of gang killing: unhurried, the "range" no further than front seat to back, the target unarmed, grabbed from the street without any escort. A Capone bodyguard, Philip D'Andrea, could reportedly drill a bullet through a quarter tossed in the air. But most of Chicago's criminals were lousy shots, city boys unaccustomed to guns used for marksmanship. New York gunmen practiced at Coney Island and Broadway shooting galleries; Capone's men would sometimes practice in various headquarters' basements. Still, Samoots Ammatuna's murder was typical: eight shots inside a barbershop, only one hit, and that one fortuitous; no one *aims* at the neck. The Italians' use of shotguns responded to the fact that these gangs really couldn't shoot straight.

Frank McErlane tried a different solution. His first crack at Spike O'Donnell marked the first recorded use in Chicago of a submachine gun. It would displace the shotgun as weapon of choice. McErlane caught O'Donnell in the open, talking with a newsboy in front of a South Side drugstore. A sedan roared up and someone called "Hello, Spike!"—a tactical error, if gratifyingly operatic, because O'Donnell turned, saw what was coming, and hurled himself and the boy to the pavement as a shotgun shattered the window behind him. Then the machine gun chattered.

It was so novel a weapon that the police were stumped. They found "a neat line of bullet holes below the drugstore window, about a foot from the sidewalk," a reporter wrote. They speculated that the array of holes came from "a machine rifle of some kind." Another reporter laid it to coincidence; he guessed it must have been a firing squad of guns in a remarkable volley.

That was pardonable ignorance. John T. Thompson had retired from the Army in 1914 to work for the Remington Arms Corporation. World War I's trench warfare turned his thoughts to a weapon that would give infantry in attack some reply to the devastating defensive fire of fixed heavy machine guns, something each infantryman could carry as he charged the enemy. Thompson founded Auto-Ordnance in the summer of 1916 to perfect such a weapon.

Any machine gun operates by harnessing the recoil power of one round's explosion to eject the spent casing on the backstroke, then chamber the next round, ready for firing, as the bolt slams back against the breech. The machine gun keeps firing as long as its operator keeps pressure on the trigger, spewing lead as rapidly as the mechanical back-and-forth action of the bolt allows. Thompson found that .45-caliber pistol ammunition was the heaviest that would still permit such continuous, rapid-fire operation in a portable submachine gun without obviously impossible constant lubrication. But that was fine, because the .45 had a tremendous wallop, and its relatively light weight meant its user could carry all the more rounds. The Thompson submachine gun was really what the Germans called their version, a *Maschinenpistole*, or machine *pistol*; theirs (modeled on the Italian Revelli) used 9-millimeter Luger ammunition. Thompson called his invention "a trench broom" that would sweep the Boche away.

The Armistice came before the company could ship any of its brooms. The "Tommy gun" never saw World War I action, so never garnered the least publicity, which meant that postwar acceptance stayed slow. In 1923 Auto-Ordnance dropped its price from $225 to $175, but by 1925 it had sold only around three thousand guns, with the sales rate declining.

The world was ignoring quite a weapon. Stripped, it weighed eight and a half pounds; fully configured, with stock and a hundred-round drum, the weight rose to a little over twenty pounds—still perfectly manageable. Its rate of fire was one thousand rounds a minute, and each round could punch through quarter-inch steel plate. The gun could stop a moving automobile and demolish it. No one then needed a permit to own or carry a submachine gun, which could be bought through sporting goods firms, hardware stores or by mail order.

When the military finally adopted the Thompson (nearly two million saw World War II service), doctrine called for firing short bursts, not stitching lead across the landscape. Either way, being able to pump many rounds virtually simultaneously into a target area made up for poor marksmanship. Equally important, it discouraged retaliation. The old way, while gunman A fired discrete rounds at gunman B—even with a shotgun—gunman C (perhaps B's bodyguard) had time to draw a bead with reasonably safe deliberation and return fire. A submachine gun in A's hands put everyone in the neighborhood of B at such obvious and immediate peril that A could expect a *roomful* of bodyguards to do little but dive for cover.

McErlane's use of a Tommy gun, even if he missed Spike on September 25, heralded a new epoch in Chicago gang warfare. By October 4, when he shot up the Ragen's Colts clubhouse, police and newspapers had caught on. The submachine gun had arrived.

Despite all the turmoil around Capone, plus a developing assault on his Cicero business by Klondike O'Donnell, he spent Christmas in New York, where he became the center of three killings.

According to his own account, it all started innocently. His little boy, Sonny, seven years old, had developed a life-threatening mastoid infection. Capone took him to the Manhattan clinic of Dr. Lloyd on St. Nicholas Place, where he offered $100,000 if the renowned specialist could save his boy. The doctor charged his usual $1,000 fee. The operation saved Sonny's life, but left him partly deaf.

Capone later said, "It was Christmas Eve when my wife and I were sent home to get some sleep. We found her folks trimming the Christmas tree for her little nieces and nephews and it broke her up." He was pretty broken up, too.

Christmas night—more accurately, the early hours of December 26, 1925—he managed to work off a lot of rage and frustration. He later dismissed the affair, saying, "A friend of mine dropped in and asked me to go around the corner to his place and have a glass of beer. My wife told me to go: it'd do me good. And we were no sooner there than the door opens and six fellows come in and start shooting. My friend had put me on the spot. In the excitement two of them were killed and one of my fellows was shot in the leg. And I spend the Christmas holidays in jail."

That account was not exactly true; it was barely approximate.

The slaughter really started in late 1923 when some of Frankie Yale's men surprised Bill Lovett almost comatose drunk, and killed him. Lovett's brother-in-law, Richard Lonergan, took over. People called him "Pegleg" (except to his face) because a boyhood accident had resulted in his left leg being amputated below the knee. Lonergan disastrously declared war on Yale. Many killings and two years later, Lonergan had almost no troops left. He persisted anyway, and after he ambushed Yale coming out of a speakeasy, only narrowly missing him, Yale determined to end the sniping for good.

Yale's chance came in early December 1925, when White Hander Eddie Lynch called to arrange a meeting. Lynch and Lonergan no longer spoke to each other. They had fallen out over a girl and, more hurtfully, over Lonergan's charge that Lynch had turned yellow. Lonergan had sent Lynch and Aaron Harms, two of his last remaining competent killers, after James DeAmato—known as Jimmy Files or just Filesy—an old friend of Al Capone's who had remained in Brooklyn and been stepping strongly on White Hand toes.

Lynch and Harms had tracked DeAmato to a suitably deserted side street one Sunday night. They were closing for the kill when DeAmato spotted them and sprinted around the corner onto busy Fulton Street. Lynch told Harms to forget it; there would be too many witnesses. But Harms hoofed it after the fugitive. Alone now, when he caught up his gun jammed, and DeAmato proceeded to pound his pursuer all over the sidewalk, a disgrace for which Lonergan chose to blame Lynch's supposed faint heart.

That did it. Lynch saw the ship sinking, anyway. He offered to trade Yale some important intelligence for a job. Lonergan planned to invade the Adonis Social Club on Christmas night, time of the annual stag party Yale threw for Black Hand gang members. The plan underscored Lonergan's desperation. He could muster only five other guns, and the only two of those with experience were Aaron Harms and Cornelius Ferry ("Needles," for his narcotics sideline). The other three, Patrick Maloney, Joseph Howard and James Hart, had been White Hand truck drivers pressed into service as ad hoc gunmen. Lonergan counted on surprise. He would strike late, when the crowd had thinned out, leveling the odds. If he could wipe out a decent number, maybe even Yale himself, the situation might yet be retrievable.

Yale conferred with Capone. He may well have summoned Capone from Chicago to return the service Yale had rendered with Colosimo and O'Banion, Capone's visit merely coinciding with his son's operation. Lonergan knew Capone's name, but no New York paper had yet pictured him. He would be a cool mind behind an unrecognized face.

The Adonis occupied the ground floor of a frame building, an apartment upstairs. A long corridor opened into a taproom with mahogany bar and brass rail. A door led to a large back room used for dancing and for managing partner Angelo "Fury" Agoglia's Lucullan banquets for which the club was famed. Charcoal gray curtains lined the walls, ceiling to floor. A beat-up piano stood in one corner; rickety tables perched on shaky legs with tattered cane-back chairs around them. For the holiday, a strip of orange bunting circled the wall, on it "Merry Christmas and a Happy New Year" in an illiterate scrawl. Some mistletoe hung about.

Following Eddie Lynch's tip, Capone arrived after midnight. Lonergan and his men arrived sometime before three A.M. and made their way to the back. The Capone-Yale stratagem confounded the invaders. Capone had a few of his own people with him, as much strangers to Lonergan as was Capone. All that greeted Irish eyes were the usual Adonis employees—including the club's singer, Helen Logan, and cigarette girl, Elvira Callahan—plus some unknown Italians sprinkled about at tables. They saw no Yale and very few known Black Handers. Instead of letting fly immediately, Lonergan and his men lounged about, guzzling. They took pains to be obnoxious with loud talk of "dagos," "wops," "Guineas," and "ginzos."

John Stabile, one of Fury Agoglia's two partners in the Adonis, was known as "Stick-em up" for his preferred vocation, confusing a reporter who wrote of a "John Stickum." Stabile came in with his inamorata, May Wilson, on his arm, her husband, Ed, conveniently doing a stretch in Elmira. She was fair, very possibly Irish; Stabile was dark, indisputably Italian. When Lonergan spied the pair he loudly scored her, urging her to stick to "white men." He may even have gone over to ask Capone his opinion; according to a survivor of the following mayhem, he had some words with one of the patrons at a table.

Shortly after three o'clock, at the nod from Capone, Fury Agoglia's brother, Sylvester, wandered over to Aaron Harms and bashed him in the head with something like a meat cleaver. At that signal, Fury hit the main light switch, leaving the White Handers illuminated by the room's only other light, a small chandelier over their table.

Out came the Capone guns. When the shooting started, May Wilson covered her ears, turned her face to the wall and shrieked until it was all over. In the taproom, bartender Anthony Desso heard Wilson, the cries of wounded men, and feet stamping past him in a panic to reach the front door. Two of the White Hand's jumped-up truck drivers, "Happy" Maloney and "Ragtime" Howard, made it out untouched. Capone and his three table companions dropped Lonergan and Ferry as the two headed for shelter behind the piano. The targets did not have time even to draw their guns. Lonergan fell first, two slugs in him, Ferry on top of him, both in a heap a few feet from the piano. So sudden was the action that police found Lonergan with a toothpick still in his lips. The pianist, George Carozza, had been playing ''She's My Baby,'' when the shooting started, the worn sheet music open on the rack.

Aaron Harms had also made it to the street; but in addition to the head wound from Sylvie Agoglia, he had been shot. He collapsed and died outside the club. James Hart had been luckier. Bullets had grazed his ear, plinked his arm; one buried itself in his thigh. He too made it out the door and was found by a policeman crawling along Flushing Avenue on hands and knees.

At 3:35, patrolman Richard Morano came upon Harms's body in Twentieth Street. When Fred Kavanaugh, acting captain of the Fifth Street Station, joined him, they followed a trail of blood to the Adonis door, which stood ajar. By flashlight they made their way through the taproom—a tangle of overturned tables and chairs—to the back, now completely dark. The lights revealed a shambles, as in the bar, with glasses and broken liquor bottles a reeking litter, along with gold-tipped cigarette butts. Bloodstains were everywhere.

Two abandoned murder weapons, .38s, lay in one corner. Police found another .38, unfired, still in Lonergan's shoulder holster, and a fully loaded automatic tucked in Needles Ferry's back pocket. The plan had worked like a dream.

Police arrested the three escaped White Handers. They also arrested seven others, innocuous Adonis employees mostly. But among them, as the Brooklyn *Eagle* explained, was ''Alphonso Capone, 'bouncer' of the club and an alleged former Chicago gunman.'' He had yet to attain much press outside of Illinois; *The New York Times* still followed the *Tribune*'s quirky spelling of ''Caponi'' in its rare mentions. The *Times* called him the club doorman.

To represent Capone and the others, Yale retained Samuel Liebowitz, probably New York's premier defense counsel before he became a hanging judge. No one would answer questions, ''on the 'advice of counsel.' '' Anna Lovett, Bill's widow and Pegleg Lonergan's sister, said, ''You can bet it was no Irish American like ourselves who would stage a mean murder like this on Christmas Day.'' Recovering, Hart claimed he hadn't even been in the Adonis; the bullets that hit him came from a passing car. The family that lived upstairs from the Adonis said they hadn't heard a thing. With the only conceivable witnesses mute, everyone arrested got out on bail, charges soon dropped for want of evidence. The White Hand gang evaporated. Capone could return to Chicago.

The South Side continued to explode. On January 24, 1926, Robert L. McCullough, on the doorstep of his mother's home for a Sunday visit, was shot three times. On February 2, dynamite demolished Ralph Sheldon's empty Cadillac; an ignition bomb may have exploded prematurely. Seven days later, Frank McErlane machine-gunned Buff Costello's bar, a Sheldon hangout, wounding one of Sheldon's top lieutenants, John Foley (always called ''Mitters'' for reasons lost to history) and a Sheldon soldier named William Wilson. The effectiveness of this attack made Detective Captain John Stege announce next day that he wanted some of those Thompsons for his own boys.

Capone was equally impressed. He immediately detailed Charles Carr, once manager of the Four Deuces, to buy Tommy guns. Carr and a companion, both in their mid-twenties, told Valley hardware dealer Alex Korecek they were bank messengers and needed a submachine gun for protection. Korecek sold them one, plus a hundred-round drum, for $210. They returned to order two more and Korecek balked; it didn't seem quite right to him somehow. Carr put his request more forcefully: Korecek would get them the guns or he would die. Doubts resolved, Korecek arranged for delivery.

Capone had to choose sides among the South Side gangs. He hated the necessity, but the choice almost made itself. Spike O'Donnell was impossible. Frank McErlane was pathological, his partner Joe Saltis unreliable, and their chief associate, John Oberta, had from childhood been nicknamed ''Dingbat'' after a comic strip. By contrast, Ralph Sheldon seemed a miracle of loyal stability, and had gladly accepted Torrio and Capone as his suppliers and allies from the first.

Not that Capone made overt breaks with Saltis and McErlane or joined directly in the fighting. None of the South Side gangs attacked his holdings or any of his people who did not stray from the outfit's territory. So Capone was willing to overlook certain liberties for the time being. For instance, Bob McCullough, who had been shot February 2, was actually one of Capone's most reliable gunmen. None of McCullough's three bullet wounds proved fatal, and he *had* been attacked on alien ground; Capone sought no retaliation. On April 15, 1926, worse happened. The bodies of Frank De Laurentis and John Tuccello were found in a sedan, parked practically at Sheldon's doorstep. They had been peddling alky in Saltis-McErlane territory for Sheldon. But, in fact, they were minor Capone operatives, helping out a good customer.

Capone had to ignore even this. He had his hands too full in other parts of the jungle.

CHAPTER 14

Who Killed McSwiggin—and Why

ON THEIR FACE, the killings that inflamed Little Italy through the winter of 1926 had nothing to do with Capone or with the murder, that spring, of Assistant State's Attorney Bill McSwiggin. In fact, they spawned the conditions that ended in McSwiggin's killing because they helped convince Klondike O'Donnell that he could muscle in on Capone's Cicero business unscathed, a folly that impelled Capone to a decisive act that would stem the hemorrhage of business and reputation.

The Little Italy fracas revolved about the collection of a second defense fund for Scalise and Anselmi when they faced trial for killing Officer Walsh. The most frightening of the collectors that Samoots Ammatuna had sent around for the first fund had been Orazio Tropea, then forty-four, widely hated and feared, known mostly by his first name or as ''The Scourge,'' a gaunt, dark, hawk-faced terror thought to possess the evil eye.

With Ammatuna dead, Orazio appointed himself head of the second fund drive, intending to keep most of it. When open-handed contributors to the first fund balked this time, The Scourge killed them. He started with Henry Spignola, killed January 10, 1926. Next came the Morici brothers, Agostino and Antonio, pasta manufacturers.

After that, the pace of donation picked up briskly, but so did Little Italy's desperation and anger. On the night of February 15, a shotgun from a passing sedan got Orazio Tropea in the head. Six days later, one of Orazio's collectors, Vito Bascone, had a bullet drilled into his forehead. Two nights after that, the body of Ecola Baldelli, another Orazio henchman, lay on a trash heap. Philip

Gnolfo, who had helped kill Henry Spignola, went into hiding after a misdirected attempt on him struck down an innocent victim instead.

Capone got back from Brooklyn to see Little Italy tearing itself apart. Klondike O'Donnell thought he saw a Capone impotent to keep order in his own backyard, let alone on the South Side, which continued to erupt in killings. He thought that signaled his opportunity to move in.

In any sweepstakes to determine the silliest drivel breathed or written about Capone over the years, a front-runner must be the story that Capone set out to murder on that April night because one of his spies reported a remark by Klondike to the effect that the O'Donnells sold better beer than the slop Capone peddled. That story pictured "Scarface," an unthinking brute, slaying on whim, animated by no more than tweaked pride. To bolster the nonsense, it cited a Cicero card dealer who, about an hour and a half before the killings, had supposedly seen Capone in "agitated" conversation with brother Ralph and other outfit gunmen in a Cicero restaurant. The dealer reportedly said that Capone snatched a machine gun from a secret panel in the wall, the others "provided themselves" with pistols, and they all rushed out.

In reality, the murders that night were the climactic battle in a war that had been raging several months.

After it was over, Henry Madigan (always called Harry) pretended that "Al Brown" had coerced him into buying the outfit's beer. *But*—said the co-owner of the Pony Inn at 5613 Roosevelt Road—"A few months ago [Jim] Doherty and Myles O'Donnell came to me and told me they could sell me better beer than the Brown beer, which was then needle beer. . . ." They would even sell it for less! "I changed," Madigan said, "and upon my recommendation so did several other Cicero saloonkeepers."

Actually, Madigan and his partner, Michael J. Windle, were ancillaries of Klondike's gang; other saloonkeepers took O'Donnell beer (as they had taken Capone beer) or *else*. James Sammons, always called Fur Sammons, was Klondike's most savagely erratic gunman. Before long, Fur Sammons and Myles O'Donnell could strut about Cicero, as one reporter put it, "flaunting the collections they had made from saloon owners whom they had compelled to buy O'Donnell beer." The gang may even have been cutting into Capone's gambling take, operating a new book that siphoned off $5,000 worth of action a day.

Granted, the O'Donnells had not yet penetrated Capone's Twenty-second Street stronghold. And even $5,000 a day in lost gambling would scarcely beggar his operation: ledgers seized later indicated that just three of his gambling joints *netted* about that much, the entire Cicero profit running to at least three million a year. But the trend was worrisome. Taken with those unanswered South Side assaults, a wide perception that Capone might supincly accept even trifling inroads by a minor menace like the West Side O'Donnells could not fail to inspire unwholesome thoughts among the rest of Chicago gangland, particularly in

Hymie Weiss, who now led the North Side gang. Weiss was a real threat. Some considered him the only person Capone really feared.

Capone had to go to war.

As always, his problem was to isolate important targets. He set two of the outfit's more obscure members scouting the territory. They were William (always "Willie") Heeney and Edward F. Moore.

The first kill came on April 3, 1926. John Ryan, a Sheldon gunman, shot Walter Quinlan, one of Klondike O'Donnell's steadier men, in a saloon at 1700 South Loomis on the border of Little Italy—out of either one's territory. It may have looked like a simple revenge killing: six years before, Quinlan had murdered Ryan's father, the redoubtable "Paddy the Bear," ruler of the Valley. Police knew it was business. Ryan doubtless delighted in the job; but the timing strongly suggests it came at Capone's urging, his scouts locating Quinlan for the boy. Police learned that Myles O'Donnell and Jim Doherty had witnessed the murder. They found both at Doherty's saloon, 4701 Roosevelt. At the same time they seized bulletproof vests that would soon figure in the killing of Bill McSwiggin. To their surprise, O'Donnell and Doherty readily rapped Ryan as Quinlan's killer.

As a bonus for Capone, Quinlan's murder indirectly eliminated another O'Donnell gunman. William McCue (always called "Rags") made strongarm sales to Klondike's prospects, then delivered the beer, for which Quinlan collected. With Quinlan dead, McCue quickly made the rounds, collected in his place, and pocketed the money, apparently figuring that the O'Donnells wouldn't notice the shortfall between Quinlan's last collection and whenever they organized a new collector. Most gunmen were not thinkers. When Klondike sent Jim Doherty around, saloonkeepers told him they'd already paid Rags McCue for that last delivery.

On April 18, Doherty and another O'Donnell strongarm dragged McCue between them to each saloon on the route, hands bound behind him, head hanging. "Is this the one you gave the money to?" Doherty asked, forcing McCue's head up. Next day police found Rags McCue in a roadside ditch.

The gangs had observed a truce for the April 13, 1926, primary. They all concentrated on electing Bob Crowe's slate of candidates, especially Joseph P. Savage, the assistant state's attorney he ran for county judge, a post that controlled the election mechanism. The gangsters had help. When an opposition poll watcher protested the beating of a policeman by the O'Donnells' Jim Doherty, Crowe's hot young assistant, Bill McSwiggin, soon appeared. "Keep your mouth shut," McSwiggin warned the precinct worker—or he'd be hauled off to the state's attorney's office. Unsurprisingly, Crowe's slate swept to victory.

Four days after the primary, Bill McSwiggin visited Capone at the Hawthorne Hotel. "If I told what that business was," said McSwiggin's father after his son's death, "I'd blow the lid off Chicago. This case is loaded with dynamite. It's dangerous to talk about it." Anthony McSwiggin was a thirty-year veteran detective sergeant. Capone confirmed the visit, but also refused specifics. He vehemently denied charges that he might have had something to do with Bill McSwiggin's murder. ''Of course I didn't kill him,'' he said. ''Why should I? I liked the kid. Only the day before he was up to my place and when he went home I gave him a bottle of Scotch for his old man.''

William Harold McSwiggin, born February 7, 1901, was only one of Bob Crowe's sixty-nine assistant state's attorneys, but now, at age twenty-five, he owned the public's eye. He had racked up seven of the eleven first-degree-murder verdicts Crowe's office had won in 1925. None, of course, involved booze gangsters. McSwiggin convicted the likes of Henry J. Fernekes—''The Midget Bandit''—and his two partners in a holdup killing; McSwiggin's latest triumph had been over Raymond Costello, a black who had killed his sixteen-year-old sweetie. Newspapers called McSwiggin ''The Hanging Prosecutor'' and ''Little Mac.'' Some colleagues called him ''Specks'' for his owlish horn-rims, others ''Billy''; at home they often called him ''Harold.'' Unmarried, he lived with his parents and four adoring sisters at 4946 Washington Boulevard, in Klondike O'Donnell's far West Side stronghold, about fifteen blocks north of Cicero's Roosevelt Road.

After parochial school, McSwiggin had attended De Paul Academy, where he boxed and pitched baseball; then he read law at De Paul University, working his way through. Of only medium height, a bit chubby with a double chin, he nevertheless was solid enough to hold down jobs as a dance hall bouncer, a truck driver and an American Railway Express security guard. But he was also smooth enough to work as a department store sales clerk. He parted his sleek black hair in the middle, had blue eyes, and liked to joke a lot. He was a sharp dresser, favoring snappy hats.

One month out of school he passed the bar; not much later he joined Crowe's office. Honors marks had helped. So had strong work for Crowe's political campaigns.

McSwiggin quickly became noted as a hard worker, a good lawyer, and the winner of tough cases that did not involve connected gangsters. He turned table-thumping tiger and spread-armed haranguer before a jury. He spoke their language. ''The average jury,'' he explained, ''is a hoodlum jury and I've always given them a line of talk they understand. By 'hoodlum' I mean they're the kind of people who eat with a knife instead of a fork, who take a bath on Saturday night, who play an occasional game of pinochle, and who cuss at the traffic cop who doesn't let them park in the Loop.''

He courted newsmen and possessed a sure instinct for getting mentioned in high-profile cases. He had interviewed Capone in the Joe Howard killing, gotten himself appointed with John Sbarbaro as Crowe's representative in the O'Banion murder, and helped try Scalise and Anselmi the second time. Newsmen gave him oceans of ink. Yet no one seemed to notice—when he unsuccessfully prosecuted Myles O'Donnell and Jim Doherty for Eddie Tancl's murder—that McSwiggin and Doherty had been boyhood chums and had kept up their friendship. By the spring of 1926, McSwiggin ranked as Crowe's ace.

* * *

On April 23, 1926, the outfit drove by Pearl Hruby's Cicero beauty parlor and hit it with ninety-two slugs from their new machine guns. They seriously wounded Hruby's boyfriend, Fur Sammons, and just missed Jim Doherty.

On April 27, Klondike O'Donnell and three companions spent the day as Crowe officials, watching a recount of the primary ballots. About seven o'clock in the evening, they put Klondike's Cadillac in a garage for minor repairs and drove away in Jim Doherty's green Lincoln. The three with Klondike were Myles O'Donnell, Jim Doherty, and either Edward Hanley—onetime policeman, onetime boxer, now Myles's driver—or Thomas Duffy, whom everyone called "Red," a Crowe precinct captain in the Thirtieth Ward, where the O'Donnells, Doherty and McSwiggin also lived. Red Duffy owned a barber shop and plied a sideline as a small-time bootlegger and gambling operator for the O'Donnells. Although no one ever firmly established who ended up in the Lincoln, the best guess is that Myles, Doherty and Hanley picked Duffy up along the way and dropped Klondike at his own home. About seven-thirty, the Lincoln pulled up in front of the McSwiggin home, and Duffy got out.

Bill had just been on the phone with a precinct captain, promising to fix a parking ticket for a friend. His father looked out the window and said, "Here's Red Duffy." Like Doherty, Duffy had also been Bill's pal since boyhood. According to various accounts, Bill announced either that he was going to Berwyn to play cards, or just "out" to get some good beer, or to see a friend. He bade his father good-bye and rode off in Doherty's green Lincoln.

Not long afterward, Willie Heeney either spotted the group or recognized Doherty's car. He flashed word of their whereabouts to Capone, who issued orders. A lead car would ram any police car that tried to interfere; two flanking cars would trail the leader but hug each curb, ready to block side-street traffic, as in the O'Banion getaway. Capone's new driver, Tommy Ross, would stay fifty feet behind those running interference, Capone and three outfit gunmen in the armored sedan. Finally, a fifth car would cover their retreat, causing an accident, if necessary, to block pursuit. Capone's group made contact and followed Doherty's Lincoln discreetly until it pulled up outside Harry Madigan's Pony Inn on Roosevelt.

The Pony Inn, two stories of cream-colored brick, stood alone in an otherwise vacant lot on the south side of Roosevelt. It had two doors, numbers 5613 and 5615, one leading to a second-floor apartment. Capone's Cadillac approached and drove up slowly from the east as McSwiggin, Doherty and Duffy stepped out of the Lincoln. Police later determined that Capone handled a machine gun himself, an example to his men. His sedan continued spewing lead as it glided past the Lincoln, then roared off. Mrs. Bach, who lived above the saloon, told of seeing "a closed car speeding away with what looked like a telephone receiver sticking out of the rear window and spitting fire."

Twenty-two bullets pocked the west wall of the building; six more splintered a small tree in front. Duffy had been riddled about as soon as his feet hit the pavement; he staggered, then crawled into the empty lot to hide behind another tree, five bullets in him. Doherty lay on the sidewalk, ripped by sixteen bullets. McSwiggin made it nearly to the 5615 entrance, then fell in a heap. Hanley, the driver, and Myles O'Donnell had not yet emerged when the firing began; they hugged the Lincoln's floorboard during the attack.

Frank J. Misek heard the tumult and ran out from an apartment across the street to see Hanley and Myles wrestle the bodies of McSwiggin and Doherty into the Lincoln. Perhaps they did not see Duffy, perhaps they thought him dead; they drove off without him. A passing motorist stopped and rushed Duffy to the West Suburban Hospital, where he died six hours later. In his pocket police found a demand for the transfer of a troublesome police sergeant and a memo book listing area speakeasies and gambling joints.

Myles and Hanley drove to Klondike's house and unloaded the bodies. When Klondike saw they were dead, he ordered them away. The bodies were dumped in Berwyn, the ravaged, blood-splattered Lincoln abandoned in Oak Park, in it five fedoras and McSwiggin's eyeglasses.

His family buried McSwiggin on May 2 with honorary pallbearers every bit as illustrious as the better grade of gangsters rated, but in addition, McSwiggin had military honors; he had been an ROTC lieutenant at De Paul. Also unlike gangsters, he merited a Mass and a grave in the consecrated ground of Mount Carmel.

Told of McSwiggin's murder, Bob Crowe said, "The shock has almost unnerved me." He promised, "We are going to get to the bottom of this." Newspapers soon made a threnody of the endlessly reiterated question, "Who killed McSwiggin and why?"

The "why" seemed to depend on the additional question of what an assistant state's attorney was doing riding around with known gangsters. Crowe first blithely assured the public it was just a "social ride." Predictably, the public found that entirely discreditable. He next hinted that McSwiggin was trying to unearth some evidence on notorious killer Martin Durkin, or maybe just doing "research into criminology." Two days after the killing, according to one paper, "it was learned" that McSwiggin had been trying to recover a bulletproof vest. Of the five that police had discovered at Doherty's saloon some weeks before, four were still impounded; but one was missing, and McSwiggin had undertaken to recover it for its owner. Crowe immediately adopted the vest theory and gratefully passed it on to a grand jury.

But why was McSwiggin killed? Some early theories held that those he'd prosecuted had sought revenge, especially Scalise and Anselmi. Or maybe his political activities had created to-the-knife enemies.

His father had the right answer. So did John Stege, by now deputy chief of detectives. Significantly it was also voiced by Joseph Klenha, village president of Cicero, who perhaps spoke for Cicero's real boss. "Those shots were never meant for my boy," said Anthony McSwiggin almost as soon as he heard the news. When someone said to Captain Stege that the gunmen were clearly the type who would kill you as quick as look at you, he replied, "And sometimes quicker." Sunset that night had been at 7:44, almost an hour before the shooting; there had been intermittent light showers. In the bad light and heat of action, Capone and his men simply failed to recognize the prosecutor.

Almost from the first, *who* killed McSwiggin held little mystery. "This machine gun slaughter," Stege announced the next day, "was done by the Brown-Torrio gang of Cicero," the nomination quickly seconded by Crowe and Bill Shoemaker, who had become chief of detectives. Shoemaker promptly obtained a warrant for Capone's arrest on a murder charge. Now all he had to do was serve it.

Capone disappeared. The murder of an assistant state's attorney—even one who consorted with criminals—generated such instant and intense indignation that Capone seriously expected the police might shoot him on sight. If the police grabbed him, at best he would be in for intense grilling, which routinely involved beatings in station house cellars—in such heat, *no one* would be immune.

The authorities punished Capone the only way they could. They raided, smashed and padlocked; they cost Capone probably a million in damage and more in missed revenues. They destroyed joints that had before been protected from even token raids. Sworn as a county deputy, Bill Shoemaker raided Stickney's Harlem Inn, supposedly long closed by injuction, but still in action. He found gunports, masked by pictures, commanding the barroom, and rooms for twenty-six whores, rooms which likewise featured secret panels and trapdoors. "It was," an observer reported, "a house in which anything might have happened."

However briefly, Greater Chicago really was shut down, and Chief of Police Morgan Collins made a personal tour to see that his orders had been followed. Such sternness heartened even the browbeaten veterans of Forest View, or "Caponeville." Capone's showcase Maple Inn—"The Stockade"—with thirty-five cubicle bedrooms upstairs, had been raided and padlocked along with the rest, but the accurate rumor swept town that when the heat died, it would reopen. On the morning of May 30, Memorial Day, three carloads of vigilantes overpowered the lone black watchman and torched the place. Berwyn firemen who answered the call merely made sure that the fire did not spread. Asked why they did nothing else, they replied there was no water at hand. Captain Stege said the police would not try to identify the arsonists. Indeed, he hoped they'd burn down more such joints; it was simpler and faster than getting court injunctions.

Judge William V. Brothers, part of Crowe's faction, impaneled a special grand jury for the state's attorney, eliciting rage from Harry Eugene Kelly, president of the elite Union League. Kelly had called for a grand jury that would investigate, along with the murders, Crowe's office and its policies. "Is Mr.

Crowe,'' Kelly asked, ''to be permitted to investigate himself and are the people to be satisfied with such an investigation?''

The people's echo of Kelly's rage convinced Crowe they were not. So he neatly sidestepped by asking Illinois Attorney General Oscar Carlstrom, once a foe but now a political ally, to shepherd the special grand jury. Carlstrom assured the public that ''Prosecutor Crowe and I will work in complete cooperation. . . .'' Well, yes.

Carlstrom's jury did not do much, though in fairness it labored under two insuperable difficulties. By state law, grand juries sat for only one month, and during most of its life this grand jury had no one to talk with who knew anything. Capone was gone. Myles and Klondike O'Donnell had driven up to the garage in a taxi early the morning after the murders, then driven off in the Cadillac they had left for repairs. They vanished for exactly one month. No one could find driver Hanley. Hardware dealer Alex Korecek, who supplied Capone's first three machine guns, sniveled to the police: ''If I tell, I'll die.'' Capone's men had vowed ''they'd take me for a ride if I ever squawked about them.'' A lawyer he did not know and who was, he claimed, ''no friend of mine,'' filed a writ of habeas corpus for his release, which the judge dismissed after Korecek *pleaded* to stay in custody. Korecek then told Carlstrom's jury what he could; but knowing that Capone owned at least three submachine guns scarcely advanced the investigation.

On May 27, just before the jury's life expired, the O'Donnell brothers arranged their surrender. After a show of silence, cajoled by threats of contempt jailings, they embraced the missing bulletproof vest story of why McSwiggin was along for the ride that night. They added imaginative details. They had been in Doherty's saloon when Duffy came to play the intermediary between McSwiggin—looking for the missing vest—and Doherty, who perhaps had it, but who was on the outs with McSwiggin because of what he considered his old pal's excessive zeal in the Tancl trial. They had all gotten in the car with two other fellows the O'Donnells did not know. But Myles and Klondike had wanted to wash up, so those two had gotten out before the stop for McSwiggin, not seeing the others again until the strangers brought McSwiggin and Doherty to Klondike's after the shooting. They knew nothing about the killings.

The obvious truth was a good deal simpler. McSwiggin wanted to go out drinking with his friends, and those friends were delighted to trot a well-known prosecutor around Cicero's saloons, showing owners who *now* had the clout and could now deliver real protection along with beer.

The court discharged the O'Donnells after two days.

Carlstrom's grand jury issued a final report that absolved McSwiggin and Crowe of any fault, and accurately judged that ''the murderers had no knowledge of the identity or position of . . . their victim.'' The report ended with the assertion that ''on the whole, a review of the years past gives no special occasion for alarm at the present moment.''

The court impaneled a second grand jury, this time led by former judge Charles A. McDonald, a man of probity and no crippling political ties. Judge McDonald got more grand juries, but got no nearer to answers. He did, though, finally kill all talk that McSwiggin had been with gangsters that night on official or even creditable business.

As the third jury wound down, Capone reappeared. He had arranged to surrender at ten A.M., July 28, 1926, three months after the murders, at the Illinois-Indiana state line—but only to *federal* agents; he still mistrusted itchy Chicago police trigger fingers.

The night before his surrender, holed up about eighty miles away, he expansively explained himself to reporters. "I'm no squawker," he said, "but I'll tell what I know about this case. All I ask is a chance to prove that I had nothing to do with the killing of my friend Billy McSwiggin.

"It's a bad time to say anything, and I've been convicted, without a hearing, of all the crimes on the calendar. But I'm innocent of everything and it won't take long to prove it. I trust my attorneys to see that I'm treated like a human being and not pushed around by a lot of coppers with axes to grind."

He also relied on his attorneys' assurance that after three grand juries had failed to indict him, the state had no case. But what about his relationship with McSwiggin and the others? reporters asked.

"Just ten days before he was killed I talked with McSwiggin," Capone told them. "There were friends of mine with me. If we had wanted to kill him we could have done it then and nobody would have known. But we didn't want to. We never wanted to.

"The police have told a lot of stories. They've shoved a lot of murders over on me. They did it because they couldn't find the men who did the jobs and I looked like an easy goat. They said I was sore at McSwiggin because he prosecuted Anselmi and Scalise for killing two policemen. But that made no difference. He told me he was going to give them the rope if he could, and that was all right with me." Very much so. As we've seen, pushing the jury either to hang or acquit got them off in the second trial.

"Doherty and Duffy were my friends, too," Capone continued, slipping into fantasy. "I wasn't out to get them. Why, I used to lend Doherty money. Bighearted Al, I was, just helping out a friend. I wasn't in the beer racket and didn't care where they sold. Just a few days before that shooting my brother Ralph and Doherty and the O'Donnell brothers, Doherty's partners, were to a party together."

He then explained that he was in a Cicero restaurant when the shooting occurred and had scooted because he feared being shot on sight. "It would have suited me better to go up and prove that I was innocent," he said, "but what could I do? What would I have got? A long time in a cell, a lot of unnecessary grilling and blame for everything. So I went away. Now I'm ready to go back.

I won't stop with the federal people. Any officer that has the right to question me can get all I know. If the police have anything but hot air they can use it. I'll answer any question that's put to me about the murder."

Next morning, Capone lawyer Benjamin Epstein, who represented him in federal matters, went with Patrick J. Roche and Clarence W. Converse, two stars of the Treasury Department's Special Intelligence Unit, to meet Capone, whose Cicero bondsman Louis Cowan posted $25,000 on a federal Prohibition indictment returned the previous May. As Capone stepped out of the courtroom, Chief Shoemaker arrested him on the warrant sworn out right after the murders. Capone would spend the night in jail, the only punishment honest policemen could visit on him for three killings.

The next day, Capone's regular counsel for state matters, Nash and Ahern, were waiting in Chief Justice Thomas J. Lynch's court. They were right about the state's case. The story that Capone had reached for a submachine gun behind a secret panel in a Cicero restaurant was denied by the terrified card dealer. When Captain Stege raided the place, he found no such panel. So much for even that slim evidence connecting Capone to the killings.

"This complaint," First Assistant State's Attorney George Gorman told the judge, "was made by Chief of Detectives Shoemaker on cursory information and belief. Subsequent investigation could not legally substantiate the information." Case dismissed. Capone posted $5,000 bond on a vote fraud charge from the April primary (later dismissed on a technicality by the Illinois Supreme Court) and walked. "They pinned a medal on him," said Sergeant Anthony McSwiggin, "and turned him loose."

"I am satisfied Capone was at the bottom of these murders," said Chief Shoemaker, "but it is one thing to be satisfied in your own mind and another to prove it beyond a reasonable doubt."

Anthony McSwiggin launched his own investigation, a driven man. The bullets that had killed his boy had killed him, too, he said. "I thought my life work was over. But it has only begun. I will never rest until I have killed my boy's slayers or seen them hanged. That's all I have to live for now." (No one asked his wife and four daughters to comment on that.) The story circulated that Capone took him up on it, handing Anthony an automatic when they met and saying, "If you think I did it, shoot me." Much later, when in prison, gangster Roger Touhy claimed that the police had set it up so that Anthony could assassinate Capone while he was in custody. Touhy had gotten the story from a detective lieutenant who was there. "Sergeant McSwiggin," he said,

> was to be in a particular room . . . when two detectives were to walk in with Capone between them. The two detectives were to step aside, leaving Capone alone, and Sergeant McSwiggin was to kill his son's murderer on the spot.

No one was afraid of any charges being brought against the sergeant. In fact, he would probably get a medal. There would just be a little story of self-defense.

So that's the way it went. They walked Capone in. But at the last minute Sergeant McSwiggin lost his nerve. He was just too decent a guy to shoot a man, even Capone, in cold blood.

Anthony McSwiggin was a good detective with good informants. So perhaps the most telling evidence against Capone was the sergeant's flat assertion that the killers had been Capone, his principal bodyguard, Frank Rio, Frank Maritote (or Diamond), and gunman Bob McCullough, with Willie Heeney and Ed Moore as the spotters. But that was only reality, not proof.

"I believe I know who killed McSwiggin and Sergeant McSwiggin probably knows," said Judge McDonald while his *fifth* grand jury rested idle. "But we need evidence. If Sergeant McSwiggin has this evidence he should bring it into . . . the grand jury room. . . . I need evidence, not rumors or hearsay."

Capone tired of the talk about the bent prosecutor. "I paid McSwiggin," he said, finally, "I paid him a lot and I got what I was paying for."

So much for McSwiggin, and so much for the O'Donnell threat. Bill McSwiggin died because he was in the wrong place at the wrong time. For that matter, so did Red Duffy. He was a minor player, not a gunman; Capone would not have wasted bullets on him, let alone risk the furor that followed. Capone had to know what the killings would cost in mandatory police raids. McSwiggin's death made them more intense, widespread and destructive. But Capone knew what had followed O'Banion's death, as well as the heat after the multiple killings of the 1923 beer wars. It didn't matter. Capone set out that night gunning for at least five humans, ready to kill them all. He could anticipate public revulsion at so many simultaneous deaths; newspapers headlined any over two as a massacre.

He *had* to attack, or accept depredations from the rest of gangland as well as further O'Donnell encroachments. It was worth the price. Whether or not Willie Heeney mistakenly reported Klondike as being one of the targets, if Capone could get Myles and Doherty (with reliable gunman Quinlan dead and ferocious Sammons incapacitated) he was sure the O'Donnell problem would dramatically diminish if not dissipate. He was right. Even with Klondike and Myles untouched, the O'Donnell gang gave him no further trouble. The strategy of dealing with his enemies piecemeal was working.

Now for Hymie Weiss.

CHAPTER 15

The Bloody Path to Peace

THE NORTH SIDE gang's leadership-by-committee barely outlasted O'Banion's funeral. Two of the projected sextumvirs had never been day-to-day members, Maxie Eisen busy with his extortion of the Jewish Chicken Killers and similar shop-owners groups, Dan McCarthy busy looting his labor unions. Louis Alterie's challenge to shoot it out with O'Banion's killers had brought down extra heat on everyone. After he then brandished a gun at the Friars' Inn in late January 1926, his own gang told him he talked too much and caused too much trouble; they exiled him to his Colorado ranch. Of the other three, Hymie Weiss radiated so much more solid competence than the slapdash Schemer Drucci and sparkled so much more than the lumpish Bugs Moran that gang leadership gravitated to him.

Weiss presented a daunting mix of intelligence, imagination, guts, pitiless brutality and vicious temper. When photographers tried to snap him, he neither courted them like Capone nor covered up like most others. He'd fix them with a glare and growl, "You take a picture of me, and I'll kill you." His brother said they'd seen each other only once in twenty years: "That was when he shot me, six years ago." When a U.S. deputy marshal interrupted a party to arrest a friend on a Mann Act warrant, Weiss ran him off at gun's point; when raiders came back in force and confiscated booze, champagne, handcuffs, guns and knockout drops, Weiss sued to recover the silk shirts and socks he claimed they had also lifted.

That's what made Weiss so dangerous: crazy unpredictability that still was controlled, cerebral and modified by a reflectiveness that argued maturity. For all his temper and relative youth (twenty-eight at the time of his death), this was not just a wild-eyed kid. When a truck-driver friend lamented his own toilsome, ill-paid lot, Weiss said, "I'd trade places with you."

Capone was right to fear Weiss; the North Side gang leader was after him.

The afternoon of August 3, 1926, two brothers, thirteen and eleven years old, were watering their horses at a cistern in what is now the Argonne Forest Preserve south of the city. When the horses shied, the boys poked in the water until they discovered the body of Anthony Curinglone, better known as Tommy Rossi or Ross, who had replaced Sylvester Barton as Capone's driver. The North Siders had abducted him over a month before, while Capone was still in hiding over the McSwiggin affair. They had wired Ross's hands and feet, beaten and tortured him, finally shooting him through the head. They then weighted him with bricks and dumped him in the cistern.

"They call me heartless, eh?" said Capone. "Ross was tortured to make him tell my business secrets. He knew nothing whatever about my affairs." Maybe not; but he knew Capone's exact daily routine, information Weiss would soon use.

With the O'Donnells neutralized and the South Side quiet because both Saltis and McErlane were in jail awaiting murder trials (for reasons we'll examine later), Capone struck first—two weeks after he emerged from hiding. That first attempt seems singularly public and improvised. Only the identity of one gunman, revealed many months later, marked it a Capone operation.

Vincent Drucci lived in a suite at the Congress Hotel, at 500 South Michigan Avenue. On the fine sunny morning of August 10, Weiss breakfasted with Drucci. At ten o'clock they strolled four blocks down Michigan. Drucci later claimed that the $13,200 in his pocket was for a real estate deal. But at Ninth Street, he and Weiss crossed to the southwest corner, site of the new Standard Oil Building, which housed executive offices of the Chicago Sanitary District, the city's most corrupt pork barrel. Morris Eller—a District trustee and boss of the Twentieth Ward—was in the office talking with John Sbarbaro.

A car swerved to the curb and out of its windows three pistols barked at Weiss and Drucci as they stood directly in front of the Standard Oil Building entrance. Weiss hit the pavement along with the other pedestrians at the busy intersection. Much of the traffic stopped as motorists cowered below their dashboards. Drucci ducked behind a mailbox, drew his automatic and returned fire. At least thirty bullets chipped chunks of concrete from buildings, burrowed into car bodies and shattered windows. Somehow only one person was wounded, passerby James Cardan, who was nicked in the leg.

As two of the three gunmen vaulted from the car and maneuvered for a clear shot, a police patrol arrived. The attack car's driver burned rubber getting away, abandoning the two.

Weiss and one of the gunmen melted into the crowd. Drucci dashed into the avenue, jumped into a stalled car, shoved his empty pistol to the head of motorist C. C. Bassett and roared, "Take me away, and make it snappy." Police surrounded him before Bassett could start.

"It wasn't no gang fight," Drucci told them, "A stick-up, that's all. They wanted my roll." Police had also collared one of the gunmen. He had dropped his pistol, but was stopped because he tried to run. He said his name was Paul Valerie and gave what turned out to be a fake address. The police did not hold him, because Drucci, when shown the man, claimed never to have seen him.

Weiss had no doubt who was responsible. He couldn't get Capone behind the Hawthorne's steel shutters, but knew from tortured driver Tommy Ross when and where Capone might be vulnerable. Posting an agent inconspicuously at hand to phone a signal was easy: the block teemed with people. Twenty-second Street was Cicero's main drag, one hundred feet wide, streetcar tracks down its center. The Hawthorne Hotel stood on the south side, the Anton Hotel next door, and both offered plenty of attractions. Shops lined the fronts of both hotels: Angelo Gurdi's barber shop, the Hawthorne Beauty Shop, Dunovsky's delicatessen and a laundry—not to mention Capone's gambling joint, the Hawthorne Smoke Shop, in the Anton. The Hawthorne Restaurant, a coffee shop, stood on the east corner of the block. Ross knew that Capone often lunched there.

Capone had been east on one of his many trips during that period, arranging for supplies of smuggled booze. He had just returned the week before, and on Saturday had grandly reopened the gambling places shuttered after the McSwiggin killing.

At 1:15, Monday afternoon, September 20, Capone and able, loyal bodyguard Frank Rio sat over coffee at the innermost of the Hawthorne Restaurant's fifteen white-tile-topped tables, as usual facing front so no one could enter without their seeing. The room, twenty-five feet wide, fifty deep, was packed, including every seat at the lunch counter, a crowd in town for the opening of the racing season at Hawthorne Park, just south of Cicero.

Both men started in alarm at the unmistakable stutter of a machine gun that sounded outside, growing louder as a car swept by, fire erupting out a window. It looked like a detective bureau car, complete with clanging gong. But as soon as it passed the restaurant, nothing.

Puzzled, Capone moved a few steps toward the front to investigate. But that "nothing"—no screams, shattering glass, or sound of rounds pounding buildings—alerted Frankie Rio to the ruse: the car must have been firing blanks, calculated to draw the curious. He sprang from the table and pinned Capone to the floor just as the real attack began.

A procession of ten sedans, one block behind the lead car, rolled unhurriedly up the street. They began firing when they reached the Anton, raking the two hotels, the shops, and the restaurant. At the restaurant, they stopped. The windows disappeared in a shower of fragments. Plates, glasses and cups on tables and counter danced and clattered and disintegrated. Slugs stitched the walls in neat lines, waist- and breast-high as Capone and Rio, along with the other patrons, hugged the floor under the tables. Chunks of wood paneling, wainscoting and ceiling plaster showered down. Rio had drawn his pistol at the sound of the blanks, but it remained mute; no one dared raise head or hand in that fire storm.

A man in khaki shirt and brown overalls appeared out of the ninth sedan, cradling a Thompson. While others from the tenth car stood guard with shotguns, the machine gunner advanced to the restaurant door at an insolently calm pace, knelt at the entrance and emptied a hundred-round drum into the already savaged room. That final salute took ten seconds.

The gunners returned to the cars, someone signaled with three honks, and the cavalcade sped off to the east, into Chicago two blocks away.

The gunmen had fired (police later calculated) at least a thousand rounds. They had put holes into thirty-five cars that lined the street—yet, miraculously, hit only four people, two of them barely scraped. Clyde Freeman, a horseman from Louisiana, sat in his car with his wife, Anna, and five-year-old Clyde junior. A bullet grazed Clyde's knee; two more sent little Clyde screaming down the sidewalk, scalp grazed, coat bullet-torn. Another hit Anna in the arm, not a serious wound. Unfortunately, a splinter from the shattered windshield lodged in her right eye.

The last casualty was Louis Barko, who claimed he was "only a lone wolf gambler," resident of the Hawthorne. A bullet nicked his shoulder and neck. Chief of Detectives Bill Shoemaker took one look at Barko and identified him as the "Paul Valerie" nabbed running away from the Standard Oil Building ambush. His residence in the Hawthorne pretty well fixed blame for the Michigan Avenue attack.

Barko, standing outside the Hawthorne, had seen everything. But he maintained gangland protocol by refusing to rap anyone at the lineup—especially not Weiss, Drucci, Moran and Peter Gusenberg, one of their top gunmen.

Capone saw one bright spot. "It has shown the authorities," he said, "that I have no corner on the machine gun market. A machine gun was used to shoot up the Hawthorne Hotel a few days ago, and they can't blame that on me. Why, I'm still paying owners of automobiles parked in front for the damage done to their cars in that raid, and I am trying to save the eye of the poor innocent woman they wounded sitting in a car in front. I have paid all her expenses and the best doctors have been engaged to care for her." Anna Freeman's eyesight alone cost Capone $5,000. He also paid for the shop damages. He did not begrudge the money. A cynic could note that it bought him sympathetic publicity. Beyond that, as one subordinate said, "The Big Fellow never wants bystanders hurt."

The tit-for-tat sniping cost Capone much more than reparations. "This is war!" screamed one newspaper, and in the show of police zeal that necessarily followed, doors of the joints Capone had just reopened slammed shut again. Capone needed counsel.

After leaving jail and Chicago, Johnny Torrio had gone to Italy, but had soon decamped when Mussolini threatened to exhibit gangsters in cages. In New

York he had become an elder statesman, helping the likes of Lucky Luciano and Meyer Lansky replicate the early Chicago-style combine said to be the pattern for national organized crime today. He split his time between there and Florida, where he had bought some property. Within days of the Hawthorne attack Capone was huddled in Florida with Torrio over the Weiss problem.

Torrio of course urged peace; and for all his hot head, Capone would not have resisted. As he put it himself, a little later, "Right after Torrio was shot—and Torrio knew who shot him—I had a talk with Weiss. 'What do you want to do, get yourself killed before you're thirty?' I said to him. 'You'd better get some sense while a few of us are left alive.' He still could have gotten along with me. But he wouldn't listen to me." Capone also declared that he had been "ready for peace . . . for some time. This fighting is useless. If the gangs would compose their differences we could all make more money than we're doing now. There's enough for all if we could get together."

He would try again.

Capone commissioned Tony Lombardo, Unione Siciliana president, to arrange a meeting with Weiss for October 4 at the Hotel Sherman. Capone would not attend—personal confrontation an unlikely lubricant for negotiations between men who had been trying to kill each other. A police official was probably invited to keep the peace talks peaceful. Lombardo would speak for Capone, the message uncomplicated. What did Weiss want for peace? Capone would agree to anything within reason.

Weiss still smoldered over the murder of O'Banion. Ratiocination could bring him to the need and wisdom of forgiving the string-puller, Capone, but the actual trigger-pullers had to go. Capone could have his peace if he would approve the killing of Scalise and Anselmi. They were still in Joliet on their manslaughter sentence, but *arrangements* could be made anywhere. Lombardo phoned the terms to Capone.

"I wouldn't do that to a yellow dog," Capone snapped. Weiss stalked out of the meeting.

Captain John Stege sneered when he heard the report, "There's no one on earth Capone wouldn't send to death if he thought his interests would be served." Chief Collins doubted the story because Weiss would never cry off without Capone's own death. The policemen did injustice to Weiss's head and Capone's heart.

This time Torrio came to Chicago to confer on what should be done and how.

Capone had already put in train contingency plans. Well before the aborted peace talk—just in case—a man who gave his name as "Oscar Lundin" took the only space available in the rooming house next door to Schofield's, a hall bedroom; but he reserved the second-floor front room for when it became free.

The next day, the building (which belonged to a popular mystery writer, Harry Stephen Keeler) changed management. On Tuesday, October 5, the second-floor front fell vacant, and new manager Anna Rotariu let Lundin move in.

The room looked hardly worth the wait. Crabbed and dingy, its ''amenities'' included a bed with a tarnished brass frame, a borax oak dresser, some chairs, a gas hot plate and a tin food box. But it was a room with a view. The rooming house, at number 740, sat just north of Schofield's at 738. Since the flower shop jutted out, its north wall shielded the shop's front door from sight. Nevertheless, the window swept a panorama of the sidewalk and the street in front of the shop and of Holy Name Cathedral, which loomed on the other side of State, its south side fronting the cross street, Superior.

About the same time, a pretty young woman who called herself ''Mrs. Thomas Schultz'' rented a third-floor room across Superior and enough down the block so that her window sighted on the alley that led to Schofield's rear entrance.

Lundin and Schultz paid rent in advance and left, never to be traced. Others moved in. Anna Rotariu remembered seeing two of them in her place. There, and in the apartment across the street, the newcomers spent most of their time in chairs drawn to the windows, the floors around the chairs carpeted with cigarette butts. Across the street, the watchers relieved the monotony by sipping wine. In 740 State, they snoozed on the brass bed, their shoes soiling the spread.

On the afternoon of October 11, 1926, the twelfth juror was seated for the murder trial (which we'll soon examine) of Joe Saltis and his driver Frank Koncil, always known as ''Lefty.'' Testimony would start the next morning. Special prosecutor Charles McDonald had hopes for a conviction despite persistent rumors that Weiss would spend $100,000 to buy Saltis an acquittal. Weiss had been an avid spectator during jury selection that day.

He and the lead defense attorney, William W. O'Brien, left court together but drove to Schofield's in separate cars, Weiss with his driver Sam Peller and Patrick Murray, a journeyman bootlegger who may have doubled as Weiss's bodyguard. O'Brien drove with Benjamin Jacobs, a small-time pol in Morris Eller's Twentieth Ward and investigator for lawyers like O'Brien, who had narrowly eluded disbarment four years before.

The two cars parked on Superior. The passengers straggled around the corner and across State toward Schofield's, Weiss and Murray in the lead, O'Brien slightly back, Peller and Jacobs well behind. Machine gun fire and shotgun blasts from the second-floor rooming-house window caught Weiss and Murray almost at the curb in front of the flower shop. Murray fell dead with seven slugs in his head and body. Weiss staggered onto the sidewalk and collapsed with ten slugs in him. He would soon die at Henrotin Hospital, never regaining consciousness.

The gunmen had sprayed the street. Standing to one side of Weiss, Lawyer O'Brien was hit four times in the arm, side and gut, though not fatally; he crawled into a basement stairwell, then made his way to a doctor's office eight doors up the block. Peller, with one minor belly wound, and Jacobs, shot in the foot, turned and hobbled back the way they'd come to a doctor's office a block away. Bullets followed them until they disappeared around the corner of Holy Name. A swath splattered the cathedral's cornerstone, chipping the inscription, "A.D. 1874—AT THE NAME OF JESUS EVERY KNEE SHOULD BOW—THOSE THAT ARE IN HEAVEN AND THOSE ON EARTH," from the Vulgate version of St. Paul's Epistle to the Philippians, so that only

EVERY KNEE SHOULD
HEAVEN AND
ON EARTH

remained.

Their work done, the gunmen ran down the rooming house's back stairs, scrambled out a rear window, and legged it through alleys to lose themselves on Dearborn. Along the way, one of them tossed the machine gun, fitted with a hundred-round drum, over a rear fence; it landed atop a doghouse.

In Hymie Weiss's pocket, police found $5,200—and a complete list of the veniremen called for the Saltis-Koncil jury. They immediately searched the room at 740 State. In addition to the arrangement of chairs, the butts and dirty bedspread, they found a pile of .45 machine gun cartridge jackets on the floor, thirty-five in all, and three empty shotgun shells, plus a gray fedora with a label from a clothier near Cicero. In the safe at Schofield's, they found the prosecution's complete witness list against Saltis.

Gunmen in the other machine gun nest, who never had to fire, simply wandered away at their leisure. Indeed, their former presence was not discovered for a week, until the woman below their abandoned third-floor apartment complained of a splotch on her ceiling, seepage from a leaky radiator above. The janitor forced the door—and called police, who found on the bed a pump shotgun with four live shells, chairs at the window ringed by cigarette butts and two empty wine bottles.

"I don't want to encourage the business," said Chief Collins, "but if somebody has to be killed, it's a good thing the gangsters are murdering themselves off. It saves trouble for the police."

Capone waxed a bit more elegiac. "I'm sorry Weiss was killed," he told reporters who gathered at the Hawthorne Hotel. "But I didn't have anything to do with it. I telephoned the detective bureau I'd come in if they wanted me and they told me they didn't want me. I knew I'd be blamed for it, but why should I kill Weiss?"

"He knows why," detective chief Bill Shoemaker growled later; and a reporter at the time pointed to the bad blood between the men and their struggle for control. But Capone wasn't in the mood for such quibbling. "It's getting to be a joke, this pinning all the murder onto me," he said. "If I stick my head out of doors without a battalion of armed men, someone tries to shoot me.'' And when he did go out with his bodyguard, the police wanted to arrest him, as in the McSwiggin case, ''All because they feel safe in calling me a murderer, knowing that I can't fight back.''

He had received the press in shirtsleeves, at slippered ease—a notable informality in those days. He handed around cigars (for the record) and undoubtedly drinks (the unreported custom with reporters during Prohibition).

''Hymie Weiss is dead because he was a bull-head,'' Capone said. ''I suppose you couldn't have told him a week ago that he'd be dead today.'' The phone kept ringing in the spartanly furnished room, some calls urgent enough to make him break off to take them, then return to the reporters. He wanted to correct an earlier statement. He had meant he was sorry Weiss had been *killed* that way, ''butchery,'' he called it. ''But don't get me wrong. I don't want anyone to think I am sorry Weiss is *dead*,'' Capone said. ''He has been out to get me—he and the [rest] of Dean O'Banion's gang of hijackers, butchers, holdup men and burglars. A fine lot of toughs for an honest man to try to do business with!''

But seriously, all this had to stop. ''I've got a boy,'' Capone said, showing the reporters Sonny's picture. ''I love the kid more than anything in the world and next to him I love his mother and then my own mother and my sisters and brothers,'' he said. ''And it's pretty terrible, too, when you think I have not been able to go home to my wife and boy for fourteen months.

''I don't want to die. Especially I don't want to die in the street, punctured with machine gun fire.

''That's the reason I've asked for peace. I've begged those fellows to put away their pistols and talk sense. They've all got families too. Most of them are kids and haven't got any children of their own, but they've got mothers and sisters.

''What makes them so crazy to end up on a slab in a morgue, with their mothers' hearts broken over the way they died, I don't know. I've tried to find out but I can't. I know I've tried since the first pistol was drawn in this fight to show them that there's enough business for all of us without killing each other like animals in the street. Competition needn't be a matter of murder, anyway. But they don't see it.

''I read in the papers,'' he continued, knowing how they relished bathos, ''that Hymie Weiss's mother was coming here from New York for his funeral. She's a wonderful mother. When Hymie was in business with us, many's the night I slept in his house and ate at his table. Why didn't he use some sense and stay out of the shooting stuff?''

He told the group about his peace offer to Weiss, and Weiss's condition. Much as he wanted peace, neither Hymie's relics nor the other gangs should suppose he'd pay *any* price: ''I am in the game to stay, and I'm backing down for no one. Let them try to kill me if they think they can. I am on the job. Any time they want peace with me, I am ready to listen.''

He didn't mind what was said and written about him, but he had agreed to this press conference "in hope that the people would get to know me right." He launched into a mix of fact and fiction. "I am a gambler and a businessman," he said, "nothing more. I never robbed a person in my life. I never killed in my life and I never broke into a building or blew open a safe." The talk of Capone as murderer was nonsense; for instance, none of the McSwiggin grand juries even indicted him. Sure, the police picked him up for questioning, but he'd never been tried for murder, obviously never convicted. Technically he didn't even have a police record!

He'd go see the special prosecutor, McDonald, or Crowe or anyone who wanted him. He wasn't hiding. But . . . "If I said too much about what I know," he said with a wicked grin, "I might embarrass someone."

No one wanted him. A dejected Police Chief Collins explained that it was a waste of time: they'd done it before and it was always the same. "Brown, of course, has an alibi. He was in Cicero when the shooting occurred." The only time it was worthwhile grabbing him was when "we have the goods on him cold, but there is no use putting him on the grill until we do." This marked the nadir of police morale. Wiseacres called Capone the "Mayor of Crook County."

The North Side remnant shared police dejection; they did not put on much of an Obsequies Chic show for Weiss's poor sweet old mother (who once had signed bail bonds for Frank McErlane after an especially frightful murder). They could muster only about two hundred of the sorrowful or curious at Sbarbaro's funeral parlor. Pallbearers came from Weiss's former classmates at St. Malachy's. Drucci, Moran and Eisen were the honorary pallbearers—no judges, this time, or big-name pols, though placards on the bumpers of the cortege did urge, in the coming fall election, support for Sbarbaro as municipal judge, for Joe Savage (Crowe's pick) as county judge, and for reelection as trustee of Morris Eller and his Sanitary District accomplices.

Church policy remained obdurate: for all his Rosaries, Weiss got no Mass and could not be buried in consecrated ground at Mount Carmel.

To the press, the jury list in Weiss's pocket and witness list in his safe came as revelation: WEISS MURDER LIFTS VEIL OFF SALTIS ALLIANCE one headline ran. But that connection could hardly have startled Capone. Among many tip-offs, Saltis and McErlane had been battling Capone's client, Sheldon, for a year. This final revelation came about because Frank McErlane, the nerve and muscle of the Saltis-McErlane partnership, found himself in jail, fighting extradition to Indiana.

On May 4, 1924, McErlane had been drunk in a Crown Point saloon, standing at the bar. His drinking buddies, John O'Reilly and Alex McCabe, had ragged him to show his marksmanship. McErlane picked a random target at the far end of the bar, lawyer Thaddeus S. Fancher, and killed him with a head shot. O'Reilly and McCabe were caught almost at once. O'Reilly drew a life sentence.

McCabe was convicted, but just before a new trial, the chief witness against him was bludgeoned to death with a hammer.

McErlane escaped back to Illinois, and thanks to Governor Small's understanding extradition policies, he was not even arrested until April 22, 1926, almost two years after Fancher's murder. McErlane would stay in an Illinois jail until August, when extradited. He showed up for one court appearance blustery drunk, his warder equally soused. A year (and much changed testimony) later, an Indiana jury would acquit him.

But in the summer of 1926, Joe Saltis had to do his own killing. In July he managed a quiet, unobserved ride for Jules Portuguese, a Sheldon adjunct, whose car had been used in the O'Banion murder. Next month, August 6, he went after one of Sheldon's better beer pushers, Mitters Foley, who had been encroaching. Saltis lured Foley from home with a decoy phone call and forced his car to the curb. Foley tripped and sprawled on the pavement, trying to run away. Saltis straddled him, shoved a shotgun to his chest and gave him both barrels. Trouble was, two witnesses saw him do it. They also identified his driver, Lefty Koncil, Saltis's political connection and partner, John Oberta, and gunman Edward Herbert.

"We've got it on them this time," exulted John Stege. It was, he said, "the first time in all these . . . beer killings that we've actually had the goods."

When it came time for his trial with Koncil in October (Oberta and Herbert would be tried later), Saltis needed help, and Hymie Weiss had given it to him, a mark of their alliance. Special prosecutor Charles McDonald had his two stalwart eyewitnesses, but two other key ones had been either bribed or scared out of town, and someone had twice rifled his files at night. Then there was that rumored $100,000 Weiss was spreading around.

So the trial held no terrors, even after the October 11 murders; Weiss had already worked his magic. But what then? With Capone suddenly freed from distractions and McErlane still in jail, Saltis figured to become a sudden dead man absent some accommodation. He turned for advice to John Oberta, who consulted Maxie Eisen, reputedly wisest of the North Siders' connections.

Eisen had returned from a trip around the world with his wife and son just in time for Weiss's murder. He was properly appalled at how matters had degenerated. Close enough to what remained of the North Siders to appear a credible spokesman, he had also stayed aloof enough not to inspire instant suspicions of a setup. Through the relatively neutral mediation of bail bondsman Billy Skidmore, he arranged a first meeting with Tony Lombardo for Saturday, October 16. Both agreed this war had to stop. "Let's give each other a break," as Eisen would later put it at the peace conference. "We're a bunch of saps, killing each other this way and giving the cops a laugh." Eisen and Lombardo agreed to meet next day, meanwhile sounding out their principals.

Capone was delighted. Peace far outweighed the satisfaction of punishing the defection of a lump like Saltis. George Moran resisted, but Vincent Drucci overruled him. Eisen and Lombardo negotiated the meeting's terms: principals only, no guns, no bodyguards. Those who had the course of Chicago gangland at their disposition met on Wednesday, October 20, 1926, probably again at the Hotel Sherman.

Eisen chaired the meeting, Lombardo by him. Capone brought Jack Guzik; his ally Ralph Sheldon came alone. Drucci and Moran attended for the North Siders. The far north, northeast and west sides contributed a mixed bag of alliances and loyalties. Pimp Jack Zuta and his lieutenant, Frank Foster, would end up siding with the North Siders. Klondike and Myles O'Donnell never again challenged Capone. Cicero's Ed Vogel was with Capone. West Sider Julian Kaufman (always ''Potatoes'' Kaufman, his front being that of a produce wholesaler) had always been close to the North Siders. Billy Skidmore, who had gambling interests in addition to his bail bond business, and Christian P. Bertsche (always known as ''Barney'') who, like Skidmore, ran casino-roadhouses in the northwest, attended as independents, but would soon combine with Zuta.

''We're making a shooting gallery of a great business,'' Capone told them all. ''It's hard and dangerous work, aside from any hate at all, and when a fellow works hard at any line of business, he wants to go home and forget about it. He don't want to be afraid to sit near a window or open a door.''

The basic treaty called for a general amnesty: no more shootings or beatings, all those in the past to be considered closed incidents. Everyone agreed to avoid and ignore ''ribbing''—challenges like, ''You aren't gonna let him get away with that, are you?'' or gossip (''Did you hear what he called you?'') often mischievously carried between gangs by police or appearing in newspapers. Gangs would punish *their own* infractors, eliminating the most probable cause of intergang grudges. Finally—the real issue—no more muscling in. Gangs would stay within assigned territories.

These reverted essentially to the original 1923 split. Sheldon would stay north and east of the stockyards, Saltis to his Back of the Yards. Klondike and Myles O'Donnell would rule in the west just north of Cicero, but with their Roosevelt Road concession intact. Drucci and Moran would have the original O'Banion territory, north of Madison and east of the Chicago River. Capone kept the near south, plus his classic south and west suburbs. He would provide for client allies like Marty Guilfoyle. Zuta, Skidmore and Bertsche would operate as before, on good behavior. Spike O'Donnell was no longer a factor: Capone would take care of him, one way or another, as he deserved.

Afterward the meeting adjourned to Diamond Joe Esposito's Bella Napoli for a celebration, which reporter James O'Donnell later dubbed ''a feast of ghouls.'' The coarse merriment of the former enemies as they relived their feuds shocked him. An observer who was not a gangster gave him a sample of the dialogue:

> ''Remember that night eight months ago when your car was chased by two of ours?''
>
> ''I sure do!''

"Well, we were going to kill you that night—but you had a woman with you."

Shouts of laughter.

Two days later a car careered south on Michigan Avenue in the afternoon, sounding a siren. A detective squad took chase and overhauled it at Thirty-first. At the wheel sat Bob McCullough, Torrio-Capone gunman since 1922, with Capone on the McSwiggin killing. "I just came to town for a shave," he said, beaming, to Sergeant John Tracey, with a nod at the golf bag next to him, "and I'm hurrying back to Burnham." The police searched McCullough and ransacked his car, sure they'd find the usual small arsenal. To their astonishment, in McCullough's pockets they found only money, in the golf bag only a mashie, a niblick, a midiron, a cleek, a spoon and other clubs, under the seats only lint. Not a single weapon was to be found. They couldn't believe it. "The peace treaty's on the square," said McCullough with a smug grin. "I wouldn't be unarmed if it wasn't." They held him for speeding.

As 1926 ended, sweetness and light suffused Chicago gangland. Scalise and Anselmi got their new trial; Druggan and Lake were acquitted of having conspired with Sheriff Hoffman when he let them out of jail anytime they wanted a jaunt. Despite the two unshakable eyewitnesses, the jury found Saltis and Koncil not guilty of Foley's murder; "I expected a different verdict on the evidence presented," said the judge.

Strife continued, but nothing that threatened peace. As a last gasp, Spike O'Donnell's people stepped out of line and two of his brothers wound up shot, though not fatally. Spike never bothered Capone or his allies again.

Just after Thanksgiving, on November 28, the unreflective might have expected all bets to come off. Theodore Anton, reputedly a particular pal of Capone's, owned both the Anton Hotel and the Hawthorne next door. Suddenly he went missing, and his body turned up over a month later. By now Capone had become so adroit at manipulating the press that his first biographer, reporter Fred Pasley, uncritically retailed Capone's affecting story of having been so devastated by Anton's abduction that he sat in a booth of the Anton restaurant through the night of his buddy's disappearance "sobbing like a child."

Then how come no new war, or even an individual reprisal? Sheriff Hoffman seemed to notice, speculating at the time that it must have been some internal disagreement. Later, the story circulated that Capone had killed Anton in a drunken rage because of some remark he took as an insult. Much later, a piano player in one of Capone's places charged that two of Capone's men beat Anton to death. The police thought, though could not prove, that young Jack McGurn, one of Capone's rising stars, killed Anton on Capone's orders. No one ever discovered exactly who did it or why, but it was *not* another gang.

Peace survived even a real intergang killing. Hillary Clements, a Ralph Sheldon soldier, disappeared on December 16, 1926. His brother appealed for his body to be returned in time for Christmas, but it was not discovered until two boys at play stumbled on it face down in a shallow hole on December 30.

Still, peace held. Many killings would in fact follow, including the St. Valentine's Day massacre of 1929; the rate of individual gang-related killings would even increase. But peace would *still* hold in the sense that gang wars had ended. Capone had become too strong and most of the others too wary. From now on he would deal with individuals who challenged him. Otherwise, he had become what his people had been calling him for some time: The Big Fellow.

Big Bill Thompson announced on December 11 that he would run for mayor again in 1927. When presented with 433,000 cards pledging support he said, "I accept, and with grateful thanks." Capone and the rest of Chicago gangland were equally happy.

Capone would scarcely have noticed the nomination on December 28, 1926, of a lawyer and master in chancery named Johnson to succeed Edwin Olson, whose term as U.S. attorney would expire January 2, 1927. Johnson seemed an ordinary man in a position that caused only minimal trouble. But his full name was George E. Q. Johnson, and he had awarded himself the "Q." to distinguish himself from all the other George Johnsons. He yearned to stand out.

For Capone, now on top of the world, that would prove an ominous ambition.

CHAPTER 16

The Big Fellow on Top

SCANDALS SURROUNDING O'BANION's murder had forced Mike Hughes out as Chicago's chief of detectives. After a while he took over as head of Cook County's highway police. In 1925, when it became clear that Mayor Dever could not dry up Chicago, Capone had booked rooms at the Metropole Hotel, 2300 South Michigan, near the old Four Deuces. With peace, he shifted main headquarters from the Hawthorne Hotel to the Metropole, and Hughes immediately started bragging of having run Capone out of Cicero.

"Chase me out of Cook County?" Capone laughed, at ease in his Metropole suite. "Well he hasn't done it and he won't do it. I moved out of Cicero three months before Hughes was even mentioned for the job of highway chief. . . . The reason I moved into this hotel was simply that my interests have expanded and I required a central headquarters."

Along with the neighborhood, the Metropole had turned seedy since Big Bill Thompson had maintained digs there. Starting with about ten rooms, including his own suite, numbers 409 and 410, Capone soon expanded to about fifty rooms on two floors, two of the rooms equipped as a gym, where he insisted his men work out. Gunmen patrolled the corridors. Politicians and city officials took turns with saloonkeepers and whoremasters waiting to see the Big Fellow. Uniformed police officers, including brass, streamed in and out, ignoring the gunmen, the open gambling, the whores, the blind pig set up in the lobby and the service bars upstairs.

As winter turned to spring in 1927, Capone's grip grew firmer without his having to do a thing. Klondike O'Donnell and Fur Sammons were about to go to jail, having been caught peddling booze. Lefty Koncil was shot to death with another Saltis gunman, Charles Hubacek—probably Ralph Sheldon's work, a payback for Hillary Clements. Saltis himself increasingly holed up at his combination farm, summer resort and estate in northern Wisconsin.

It cannot properly be termed a ''comeback,'' because Bill Thompson had never really been away. ''The most notorious wowser of the day'' could always euchre the press into covering him. A scant year after voter disgust had forced him out of the 1923 mayoral election, he constructed a yawl that cost $25,000 beyond all the services and materials (''tide water cypress'') his clout could still squeeze from suppliers. He called it the *Big Bill*, his bust its figurehead. He promised to sail it to the South Seas to capture for Chicago the fabled tree-climbing fish (exhibits of which, no one bothered to point out, already graced the Field Museum). Big Bill added that *Big Bill's* voyage would illuminate the need for a commercially navigable waterway from Chicago to the Mississippi, which meant dredging sixty-three miles of the Illinois River. He offered to bet anyone the cost of the schooner that he'd successfully complete the expedition. After being towed over the Illinois shallows to the Mississippi, Big Bill delivered headline-grabbing speeches almost the length of the mighty river—then jumped ship before New Orleans and returned to Chicago. But, significantly, no one had taken his bet.

No sensible person would have bet against his return as Chicago's mayor, either. He had totally broken with Fred Lundin, who sponsored a new candidate, John Dill Robertson, MD, a genial quack whom Thompson, as mayor, had made city health commissioner to universal indignation. On April 6, 1926, a year before the mayoral election, Thompson had destroyed them both at what became known as ''The Rat Show,'' a noon rally in a Loop theater, the Cort. Thompson displayed two large gray stockyards rats, whom he introduced as Fred Lundin and Old Doc Robertson. ''This one on the left here is Doc,'' he told the tittering audience. ''I can tell him because he hadn't had a bath for twenty years. . . .'' Thompson flashed his perennial-boy grin, then addressed the cage on his right. ''Fred, let me ask you something. Wasn't I the best friend you ever had? Don't hang your head like that. Isn't it true that I came home from Honolulu to save you from the penitentiary?'' Well, what gratitude could you expect from rats?

''Do you remember, Doc,'' he said, turning back to the left, ''how thousands came to me to protest against the appointment of yourself as health commissioner? . . . Do you remember how I stood up against them and honored you with that great office? If you do remember, you know why I now call you a rat for turning against me with Lundin.''

Robertson withdrew from the February 1927 primary, and Thompson faced Dever in April. Dever did not want to run again, but the Democrats needed him and *someone* had to try to stop Thompson.

Big Bill's campaign startled the entire nation. ''In some mysterious way,'' a *New York Times* editorial marveled, ''he is running against George V''—the King of England! Thompson attacked Dever's superintendent of schools as ''King

George's stool pigeon,'' leader of a plot to undermine patriotism in Chicago's children. That wasn't all. ''I shouldn't be surprised,'' Thompson declared, ''if the King had something to do with slipping over the Volstead Act on us so that all their distillers can make fortunes selling us bootleg liquor.'' But Cowboy Bill would ride to the rescue. ''If George comes to Chicago,'' he promised, ''I'll punch him in the snoot.''

Poor Dever. ''I try to get my opponent to talk on issues,'' he wailed; ''I ask him questions. His only answer is—'America First.' How can I campaign against a brain like that? What kind of a mind has he? I don't know what this King George stuff is all about.'' *That*, a Thompson supporter retorted, was precisely Dever's problem. Or as the Peoria *Star* put it, ''Bill knew his electorate better than his opponents [did]. . . .''

What hurt Dever most with that electorate was his insistence on at least *trying* to enforce Prohibition. The fact that he could do only very ineffectively what the vast majority did not want done at all endeared Dever to no one, wet or dry. Against Thompson's hurrah, Dever's backers could ask voters only if they didn't admire the ''clean, wholesome, adequate schools'' and scandal-free administration their candidate had provided—all under the thrilling campaign slogan, ''Dever and Decency.'' As one old pol commented, ''Who the hell is attracted by decency?''

Only a single issue mattered to Capone and other gang leaders. Big Bill had boasted of his campaign platform, ''You can't find anything wetter than that in the middle of the Atlantic.'' Under Dever's eye, police had been forced to padlock thousands of joints. ''When I'm elected,'' Thompson trumpeted, ''we will not only reopen places these people have closed, but we'll open ten thousand new ones.'' His literature assured the thirsty, ''I will discharge any Chicago policeman who crosses the threshold of anyone's home or place of business, without warrant of law.'' On the stump, he rendered that into ''no copper will invade your home and fan your mattress for a hip flask.''

In his Metropole office, Capone added a portrait of Big Bill to those of Washington and Lincoln. His men collected $40 for Thompson's campaign from those joints without slot machines, $250 from the more fortunate. More money came from Daniel A. Serritella, a particular friend of Capone's, and the man who would become city sealer in Thompson's administration. In all, Capone contributed at least $100,000; Frank J. Loesch, then head of the Chicago Crime Commission, thought it might have amounted to $260,000. Superpimp Jack Zuta kicked in at least $50,000 of his own, and said, ''I'm for Big Bill hook, line and sinker and Big Bill's for me hook, line and sinker.''

Boss Brennan sneered that all the hoodlums in town were for Thompson—who turned this remark to his advantage, calling all his audiences ''fellow hoodlums,'' including one gathering of titillated society matrons. When Thompson and Crowe left the Sherman one night to find the tires on their car slashed, Crowe said, ''The Dever decent element is giving rough treatment to us Thompson hoodlums. I suppose some silk-hat highbrow from the Gold Coast did this uplift deed with his umbrella.''

Since the hoodlums were indeed for Thompson, the Democrats ran scared and wary. Dever demanded police surveillance of gunmen. When North Side thugs beat a watchman and trashed the office of a Dever alderman in the heart of the gang's Forty-second Ward, Chief Collins ordered his men to pick up the North Side leaders. On the day before the April 5, 1927, election, a detective squad caught Vincent Drucci and two companions coming out of the Hotel Bellaire. Detective Daniel Healy, though only in his early twenties, had killed one holdup man and had once slugged Joe Saltis silly. He did not like criminals. Almost at once, he and Drucci began snarling at each other. Drucci objected to Healy's holding his arm, cursing the detective. Healy swatted Drucci, pulled his gun and said, ''You call me that again and I'll let you have it.''

In the squad car, on the way from the detective bureau to court for a habeas corpus hearing, the squabbling, maybe scuffling, continued. Healy's report to superiors, sanitized in the press was: ''When Drucci got into the car he said, 'You———, I'll get you. I'll wait on your doorstep for you.' I told him to shut his mouth. Drucci said, 'Go on, you kid copper, I'll fix you for this.' I told him to keep quiet. Drucci said, 'You take your gun off and I'll kick hell out of you.' He got up on one leg and struck me on the right side of the head, saying, 'I'll take you and your tool (revolver).' ''

When the car stopped, Healy got out. His account continued: ''He put his arm through the right side [window], through the curtain and said, coming toward me, 'I'll fix you,' grabbing hold of me by the right hand. I grabbed my gun with my left hand and fired four shots at him.'' They struck in the arm, leg and gut; Drucci died before the squad car reached the county hospital.

When Drucci's lawyer, Maurice Green, used the word ''murdered,'' Chief of Detectives Bill Shoemaker snorted, ''I don't know anything about anyone being murdered. I do know Drucci was killed trying to take a gun away from an officer.'' He later added that they were striking a medal for Healy. The coroner's jury held it justifiable homicide.

One of Thompson's supporters later wrote that the ''Capone men . . . were enlisted against the 'Dever and Decency' gunmen. . . . Steal the election from Thompson? Not where the Capone mob was marshaled!'' In fact, the election was quiet and lopsided. ''They was trying,'' explained Will Rogers, ''to beat Bill with the better-element vote. The trouble with Chicago is that there ain't much better element.''

That was not fair. While Dever appealed too much to the portion that contained merely bigoted bluenoses and the rapacious, much of Chicago's *real* better element—like Professor Charles E. Merriam, social worker Jane Addams and reformer Harold Ickes, who later joined Franklin Roosevelt's Cabinet—stuck by Dever for his personal honesty and intentions. On the other hand, an astoundingly large number of others could not stomach the Brennan Democratic machine. A contemporary academic, visiting Chicago that spring, expressed his astonishment "to find that some of his most intelligent and progressive friends proposed to vote for 'Big Bill.' " When he asked what motivated "such an apparently incredible decision, the answer came forth unhesitatingly that Thompson and his crowd grafted for the common man, while Dever and his polished associates grafted for the great traction interests and the public utilities."

Mack Staley voted for Thompson in that election. Today, nearing ninety, he still works as a salesman for a stationery store. "Big Bill was pretty good," says Staley, who still remembers fondly how Thompson annually treated the kids to a day at Riverview, a city amusement park.

"Tell 'em cowboys; tell 'em," shouted the Intellectuals' and Common Man's Choice through a megaphone, as he stood on a table in the Louis XIV Room of his Sherman Hotel headquarters, waving the sombrero he had affected since his youth on the range, "I told you I'd ride 'em high and wide!" He had won a landslide: 512,740 to Dever's 429,668. Pundit Elmer Davis asked himself if there was any mystery about the returns, and answered, "Well, one mystery, perhaps—in spite of everything, 430,000 people voted for Dever."

The mayor-elect led his celebration to the Fish Fans clubhouse, a hulk moored in Lake Michigan off the coast of Lincoln Park. Founded by Thompson in 1922, ostensibly to promote the stocking of state waters with food fish, the club served as a convivial Thompson political hangout, its members' lockers awash in booze. Fifteen hundred trooped aboard that night, ignoring a warning from the caretaker that the rotting hull wouldn't take such numbers. Sure enough, as the members stomped their glee, the Fish Fans club settled majestically in six feet of water to the bottom. Miraculously, no one was hurt; Thompson didn't even get his feet wet.

Capone and his people must have thought that auspicious. A few months after Thompson's inauguration, federal agents Pat Roche and Clarence Converse mounted a raid on a Capone brewery. They kept their plans so secret that no Prohibition officials or police got wind—which meant Capone didn't know about it either. "Our observation convinced us," Roche said, "[that the brewery] was immune so far as the police were concerned. There were always police flivvers around it and the trucks would go out with a guard of motorcycle cops. They were there tonight when we raided it, but made no attempt to interfere once we had established our identities."

One of the employees knew what to do. He eased up to the feds. "Al Brown owns the place," he whispered, "everything will be squared." He gaped, dumbstruck when the magic name and usual gracious offer failed to dispel the problem. *Nobody* turned down the Big Fellow!

On the day after Thompson's election, a tiny tragedy occurred in the lake behind Roosevelt Dam in Arizona. The *Santa Maria*, a four-man amphibious airplane piloted by Commander Francesco de Pinedo was making a world tour for the glory of fascist Italy. It had set down to refuel. Spillage had created an oil slick all around the plane. One young lout among the gawkers lit a cigarette, tossed the match into the water, and—*whoosh!*—the *Santa Maria* went up in flames.

Mussolini ferried de Pinedo his craft's sister plane, *Santa Maria II*, and the commander continued on, his last stop in the U.S. scheduled for Chicago on May 15, 1927. On hand to greet him as he splashed down on Lake Michigan at the foot of Monroe were Chicago's most luminous Italians, foreign and domestic: Italy's consul general, Leopoldo Zunini; the president of Chicago's Italian Association of Commerce, Italo E. Canini; the city's premier fascist, Ugo M. Galli; and Mayor-elect Thompson's personal representative, Judge Bernard P. Barasa. In addition, Anthony Czarnecki, who had become the collector of customs, represented the federal government and Alderman Dorsey R. Crowe led a committee of one hundred appointed by the outgoing mayor. Sirens sounded, flags waved. U.S. Army and Navy fliers flocked about. One reporter estimated that one thousand of Chicago's less luminous Italians waited in hopes of kissing the hero. (Coincidentally, six days later, May 21, 1927, Captain Charles A. Lindbergh landed in Paris.)

Despite the throng of greeters, one of the first embraces came from Al Capone, there at *police* invitation. The brass feared anti-fascist riots and concluded that Capone's presence would deter disorder more surely than squads of patrolmen. Such had become the Big Fellow's perceived clout. He had even been "taken up" by an element of Chicago society. He had awaited de Pinedo's arrival aboard the yacht of the millionaire founder of the Zenith radio company.

Our decades did not invent "celebrity." Society found the same frisson in consorting with Capone that one might in having a tame and amiable tiger as a house pet. Here was a certifiably dangerous man, whose very bulk radiated power, a constant lawbreaker who had unarguably killed personally, whose fiat had certainly slaughtered at least dozens, maybe hundreds, slaughter that *continued*. Yet the respectable could consort with him in safety, even with unfeigned pleasure. Capone dressed, spoke and acted well, if somewhat floridly. One contemporary called him "a fervent handshaker, with an agreeable, well-nigh ingratiating smile." He flashed it often and easily.

He was almost embarrassingly generous. His Christmas shopping bill for friends and acquaintances ran to over $100,000. As lagniappe he might give special friends a belt buckle with their initials in diamonds. He once ordered thirty buckles at a time, each costing $275. (A whitefish dinner with french fries and salad then cost 65 cents at the North American Restaurant, short ribs 40 cents; a pound of coffee at the A&P 45 cents.)

Capone's loyalty was legendary (as in "I wouldn't do that to a yellow dog"). So was his sense of honor. "If he gives you his word," admitted one of his sharper critics, "you can believe him." He made an interesting companion, spoke knowledgeably, with ponderable opinions and even wit on sports, politics, current affairs, the theater, movies, jazz, Italian opera (admired *Rigoletto* and *Il Trovatore*, adored *Aïda*). He once commented on a predecessor: "I'll have to hand it to Napoleon as the world's greatest racketeer. But I could have wised him up on some things.'' The *figlio puttana* had gotten a swelled head. Elba should have been a warning. ''But he was just like the rest of us. He didn't know when to quit and had to get back in the racket. He simply put himself on the spot.''

Of course Prohibition remained Capone's best, most resonant subject. His message reassured those who needed permission to break the unpopular law. ''Sure,'' he'd say when asked if he was indeed a bootlegger, ''and some of our best judges use my stuff.'' He'd point out that Cook County harbored some seven thousand saloons when the Eighteenth Amendment passed, spent about $70,000,000 a year for beer and booze, and voted five to one wet. ''Well,'' Capone would declare, ''you can't cure a thirst by law.'' He neatly blurred any doubts about his patrons' moral supremacy. ''They call Al Capone a bootlegger,'' he'd say. ''Yes, it's bootleg while it's on the trucks, but when your host at the club, in the locker room, or on the Gold Coast hands it to you on a silver tray, it's hospitality.'' What did he do that was so awful? He ''supplied a legitimate demand. Some call it bootlegging. Some call it racketeering. I call it a business. They say I violate the prohibition law. Who doesn't?'' Technically, of course, the buyer committed no crime; but his moral point made sense.

Few of Capone's other pronouncements on society would have ruffled the most stolid burgher. ''People respect nothing nowadays,'' he'd complain. ''Once we put virtue, honor, truth, and the law on a pedestal. Our children were brought up to respect things.'' But now, ''look what a mess we've made of life!'' He deplored birth control, which sapped America's vitality. He loathed homosexuals. He execrated the flapper's bobbed hair, showcase clothing and brashness. ''The trouble with women today,'' he'd pontificate, ''is their excitement over too many things outside the home. A woman's home and her children are her real happiness. If she would stay there, the world would have less to worry about the modern woman.'' He deplored the twenties' loosened moral climate with a vehemence that displayed more than mere chagrin over competition for his whorehouses. As for the end of tolerated vice, he said, ''Reform did not end prostitution.'' It spread it around and made it dangerous. ''Now the girls no longer are inspected once a week by health department doctors. Now they are not concentrated down on the Levee. Instead, they are living in the swank apartment houses, associating with the wives and daughters of the best people in town. They simply went underground.''

Capone had transformed himself from ''Scarface Al,'' the ineffably bad, into a popular public figure of romance. To those he favored, he was ''Snorky'' (slang then for ''stylish'' or ''up-to-date''; one member of the outfit was told it meant ''gorgeous,'' and thought people used it to butter up the boss). Capone had become a tourist attraction. Busloads would drive by the Hawthorne, known as ''The Capone Castle,'' and the Metropole, agog for a glimpse of anything or anyone connected with the man.

Even in routine movements about the city he lived up to billing, a dazzling show in his seven-ton rolling fortress, a flivver scooting a half block ahead like a pilot fish, a consort sedan trailing, both packed with bodyguards. The cry "There goes Al!" would rise and the curious would line the curb, rewarded by a vision: Capone lolling in the back seat snorkily decked out in one of the twenty or so $135 custom sack suits he ordered each year. (At this time, Maurice Rothschild in his upscale State Street store was glad if his top-of-the-line $85 suit fetched $50 on sale; a full-length raccoon coat went for $195.) Snorky favored green and canary yellow; his custom shirts (ordered by the dozens) cost up to $27, collars and sleeve monograms extra; his tie and handkerchief of course matched (he once ordered twenty-eight sets, at about $7 apiece). Under the ensemble he snuggled in underwear knit of Italian glove silk ($12). On his head sat one of his wide-brimmed signature hats, a *borsalino* the hue and texture of Grade A cream. As he acknowledged the plaudits with a wave of his fat black wand of a cigar, the crowds might catch a gleam of his $50,000, 11.5-carat Jagersfontein diamond ring.

For strolls around town, out from behind the armor plating, Capone had a *real* armed entourage. As many as eighteen bodyguards surrounded him, depending on location and current state of gangland tension. Routinely, four patrolled ahead, four guarded his rear, two his flanks. At the theater or opera he always bought extra seats so bodyguards surrounded him and his guests. One patron, seated directly behind Capone, nervously calculated that if the human shield should prove necessary, *he* sat in the line of fire.

The menials in Capone's organization—bodyguards, drivers and the like—could count on salaries that ranged from $100 up to $500 a week.

Naturally, this opulence and payroll required a princely income. Despite the turmoil of 1926, U.S. Attorney Edwin Olson estimated Capone's gross for that year at $70,000,000. A newspaper pegged his personal net at $3,000,000. Before vigilantes burned it, the Stockade alone turned about $100,000 a month. Records the government seized showed that a single Capone gambling joint, the Hawthorne Smoke Shop, cleared a $587,721.95 profit in just under two years. With peace and Capone's ascendency, 1927 receipts exceeded $100,000,000 according to a later Internal Revenue estimate. Sixty million of it came from beer and booze, beer by far the larger contributor: Capone claimed that in the lushest months under Thompson's benign eye, he was lucky to move five thousand cases of booze; with overhead, payoffs, raids, bail bonds and such like, the net was maybe $10 a case for a trifling $50,000 a month, $600,000 a year. Twenty-five million more of the total came from gambling, $10,000,000 from whores, dance halls and roadhouses, $10,000,000 "miscellaneous," including labor racketeering and enforced tribute from other gangs.

Of course, expenses ate up a lot of the gross. By 1929, Capone would claim to be paying $30,000,000 a year just in political and police protection.

No one knew—probably including Capone—exactly how much he raked off as his share. As noted, the government could persuasively *detail* $218,057.04 for 1927, and that could scarcely have accounted for a tenth of reality.

He needed every penny to support two outrageously expensive habits, gambling and generosity. ''Al,'' said one observer about both, ''is really a bigger boob than any small towner who comes in from the sticks to do Chicago in one night.'' He often carried a $50,000 whip-out wad, ready for action. He loved shooting craps, never for less than $1,000 a pass except to spare embarrassment to friends not affluent enough to fade such bets. On occasion he'd bet $50,000 or $100,000. He lost with stupefying regularity.

Fifty or a hundred thousand might also go on a horse's nose. He had started plunging on horses in 1924 when he got to know some jockeys and trainers at the Hawthorne track. With their tips and inside information, he beat the books for over $50,000 in one week and immediately indulged his second habit of munificence, throwing lavish dinner parties for the trackmen. His luck (or their accuracy) changed, and he dropped about $200,000 by the end of the meeting. But they remembered his generosity, and at the next meeting he *rocked* the books for a half million in a few weeks' time. He was hooked.

With a select few he would generously share even near-sacred inside tips about fixes. One day he spotted a young news photographer he liked at the finish line. ''Hey, kid,'' Anthony Berardi remembers Capone saying, ''why don't you bet number five?''—a dog of a steeplechaser that had never run on the flat; the morning line had it at ninety-nine to one. ''How the hell can this horse win?'' Berardi said. ''Forget it.'' Before the start, one of Capone's gunmen strode up and tucked a $5 win ticket in Berardi's pocket. The horse went off near sixty to one, and Berardi, pocketing over $300, kicked himself for not having *listened.*

But fixes, tips and inside dope could not keep pace with Capone's gambling urge, and his handicapping, like his craps shooting, was horrible. He invariably ended each meeting, as then-current slang put it, ''with the bag over his head.'' He unfailingly paid the bookies what he owed. In the spring of 1927 he told friends, ''I've lost a million and a half on the horses and dice in the last two years. And the funny part is, I still like 'em, and if someone handed me another million I'd put it right on the nose of some horse that looked good to me.'' Two years later he told a friendly newsman that he had ''fooled away'' $7.5 million since he had arrived in Chicago. No one who heard it doubted the figure.

Capone's parties for the horsemen were typical. His parties generally achieved Saturnalia. At their most demure, he and a dozen henchmen might wander into a cabaret or jazz joint—one he owned or someone else's—have the doors closed, and announce that drinks were on him for the rest of the night. He usually tipped the waiters and musicians $100 and never gave hat check girls less than $10. He and his men would sit in a corner, one musician recollected, ''very gay and noisy,'' but with the bodyguard continually scanning for danger. ''Al's big round face had a broad grin plastered on it and he was always good-natured,'' the musician noted, ''which didn't annoy me at all.''

Capone also might take over a joint for a few hundred of his friends and associates, tossing $5,000 to the owner, keep the change. Banquets at the Metropole could last two days and cost $1,800. Once he threw himself a three-day champagne birthday party, Fats Waller providing the music. When Scalise and Anselmi's new trial resulted in a final acquittal on June 9, 1927, Capone threw them the grandfather of all gangland parties, costing some $25,000, and described by one guest as "a deluge." It featured a climactic shoot-out with corks from bottles of Piper Heidsieck and Mumm's, the champagne forming a pool on the floor. Later in the year, before Jack Dempsey fought the famous "long count" match against Gene Tunney, Capone outdid himself with an all-week, $50,000 gala. This was a peculiarly handsome gesture, since that's the amount he would lose on Dempsey, his pal. He had planned to spread a little more in the right circles to help ensure victory. Dempsey claimed to have dissuaded his number one fan from fixing the fight, pleading that it would be unsporting, whereupon Capone sent a mass of flowers to Dempsey and his wife with a card, "To the Dempseys in the name of sportsmanship."

His benefactions cost him only marginally less than his hospitality, and doubtlessly garnered him even more good will. Before they fatally fell out, Theodore Anton liked to tell of the rainy evening a ragamuffin dragged himself into the restaurant, selling newspapers.

"How many you got left, kid?" Capone asked.

"About fifty, I guess."

Capone peeled off $20. "Throw them on the floor," Capone said, forking over the bill. "Run home to your mother."

Panhandlers, grifters, and especially ex-cons—but really anyone with an authentic or at least plausible hard-luck story—knew him as an easy and open-handed touch. One ex-policeman still remembers the time Capone heard of an old woman turned out of her apartment for nonpayment of rent. Capone instantly dispatched a truck to gather up the crone's valueless possessions, heaped on the curb. He ordered his men to treat the junk like treasure and to cart it to a new apartment he'd paid for. Not that his personal emotions had to be engaged. Merchants in Cicero had standing orders to give the needy coal, come winter, groceries and clothes any time, charging Capone. When the Crash came in 1929, he was first to open soup kitchens. He may have clouted much of the makings from produce companies; maybe some of those Cicero merchants did not get paid in full, either. But years later, when A. J. Liebling visited Chicago, a cab driver still remembered Capone as "a good guy" because he fed so many.

"You can say what you want about Al Capone," says a woman who was sixteen in 1927. "If people were desperate and needed help, he was there to help them. As long as you were on the up-and-up. He didn't expect anything in return, and he never expected you to pay him back." The woman's son, a detective sergeant with the Cicero police force, had absorbed the stories all his life. "My people," he insists, "thought of Capone as Robin Hood." And the policeman's mother echoes, "That's right, just like Robin Hood."

Actually, he was both better and worse. On the creditable side—unless one counts racketeering and extortion from other bootleggers and gambling operators—Capone did not steal. To the contrary, he prided himself on turning burglars, robbers and stickup strongarms into useful servants of public desire. He tolerated no free-lance thuggery. Of course, what he took did not come largely from the rich, but from the middle class and the better-paid workers who could afford a speakeasy stein of beer or a flutter at the tables. He got virtually no action from the poor; they could barely eat. "I never knew what gravy was," remembers Santa Russo Baldwin, one of nineteen children whose father was a laborer. "And steak, I didn't know at all what that was. We had neck bones. We had a lot of spaghetti—with peas, with beans, with oil. We never went hungry; but we had a lot of neck bones." Her father made his own wine. People like that never saw the inside of a speakeasy, let alone a cabaret or gambling joint.

On the other hand, what Capone *gave* to the poor as a regular program (distinct from generous impulses) did not represent any enormous percentage of his take. He was not in the Robin Hood business.

His personal business, especially after 1927, was the administration of a huge and growing empire, using the "real gift for organization" one critic admired. Visitors to his Metropole office would usually discover him in shirt-sleeves, desk buried in papers, nine phones jangling, conversation continually interrupted by urgent requests for instructions, decisions. One time Tony Berardi secured an appointment to take another photo. Berardi often boxed in the Thursday-night amateur bouts at Sylvio Feretti's gym, one of the reasons passionate fight-fan Capone liked him. Before Berardi could start shooting, Capone looked up from the pile of work and said, "Tony, you think you can lick me?"

"Al, for Chrissake, I came here to get your picture, I didn't come here to fight you. But you want to find out, let's go to Sylvio's and put the gloves on."

"Kid," sighed Capone with palpable regret and a grimace at his laden desk and at the three or four men waiting to claim the boss's attention, "I don't have time."

Those men hugely admired their Big Fellow. His cachet marked them, too. They felt like an elite just for having been recruited. "Capone hires nothing but gentlemen," boasted Harry Doremus, one of the ad hoc gentry. "They must be well dressed at all times," Doremus said. Also well-spoken, "yes, sir" and "no, sir," ready to their tongues without notably thuglike growl-accents. "He hires his men with great care and takes pains that they are of his own type in dress and conduct."

Birdseed aside, that's how they thought of themselves and of Capone. They gloried in his power, and chorused examples of it. Like the time he had one of his clerks get a judge on the phone. Snatching the receiver, without preamble he uttered the name of a law officer who had caused some trouble, then snapped, "I thought I told you to dismiss that fellow." The clerk heard some squeakings on the other end; Capone bellowed, "Forgot it? Well, don't forget it again!"

Out of the office, when he moved among the people, Capone was gracious, a model for his gentlemen. "He always tipped his hat to us," says a woman who grew up in a neighborhood Capone frequented on inspection tours of alky cooking, "and he was always polite." She got to see him in the gold-rimmed spectacles he sometimes wore, but in which he never permitted himself to be photographed.

His politeness included police. Joseph A. Refke was a rookie in 1927, the first time he saw Capone. "Hey Joe," said his partner, pointing to an enormous parked Cadillac as their patrol car rolled by, "that's Al Capone over there." Refke thrust his hand out the window. "Hello, Al!" he called, waving furiously. "He didn't know me from Adam," Refke recalls, but Capone waved back, flashed a huge smile, "Hi, officer!" He seemed almost like a police buff in his feelings for street cops. "I got nothing against the honest cop on the beat," he once said. The ones you couldn't buy, "you just have them transferred someplace where they can't do you any harm. But don't ever talk to me about the honor of police captains or judges. If they couldn't be bought, they wouldn't have the jobs."

He had a point. Once, a squad of young overreachers, acting on a tip that an escaped prisoner was at a certain South Side address, burst in on the headquarters of a Capone client gang. The escapee wasn't there, but many gang members were. When they heard the raiders they threw their weapons in a pile. The policemen proudly bore the confiscated artillery to their captain.

"Who gave you such orders?" he roared. "Take that stuff back." In those days, the standard internal threat to erring policemen was: "One more like that and you'll be transferred to Hegeswich"—a precinct so barren and forlorn Leopold and Loeb left Bobby Franks there, assuming no one would stumble upon his body. The eager-beaver policemen figured they were halfway to Hegeswich. Desperately, they went to make their peace with the Big Fellow. "I understand your captain wasn't to blame," he said, "and that you boys just made a mistake. After this, don't pull another boner."

Incidents like that confirmed his men's awe for Capone. What's more, he took care of them. One who was shot in an intergang scuffle glowed when Capone sent him, all expenses paid, on full salary, for a six-week recovery in Miami. Even when he reflected that the purpose probably was to keep him out of the way while he was still hot, the gesture sparkled in his memory.

Capone's very excesses helped him win the outfit's admiration and affection. Unlike the reproach implicit in Torrio's asceticism, the gusto of Capone's debauches gave license to his men's preferred amusements. And he outdid them in all. His gambling would have awed Nick the Greek. The wife of one gang member recalled seeing Ralph trying to stuff Capone, inertly drunk, into a car—an unremarkable sight, despite the line Capone had handed Robert St. John about selling, not drinking the stuff. As for sex, despite his abiding affection for Mae, Capone was routinely unfaithful. At the Roamer Inn, one of the earliest Torrio-Capone suburban roadhouse brothels, romance was visibly budding between a resident whore named Marcelle and a musician until a bartender blighted it. He told the musician, "I'd lay off if I was you, bud—that's Al Capone's girl." Not that Capone lacked generosity in this area, too, if personal property was not involved. A young caddy at the Burnham Golf Course once heard talk of some kind of party Capone was throwing for his boys at the clubhouse that night, an "orgy" they called it. Curious as to what that word meant, the caddy skinned up to a second-story balcony. Through the window he saw about twenty couples, most of them nude, Capone fully clothed, standing to one side, the gracious host, just laughing and enjoying his guests' pleasure.

Capone's standards and conduct differed drastically at home (except for lavishness: a later tax issue would be where he got the money to pay for, among much more, about $1,500 worth of Chinese rugs delivered to his Prairie Avenue home in November 1927). With his family he turned determinedly domestic. Early in 1927 he met a reporter at the door, decked out in a pink apron, carrying some spaghetti sauce he'd been cooking. He liked to lounge in bathrobe and slippers, playing games with his idolized little Sonny (whom the family also called, sometimes, "Jiggs").

Capone's sister, Mafalda, twelve years younger, adored him. He sent her to the Richards School, an expensive and exclusive institution for young ladies. At Christmas he showered students and faculty alike with gift baskets. Mafalda would later complain about all those mean headline stories that traduced her brother. "I wish they would say how good he is. Everyone who really knows him says it."

Even with his brother John, only two years his junior, he played the protective paterfamilias. (Though still only twenty-eight in 1927, Capone looked, one observer thought, fifteen years older, and had an air as if—Capone himself often said—he had "lived a thousand.") "Mimi," as the family called John, was a scapegrace, fined for disorderly conduct at age eighteen, trusted with only the most scantling jobs. One was escorting beer deliveries to the suburban cabarets and whorehouses. At Burnham's Arrowhead Inn, he fell in love with Lillian, a singer. Capone thought it a bad match and knew how to break it up. "Fire that girl," he ordered the bandleader, Milton Mezzrow. "Get her out of here. If I hear any more stuff about her and Mimi you're booked to go too."

"Mezz" Mezzrow, one of jazz's near-greats, was new to the job of bandleader and took it seriously; besides, he had hired Lillian. He said, "I won't fire her. She's one of the best entertainers we got around here. Why don't you keep Mimi out of here, if that's the way you feel about it?"

Well . . . "She can't sing anyway," Capone grumbled.

"Can't sing! Why, you couldn't even tell good whiskey if you smelled it and that's your racket, so how do you figure to tell me about music."

The five or six gunmen with Capone started to laugh and Capone positively guffawed. "Listen to the Pro-fes-sor!" he singsonged to his men. "The kid's got plenty of guts." But, he added, he better not catch Mimi fooling around with the girl anymore.

Such evidences of fair-mindedness added to Capone's stature in his men's eyes—though jazz musicians could get away with much, like boxers, respected and indulged as pets by most gangsters.

With Capone at his peak, firmly supported by the largest, toughest, best-organized aggregation of muscle among the gangs, with complaisant Bill Thompson mayor, 1927 should have been Capone's wonder year. He should have appeared unassailable. Instead, from the spring on, he faced repeated attempts on his life and ended the year in exile.

CHAPTER 17

Insanities

JOSEPH AIELLO COMMISSIONED the attempts on Capone. Dean of a far North Side clan of nine brothers, with numberless cousins, Aiello led a gang composed of family, hired guns and the detritus of the Genna gang.

How could Aiello have hoped to displace the Big Fellow? "Those guys," says Mark LeVell, indefatigable researcher into the ways of Chicago's gangland, "thought if you stick a forty-five under your arm you can conquer the world. As long as you have a gun you have a chance." As with O'Banion, the issue turned less on money than ego.

The start of Prohibition had found Aiello a partner with Tony Lombardo, already successful in the produce business. They quickly made a fortune selling sugar and other alky-makings to the Gennas. By 1927, Aiello had snared a respectable share of the bootleg trade, including a fragment of the Genna's alky-cooking business. The family also owned a large commercial bakery. Aiello lived in a three-story mansion in tony Rogers Park, built to order. He was doing fine.

But Capone had elevated Lombardo to presidency of the Unione Siciliana, for which post Aiello considered himself the prime candidate. Aiello couldn't abide his partner's glory; they quarreled and split up. And when he thought he stood a chance, Aiello came after Capone, author of his disappointment.

Aiello offered the chef at the Little Italy, one of Capone's favorite restaurants, $10,000 to spike Capone's food with prussic acid. The chef told Capone, who said, "If I had known what I was stepping into in Chicago I never would have left the Five Points outfit."

Aiello offered $50,000 to *anyone* who would kill Capone. Starting in May of 1927 at least four hoods came to try for the bounty—one each from New York and Cleveland, two from St. Louis. But Capone's intelligence network blanketed the city; nearly every waiter, bookie, street hustler, cabbie, newsboy, grifter—and most policemen—were his spies. The hoods barely hit town before Capone knew it and they hit the pavement, dead. The first, New Yorker Antonio Torchio, was found May 25, 1927, shot five times. Three days later, two hundred machine gun slugs shredded the Aiello bakery, wounding a baker and brother Tony, who had recently been acquitted of killing Joseph Nerone, "the Cavalier," killer of Tony Genna.

When Aiello ran out of imported talent, he tried old Genna killers—with the same result. Corpses piled up throughout the summer, ten in all. A myth flowered that some were found with nickels pressed into their palms, the killer's valuation of his victims' worth. A. A. Dornfeld, long chief of Chicago's City News Bureau, insists that he never ran across an authenticated instance. The myth probably started because of the history of the man largely responsible for the deaths, "Machine Gun" Jack McGurn, one of Capone's favorites. He was the nearest thing in the organization to Scott Fitzgerald's description of Gatsby: "an elegant young roughneck."

Jack McGurn was not his real name; neither were his other two supposedly real names. Born 1903 in Licata, Sicily, Vincenzo Gibaldi was brought to Brooklyn by his parents, Thomas and Josephine, at age one. When Vincenzo was a young adolescent, two members of the Irish White Hand gang blew his father apart with shotguns in a case of mistaken identity, Thomas looking fatally like Willie Altierri, a Frankie Yale gunman. Thomas had just finished getting a shoeshine when Bill Lovett's men struck. He held three nickels to pay and tip.

His widow married grocer Angelo Demory and moved to Chicago when her son was in his mid-teens. Vincenzo Gibaldi grew to a trim welterweight, boxing successfully as an amateur but less so when he turned pro. His manager dubbed him Jack McGurn, and eventually advised him to quit, charging he lacked heart in with the tougher opponents. The lack was not evident outside the ring.

He had been teaching himself to shoot. Just after his nineteenth birthday he returned to Brooklyn, killed the two White Handers who had pulled the trigger on his father, and just missed getting Wild Bill Lovett, wounding him severely. He put a nickel in his victims' hands, tossed one at the wounded Lovett. The three nickels in his father's hand had "paid off" his killers.

In 1923, gunmen murdered his stepfather—this time on purpose; Demory had been tied in with the Gennas. Gibaldi got those killers, too. By then he had joined Torrio and Capone. As "Al Brown" and many other criminals did to frustrate identification and protect the family name, Gibaldi adopted an alias. He changed two letters of Gibaldi to become Gebardi. Others in the family also called themselves that, though some used Demory or its alternate spelling, "Demora." For Vincenzo, though, his ring name stuck, newspapers making it "Machine Gun" Jack McGurn (though in fact he favored pistols) when he became notorious for at least twenty-two killings. Thereafter, many would explain that his *real* name was Demory or Demora . . . or maybe Gebardi.

Capone doted on McGurn, who indeed cut an attractive figure. Except for a large and shapeless nose, he was handsome, with sleeked-back hair and soulful eyes. He was a spiffy dresser, a sinuous and athletic dancer who could beguile a flapper with his Charleston and tango or amaze a barroom with full, to-the-floor splits. He shot tournament-level golf, rode horseback like an Apache, displayed lithe mastery of every other sport he tried. ''McGurn was a *real* nice guy,'' says Tony Berardi, who shot McGurn's wedding picture. ''If you didn't know who he was, you'd want to become his friend.''

The new chief of detectives, William O'Conner, ordered that ''known criminals'' be tested for insanity. First tested was McGurn. ''There is something wrong with McGurn,'' said the tester, though she allowed it might be just befuddlement at the questions asked. Or, as an attendant detective remarked, ''If he isn't nutty now, he soon will be.'' Frank McErlane wriggled out of the psychiatrists' grasp on a writ, and if he was no psychopath, neither was Jack the Ripper. When Dingbat Oberta showed up waving a certificate of mental health from a private psychiatrist, the program died.

O'Conner should have tested Joe Aiello, who kept at it, fantasizing that he had strong support from stalwart allies. Moran's gang did back him, but was in tatters. Aiello also counted on a pact with Barney Bertsche, Jack Zuta and Billy Skidmore, all chafing at having to cut Capone in. But not unlike that greater racketeer, Napoleon, Capone was adept at isolating enemies so he could defeat them seriatim. He did it to Aiello. Since Moran was both intractable and relatively powerless, Capone did not waste time on him. He invaded the near North Side headquarters of Skidmore's gambling and bail bond operations, backed by ten convincing henchmen including Frankie Rio, McGurn and Jack Heinan, the only one of Capone's people ''reputed tougher than McGurn.'' Capone bore only a message of peace—at a price. ''Listen, Billy . . .'' he reportedly said. His quarrel was with Aiello, the man gunning for him. The others would have no problem if they kept their noses clean. ''Either you quit playing ball with the Aiellos—or . . .'' Skidmore promised to relay the message. Capone's opposition never learned to coordinate their ambitions or discontents.

Aiello continued his mad pursuit anyway. About November 13, 1927, police Sergeant Bernard Smith, acting on an anonymous tip, raided an apartment in Rogers Park, Joe Aiello's neighborhood. He found thirty-seven sticks of dynamite plus a list of names and addresses which led to a room in the Rex Hotel on North Ashland (not the one on South State, headquarters for Capone's whoremaster, Duke Cooney). At the Rex, police surprised Angelo La Mantio and four other Aiello mobsters with three rifles and a box of ammunition. La Mantio's pockets disgorged rent receipts for an apartment way west on Washington Boulevard, and a key to room 302 in the Atlantic Hotel at 316 South Clark Street.

The apartment on Washington faced number 4442, home of Tony Lombardo; it was a machine gun nest. In the otherwise empty Atlantic Hotel room, police found rifles clamped by vises to stands at the window, binoculars lying beside them. The rifles were trained on the entrance of a cigar store across the street, at 311 South Clark. It was owned by Michael Kenna. Hinky Dink's store remained a focus of Chicago politics, and Capone appeared there almost daily.

At the detective bureau, Capone denied ever having clapped eyes on any of the men in custody. But he reduced La Mantio to a quivering heap with the Look. That, plus a session in the "goldfish bowl," the soundproof basement room for *intense* detective interrogation, convinced La Mantio to admit to Joe Aiello's generalship of the plot. The police brought Aiello in, too.

It was now Capone's turn to act crazy. He sent six cabs full of gunmen to the detective bureau. One policeman saw them through a window and assumed they were detectives bringing in prisoners. But instead of coming in, the group fanned out, some patrolling the streets and adjacent alley, others watching the bureau's front and back entrances from doorways. The dumbfounded policeman called a colleague to come see. Finally three of the watchers made for the bureau's entrance, one of them transferring his .45 from shoulder holster to overcoat pocket. The incredulous policemen recognized one of the three and suddenly realized these were Capone men! A hastily assembled posse sortied out and surrounded the three—Louis Campagna, Sam Marcus and Frank Perry, all armed. Shoved in a cell near Aiello, Campagna started snarling at him. A policeman who understood the Sicilian dialect posed as a prisoner and listened.

"You're dead, friend, you're dead," Campagna said. "You won't get up to the end of the street still walking."

"Can't we settle this?" Aiello pleaded. "Give me just fifteen days and I'll get out of town and stay out."

"You have broken faith with us twice now," came Campagna's answer. "Now that you have started this, we'll finish it."

Upon release, Aiello asked O'Conner for an escort. "Sure I'll give you police protection," the chief of detectives replied, "all the way to New York and onto a boat." But police did see Aiello safely away in a cab with his wife and little boy, who had appeared with his lawyer to post bail. The next day his lawyer presented the court a doctor's certificate excusing Aiello's failure to show up for arraignment on grounds that he was suffering a nervous breakdown. Capone's men believed it.

Joe, Tony and Dominic Aiello cleared out of town, going to ground in Trenton, New Jersey. Capone announced that he accepted Joe's plea for peace. "I'm willing to talk to anybody any place to bring about a settlement. But I'm going to protect myself. When somebody strikes at me, I will strike back." When Dominic returned the following January, two of Capone's men, Lawrence Mangano and bodyguard Phil D'Andrea, phoned him a warning, then shot up the bakery again for emphasis. Joe Aiello stayed quiet for a while, but was not through. With that .45 still under his arm, he would be back at it a year later.

Meanwhile, Capone looked and felt invincible. He gathered his key people at Lawrence Mangano's Minerva Athletic Club (a speakeasy and gambling joint). "I'm the boss," he told them. "I'm going to continue to run things. They've been putting the roscoe on me now for a good many years and I'm still healthy and happy." No one could chase him out.

This hubris would be immediately punished by the least likely nemesis in town.

Craziest of all was Big Bill Thompson's conception of himself as logical replacement when Calvin Coolidge did not choose to run in 1928. The Omaha *World-Herald* noted that "Mayor Thompson . . . has made no bones about having his eye on the Republican Presidential nomination." He retained just enough grip on reality to understand that he had no hope of national office while he presided over the city of Al Capone, by now symbol, around the *world*, of the law's failure.

The Siege of the Detective Bureau—as newspapers dubbed the try at Aiello—triggered Thompson's action. It had generated national headlines and made Thompson look out of control of his city. Thompson had brought Mike Hughes back as police chief, and Hughes delivered the message: Capone had to leave town. Capone knew that for all the protection money he paid, those he bought did not have to stay bought, and *they* had the real clout anytime they chose to exercise it.

December 5, two weeks after his triumph over Aiello, Capone announced, "I'm leaving for Saint Petersburg, Florida, tomorrow." He had some property there he wanted to sell. "I don't know when I'll get back, if ever," he said. "But it won't be until after the holidays, anyway." Chief Hughes said, "He isn't going to come back."

Capone welcomed the press into his Metropole office for a valedictory and apologia. Just back from an eight-day hunting trip in Wisconsin, stubble on his jowls, he was dressed in an "ultra-nifty hunting suit," having outfitted himself and about twenty companions at Marshall Field's for the trip. "Let the worthy citizens of Chicago get their liquor the best way they can," he said. "I'm sick of the job. It's a thankless one and full of grief."

What had he done to deserve such persecution? "I have never been convicted of a crime," he said accurately, before his imagination took over, "nor have I ever directed anyone else to commit a crime. I have never had anything to do with a vice resort. I don't pose as a plaster saint, but I never killed anyone. I never stuck up a man in my life." No one knew about that Brooklyn crap game. Then Capone moved back to the truth. "Neither did any of my agents ever rob anybody or burglarize any homes while they worked for me. They might have pulled plenty of jobs before they came with me or after they left me, but not while they were in my outfit."

What did his career of rehabilitating hoodlums earn him? Constant public abuse. Being called a killer. Headlines with "Scarface" in them. "I could bear it all," he said, "if it weren't for the hurt it brings to my mother and my family.

They hear so much about what a terrible criminal I am. It's getting too much for them and I'm sick of it all myself.''

He was known everywhere as a gorilla. Recently, a man desperate for $3,000 had told Capone that if he gave it to him, the man would insure himself for $15,000, name Capone as beneficiary and commit suicide. The maniac wouldn't accept ''no''; they had to toss him out bodily. ''Today,'' Capone continued, ''I got a letter from a woman in England. Even over there I'm known as a gorilla. She offered to pay my passage to London if I'd kill some neighbors she'd been having a quarrel with.''

That's what he had to put up with. Why? ''Because I've given the public what the public wants. I never had to send out high pressure salesmen. I could never meet the demand.'' Sure he broke the Prohibition laws. ''Who doesn't? The only difference is I take more chances than the man who drinks a cocktail before dinner and a flock of highballs after it. But,'' he added, as always hitting the moral if not legal truth, ''he's as much a violator as I am.''

What was Capone? ''I've been spending the best years of my life as a public benefactor.'' He showed the public a good time, that's all. ''Public service is my motto. Ninety percent of the people of Cook County drink and gamble and my offense has been to furnish them with those amusements.'' And always on the up-and-up, a point of pride. ''My booze has been good and my games on the square.'' He did not mention prostitution.

But now, good-bye to all that. ''I guess the murder will stop,'' he said. ''There won't be any more booze. You won't be able to find a crap game even, let alone a roulette wheel or a faro game.'' Chief Hughes wouldn't need the extra three thousand police he'd been asking for. ''Say,'' the thought struck him, ''the coppers won't have to lay all the gang murders on me now. Maybe they'll find a new hero for the headlines. It would be a shame, wouldn't it, if while I was away they would forget about me and find a new gangland chief?''

He bore no grudges despite all. ''I leave with gratitude to my friends who have stood by me through this unjust ordeal,'' Capone finished his remarkable performance, ''and forgiveness for my enemies. I wish them all a Merry Christmas and a Happy New Year. That's all they'll get from me this year. I hope I don't spoil anybody's Christmas by not sticking around.''

It might, at that, prove a bleak season for those public servants accustomed to a place on his $100,000 gift list.

Talk of Florida just lent protective indirection to his itinerary. Leaving family at home, Capone entrained for the West Coast with only two guards. He headed first for the races and whoopee of Tijuana, Mexico, then north of the border to San Diego. ''They treated me fine. I was invited out to some big ranches by prominent people, too.'' Next, on to Los Angeles.

He lounged in his suite at the Biltmore, snorky in one of his green ensembles, pinkie diamond flashing, just another tourist. ''This gang war stuff is greatly overdone,'' he said, ''and I get tired of it. I'm strictly against gang wars of any kind and I just want to get along with everybody, but every time there is a gang battle they just seem to deal me a hand whether I know anything about it or not.''

When he noticed a reporter eyeing his two companions, he blandly passed them off as his cousin and a friend. ''They made the trip with me for company,'' he explained. ''All this talk about my having a bodyguard with me is just more bunk. I don't need one.'' He wasn't looking for trouble and didn't expect any.

It found him, anyway, in the person of Los Angeles's toughest detective, ''Roughhouse'' Brown. When the story of Capone's arrival appeared in the newspapers, Brown came by with a message from Police Chief James E. Davis, who said, ''We have no room here for Capone or any other visiting gangsters whether they are here on pleasure tours or not.''

Capone gave up at once. ''I'm just tired of all this rousting around,'' he said. ''If they don't want me here, that's good enough for me.'' But when he reached the Santa Fe Railroad's station, ready to board *The Chief*, Chicago-bound, Capone asked reporters, ''Why should everybody in this town pick on me? I wasn't going to do anything here.'' He and his pals came as tourists. ''I thought that you folks liked tourists. I have a lot of money to spend that I made in Chicago. Whoever heard of anybody being run out of Los Angeles that had money?''

He was pretty burned up about it all, but he'd be back, because he liked the place. ''We all had a fine time today before the police told us that we had to go back home. We went through a film studio. I never saw them make pictures before. That's a grand racket.'' He'd toured some stars' homes. ''I think Mary Pickford's old home was nicer than the one she lives in now,'' the fan confided.

He knew what to do before his next visit. ''When I get a little business done in Chicago I am going to send out a lot of money here and have some real estate man buy me a large house. Then I will be a taxpayer and they can't send me away. Anyway, when the real estate men find out I've got money they won't let me go, even if I want to.'' Piled on the other indignities, someone had stolen a jug of wine he'd laid by for the trip. ''Now I won't have a drink between here and home,'' he said, waving farewell. ''Pretty hard on a fellow like me that didn't mean no harm and only wanted a rest from business.''

When the news flashed to Chicago that Capone was on his way, Chief Hughes said, ''We will have a reception committee at the station to meet him.'' Capone laughed. ''He's always making jokes,'' Capone said. ''I am a property owner and taxpayer in Chicago, and certainly I can return to my home.''

But this jolly mood soon darkened. It was not only a dry trip, but an uncomfortable one. ''Privacy,'' Capone said, ''there ain't any.'' He'd look out the window at every whistle stop and see sheriffs guarding against who knew what diabolical schemes the archcriminal might hatch. ''At Kansas City the police in the station were so thick that some of them were pretending to sell apples.'' He was the one who needed protection. ''People were trying to get in all the time to look at me.''

Mike Hughes had not been joking. His chief of detectives William O'Conner confirmed that police would meet the train and promised that if Capone didn't get right back on, outbound, they'd hustle him straight to the hoosegow. ''The idea,'' he explained, ''is to show Capone that for him the 'no parking' sign is already out.''

''It's pretty tough,'' Capone moped to a reporter as the train crossed Illinois, ''when a citizen with an unblemished record must be hounded from his home by the very policemen whose salaries are paid, at least in part, from the victim's pocket. You might say that every policeman in Chicago gets some of his bread and butter from the taxes I pay. And yet they throw me in jail—for nothing—when I seek to visit my own home to see my wife and my little son.

''I am feeling very bad—very bad. . . .''

He did not intend giving in so easily as he had in Los Angeles. ''I am going back to Chicago,'' he vowed. ''No one can stop me. I have a right to be there. I have property there. I have a family there. They can't keep me out of Chicago unless they shoot me through the head.''

What about reports of Capone's attempt to have the chief of detectives killed?

''You want to know the truth?'' Capone said. ''Why, it's ridiculous. I'm not a fool. Two of my men were picked up. They had guns—for their protection. They were dragged into the chief's office and one of them was punched in the stomach so that the gun in his belt slipped. It was about to fall, and naturally he reached to catch it. So O'Conner calls that an attempt to kill him in his own office.'' Capone threw up his hands in disgust.

Squads of police waited at every Chicago terminal. Just in case, they had posted Daniel Healy—a sergeant since killing Vincent Drucci—at one where the train did not stop. Capone planned to finesse them. He had phoned Ralph to meet the train in Joliet on Friday morning, December 16. The crazy, crude son of a bitch arrived an hour early with two sedans and four henchmen. A deputy noticed them cruising and checked the registration. Joliet police arrested them, their pistols bulging, a shotgun in their possession, as they lounged about the depot.

''You're Al Capone,'' said John J. Corcoran, chief of the Joliet police, when Capone alit at 9:45 that morning.

''Pleased to meet you,'' said Capone.

Then he saw himself surrounded by six of Joliet's finest, shotguns leveled. ''Well, I'll be damned. You'd think I was Jesse James and the Youngers, all in one. What's the artillery for?'' He gingerly handed over a .45 and a smaller revolver, then added two extra clips, saying, ''You may want some ammunition, too. These are no good to me now.''

Shoved in a cell with two fragrant derelicts, he got rid of them by paying their $22 fines from the $2,447 wad the police had found on him. Eight hours later, freed on $2,400 bond, each, the group drove to Chicago Heights for a welcome-home banquet.

''When I come back for my trial,'' Capone had said as he left, ''I am going to make a good big donation to the worthy charities of Joliet. I am not mad at anybody.''

He was back the following Thursday, the whole matter settled expeditiously in judges' chambers. Total concealed weapons fines and costs amounted to $2,601 in local magistrate's and state circuit courts. The latter charge came to $1,589.80.

''Maybe this will be a lesson to you,'' said Circuit Judge Fred D. Adams.

''Yes, judge, it certainly will,'' said Capone with a smile as he whipped out the wad and leafed off a grand and six bills. ''I'll never tote a gun again in Joliet.'' He waved away the $10.20 change proffered by the county clerk, Albion F. Delander, father of a Miss America. ''Keep the change or give it to the Salvation Army Santa Claus on the corner and tell him it is a Christmas present from Al Capone.''

Meanwhile, in Chicago, police surrounded the house on South Prairie. He would be arrested every time he poked his nose out of it. ''We don't want Capone here, so we are trying to make it uncomfortable for him,'' Hughes explained. ''Of course we might let him out if he were going to catch a train that went far enough from Chicago.'' In Italy, coincidentally, Mussolini had just instituted a policy of ''forced domicile,'' house arrest for dissidents, Sicilian Mafia suspects and other state enemies.

Capone knew where power resided. He left town again, this time without fanfare and this time really for Florida.

Police wait for Al to appear outside Capone home on Prairie.
CHICAGO SUN-TIMES

GODDARD COLLECTION

Once allies, the Genna brothers were eliminated when they proved trouble-some and greedy. Angelo (above left) *crashed his car when he was shot; Mike* (above right) *was killed by police; Tony was shot in the back.*

ANTHONY BERARDI

Samoots Ammatuna followed the Gennas.
MICHAEL GRAHAM

Gabby Hartnett and fans: (left to right) *Roland Libonati, Sonny and Al*
ANTHONY BERARDI

GANG LEADERS, Allies and Foes

Terry Druggan
ANTHONY BERARDI

Frank Lake
CHICAGO SUN-TIMES

Joe Aiello
CHICAGO SUN-TIMES

ANTHONY BERARDI

Jack Zuta

Joe Saltis

ANTHONY BERARDI

Frank McErlane

Danny Stanton

Claude Maddox

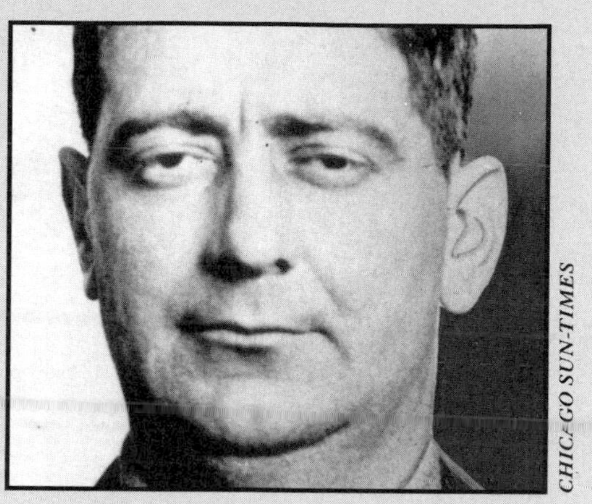

Klondike O'Donnell

CHICAGO SUN-TIMES

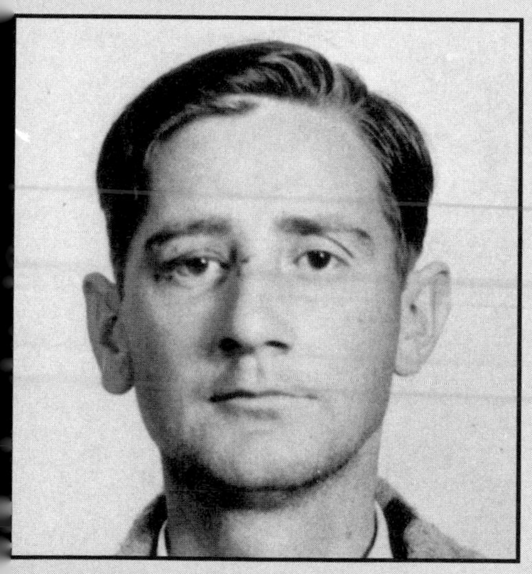

Myles O'Donnell

Hymie Weiss took over the North Side gang after O'Banion's murder. Weiss was said to be the only one Capone feared.

GODDARD COLLECTION

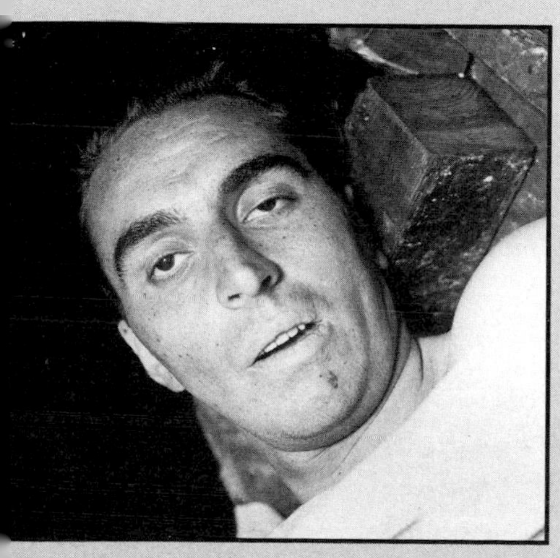

Vincent Drucci succeeded Weiss briefly—until Detective Healy shot him.

George "Bugs" Moran, last of the North Side leaders

CHICAGO SUN-TIMES

CAPONE'S MEN

MICHAEL GRAHAM

Bodyguard Frank Rio

Jack Guzik, business manager, bagman

MICHAEL GRAHAM

Jack McGurn, his shoulder and arm scarred by bullets from the North Side Gusenbergs, with Louise Rolfe

John Scalise and Albert Anselmi, killers

CHICAGO TRIBUNE

Frank Nitti with moustache he grew while trying to hide from a tax indictment.

Tony Lombardo, Capone's choice as president of the Unione Siciliana
ANTHONY BERARDI

Capone and some of the boys: McGurn (seated left), *Rocco De Grazio* (standing left), *Louis Campagna* (standing, second from left). *Tony Berardi remembers all three standing* (right) *as detectives.*

ANGELO DI IACOVA

Bill McSwiggin asking jury for conviction of cop-killers Scalise and Anselmi
ANTHONY BERARDI

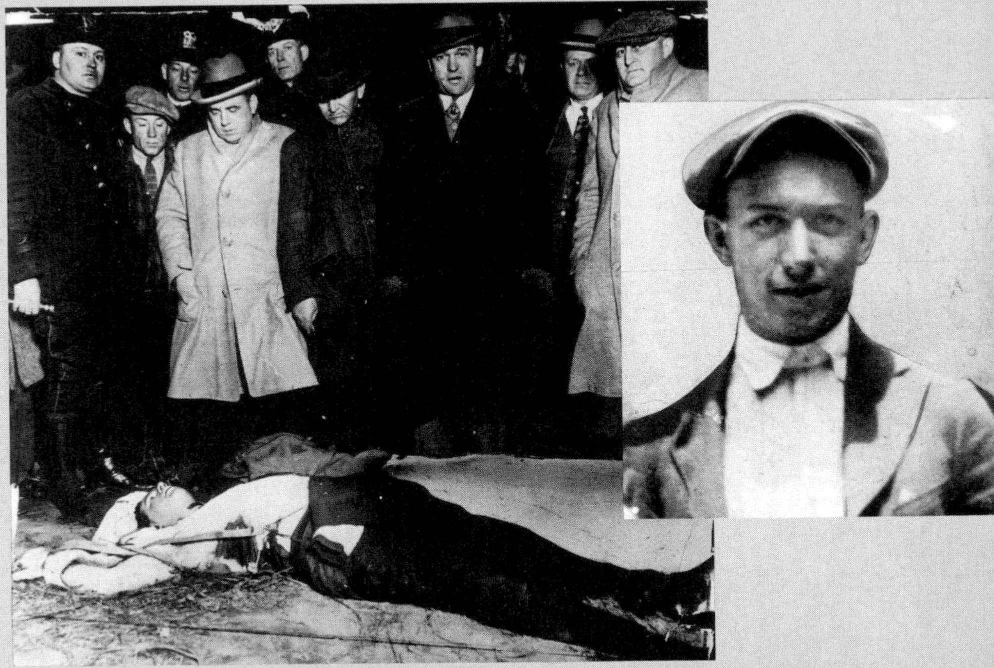

McSwiggin was shot when he went out drinking with hoodlum pals. The O'Donnells's ace gunman, Jim Doherty (inset), *was killed with McSwiggin and Red Duffy.*
CHICAGO TRIBUNE

Palm Island estate
CHICAGO TRIBUNE

CHAPTER 18

Elba—and the First Step to Waterloo

CAPONE CALLED IT ''the garden of America, the sunny Italy of the new world, where life is good and abundant, where happiness is to be had even by the poorest.''

Miami sits on Florida's east coast, a scant subtropical fifty miles from the tip. Across Biscayne Bay lies Miami Beach, nature's breakwater, called by one admirer, the ''Billion-dollar Sandbar.''

Capone loved Miami with its bright midwinter heat, Mediterranean colors and air of wide-open, anything-goes indulgence. Casinos ran without even pretense of concealment; slot machines whirred and jingled in nearly every store and hotel lobby, and police moonlighted as escorts delivering the slots' proceeds. On Miami Beach, the city manager, Claude A. Henshaw, had the concession for giving operators approval. One complained bitterly when private detectives raided and hauled away his gambling paraphernalia. He paid $1,000 a week for protection, he told the Dade County sheriff, and would see someone in jail before he'd stand for a double cross. The equipment was back in place next day. As for Prohibition, when one bootlegger had three hundred cases of whiskey hijacked by sheriff's deputies, he managed to track it to a barn in nearby Hialeah and was loading the lot onto trucks when the local police descended to arrest him, taking the booze to their station house. The sheriff's office graciously ceded two thirds of the loot to its rescuers, selling the one hundred cases they kept to the exclusive Biltmore Hotel in Coral Gables.

This was Capone's kind of town. He wanted a home here, and figured lots of his friends would want to join him. ''Furthermore,'' he said, ''if I am permitted, I will open a restaurant.'' He'd also join Rotary if they wanted him.

Since Miami's chief of police, Leslie H. Quigg, could see no legal way to bar Capone, he merely shrugged, saying, "If he's just here to have a good time, and doesn't start any rough stuff I won't bother him."

With three guards, Capone registered as "A. Costa" at the Ponce de Leon, a downtown Miami hotel whose silver service bore its engraved logotype, a plushness considered locally the acme of class. Capone took the penthouse, suite 804, whose walled, awninged patio overlooked busy Flagler Street. This served nicely as headquarters. For Mae and Sonny, and his more formal entertaining, Capone rented what the government later characterized as a "palatial home," paying $2,500 for the six-month season; it sat on Miami Beach's Indian Creek Drive, the sandbar's ritziest address.

The Miami Beach chamber of commerce vowed to drive Capone out; the Women's Club and an uplift group that called itself the Committee of One Hundred began to badger Miami Beach mayor J. Newton Lummus, Jr., to *do* something.

The city manager and mayor summoned Capone to conference. "We have talked over the situation with him," Henshaw later announced, "and explained that a majority of the citizens do not want him here." The mayor burbled, "Mr. Capone was one of the fairest men I have ever been in conference with." The paragon played along, telling reporters, "If I am not wanted here I will leave immediately. Where I will go from here I have not decided."

That might have posed a problem if he'd had any intention of leaving, Lummus and Henshaw any intention of banishing him. Before Miami, Capone had gone to St. Petersburg, as announced, and police had dogged him; hearing he might buy a house in the Bahamas, the governor declared him non grata. On a scouting trip to New Orleans, brothers Ralph and Albert ended in jail when police found three revolvers on them; released, they were threatened with rearrest if still in town after noon the next day.

Back in Chicago, when Zuta, Bertsche and Skidmore forgot Capone's warning, the outfit bombed their places, with shooting to point up the lesson: Isadore Goldberg, one of Zuta's chief lieutenants, got it. Yet all this discipline barely served to keep the dissidents in line. Sooner or later Capone would have to act decisively.

In Miami, he lay low until sensation over his presence passed. He cultivated the owner-manager of the Ponce de Leon, Parker A. Henderson, Jr., son of a former Miami mayor, a pudgy, susceptible young man with a taste for the louche—the Pet Tiger Syndrome. Capone called Henderson to the penthouse, introduced him to real gunmen, invited him to Indian Creek for a civilized dinner with charming Mae and houseguest Duke Cooney, Capone's chief whoremaster. Henderson was enthralled. Here he was, *pals* with the world's most dangerous criminal! He became Capone's usefully respectable dogsbody.

Eighteen times from January 14 to April 2, 1928, at Capone's request,

Henderson took along Nick Circella—an armed bodyguard—and picked up money orders for "Albert Costa" sent from Chicago to Western Union, a total of $31,000. Henderson had to endorse the checks; he cunningly disguised his handwriting as he signed "Albert Costa." How exciting! What fun!

And what an opportunity. When real estate agents pestered Henderson for an introduction to Capone, Henderson told his friend, Newt Lummus, about it. "If anyone sells property to Capone, you and I should do it," replied the mayor of Miami Beach, who was also a real estate agent. They showed him a number of estates, and in March he was smitten by the one at 93 Palm Island.

Palm was a manmade island in Biscayne Bay, a narrow construct hardly three quarters of a mile long, just off the original causeway (then called "County," now "MacArthur") that linked Miami with its sandbar. Palm Island lay in Miami Beach's jurisdiction. Palm Avenue ran the spine of the tiny island, villas lining the road, north and south, their backs opening onto the surrounding bay. St. Louis brewer Clarence M. Busch had built the fourteen-room, two-story, white-stucco, Spanish-style number 93 in 1922. Its lot had a hundred-foot frontage, and ran three hundred feet back to the bay on the north. A three-room gatehouse stood on the road.

Capone gave Henderson $2,000 earnest money to bind his $40,000 offer to the current owner, James Popham, and then $8,000 to close on March 27, 1928. At the suggestion of Lummus, who could anticipate the better element's wrath, Henderson fronted the deal, signing the notes and mortgage papers; about six months later he made the deed over in Mae's name. Capone ordered a start on improvements that in time would cost roughly $100,000, including the largest private swimming pool in Florida, its filtration system able to handle both sea and fresh water. But he couldn't enjoy his new home just yet; he had to return to Chicago for the April 1928 primary.

Mayor Thompson kept denying he wanted the presidential nomination, and with each denial made another cross-country speaking tour. The primary was crucial. With three years to go as mayor, Thompson was not running, but to prove his clout before the summer nominating convention, he had to demonstrate that his endorsement could shoo in to reelection his principal allies: Bob Crowe as state's attorney, Len Small as governor, and Frank L. Smith as U.S. senator. (The latter had been elected before, but the Senate had *twice* refused to seat him on grounds that the election had been corrupt.)

Thompson and his candidates had never lost a primary and had concluded a pact with Senator Charles S. Deneen, head of the only other major Republican faction. But Bob Crowe insisted on running his friend, Bernard Barasa, for a place on the Board of Review, a tax-setting plum. The incumbent, Edward R. Litsinger, was one of Deneen's few supporters in any Cook County office. That meant a primary battle despite the pact; Deneen ran candidates for all offices.

While no one ever fixed specific responsibility for the bombings, the first four targeted Thompson men. Explosions rocked the homes of Charles Fitzmorris—once Thompson's police chief, now city controller—and Dr. William H. Reid, commissioner of public service; next came John Sbarbaro's mortuary and the apartment building of Lawrence Cuneo, Crowe's brother-in-law and secretary. No one was hurt, but the mayor and state's attorney got police guards around their homes.

With less than a month to go, the Deneen faction looked already routed. Thompson commanded patronage, a force of 100,000 campaign workers, and a faction united behind his campaign vaporings about no World Court, America First, flood control and King George. It got worse for Deneen. On March 14, 1928, his supporter, Joseph H. Haas, county recorder, succumbed to pneumonia at age seventy-one, and control of the six hundred jobs Haas had in his gift reverted to Thompson. One week later, March 21, someone murdered Diamond Joe Esposito, Deneen's most effective street politician and the opponent of Crowe supporter Joseph Savage for Twenty-fifth Ward committeeman.

Some eight-thousand braved a storm to watch Esposito's Mount Carmel funeral (with full rites) on March 26. That night, about 11:20, a blast ripped Deneen's home. Five minutes later, another tore into the home of Judge John A. Swanson, Deneen's candidate opposing Bob Crowe for state's attorney. The bombs wrecked the homes but hurt no one. Deneen was on a train to Washington at the time. Swanson, however, only narrowly escaped, having just driven past the spot where the bomb exploded.

Of course, Thompson's side *had* been hit first. And bombings had become almost a given of Chicago life. These numbered the sixtieth and sixty-first in about five and a half months ("I see you got out alive," a Washington hotel clerk remarked when he saw a traveler register from Chicago); eight insurance companies refused coverage on halls that Swanson wanted to rent for rallies, and a church turned him down unless he would personally indemnify the building. So the city might have dismissed this latest incident. But Crowe blundered. His faction offered rewards totaling $65,000, and he predicted that the bombs would prove to have been set by the Deneen side itself. "After having bombed the homes of friends of mine," Crowe said, "and having made no headway, they are now bombing their own homes in an effort to create the impression that the forces of lawlessness are running the town." Thompson enthusiastically embraced Crowe's theory.

"That's plain poppycock," Judge Swanson replied. "Why, this was within two seconds of being a direct attempt on my life. The bomb dropped fifty feet away from me, and it is ridiculous to think any of our people would endanger the life of one of our candidates. . . ."

In fact, Chicago did think the forces of lawlessness were in charge, and had grown sick of it, especially since Thompson and his police chief, Mike Hughes, had so recently and loftily proclaimed victory over crime, supposedly redeeming Thompson's campaign pledge to "drive the crooks out of Chicago in ninety days." The cynical manipulativeness of Crowe's statement was too much. At a rally, Deneen candidate Litsinger said Thompson spent $243,000,000 to run Chicago. "And what are we getting?" he asked.

"Bombs!" the answer reverberated. "Pineapples!"—slang for tossable bombs, like hand grenades. The newspapers dubbed it "The Pineapple Primary," a name that immediately caught on. Litsinger's opponent, Bernard P. Barasa, became Bernard Pineapple Barasa. Chicago had again amused, amazed and scandalized the nation. President Coolidge had stationed Marines in Nicaragua to protect property owned by U.S. corporations from rebel Augusto César Sandino. "It seems that American property," said Senator George W. Norris, "is safer in Nicaragua than it is in Chicago."

Despite the exile, Thompson remained Capone's candidate, and he did his considerable best for Thompson's slate, personally directing his field forces in their usual election duties. In places Capone controlled, the Thompson ticket squeaked through. In Cicero, Crowe won 5,180 to 4,923, not vaguely enough. He lost the state's attorney nomination to Swanson 265,371 to 466,598. Pineapple Barasa lost to Litsinger 307,941 to 417,527. Even Thompson's choice for sheriff lost—to a political unknown! Statewide, Governor Small and the seatless Senator Smith both lost big.

Thompson, crushed, sulked in Wisconsin all summer, leaving Corporation Counsel Samuel Ettelson to run the city. The mayor had said before the election he would definitely resign if Swanson won. "Well," he grumped, when asked about that, "I'll say definitely now that I'm not getting out." On Ettelson's advice, he did replace blowhard Mike Hughes as police chief with William F. Russell, who as deputy chief had once explained why he did not move against the numbers racket in his territory: "Mayor Thompson was elected on an 'open town' platform. I assume the people knew what they wanted when they voted for him."

Comment across the country, even in Europe, hailed the primary as the birth of "moral Chicago." They overlooked the character of its midwife. Anton Cermak, who succeeded as Democratic boss when George Brennan died in August, was a much more efficient and imaginative boodler than Thompson.

The heat naturally came off Capone, since Thompson's presidential hopes were no longer even a joke. But Capone hurried back to the sunny Italy of America anyway, eager to supervise refurbishment of his Palm Island home. While Capone busily ordered furniture and furnishings, crews of contractors, workmen and landscapers labored. Capone added a new dock for his speedboats, new garages, a boathouse, mosaic walks, rock gardens, fountains. He paid for the best, and did his best to see he got it. One crew of tile layers, working outside, set their lunch pails to the side only to find them missing at noon. A Capone gentleman called from the house, "C'mon in." The workers entered to find an enormous feast spread for them, a treat repeated each day. "Mr. Capone wanted," one of the workers said, "to make sure we were treated like kings because he wanted a good job on his house.'' It was practically a policy. Capone's first employee at the house was handyman-butler Daniel J. Brown, ''Brownie'' to everyone. ''Mr. Capone treated me good,'' said Brown long after the fact. ''If there were more people like him there wouldn't be so many poor people.''

Capone got the results he wanted and was engagingly proud of his accomplishment. ''Honestly,'' Capone would ask visitors, ''how do you like my place? It's all my own work. It was only a bare home when I got it and I really supervised all the improvements myself.''

Capone could afford the time because he had left the outfit in capable *and* loyal hands. Frank Nitti, onetime barber and fence, had emerged as Capone's chief line operating officer and tactician. Jack Guzik remained his primary numbers man and business strategist. Ralph Capone, Lawrence Mangano and Charley Fischetti had charge of beer distribution; Joe Fusco specialized in liquor, which would remain his post-Prohibition field. Burnham ''Boy Mayor'' Johnny Patton oversaw brewery operations and political fixes (Guzik functioning as bag man). Frank Pope, with Anthony Volpe (known as ''Mops''), ran gambling, concentrating on handbooks, while Peter P. Penovich, Jr., managed the floating casinos, James V. Mondi overseeing the Capone take from independents, with Hyman Levin (known as ''Loud Mouth'') as chief collector. Duke Cooney had charge of brothel operations. George Barker (''Red''), William J. White (''Three-Finger Jack'') and Murray L. Humphreys (''The Camel'') became labor specialists, Humphreys at one time running sixty-one locals.

Jack McGurn headed the outfit's gunmen, along with Frank Milano; Frank Rio (who often used the name Cline) and Frank Maritote (Diamond) were chief bodyguards, along with sharpshooter Phil D'Andrea. James Belcastro and Joseph Genaro led the bomb squad, always ready to add a convincing closing argument to a sales pitch. Some of the rising young stars were Sam Hunt (called ''Golf Bag'' because that's where he once secreted a shotgun), Anthony Accardo (his usual alias ''Joe Batters''), Joseph Aiuppa (''Joey O'Brien''), Sam Giancana (''Mooney''—who later reputedly shared a girlfriend with John F. Kennedy), Paul Ricca (''the Waiter,'' his real name Felice de Lucia). All would become forces in organized crime after the Capone era.

In one way the outfit was not monolithic, but a conglomeration of entrepreneurs. Even at the most minor executive level, members owned hotels, speakeasies, restaurants, cabarets, and interests in various rackets, the outfit a profit participant. Only the mass stayed on salary as hired help, but they too could aspire to greatness. One $500-a-week driver-gunman hankered after a handbook to run, certain his take would zoom to $3,000 a week.

Times continued lush. A New York Stock Exchange seat now sold for a record $335,000, and volume mushroomed in a raging bull market. The nation's stock frenzy left Capone unmoved; he gambled compulsively on nearly anything else, but labeled the stock market ''a racket.'' Yet except for inflated beer and booze prices, good living stayed relatively cheap. One person could luxuriate in Chicago's lakeside Drake Hotel as a permanent guest for $100 a month, a family of four for $297.50; the table d'hôte dinner in its Italian Room cost $1.50. Smokers got two packs of Old Gold, Lucky Strikes or Chesterfields for a quarter in the Loop; an evening of roller skating cost 30 cents for ladies, 40 cents for gents. Model A Fords, replacing the model T, ran from $385 to $570.

Capone did not expect good times to last, as is, for the outfit. Not that he foresaw the '29 Crash or the Depression. But unlike many who were resigned to Prohibition as a fixture (Clarence Darrow asserted in *Vanity Fair* that the Eighteenth Amendment would never be repealed), Capone accurately predicted it would end within four or five years. The outfit would need new business.

The prototype already existed in their labor racketeering. That had never been a matter of gangsters taking over free associations of laborers. Most unions were "owned" by gangsters who either had helped organize them, like Big Jim Colosimo and his white wings, or had muscled in and taken them over. The racketeer owners paid benefits, negotiated, and called strikes for higher wages and better conditions. But they also bled off much of the dues, pilfered pension funds, and extorted money from employers to avert strikes or for sweetheart contracts.

The membership generally stayed docile. "The members," Capone once explained, "will always vote for the loudest talker, the guy who promises them the most in the way of more pork chops. You give it to them with one hand and you take it away with the other. As long as their take-home pay is higher this year than it was last year they don't care how much you take from them in the way of dues." They considered dues the same as taxes, said Capone, "just something you got to pay the thieves who run things." Dissidents were first offered executive posts; if that didn't shut them up, they got a beating. "When they get out of the hospital," said Capone, "if they still want to squawk you get rid of them."

Business racketeering would be another area for growth. It was no Capone innovation. As noted, the *Tribune* and the Hearst papers had employed circulation sluggers before Prohibition; garage owners hired members of a West Side gang to puncture tires, compelling car owners to seek monthly parking. Maxie Eisen exacted tribute from most West Side food stores and peddlers. Capone had seen the "protective associations" back in Frankie Yale's Brooklyn.

In late 1927, the Employers' Association of Chicago issued a list of twenty-three businesses manipulated by racketeers. They included the expected, like laundries, shoeshine parlors, garbage collection, fish, poultry and meat retailing, grocery stores and delicatessens; but racketeers were also extorting from doctors, photographers and dental laboratories. By 1929 the list would grow to ninety-one, and one observer guessed that at least 70 percent of the rackets were controlled by Capone or close affiliates.

Strangely enough, Capone had gotten into one business honestly and by invitation. Back in May 1927, Morris Becker had been approached by a representative of the Master Cleaners and Dyers Association, a racket run along classic protective-association lines, making up to members their dues, assessments and other exactions by the enforcement of high prices. The Employers Association estimated that by 1929 such extortion-driven prices, passed along to the public, cost Chicago $130,000,000 a year; two years later the Chicago Crime Commission put the figure at an even $200,000,000.

On the South Side, only Morris Becker held out against the Master Cleaners. After forty-two years in the business, he could turn a tidy profit doing men's suits at $1.25, dresses at $1.75, versus the Master Cleaners enforced schedule of $1.75 and $2.25, respectively. In May 1927, the Master Cleaners got around to Becker's case. They sent in Sam Rubin. When investigators asked Rubin how he, who had never worked in the industry, could function as head of the Master Cleaners' captive retail union, he answered, "I am a good convincer."

"Oh," Becker said when his plant foreman introduced him to Rubin, "you are the Mr. Rubin I hear so much about."

"Yes," said Rubin, "and you will hear a great deal more. I want to tell you something—you are going to raise prices."

"The Constitution guarantees me the right to life, liberty and full pursuit of happiness."

"To hell with the Constitution," Rubin said. "I am a damned sight bigger than the Constitution."

A bomb exploded in Becker's plant three days later. When he told another Master Cleaners official he was sticking to his prices anyway, the man said, "If you do, Becker, you're going to be bumped off." The association's secretary then came to demand a $5,000 donation. A strike suddenly emptied Becker's plant. "See Crowley" union officials told him when he asked the cause. F. W. Crowley, head of the Master Cleaners, offered him a final chance before being wiped out: he could join the association for the $5,000 payment—in cash, though Crowley would allow installments, since Becker said he could scratch together only $3,000.

Instead of paying, good citizen Becker went to the state's attorney's office. A grand jury indicted fifteen Master Cleaners officials, who hired Clarence Darrow, an extravagance, since Crowe's office presented a most languid prosecution. They called only Becker and his son as witnesses. Becker asked where the others were. "If you want witnesses get out and get them," an assistant state's attorney snapped. "I'm a prosecutor—not a process server." The jury took fifteen minutes to acquit. "Every complaint made to the state's attorney by Becker has been presented to the grand jury," Crowe blandly replied to criticism.

That was enough law and order for Becker. He went to see Capone, talking his way past all the Metropole bodyguards, the sort of gutsy "listen to the Pro-fes-sor" behavior that won Capone's admiration. Becker wanted him as a partner in a new cleaning business. As Becker said later, "I knew the money that could be made in the business by giving the public a fair deal and I went to Capone. I told him of the risks of dynamiting and other attacks which would be entailed through starting an independent concern. But—well, you know Mr. Capone. And so does the Master Cleaners and Dyers Association."

Bombings? Beatings? Burnings? No more of that stuff, Becker exulted. "My partner, Capone, knows how to give better than he gets and the cost of cleaning and pressing is coming down." The Master Cleaners' lawyer tried to discuss that with the titular chief of Sanitary Cleaning Shops, Inc. "Get the hell out of my office," said Capone. "You try to monkey with my business and I'll toss you out of the window."

With pardonable smugness, Becker said, "I now have no need of the state's attorney, the police department, or the Employers Association. I have the best protection in the world."

A usually less starry-eyed critic wrote of this episode, "Al Capone did not voluntarily impose himself upon legitimate business in Chicago. He was invited into it by Morris Becker. . . ." Except for this one instance that simply was not so. Capone gravitated to business racketeering naturally, finding it a sensible fit with his union racketeering. For example, twenty-five Loop skyscrapers once paid his organization $1,000 apiece to avert a walkout of the Elevator Operators' Union. Capone had self-set limits. He would not touch drugs, which he thought could destroy the outfit; he hated kidnapping, which he thought a barbarous blow to the victim's family; he would have nothing to do with pickpockets. But he would enter any other promising racket.

At the same time, he never double-crossed *anyone* who stayed straight with him. When he pulled out of Sanitary, a couple of years later, it was because the money had been relatively meager, and Becker had reached an accommodation with Master Cleaners. But long after Becker had anything to hope or fear from Capone, he declared, "Al Capone was scrupulous in living up to his bargain. I was equally scrupulous. . . . If I had it to do over again I would never ask a more honest partner in any business."

Even some of the better element in Florida agreed. Consider, for instance, the villa Capone took for six months before he bought a house. Since "Al Brown" had leased it, the owners did not know who their tenant was until they heard a ship's radio news flash on their way to the Riviera. They returned at the end of the lease expecting to find their property a blasted battleground. In fact, they discovered not a scrape, scuff or tear, let alone any breakage or anything missing. Quite the contrary. The house's silver and china had proven inadequate for the scale of the Capones' entertaining, so they had doubled most of it, leaving the new pieces behind.

To the owners' mind that more than offset an unpaid phone bill of about $400. But soon Mae showed up, all pretty apology for having neglected the bill, and handed over a $500 note to settle up. When the owners could not make change, Mae waved it off. That would take care of any incidental damage.

After Ammatuna's murder in late 1925, Frankie Yale had favored Joe Aiello as president of the Unione Siciliana (which changed its name to the "Italo-American National Union," though for clarity's sake we'll stick with "Unione''). Even so, Yale had not balked when Capone maneuvered Tony Lombardo in. But Yale soon resented Lombardo's independence and the diminished share of Unione proceeds remitted to him as head of the national.

In the spring of 1927, Yale started making up the deficit at Capone's expense. His Black Hand gang was supposed to oversee safe landing of whiskey smuggled into Long Island for Capone and safe passage for Capone's trucks through New York. Suddenly, many of Capone's trucks were being hijacked before they cleared Brooklyn, enough of them so that his supply of ''the real quill,'' as it was then called, dwindled alarmingly. Capone voiced his suspicions of a Yale double cross to his old Brooklyn friend, Filesy DeAmato, and asked him to check.

In June DeAmato reported back—incautiously from a phone booth—that indeed Yale was arranging the hijackings. Too late, DeAmato realized he had been overheard. On July 1, 1927, he tried a preemptive ambush of Yale. His seven shots missed. Six nights later, a black sedan drove by DeAmato as he stood on a Brooklyn sidewalk. One of the occupants squeezed off three pistol shots; two of them, to the neck and chest, killed DeAmato.

The hijackings continued, as did Capone's fury at both Yale's betrayal and his murder of Capone's old friend (and spy). But Capone was busy on other fronts; besides, revenge takes time—if only to let extra wariness in its obvious target slacken. When Capone returned to Florida in 1928, he had the leisure to plan retribution.

Jack Guzik, Dan Serritella and Charley Fischetti visited Capone in Florida in late June. Jack McGurn soon joined the planners, along with Scalise and Anselmi. All left on a Chicago-bound train on June 28, but the killers got off in Knoxville, Tennessee, where one, who used the name ''Charles Cox,'' plunked down $2,400, cash, for a black Nash sedan. They drove to Brooklyn, McGurn their guide to the streets where he grew up.

Around two P.M. on July 1, 1928, Yale left home in his new coffee-colored Lincoln. By four o'clock, he and his usual business driver, James Caponi—no kin to Al, so nicknamed ''Sham Brown''—were drinking in Yale's Sunrise Cafe at Sixty-fifth Street and Fourteenth Avenue. A bit past four a phone call urgently summoned Yale home. Something was wrong with Lucy, his new wife (Yale had ditched Maria and their two daughters). He dashed out, declining Caponi's offer to drive him. He now lived on Manhattan's West Eighty-first Street and so picked up New Utrecht Avenue, a broad artery cutting due north. At some point he spotted the black Nash trailing him, because he suddenly swerved west onto narrow, residential Forty-fourth Street. He crossed Tenth Avenue. The Nash caught up and artillery sounded, pistols, shotguns, machine guns. Yale's Lincoln hopped the curb and smashed into the front stoop at 923. One gunman vaulted from the car, ran over and blew out whatever was left of Frankie Yale's brains with a .45.

The killers took off. They turned north on Ninth Avenue, then again west on Thirty-ninth, and abandoned their car about three long blocks from the foot of Thirty-ninth Street, making their way to the ferry for Staten Island, where a car waited to whisk them over one of the new bridges to New Jersey and a clean getaway.

In the Nash, police found a sawed-off pump shotgun, a .45 Colt automatic, a .38 Smith and Wesson and a machine gun with nine rounds fired from the hundred-round drum. They traced the machine gun to Peter von Frantzius, a Chicago sporting goods dealer who would soon loom even larger in Capone's story. They traced the handguns to a batch of twelve that Parker Henderson had bought in Miami on Capone's instructions, and had left, as ordered, in an empty room at the Ponce de Leon. Police also saw Yale's initials gleam in diamonds on his belt buckle, one of Capone's gifts to special buddies.

Yale's more permanent buddies gave him a glorious send-off, with a $15,000 silver coffin and a cortege of 104 cars filled with mourners, another 38 with flowers. One wreath of roses and orchids had no card attached; but its white ribbon delivered in gold letters the promise of revenge, "We'll See Them, Kid."

Back in Miami, Dade County solicitor Robert R. Taylor summoned Capone for questioning, and learned nothing beyond Capone's assertion that "I'm in the cleaning business, in Chicago," and admission that he also was a gambler with "an interest in a dog track in Cicero."

Morris Becker's partner returned to Chicago in late July in time for another raid on his Metropole headquarters. Disgusted, on July 30 he moved a block north on Michigan to the Hotel Lexington. Like the Metropole, the Lexington had deteriorated considerably since the days when, newly opened, it had hosted President S. Grover Cleveland when he came to open the Columbian Exposition.

"What!" cried the Lexington's manager, asked if he now hosted Capone's new headquarters, "Al Capone in my hotel? Why, the very idea!" Eventually Capone and his people would occupy as many as fifty-four rooms, amounting to the entire fourth floor, much of the third, and scattered other rooms for their women.

Capone installed his woman on the fifth floor. This current replacement for Marcelle from the Roamer Inn was another of his brothel inmates, a teenage blond Greek. Troubled by a vaginal lesion, she went to Capone's doctor, David V. Omens, who doubled as a partner in that Cicero dog track of Capone's, the Hawthorne Kennel Club. A Wassermann test revealed syphilis in the girl. As we'll see later, it's unclear how much treatment Capone received, if any, and how soon he got it. But, as we'll also see, it was already too late.

One of Capone's earlier visitors at the Lexington was Frank J. Loesch, head of the Chicago Crime Commission. Born in 1852, the son of German immigrants, Loesch had worked his way through law school at Northwestern, founded his own firm and became counsel for the Union Station and the Pennsylvania Railroad west of Pittsburgh. Concurrently, he rendered public service, acting as a special counsel to prosecute voting frauds, heading the bar association, helping found the Crime Commission, which he now led—a tall, straight, white-haired but vigorous seventy-six. Loesch was also "a xenophobic bigot," as a later observer characterized him. "The real Americans are not gangsters," he once said. "Recent immigrants and the first generation of Jews and Italians are the chief offenders, with the Jews furnishing the brains and the Italians the brawn." He much admired Mussolini's methods of dealing with the Mafia.

After 1928's Pineapple Primary, Loesch decided that only a brainy Italian could rescue the general election in November from the same chaos. Besides, he wanted "to remonstrate with him about his conduct" and to explain why "a man of his ability ought to be a better American." The interview took place in late August.

Loesch told a Senate committee four years later that he saw twenty-five or thirty of Capone's men about the hotel, "evidently all dark-skinned fellows, probably none who could talk English." In any case, they didn't try. The half dozen in Capone's office just fingered their guns as Loesch examined Capone's portraits of Lincoln, Washington and Big Bill Thompson.

For small talk, Loesch asked how Capone expected to beat *both* sides of the law. Capone deprecated the law, but said he expected rivals would gun him down some day—"but they'll only get me when I'm not looking," he added. Pleasantries over, Loesch said, "I am here to ask you to help in one thing. I want you to keep your damned Italian hoodlums out of the election this coming fall."

Capone found the plea irresistible. His candidates had lost in the primary. It really didn't matter which set of winners, Democrats or Republicans, he would have to suborn next, but the plea itself flatteringly ratified his position. *This* was clout, acknowledged. He said he'd help.

That was the version Loesch told under oath to a Senate committee. More informally, before the Southern California Academy of Criminology, he claimed that Capone had said he could of course guarantee the South Side, and take care of the North Side too, "because they are all Dagoes up there." For the Irish gangs out west, though, he would instruct the police to intervene. The night before the election, he had the police fan out with seventy flivvers to pick up and hold hoodlums he thought might cause a ruckus. The problem with that version is that, under oath, Loesch had the chief of detectives worrying about demographics and dispatching squad cars. In either case, it turned out to be, Loesch said, "the only honest election that we had had in thirty years."

Before it took place, Brooklyn's Black Hand gang, in cahoots with Joe Aiello's regrouped force, paid off a first installment on that funeral promise, "We'll See Them, Kid." They couldn't get to Capone, of course. He had returned to Florida for the rest of the summer. But there or in Chicago, his caution and his loyal bodyguard rendered him virtually invulnerable. He notoriously was *not* looking the other way. Tony Lombardo presented an easier target. He had come to America as a teenager, arrived in Chicago with $12 in his pocket, and prospered mightily. He was thirty-six years old and not especially wary.

Lombardo finished some routine work at Unione headquarters, 8 South Dearborn, and left the office about four-thirty the afternoon of Friday, September 7, 1928. Flanked by bodyguards Joseph Ferraro and Joseph Lolordo, Lombardo strolled up to the corner of Madison, turned left and crossed the street. A minor spectacle held the attention of the Friday rush-hour crowd. The Boston department store planned an exhibit around a real airplane; workmen were hoisting it with block and tackle up the side of the building to an upper story. Distracted with the rest, Lombardo and his two guards did not notice the two men loitering in the entrance of a busy chain restaurant, the Raklios. Lombardo had just passed the restaurant when a voice cried, "Here he is."

The waiting gunmen darted out of the restaurant's doorway a few steps to the marked trio, and fired. Dumdum bullets shattered Ferraro's spine and he dropped, mortally wounded. Two more dumdums demolished the back of Tony Lombardo's skull, killing him outright. Though not a shot hit Lolordo, a policeman nabbed him as he chased one of the killers—which meant that both got away.

Next day came one of the Church calendar's greatest feast days for Italians, that of Our Lady of Loreto. Father Louis M. Giambastiano had seen the killing in Little Sicily during the D'Andrea-Powers fight; now it was this Capone-Aiello business with more killings. Back then he had urged his flock to tell police what they knew; none had dared to. The best the priest could do, this time, was post a sign in sorrow outside San Filippo Benizi:

FRATELLI
per rispetto a dio in cui credete
per onore della patria e dell'umanita
PREGATE
perche cessi l'indegna strage
che disonora il nome italiano
dinanzi mondo civile

He urged his brothers—for the respect they owed the God in Whom they believed and the honor of their country and humanity—to pray for an end to the horrid slaughter that dishonored the Italian name before the civilized world.

The slaughter had just begun. Capone came north for Lombardo's splendid funeral, face bewhiskered as the usual badge of grief. At the bronze and silver coffin stood a floral heart eight feet tall with a card, "From Al Capone." It might have read, "We'll See Them, Kid." He had an arrangement in Brooklyn, a man detailed to ensure no more hijackings; and maybe they had a legitimate beef about Yale's killing. The Aiellos were different. Joe Aiello remained holed up. Until he could be located and set up, the outfit hit other targets, shooting up more Aiello property, wounding brother Tony again. Meanwhile, Capone went back to Florida once more.

Pasqualino Lolordo, older brother of Tony Lombardo's bodyguard, Joe, and another friend of Capone's, became Unione president. With Capone away, Joe Aiello snuck back to Chicago and, though still lying low, revived his lunatic project of clearing the way for his own succession to the Unione presidency.

Patsy Lolordo and his wife, Aleina, arrived home from downtown about two-thirty the afternoon of January 8, 1929. Home was the lushly furnished top floor of a decrepit, three-story building Lolordo owned at 1921 West North Avenue. At the door waited two men Aleina had seen many times but whose names she professed not to know. She laid out a spread for her husband and guests: sandwiches, pastry, a box of cigars and four bottles of wine. Busying herself in the kitchen, she heard the two men leave at about three o'clock. Five minutes later came a knock at the front door; she heard her husband answer and greet his new visitors effusively. As they settled down to talk, Lolordo closed the door to the kitchen, where his wife stood ironing while their maid scrubbed the floor.

After hearing about an hour's worth of conviviality, Aleina suddenly heard a volley of shots. She rushed through the door, brushing past three gunmen to her husband, crumpled where he had stood before the ornate fireplace. As she slipped a velvet pillow under his bleeding head, the gunmen unhurriedly left, one dropping his .38 to the living-room floor, another tossing his aside when they reached the second landing on the stairs. They had fired eighteen shots, eleven of them obliterating Lolordo.

Early reports said that Aleina had identified a picture of Joe Aiello as one of the three visitors. "She didn't identify anyone," said John Stege, by now deputy police chief. "I don't know how the report got started. It was the same in this case as in other cases—no identification, no aid."

This nonsense had to stop. Patsy Lolordo made the fifteenth person connected with Capone to be killed in the past three months. Aiello was hard to pinpoint, and had a following among the alky cookers Capone still needed. Anyway, Moran was the key. Without his backing, Aiello could not mount further defiance. Two of Lolordo's killers had been Pete and Frank Gusenberg, the best of Moran's few remaining guns; the third was James Clark, another Moran man. Furthermore, the North Side gang, for all their weakness, had been asking for it. In the last year, the Gusenbergs had shot Jack McGurn twice, the second time trapping him in a phone booth, and almost killing him with pistol fire. When the outfit interfered with traffic to Moran's dog track, the Fairview Kennel Club, he retaliated by setting fire to Capone's Hawthorne Kennel Club.

Without Moran the North Siders would be through. After him, no one of even his own modest leadership abilities remained.

Just past the New Year, Capone had taken to bed on Palm Island with pneumonia—an illness that would help send him to jail. But sick or no, Capone started plotting.

CHAPTER 19

Hearts . . .

AFTER THREE YEARS as copy boy for the City News Bureau, Walter Spirko, just turned eighteen, had become a cub reporter, his beat the drearier court cases plus the coroner's office—which meant much more. ''You check that Coroner's office, Walter, *every hour*,'' city editor Isaac Gershman had cautioned. At that office were recorded all ''unnatural deaths,'' the essence of police-news filler, which is what the bureau collected for Chicago papers.

Gershman added wise counsel. ''Some smart alec's gonna say, 'C'mere kid, I got a scoop for you that's gonna set this town on its ear.' Don't get all excited and run to the phone to call us up. They're just pulling your leg; they always do that to new guys.''

Sometime after ten-thirty on the morning of February 14, 1929, Saint Valentine's Day, Spirko made his hourly check at the cubicle where deputy coroner William Bauman logged in new deaths as he received word of them on the phone.

Before Spirko could inspect the book, Bauman's phone jangled. The first words galvanized the deputy. ''What!'' Spirko heard Bill Bauman exclaim. ''Seven men killed . . . two-one-two-two North Clark . . .''

Spirko peered out of the cubicle and scanned the other coroner's assistants busy in the office. ''There's lots of them on the phone,'' he recalls, ''and I'm wondering which one's on the other end of that line.'' Bill Bauman kept scribbling information furiously between exclamations, then slammed the earpiece back on its hook. ''Walter,'' he cried, ''you better call your office. You hear what I—?''

''Sure, I heard you,'' Spirko interrupted. ''Seven men shot and killed at two-one-two-two North—''

"Well? *Get going!*" Bauman roared as he leapt up and rushed to the private office of Cook County Coroner Herman M. Bundesen. Spirko followed and leaned against the doorjamb in time to hear Bauman blurt, "Coroner, we got seven men shot and killed at . . ."

Suddenly Spirko thought that this might be for real. Bundesen effused dignity. Surely a figure of such grandeur would disdain part in so silly a prank on a mere cub. Spirko darted to a desk and snatched up a phone.

"Gersh, there's seven men . . ."

"*Wal*-ter! We just got through *telling* you. Forget the scoops and go about your business." The city editor hung up.

By now the coroner had donned his chesterfield and derby and was making his stately way to the door, saying, "Well, I'd better get out to the scene." Spirko called in again. "Gersh," he said, desperate, "on that seven men? Do you think Doctor Herman M. Bundesen would stoop so low to play a joke like that? Well, I stood there when he was told about it, and he's on his way to the scene."

"I still don't believe it."

Even those already on the scene at 2122 North Clark had trouble believing it.

Though only a short block west of the fashionable residential street, North Lincoln Park West, in 1929 that part of Clark harbored an undistinguished mélange of stolid rooming houses, shops and light industry. Number 2122 fronted Clark about twenty-five feet on the west side of the street, one story of dingy terra-cotta brick squatting between three-story rooming houses; it ran 150 feet back to an alley. A small sign on the roof offered the building for sale, but since October it had been leased to one Frank J. Snyder.

A double window faced Clark, the top half painted black so passersby could not see in, the bottom blocked by a large placard reading "S.M.C. Cartage Company." Not that a peek in would have rewarded the curious: A partition turned the front of the building into a small office space, shielding whatever was going on in the rear. The office had a raw, unused look, containing a small counter, a double desk in the middle of the office, a smaller one against the partition, some chairs and a couple of telephones. The words "MOVING" and "EXPRESSING" ran vertically down on either side of the building's street door. S.M.C. Cartage appeared to be a modest-sized furniture hauling company. Trucks entered in the rear, where double doors opened to the alley.

In fact, 2122 North Clark was occupied by George Moran's North Side gang. Their business headquarters was in the Loop at 127 North Dearborn, and their main hangout was the Wigwam, a bar in the Marigold Hotel about two miles north. The garage served as their main booze warehouse, complete with a storage area so cunningly tucked into the rafters that the platoons of investigators who would soon invade the place never found it.

Moran's relations with Capone had always been at best testy. He had argued with Weiss and Drucci against Capone's peace proposals. Despite a large head and thrusting, deeply dimpled jaw that made him look large in pictures, Moran stood barely at middle height. In public, he delighted in referring to Capone as "the Beast" and "the Behemoth," scorned his use of bodyguards and dismissed his merchandise as "rot-gut alcohol and green beer."

Nevertheless, after the peace of '26, and despite the year's many fractures of it, Moran got a major portion of his carriage-trade booze supply through Capone, real whiskey smuggled in by Detroit's Purple Gang. The favored brand was Old Log Cabin, but Moran thought the price he paid Capone left him too little profit. He found a supplier for a cheaper brand that would leave him greater margin.

Capone didn't mind. Paul Morton, brother of the martyred Nails, gladly took consignment. But Moran had misjudged his market. Customers rejected the new brand and demanded Old Log Cabin again. When Moran reapplied for his franchise, Capone turned him down. Why should he cut off a friendly good customer in favor of such a nettlesome one? Moran's idiot solution was to hijack what he needed. Soon, nothing coming from Detroit was safe, the final outrage.

In what had become a trademark Capone prelude to murder, men applied for rooms across the street from the garage, in rooming houses managed by Mrs. Michael Doody, 2119 North Clark, and Mrs. Frank Arvidson, at 2135. To explain their odd-hour comings and goings, they said they were cab drivers who drove at night and slept days—despite which they insisted on front rooms, overlooking Clark, rather than quieter back rooms.

Next came the elegant setup. Through free-lance hijackers, never identified, Capone's people arranged for at least one load of Old Log Cabin to be sold to Moran, with promise of more to come. The free-lancers then offered a *big* stolen shipment at the excellent price of $57 a case. They would deliver to Moran, personally, at the garage around ten-thirty, Thursday, February 14.

St. Valentine's Day dawned bleak and cold, threatening snow flurries. Chicago expected killings: that midnight, Charles Walz, eighteen, and Anthony Grecco, nineteen, were scheduled to be the first put to death in Cook County's new electric chair; they had murdered policeman Arthur Esau during a holdup. A third murderer, David Shanks, killer of a Peoria schoolteacher, was supposed to join them, having missed the honor of going first when he had received an eleventh-hour stay six days before.

By ten o'clock that cloudy morning, the temperature had "warmed" to eighteen degrees, up from the day's low of thirteen degrees two hours earlier, headed for a high of twenty-six degrees at two o'clock that afternoon. Shifting gentle-to-moderate winds swirled only a modicum of snow—officially a "trace of precipitation."

On Palm Island, Florida, at that moment—by the clock an hour later, eleven A.M., Eastern Standard Time—Capone was, exceptionally, up and about. He had an appointment in downtown Miami scheduled for twelve-thirty.

* * *

From across the street, the watchers on Clark saw a total of seven men drift into the garage.

John May, age about thirty-five, father of seven, arrived early. Later, some thought him an entirely innocent victim, "an ordinary mechanic." Actually, he had been a failed safecracker, charged twice with robbery and larceny. Up to a month and a half before, he had driven loads of beer and booze for Moran, then had been assigned at $50 a week to maintain the gang's trucks. He had brought with him his Alsatian, Highball, whom he leashed to a wheel of the jacked-up truck under which he was working at ten that morning.

"Frank Snyder," the garage's leaseholder, also got there early. His real name was Adam Heyer, alias Arthur Hayes, a furtiveness that colored even his private dealings. Heyer's son by a first marriage had no idea where his father lived, though they met for dinner every few months; he didn't even know if his father was forty or forty-one. Heyer's bride of seven months had no idea where he worked or what he did. An accountant by training, his first jail time—for robbery—had come twenty-one years before, his latest six years ago for parole violation on a confidence-game conviction. As the gang's business manager, he ran Moran's dog track.

Albert Kachellek would turn forty in eleven days. Born in Germany, brought to Chicago at age five or six, he had first done time in 1905 at age sixteen for robbery—four months in the workhouse. That same year he drew four years for burglary. From then on, regularly in one jam or another, he called himself James Clark; "he did not want to hurt my mother's feelings," explained his sister, who indignantly protested the newspapers' repeated reference to him as Moran's brother-in-law. "Where do they get that stuff from?" she demanded. Clark's estranged wife, Dot, was no kin to Moran. Clark was one of the three gunmen Moran had lent Aiello for Lolordo's murder.

The other two, the Gusenberg brothers, Peter, forty, and Frank, thirty-six, were the gang's main muscle. "Tough sons of bitches," Howard Browne recalls their fearsome reputation, "who wanted you to know it, so they'd slap you around for nothing." Pete had done time in Leavenworth for mail robbery. Frank had managed to beat most raps, including murder, except for one ninety-day disorderly conduct sentence. Their kid brother, Henry, whom they had muscled into a job as a $175-a-week, no-show Loop movie projectionist, would join them at the garage a little later. At noon they would drive several trucks to Detroit to pick up more smuggled booze.

Reinhart H. Schwimmer, twenty-nine, recently divorced for the second time, had no business in the garage, literally or figuratively. Although then called an optometrist, and calling himself "Doctor," he'd had no medical training, but had come to his vocation of fitting eyeglasses by inheriting his father's business ten years earlier. Today he'd be called an optician. He had recently closed his failed office, though he hoped to open a new one soon. Meanwhile, he was supported by his mother, Josephine—a self-described "scalp specialist," who may have had a sideline in fortune-telling. She gave him the $70 his room at the Parkway Hotel cost each month; she also gave him walking-around money, $15 just the day before. Schwimmer was another who found gangsters irresistible. He had been friendly with O'Banion—had been photographed prominently at his funeral—and had insinuated himself with each of O'Banion's successors. Lately he had been boasting inaccurately that he shared in the gang's profits and could, at will, have anyone he liked rubbed out. This morning, Schwimmer had dropped by for a cup of coffee and a chat, planning to meet his mother at two o'clock that afternoon.

Probably the last to arrive was Albert R. Weinshank, thirty-six, proprietor of a speakeasy and Moran's point man for efforts to infiltrate Chicago's cleaning and dying racket.

The timing of Weinshank's arrival is inferential, but persuasive. He bore considerable resemblance to Moran, especially in his chunky, middle-height build. That morning he wore a tan or olive-brown hat and a gray overcoat that matched Moran's usual winter costume. From across the street, the watchers could easily have imagined, with Weinshank's arrival, that it was time to telephone the signal.

Police later insisted that the car came from the Circus Café, headquarters of the gang led by Claude Maddox, Capone's ally in the northwest. The Circus stood on North Avenue, about two miles southwest of the garage. If the police were right, the car's driver probably headed east on North Avenue (the 1600 block), and turned north on La Salle, which ran into Clark; but then he followed Lincoln Park West as it forked away from Clark just below Dickens Avenue (2100). He did that in order to turn left at Webster (2200) for the short block west, back to Clark. This maneuver would let the car pull up—without an awkward U-turn on that fairly narrow busy, slippery, snow-swirled street—directly in front of the garage.

The car was a seven-passenger Cadillac, just like the detective squad cars, complete with a siren, as well as a gong on the running board, and a gun rack on the back of the driver's seat. In it rode five men, two of them in police uniform. As the car headed west on Webster, nearing the intersection of Clark, a truck, heading north on Clark, also approached the intersection.

Elmer R. Lewis, delivery man for the Beaver Paper Company, searched vainly in the gloom of gusting light snow for the address of his next delivery. He wasn't paying much attention to his driving. The occupants of the Cadillac, adrenaline pumping, were paying even less. Their driver impatiently rushed the intersection instead of waiting for the crawling truck to clear it. He swerved left into Clark as Lewis, craning to spot house numbers, entered. The rattly old Nelson–Le Moon's left front bumper struck the Cadillac's left rear fender.

The sedan pulled up opposite 2156 North Clark; Lewis also pulled to the curb, apprehensive over having hit what looked like a police car, regardless of fault. As he got out, he saw someone he took to be a detective emerge from the

Cadillac, a man about five-foot-ten, Lewis later told the police, maybe 165 pounds, about thirty-five years of age, clean-shaven, with a light complexion, "American" rather than "Italian." The man wore a blue suit and—rather elegant for a detective, Lewis thought—a chinchilla overcoat. Lewis could see no one else in the Cadillac very clearly: the top was up, side curtains in place. He did notice that adhesive plaster covered a crack in the isinglass of the rear window.

Before Lewis could approach, the man in chinchilla laughed and waved him on, no harm done. Hugely relieved, Lewis forgot his delivery for the moment, continuing north a half block before he remembered to return. By then the sedan had proceeded down Clark to stop at 2122, S.M.C. Cartage. It was just about ten-thirty.

Like Schwimmer, George Moran lived at the Parkway Hotel, around the corner from the garage, at 2100 Lincoln Park West on the corner of Dickens. Although no witnesses saw him, the talk was that Moran arrived a little late that morning for his appointment with the presumed Old Log Cabin hijackers. With him walked Ted Newberry, one of his key people in the booze trade. About the same time, Willie Marks, Moran's specialist in labor racketeering, swung off a streetcar coming from the Loop. All three supposedly saw the Cadillac pull up and four of its occupants get out, including the two in police uniform, leaving the driver at the wheel.

If the talk was right, none in the Cadillac could have worn familiar faces. Neither Moran nor the other two lacked courage. They would not have stood by while known gang rivals invaded their depot.

Police were another matter. No bootleggers, current with their payoffs, would much fear even an unscheduled raid; fixing it would constitute another overhead item. But why let themselves in for a pinch? The three would naturally have scooted, tradition also saying that Moran returned to the Parkway to warn Henry Gusenberg not to show up at the garage.

Inside, evidence showed that the seven men reacted to the strange faces in uniform the same way Moran did—with disgusted resignation. As a police statement put it later, "The seven men thought they were facing only arrest. . . . Otherwise [they] would have sold their lives dearly."

When the invaders passed through the door in the front office partition, they found themselves in a makeshift passageway. Perhaps as many as ten canvas-covered trucks plus some automobiles jammed the garage in back. They had been parked to form an aisle from the office, which opened into the only clear space left in the back, an area near the building's whitewashed, though dingy, north wall. A table stood under a bright bare bulb that flooded the area with light. A pot of coffee perked on a hot ring, next to it an open box of crackers. Six crowded around it, with May still working under the truck.

The two in police uniform undoubtedly went first into the open space, barking their announcement that this was a raid, their orders to put hands up and face the wall, May joining the rest. The two in mufti likely lagged behind in the passageway until the gang members had grumblingly obeyed. In any case, the seven apprehended nothing more than inconvenience. Except for Schwimmer, they'd been through the drill before. They may even not have anticipated arrest, since the garage was currently innocent of alcohol, save for any hidden in the rafters.

As commanded, the seven lined up against the brick wall, faces to it, hands high in the air. Frank Gusenberg stood on the left of the line, then Pete, next Weinshank, Heyer, Clark, May, with the hapless optician on the right. All wore their overcoats except Clark—and May, who had on brown overalls and a jacket. Schwimmer sported a carnation as boutonniere.

One of the raiders went along the line, patting down the victims, stripping them of weapons: certainly from the Gusenbergs and Clark, possibly from all of them, even Schwimmer. In the rush, Frank's pistol, fully loaded, dropped to the floor and lay there.

Among them, the raiders carried two Thompson machine guns, one with a twenty-round box clip, the other with a fifty-round drum. The machine gunners took position at either end with the shotguns in the center. In the horrid instant of the first shot, the seven must have grasped their doom, but it happened so fast, the first bullets found all of them with their hands still thrust in the air.

The first sweeps of the machine guns sprayed .45 slugs across the victims' lower backs and at the level of the upper shoulder, neck and head. May started to turn, perhaps to protest that he wasn't *really* a target. Buckshot blew the left side of his skull away. Another shotgun blast filled Schwimmer's torso with buckshot.

Four of the victims fell straight back to lie supine at right angles to the now pocked wall. Clark staggered sideways and fell parallel at the base of the wall. Pete Gusenberg slumped into a chair, not sitting, but rather with his chest over the back of it, body sideways on the seat. One machine gunner squatted and sprayed the line of bodies a last time, putting holes in the tops of heads, his misses ripping through upturned feet.

In a brilliant touch the killers covered their escape with a mock arrest, in case anyone wondered about all that shooting.

Mrs. Joseph Morin, across the street, on the third floor of number 2125, was about to clean the front room of her apartment. As she raised the blinds, she noticed two men come out of the garage across the street. Oddly, they had their hands in the air. Immediately behind them came two men in police uniform, carrying, Mrs. Morin remembered next day, "long guns which they were pointing at the backs of the men."

Samuel Schneider heard nothing special in his tailor shop on the ground floor of the next building north of the garage at 2124. Certainly he heard no shooting. But he saw out his window the men crossing the sidewalk. They "took it so easy," he told a reporter, "I thought they were police officers."

In the third floor back of the same building, Mrs. Max Landesman stood at her ironing board. She did hear the muffled rataplan and blasts and thought she knew what it was. "Of course," she said later, "I got a little curious." She reached the front windows of her apartment in time to see a man get into the Cadillac, which looked to her like a police car.

Before it pulled from the curb, heading south, a streetcar clattered past in the same direction. In their rush, the raiders, finding their lane congested, swerved around the wrong side of the trolley, into the oncoming lane. Police car or no, that seemed to Jeanette Landesman like a lot of reckless nerve. Besides, what was all that noise about? Certain it had come from the garage, she marched downstairs to investigate.

When the garage's front door seemed stuck, she returned to her building and enlisted one of the roomers, Charles McAllister, who was about to leave for work. McAllister forced the door and disappeared into the office.

He emerged ashen. There were dead men in there! And one who was alive, who had called, "Who's there?" as McAllister came upon the bodies. Landesman immediately called the nearest police station, the Hudson precinct.

The first policeman in was Lieutenant Thomas Loftus, who saw Frank Gusenberg painfully crawling toward him. Loftus picked up Frank's revolver from the concrete floor.

"Do you know me, Frank?" asked Loftus, who had indeed known the gangster for years.

"You're Tom Loftus," Gusenberg croaked.

"Who did it?" the policeman asked. "What happened?"

"I won't talk."

"You're in bad shape, Frank—"

"For God's sake, get me to a hospital!"

This was too good an opportunity. Loftus kept pressing Gusenberg for an answer before he would call for help. "I want you to explain this shooting," he said.

"I refuse to talk."

Loftus gave up. He called for an ambulance. On the way to the Alexian Brothers' Hospital, a little over a mile west, the lieutenant tried again. He said he had seen Gusenberg's brother there, too, dead. Who did it?

"I won't tell," breathed Gusenberg, adamant though far gone. He died about 1:40 that afternoon, never telling.

The scene of slaughter stunned even hardened police. Heyer, who had stood in the middle of the line, had been so riddled in the cross fire that the autopsy would be unable to trace the individual bullet tracks in him. Bullets had flown so thick they tore Pete Gusenberg's trousers in the seat—"rather obscenely," as one account put it. Seven victims were so outrageously many, and the enormity of it so disoriented the police, that they neglected to notify their desk sergeant, who would have passed word immediately to headquarters. For about thirty minutes, police headquarters knew little about what would soon be called "the massacre."

Of course, the coroner's office knew because of the ambulance call. When cub newsman Walter Spirko called his city editor the second time to insist that something *must* be up, Isaac Gershman phoned the bureau man at headquarters. He was told only that the police had received a complaint about some disturbance at 2122 North Clark—a brawl or something—but no word of anyone dead. Gershman, who had a clear half-hour beat on the world, sent out a routine bulletin: Men Injured in Brawl.

Before long, Gershman could hear police sirens wailing through the Loop The word was out and *everyone* had the story. He got Spirko on the phone. ''Walter, you son of a bitch, why didn't you *convince* me!''

CHAPTER 20

. . . and Flowers

AT 11:25 POLICE SIRENS still howled through the Loop and throngs began to choke all streets surrounding the garage. At that moment in Miami—12:25 local time—a limousine, light blue and polished to a dazzling glare, pulled up to the Dade County courthouse. Capone stepped out, a tropical delight in checkered sports jacket, white flannel trousers, sport shoes and his trademark white fedora. The uniformed chauffeur and one bodyguard stayed with the car; another bodyguard took up station at the elevator inside the courthouse. Both guards were bulky but agile men, eyes always on the move, both dressed almost as snazzily as Snorky.

On the dot of 12:30, Capone breezed into the office of Dade County Solicitor Robert Taylor, who had quizzed him the summer before. Taylor had invited Capone in to talk with Louis Goldstein, an assistant district attorney from New York, there to ask more questions about the murder of Frankie Yale. Later, many assumed that Capone had set up an unshakable alibi for himself, but neither the idea nor timing was his. Nor did he need an alibi, his presence in Florida clearly established by his almost daily public appearances about town.

"Well," Capone smiled after shaking hands with Taylor, Goldstein and the Miami officials gathered for the show, "what can I do for you?"

Afterward, no one would reveal what Capone was asked or what he answered about Yale—an ongoing murder investigation. As for the rest of the conversation, Capone coyly said when asked, "All I did was answer a lot of questions that probably wouldn't interest anybody." Two years and nine months later he would learn he was wrong, but at the time most of the Dade County solicitor's questions seemed indeed toothless. Capone neatly parried the rest.

For instance, Taylor asked when Capone had first met Parker Henderson. About two years before. But Taylor followed up with, "Can you tell me the names of the men who were with you?" Capone bridled. "I don't like to give names unless I have an idea what it's all about. Will all the questions I answer be put down and used?"

"Well," said Taylor, "we intend to keep a record and anything you say will be used for or against you." Stenographer Ruth Gaskin sat taking it all down.

"That's just what I wanted to know," said Capone. He couldn't remember those names.

When some of his answers figured in Capone's tax trial, a few critics hooted at his boobery for continuing after that warning. It made you wonder about "the mental capacity of Capone." But the first major tax trials of gangsters—his brother Ralph and Druggan and Lake—lay nine months in the future, and were possible only because the government could tie those three to *provable* income. For reasons we'll explore later, few realized when Capone answered Taylor that illegal earnings were taxable. Actually, the IRS had been looking at gangsters' tax situations for a couple of years and had specifically targeted Capone on October 18, 1928, the previous fall. Capone did not know that and could not have foreseen any reason not to answer what seemed routine questions, though he did so warily.

"What is your business?" Taylor asked.

"I'm a gambler," Capone said. "I play the race horses." Perfectly legal in Florida, Illinois, and, if the New York assistant D.A. cared, his state, too.

Taylor, who had checked Capone's telephone records, asked, "Who is Mitchell of Oak Park, Illinois? He called your home three times on the twentieth."

"He commissions money on race tracks for me."

Again, this was nothing illegal. In fact it implied that Capone didn't even deal with illegal bookies, but had bets placed at the tracks for him. But Taylor also probed for damaging admissions. "Besides gambling, you're a bootlegger, aren't you?" he asked the man whose name was a world synonym for the art.

"I never," said Capone, not under oath, "was a bootlegger in my life."

When it came to potentially embarrassing relationships, he deflected the county solicitor's questions.

"How long has Dan Serritella been living with you?" Taylor asked.

"He don't live with me. He visits for a few days now and then," distancing himself from Thompson's city sealer and a future state senator.

"Do you know Jack Guzik?"

"He's a friend of mine in Chicago."

"What is his business?"

"He's a fighter," Capone answered, joking about the roly-poly ex-pimp.

Capone also distanced himself from the suspicious use of an alias. When Taylor asked, "Do you know Al Brown?" Capone answered, "People call me that name but I never use it. My right name is Capone."

He answered questions about money with at best restrained candor. He denied that he'd been wired money under the name "A. Costa"—perhaps because he thought that looked fishy. But Taylor changed the question: "You got money at the Western Union office here from Chicago?"

No harm in that, even in the context of a tax inquiry: receiving money is not the same as receiving provable income. "That's correct," Capone answered, "all of it comes from Chicago."

"In connection with the gambling business?"

"Yes." According to popular thought at the time, illegal gambling winnings were not taxable income.

"Did Parker Henderson ever negotiate any money transactions for you?" asked Taylor.

"Only when he bought my home."

"How much did you give Henderson to buy the home?"

"Fifty thousand dollars."

"Cash or check?"

"Cash."

"Did you always find him honest in his dealing with you?"

"Absolutely, yes." Thought of the beguiled puppy having dared be anything else must have tickled Capone.

The whole exchange brought up a point. "Do you keep a record of your money transactions?" Taylor asked.

"Positively, yes."

"Where is that record?"

"That's my personal affair."

Capone even engaged in a little high-level poor-mouthing when the county solicitor asked if he was, as reported, going to buy a Bahama island called Cat Key, recognized as a marvelous staging point for running booze to the Florida shore.

"I don't know. Possibly."

"What are you going to do with the Key."

"I don't think I'll get it. I haven't got the money."

"How much do they want for it?"

"Half a million."

Then Taylor cut too close to the knuckle. "Did you get any money from Charley Fischetti while you were staying at the Ponce de Leon?" he asked. "Henderson says you received various sums, from one thousand dollars to five thousand dollars."

"What has the amount of money got to do with the question?" was all the answer Capone would give. Bringing up Charley Fischetti specifically suggested income from operations.

Considering how little he knew about the government's intentions, it was a cool and circumspect performance.

* * *

Back in Chicago, honoring tradition, police chief Bill Russell talked tough. "It's war to the finish," he declaimed, adding the time-hallowed promises: "I've never known a challenge like this—the killers posing as policemen—but now the challenge has been made, it's accepted. We're going to make this the knell of gangdom in Chicago."

His police did become a frenzy of activity. They raided the rooms and apartments of the victims. They found a loaded gun in Schwimmer's Parkway Hotel room, but could not locate Moran anywhere in Chicago. At Pete Gusenberg's apartment they found Myrtle, who said she had divorced a man named Coppelman the year before and was now married to Pete, except she thought his name was Gorman and she had no *idea* he was a gangster. At brother Frank's hotel room they found two fringe gangsters, including a brother-in-law, Paddy King; the two were probably there to retrieve a machine gun, which a janitor had found behind a loose wall board in a closet. The police also found the witnesses on Clark Street, including the two who had rented rooms to the watchers. But the landladies could identify no mug shots as belonging to the fake cabbies.

In short, the police hadn't a clue.

The day after the massacre, Frederick D. Silloway, assistant administrator of the Prohibition forces in Chicago, caused considerable fuss with charges that the men in police uniform really were police, not masquerading gangsters. When questioned, he claimed newsmen had misquoted him. All he meant to say was that he believed the massacre "an outgrowth of a hijacking job on Indianapolis Boulevard" some weeks before; on that job, uniformed police and, alas, a Prohibition agent had been involved. Silloway's thesis was goofy: that five hundred cases of booze had been hijacked from Moran and that Moran consequently stopped all protection payments "and the result of this move was the machine gun killing of yesterday." Chief Russell said, "If Major Silloway has any evidence in this case he ought to come to us with it." No matter what, "we are going to find out who committed this crime. As far as Major Silloway's statement is concerned, I would just as soon prosecute policemen, if they are guilty, as anyone else."

Legend grew that Russell's chief of detectives, John Egan, had been even gruffer about the possibility of errant police, threatening to "toss 'em by the throat into a cell." In fact, he rejected the Prohibition enforcer's charge even more tartly than his boss. "It is all very well for Major Silloway to hatch theories and give them out for publication," Egan said, "but we will pay more attention when he can produce facts." To be safe, though, 255 detectives had to account for their whereabouts at the time of the massacre. Silloway was transferred, then fired.

Another enduring legend had Moran saying something like, "Only Capone's gang kills like that." The story appeared in Miami papers; in Chicago they knew that Moran had sent word to Chief Egan accusing no one and professing bewilderment. "We don't know what brought it on," he let Egan know. "We're facing an enemy in the dark."

The more new information that came in, the less light was shed. H. Wallace Caldwell, president of Chicago's board of education, should have made an unimpeachably reliable and responsible witness. At first report, he had seen the murder crew moments before their car hit Elmer Lewis's truck, his attention attracted to them because they had just run a red light. He added the fascinating clue that one of them was missing a front tooth. Dentist Loyal Tacker, whose office was up the street at 2530 North Clark, said he'd extracted the front tooth of a rough-looking putative hoodlum the night before the massacre. Soon, a prime suspect would turn out to be missing a tooth.

Trouble was, Caldwell placed the murder car driving north on Clark, which it certainly never did, then turning around. Furthermore, it seemed unlikely that from cars passing head on—with snow swirling however lightly—even the most gap-toothed grin would be readily visible. A later version amended the claim. It was Caldwell's chauffeur who had spotted someone missing a tooth, and the fellow had been loitering outside the garage. As for the dentist, although his name had not been published, Tacker turned up in Detroit several days later, wandering dazed, his head bruised, with a tale of having been abducted the day after the massacre. He could not guess why he had then been set loose. The police mug shot he picked out as being his patient was of a prisoner penned in the county jail since November.

George Arthur Brichetti was another matter. Described as "a youth," he claimed to have seen a car pull up to the garage's back doors in the alley, two of its four passengers in police uniform. So far, pretty plausible. But young Brichetti also insisted he saw three of them go in the rear doors, at which point he ran around front, hoping to see more, and got there in time to watch the killers file out, one of the "civilians" missing a finger and saying, "All right, Mac," as they got into the car. On the strength of that, the police started looking for Capone's chief bomber, James Belcastro, who had a finger missing, and for Jack McGurn. No one called him "Mac," but Brichetti was later said to have identified his picture.

Meanwhile, theories about the massacre's motive abounded. The Gusenbergs had stiffed some hijackers, and this was revenge. It was somehow connected with Moran's moves on the cleaning and dyeing trade. It was the Purple Gang, tired of the hijackings. It was an intramural Moran gang killing. J. Thomas Heflin, senator from Alabama (and uncle of current Alabama senator Howell Heflin), explained to his colleagues that Mussolini's agents had killed gangsters who "wouldn't swear allegiance to Fascism" once Mussolini had sold out to the Pope in the recent Concordat.

With so much heat, State's Attorney John Swanson announced the start of Prohibition in Chicago, telling police deputy chiefs and captains that this time *all* speakeasies had to close, for real. Casinos had to close, too, and horse parlors and brothels—the idea being that the "means and sources of [gang] revenue must and will be stopped."

Two days later, it had become uncommonly hard to find any kind of action in the city. "How is it?" one saloonkeeper asked another.

"Dead."

"Dead is right!" said the first.

Not in Miami. Two nights after the massacre, Capone hosted a mammoth press party at Palm Island as part of the hoopla surrounding the championship fight between Jack Sharkey and William "Young" Stribling. He visited Sharkey's training camp and was photographed standing between Sharkey and Bill Cunningham, a former football all-American. He played golf at the Hollywood Country Club—a memorable event for his caddy because Capone played only the front nine and tipped $1 a hole, whereas $1 was thought generous for a full round.

Altogether, except for the first half of January, Capone had been having a marvelous time in Florida. The duration of his January illness would become subject of a federal trial, but the condition was real and fairly serious. Dr. Samuel D. Light had come on January 5, 1929, to find that Capone's influenza had degenerated into double pneumonia. As we'll see later, a young local doctor, Kenneth Phillips, had taken over the case and called Capone family doctor David Omens down from Chicago for consultation. But by mid-January, Capone had been on the mend. Although the exact date would be disputed, policemen on duty at the Hialeah racetrack had seen Capone there soon after the meeting opened on January 17. Charter pilot Edward Nirmaier had flown Capone and his small party on a day's picnic to Bimini on February 2. Capone had led a longer expedition by steamer to the Bahamas on February 8, staying until the 12th, two days before the massacre.

Back in Chicago, the massacre investigation followed two lines. Herman Bundesen kept his coroner's jury sitting after hearing testimony from doctors about how the victims died and from relatives about how they had lived. The jury would investigate the murder weapons. Meanwhile, the police kept up the heat.

When Chief of Police Russell announced on the afternoon of February 21 that, in the drive to dry up Chicago, police would start looking in all garages, barns and sheds that fronted on alleys, searching for stills, the threatened diligence paid off. That night, one week after the massacre, neighbors saw smoke drifting out of a garage at 1723 North Wood Street. The answering firemen took one look inside and called the police.

Someone in the rented garage had been using a hacksaw, ax and acetylene torch to dismantle a 1927 Cadillac seven-passenger sedan. Bit by bit, so as to avoid detection, they had burned the canvas top and side curtains, the woodwork and upholstery. Captain John Stege deduced that Russell's announcement had hurried them, and "they probably used a little gasoline to speed up their work." It evidently flared, singeing one of the workers, whose charred coat and hat were in the garage. An unidentified stranger had presented himself at a nearby doctor's office for treatment of burns he claimed had resulted from a still explosion, but had run off when the nurse told him he'd have to wait for the doctor to finish with another patient. The nurse guessed that he feared a report to police. A police siren dismounted from the car lay in a corner of the garage. A Luger and grips from another pistol were found, but so handled as to be useless for fingerprints.

As soon as one deputy chief saw the charred, dismantled remains in the garage, he announced, "The murder car has been found." Assistant State's Attorney Harry F. Ditchburne added, "There is no doubt in our minds about the car," then generously asserted that since the wreck certainly did *not* belong to the police department, "the detective squads are freed from suspicion."

The garage had been rented on February 12, two days before the massacre. The tenant called himself "Frank Rogers" and took the garage for a month, paying $20 in advance. He foolishly gave the blameless owner a real address: 1859 West North Avenue, practically right around the corner. Although police found the place deserted, they knew all about it. Next door sat the Circus Café, Claude Maddox's headquarters. A month before, police had raided 1859 West North and had discovered Maddox, who gave his real name of John E. Moore. They found him hunkered down in a rear, first-floor room, bare of all furnishings except a fully loaded hundred-round machine gun drum and a dozen overcoats, one with a loaded automatic in its pocket. It was this sort of thing that accounted for Maddox's nickname, "Screwy."

Police traced the car's insufficiently obliterated serial numbers through a couple of legitimate trade-ins to a cash sale for $850 in late December. The buyer gave his name as "Morton" and his address as "Los Angeles." There, the trail ended. No one could find "Frank Rogers" either. Witnesses had seen Maddox and two henchmen in grease-covered overalls near the Wood Street garage the previous Monday; but even had they been seen at it, no law prevented citizens from demolishing Cadillacs—even those that resembled otherwise unidentifiable murder cars. Even if they could find him, police had no case against Maddox.

Chicago had been here before. Rewards totaled $50,000: $10,000 from citizens, $20,000 each from the city council and the state's attorney's office, with hopes for another $50,000 from the Association of Commerce. When some on the city council fretted about the funding of their share, Alderman Jacob M. Arvey, who in time would sponsor the political career of Adlai E. Stevenson, said, "Why worry? We'll never have to pay the reward, because the men will never be arrested."

Not *the* men, maybe, but tradition demanded the arrest of some. Two officials checking out leads in St. Louis learned that the ploy of donning police uniforms to lull victims had been standard with two former members of a local gang called "Egan's Rats"; they had shot up a rival gang in Ohio using it. One of them even had a front tooth missing! Police announced the Egan's Rats connection but reserved the names until they could circulate photos. Then they put out a wanted list of seventeen names, including Maddox, McGurn, Belcastro, Joseph Lolordo, Frank Maritote, and just for luck Joe Aiello and his brother Sam.

About one-thirty next afternoon, February 27, they found McGurn registered in the Stevens Hotel as J. Vincent D'Oro. Separated from his wife, he had been cuddling since January 31 with Louise Rolfe, an extravagantly blond lovely, like him a tournament-level golfer, who had first come to public notice at age fifteen when she had caused a fatal, three-way car crash also involving the Illinois attorney general, Edward J. Brundage. Three floors above their room the state's attorney's men had set up a headquarters to investigate the massacre. "We've got him cold bang!" a police lieutenant exulted when George Brichetti and an unnamed woman who claimed to have seen the getaway car picked McGurn out of a lineup.

That night, in Miami, Sharkey won a decision over Stribling in ten uncombative rounds. Capone was there with his pal Jack Dempsey, who dusted off his box seat for him.

Earlier in the day, a government process server had delivered a subpoena to Capone, returnable on March 12, 1929, before a federal grand jury in Chicago. The jury wanted to ask him about Prohibition violations in Chicago Heights. "I have nothing to fear," announced Capone, "and I'll be there, but I'd like to end my vacation first."

Chicago police and prosecutors had shown pictures to witnesses, pumped informers, and released the names of the two former Egan's Rats they now concluded had been masquerading as police. The first was gap-tooth Frederick R. Burke, fugitive since he had been identified with an April 2, 1925, bank robbery in Louisville; he had skipped bail after indictment in 1927. The other suspect was Burke's most usual accomplice, James Ray. Both were nearly six foot and were meaty enough to make convincing police; neither would have been recognized by Moran or his men. For a bonus, police proposed Joe Lolordo as one of the machine gunners, presumably avenging his bother's murder. Soon, New York police let Chicago know they suspected Burke and Ray in the Yale killing. Another assistant state's attorney, David N. Stansbury, divined that the two had received $10,000 each for St. Valentine's.

Thanks to witnesses and informers, police also started looking for Capone's murder twins, Scalise and Anselmi, though prosecutors ended up with enough evidence to indict only Scalise at the same time they indicted McGurn. By then—mid-March—they had given up on all other suspects, and had concluded that McGurn had only helped plan the massacre. His alibi against being at the scene was too good, literally a *beautiful* one. Reporters dubbed Louise Rolfe "the Blonde Alibi," a stunner if you ignored somewhat thick ankles and discounted assurances by a woman reporter that her hair was "too yellow to be natural."

Rolfe maintained that she and McGurn had spent that Valentine's Day in bed until one-thirty. Indeed they had hardly peeped out of the hotel room since their arrival two weeks before, meals and newspapers being sent up. ''When you're with Jack,'' she explained, eyelids aflutter, ''you're never bored.''

The police part of the investigation had played out. ''What does he say?'' a reporter had asked after McGurn's interrogation. ''What did he ever tell a policeman?'' had come the answer.

The authorities had no interest in Capone, either, but for no evident reason other than contrariness, Capone decided to end his vacation on his own schedule. On March 5, Dr. Kenneth Phillips—twenty-nine years old, young, struggling, and glad for the business—completed an affidavit that pneumonia and pleurisy had confined Capone to bed until February 23 and had left him too debilitated to travel. Capone had written to William Waugh, who as a federal prosecutor had turned down $50,000 from John Torrio's representative, saying, ''You're talking to the wrong man.'' Now in private practice, Waugh represented Capone, whose letter read, ''If a federal judge says a doctor's affidavit is not a legal reason for my staying here, I will take an airplane for Chicago at the earliest possible moment.'' Capone added virtuously, ''I don't want to be in contempt of court.''

Judge James H. Wilkerson had his suspicions. When he saw in Capone's letter the statement ''It would be dangerous for me to come to Chicago now,'' he rumbled, ''I wonder what kind of danger he means.'' But he granted an eight-day stay to March 20, remarking that it would give the U.S. district attorney ''time to satisfy himself as to Capone's condition of health.''

When Capone duly appeared on March 20—excited stenographers and court factotums clustering about for a glimpse—anticlimax reigned. The grand jury kept Capone waiting for hours, quizzed him briefly in the afternoon, then put him over for a week; that entitled Capone to collect $3 a day from the government in witness fees. The question of whether local police could touch someone under federal subpoena had much exercised the newspapers. The point turned out to be moot. ''Why bother with questioning Capone?'' Captain Stege said when asked if he had the Prairie Avenue house staked out. ''Think he's going to tell anything? Ho! Ho! Not Capone.'' Besides, the deputy police chief pointed out, Capone *was* in Florida at the time of the murders.

When they got around to him again, the grand jury seemed more interested in Capone's tax status than his knowledge of booze in Chicago Heights. Reports circulated that Capone ''was even willing to pay an income tax, if he owed one,'' and would split any difference with the government over how much.

Meanwhile, the government had indeed looked into Capone's state of health when he asked for that delay. After his second grand jury appearance, Capone was arrested on a federal warrant charging him with contempt, claiming he had been perfectly able to appear on March 12. Capone posted $5,000 bail, and walked away. The trial would not take place for almost two years.

* * *

Inquiry by the coroner's jury in the massacre seemed to promise results. The first sessions, February 23 and March 2, contributed little but pathologists' reports, vague replies by often reluctant relatives of the victims, and professions of mutual admiration among the offices of the coroner, state's attorney and police. An assistant state's attorney hazarded the opinion that no "definitive conclusion concerning the identity of the killers could be reached today," an understatement, "or possibly for another ten days," manic optimism.

But the coroner had an idea, which the *Chicago Daily News* later claimed came from them. He would engage the country's premier expert in the new forensic science of ballistics to look at the shell casings and bullets found in the Clark Street garage and taken from the victims' bodies. A day or so after his jury's second adjournment, Bundesen announced the hiring of Calvin Goddard, then working out of New York. A self-described "gun crank" since childhood, Goddard had been an ordnance officer during the war, and had established the police ballistics bureau in New York.

By the time the coroner's jury met for its third session, April 13, Goddard could tell them convincingly that the St. Valentine's Day killers had used two machine guns—firing, respectively, twenty and fifty of the seventy shell casings found in the garage. They had also used two shotguns, both pump automatics, since double-barreled shotguns are loaded by hand, so leave no ejector marks on their cartridges. He also assured them that the machine guns and shotguns owned by the Chicago police and by police of the more suspect suburbs, like Cicero, had been tested and were not the murder weapons.

Goddard and ballistics impressed the jurors mightily—especially when the coroner interrupted proceedings with a case unrelated to the massacre: Goddard demonstrated to the jurors that a gun taken from George Maloney had been used in a hitherto unsolved murder that had occurred nine months before. He similarly pinned two other killings on Maloney. The evidence persuaded the jurors to vote murder indictments. It also moved two of the jurors, Burt A. Massee, vice-president of Colgate-Palmolive-Peet, and Walter A. Olson, president of his own rug company, to help fund the Scientific Crime Detection Laboratory in affiliation with Northwestern University, with Goddard as its director. If suspect guns ever showed up, Goddard could determine whether they were used in the massacre.

No one knew where those guns had gone; but Coroner Bundesen tried to find out where they had come from. The attempt hardly displayed certain elements of corporate and civic Chicago at their best. The trail first led to an old-line "respectable" sporting goods firm, Von Lengerke and Antoine, that would sell machine guns to *anyone*, including seven by mail to "Gopher State Mines," a company that existed only on a letterhead accompanied by a check; the firm also sold four guns to Capone's dog track.

The star of the inquest, though, was Peter von Frantzius, a wispy, sharp-featured sporting goods dealer who specialized in gun sales—including the

Tommy gun found in the Nash abandoned by Yale's killers. Second billing went to Louis Wisbrod and Frank Thompson, both short and rotund, who acted at different times as arms dealers themselves and as sales agents for von Frantzius.

At one point, after other testimony had given von Frantzius the lie, Bundesen recalled him. "Do you wish an opportunity to change your testimony?" the coroner asked. "Yes," came the bland answer, "if the circumstances demand."

When Thompson started selling for von Frantzius, his customers, by his own account, seemed of doubtful respectability—not least because they gladly paid him more than double legitimate retail for the guns he obtained. His best customer, whom he met in a Detroit bar, called himself "Joe Howard," gave his address as "General Delivery," and once arranged to take delivery of machine guns on a Chicago street corner. "Howard" turned out to be James Shupe, known as "Bozo," a small-timer with ties to the outfit through Scalise and Anselmi.

At one point, Bundesen charged that von Frantzius really didn't care where the guns he sold went, did he? "No," came one of the man's few certifiably honest answers, "we are in the business of selling firearms."

Despite promises to look at mug shots, the arms dealers somehow never were able to identify their customers. But selling machine guns even to criminals then constituted no crime, so Von Lengerke and Antoine, von Frantzius, Thompson and Wisbrod paid no price. All stayed in business.

The two St. Valentine's Day machine guns showed up exactly ten months after the massacre. On Saturday night, December 14, 1929, two cars bumped in a trifling accident outside the city hall of St. Joseph, Michigan, a fashionable Lake Michigan resort town. The motorists argued vehemently over what turned out to be $5 damage. Policeman Charles Skelly came along and insisted they drive to headquarters to sort out blame. Skelly swung up on the running board of one car. Its occupant hauled out a pistol and shot the officer three times. Skelly died in the hospital three hours later.

The killer tried to drive off, but his wheel broke when he banged into a road obstacle. He abandoned his car, commandeered another at gunpoint, and got away. Papers in his car led police to a bungalow south of town belonging to Frederick and Viola Dane, who had lived there only two months. A search turned up two machine guns, revolvers, ammunition, two bulletproof vests, and $319,850 in bonds, $112,000 of them stolen the month before in a Wisconsin bank robbery. A laundry mark, the initials FRB, set one investigator thinking about those wanted circulars on Fred R. Burke. "Dane's" description went to Chicago.

Ballistics expert Goddard tested the seized machine guns. Bullets from one matched those pumped into Schwimmer; bullets from the other matched those taken from Clark's body. At that point, Goddard stopped testing; these were unarguably the massacre guns. What's more, one of them had fired bullets into Frankie Yale a year and a half before in Brooklyn! And the bulletproof vests found in the bungalow matched those sold, as a sideline, by Wisbrod, the machine gun dealer.

In any good detective story, a break like that would have unraveled the mystery. In this case, identification of the murder guns represented the last *fact* any outsider ever learned about the massacre.

December 2, 1929, not long before Skelly's murder, Assistant State's Attorney Harry Ditchburne had to admit in court that the state still found itself unready to try Jack McGurn despite five previous continuances. Since Illinois law generally required dismissal after four, Ditchburne dropped the case. They had no stronger case against Scalise, Anselmi, Joseph Lolordo, or any other suspect.

The federal government would eventually indict McGurn under the Mann Act. He had taken Louise Rolfe across state lines for exuberantly plain immoral purposes—a sexy vacation in Florida, unmarred by wedlock. They also charged her as coconspirator without explaining how the alleged victim could enter into a conspiracy to debauch herself. McGurn married Rolfe, supposedly on advice of a newspaper editor who pointed out that she could not then be compelled to testify against him. No matter; they were convicted anyway, McGurn getting two years, Rolfe four months. Their convictions would be overturned on appeal.

Burke was captured. Michigan, however, would not turn him over to Illinois. They put him in prison for Skelly's murder. He died there in 1949 of a heart attack, and was never tried for the massacre.

Would he have been convicted? A later writer declared that only Burke was, with "moral certainty," one of the killers. Was he? Gangster Alvin Karpis, thirty-three years in one prison or another, told the writer that Burke did take part, and named the other participants, including Claude Maddox, George Ziegler (who planned it, Karpis said), Gus Winkler and Raymond Nugent, known as "Crane Neck," with Byron Bolton as a lookout (Bolton had indeed been suspected, along with Fred Weston, whom Karpis failed to name). Karpis said his information had come from Nugent. Former outfit member George Meyer knew Karpis in prison; the cover of Meyer's book suggests that he might have been the unidentified driver of the St. Valentine's Day car, though he refuses to discuss it, there being no statute of limitations on complicity in murder. "If you believed half what Karpis told you," Meyer says, "he knew everyone and all their business. Karpis didn't know shit!"

Was Burke there? Writing closer to the event, Walter Burns pointed out that Burke had never been known to tie in with Capone's outfit before or to take part in Chicago crime. He certainly had possession of the machine guns used; how he got them was another matter. Burns also insisted that Burke was identified only "dubiously by his photograph as one of the two men disguised as policemen." Bank robber J. Harvey Bailey said, when he had nothing to gain from it, that he had been with Burke in Calumet City at the time of the massacre.

Logic also argues against Karpis's nominations, especially Burke. Like Bolton, who was from Detroit, Burke operated as a free-lance stickup artist with no ties to any gang once he had cut loose from Egan's Rats. Using out-of-town members of correspondent gangs was standard; but Burke and Bolton were under no such discipline, their loyalty unmonitored. Would Capone have countenanced the use of such men? Particularly a flake like Burke, who had the kind of dippy judgment that would let him murder a policeman—the unfailing trigger of an all-out manhunt—rather than run the minuscule risk of being recognized as a fugitive by someone at a small-town police station when he wandered in with Officer Skelly as party to a minor auto accident?

From another angle, would Capone have mounted such an operation without his most reliable killers, Scalise and Anselmi? Would he have trusted planning to someone besides McGurn, whose tactics had confounded the assault on him from Aiello's assassins? Of course, if Scalise and Anselmi were there, they could not have masqueraded as policemen; by then, the North Siders knew them by sight (that also would gainsay the story that Moran saw the car pull up and its occupants get out). But Scalise and Anselmi could have been in plain clothes, waiting to appear until the "police" had the victims facing the wall. And the "American" that truck driver Lewis saw could have been the driver.

For what it's worth, George Meyer says flatly, "Those two policemen were real cops—out of the Sheffield Station." Howard Browne says no: the firm consensus of police and newsmen made them two unidentified Egan's Rats on loan from St. Louis, the two civilians Scalise and Anselmi. Burke, Browne insists, "was never near that garage." Capone would never have had him.

Was it a Capone operation?

"The St. Valentine's Day massacre never really made sense," says Mike Graham, researcher into Chicago's past and proprietor of a twenties museum. Could the author of this most shocking of killings be the man who "didn't want any trouble"? Capone surely would have anticipated the public outcry such slaughter would evoke and realize it must impel even greater losses than he had suffered after McSwiggin. For what? Well, as Mark LeVell points out, "The beer wars were over when Moran lost the best of what was left of his troops." True; but the wars would have been equally over with Moran's solo death.

Some theorists today speculate that Capone's routine protestations of innocence might have been accurate. Perhaps, after all, it was police miffed at seeing their payoff suspended or some intragang squabble. One theorist postulates an outside group of "hillbilly gangsters." But these explanations ultimately make no more sense.

Besides, every wrinkle in the planning spoke of a Capone operation: the lookouts planted far ahead, the untraceably purchased "squad car," the garage rented a scant fifteen-minute drive away where the car could be dismantled, the setup to bring Moran to a known place at a known hour, the charade "arrest" that covered the killers' getaway. Every detail that could be *planned* bore Capone's signature.

No one could plan the reaction or thought processes of the killers. That was why Torrio and Capone sent for Yale on the Colosimo and O'Banion kills, and why Yale left the Lonergan kill to Capone. Put a cool head in charge to modify plans if necessary. No Yale, no Capone, and no McGurn went into the garage that morning. While Scalise and Anselmi were marvelously useful killers, their performance at the Genna-Walsh-Olson killing, especially when panic marred their escape, showed them no thinkers.

Some writers made much of the volume of phone calls traced from Chicago to Palm Island before the massacre, especially McGurn's call on February 11—ignoring the government's proof that Capone wasn't there on that date, but was in the Bahamas from the eighth to the twelfth. This is sheer speculation, but if one of the killers had picked up a front office phone at 2122 North Clark Street to call Palm Island at ten-thirty, when Capone *was* there, and if he had said, "Snorky, we've got them lined up, and one of them we think is Moran, but there are *six* others; what now?," it's not difficult to believe that the man who had lost millions from past multiple killings would have told them to pretend it was a police shakedown or mistake and to get out fast. He could always isolate his enemies again, as he had in the past, for more surgical elimination. At *most* he might have had them "arrest" the person they thought was Moran, conceivably the Gusenbergs, too, and abduct and kill them. But a *massacre*? More likely, Capone's plan to get Moran got out of hand.

Throughout the country, it caused unprecedented revulsion. The New York *Sun* editorialized that "crimes like this constitute the underworld's supreme defiance of society." The *Boston Globe* saw in it "many reasons to believe that the rest of the United States is growing more like Chicago." And the *Record* admitted, "It's not impossible in Philadelphia."

If Chicago would do nothing about Capone—and clearly Chicago, even after the massacre, would not or could not—others would.

CHAPTER 21

Brotherly Love

ALTHOUGH THE SPECIAL Intelligence Unit of the IRS did not win permission to target Capone until October 18, 1928, government agents and prosecutors had cherished the idea since 1927, when the Supreme Court decided a case that had started six years earlier. The IRS had always claimed that any income was taxable, even if illegally earned; but IRS claims are not law. In 1921 the government had prosecuted two-bit bootlegger Manley Sullivan for evading taxes on his booze profits. Sullivan's lawyers argued that it ill became Uncle Sam to share in illegal money. Moreover, to compel declaration of illegal earnings on pain of prosecution for tax evasion invaded their client's Fifth Amendment right against self-incrimination. The Supreme Court rejected both arguments.

On the strength of the *Sullivan* decision, IRS agents in Chicago began keeping scrapbooks about gangsters' newsworthy expenditures, like Capone's Florida mansion and lavish parties. Such spending argued command of much money. Still, the government could not tax money people *had*, only money they *earned*—income. To nail Capone or any other tax evader, the government needed proof of income.

Early as 1928, the IRS compiled what it needed on Terry Druggan and Frankie Lake, who in March were indicted for evading taxes on income from ownership of illegally operating breweries.

The IRS had an equally promising hook into Ralph Capone. In 1926, agent Eddie Waters had convinced Ralph he could avoid trouble by declaring income and paying tax on it, and had offered to fill out the forms when Ralph complained of their complexity. Ralph agreed and admitted to income totaling $55,000 for the previous four years. Although laughably low, it *was* admitted income. Waters told Ralph he owed a tax of $4,065.75. Ralph inexplicably did not pay, and in January 1927, the government moved to attach his property, including valuable racehorses.

Ralph tried to palter, claiming that he was too broke to pay the entire sum, that he had sustained heavy gambling losses, and that some horses had died. All he had left, he said, was half interest in two hayburners, but he would borrow $1,000 if the government would settle for that. The local tax collector recommended acceptance—after insisting that Ralph put his affecting tale in writing and swear to it. Washington turned down settlement anyway, and the head of Treasury's Special Intelligence Unit, Elmer L. Irey, set an agent to work looking into Ralph's finances.

By July 1928, the IRS could demonstrate Ralph's interest in four respectable Thoroughbreds. What's more, he had cleaned out a safe deposit box the day before he had started pleading poverty to the government. In November 1928, Ralph upped his offer to $2,500, then to the full amount, $4,065.75—but, incredibly, refused to pony up the additional $1,000 or so in interest and penalties the government now demanded, which would have settled the matter.

If the IRS could prove greater income than Ralph's admitted $55,000, and greater assets, the government might put him away for a significant amount of time.

In Ralph's case, as with Druggan and Lake, prosecution success turned on being able to fix ownership of property that had provable or inferable income flowing from it. Al Capone had been more cunning. He seemed to own nothing in his own name and showed no visible income.

The IRS could bring two kinds of cases. The more straightforward was called a "specific item" prosecution: the evader declared a $50,000 income and the government could document receipt of another $25,000. The other kind of prosecution, "net worth," was trickier. If the government could show that as of a certain date the target had only, say, $10,000 in assets and a $5,000-a-year income, then two years later owned a $20,000 house and $10,000 car—yet declared and paid tax (if at all) only on that same $5,000 income—prosecutors could reasonably expect a jury to find that the rascal really had enjoyed an income for those two years equal at least to the difference between the earlier and present assets, or to the difference between starting assets plus income and proven expenditures.

Merely having the higher assets or spending the money was not proof of taxable income, and of course defendants did not have to prove where they got the money. But realistically, unless they could demonstrate a convincing nontaxable source for the excess—anything from inheritance to loans to the kindness of strangers—most juries would vote that it was indeed income and that nonpayment of taxes on that income was criminal.

First, prosecutors had to establish what they today call the "starting point"—that admitted or provable initial level of assets or income (or combination) with which juries could compare later assets and expenditures. Otherwise the accused might have owned the assets or the monies expended all along. Tax declarations for earlier years did nicely as starting points. Prosecutors particularly treasured statements of assets in divorce proceedings, since the wretches often swore to understated assets hoping to avoid big settlements.

Not until 1954 would the Supreme Court unmistakably affirm the legitimacy of net worth prosecutions. Today's prosecutors seldom traffic in them, since proof is necessarily all circumstantial, something juries view with suspicion. In Al Capone's case, net worth looked impossible. His dealings seemed to yield no discoverable starting point. He had never filed an income tax return, owned nothing in his own name, and had never made a declaration of assets or income. The IRS kept looking.

After admirably painstaking digging that yielded nothing in the search for more of the goods on Ralph, the IRS happened on the key exhibit fortuitously.

Conditions in Chicago Heights had become scandalous beyond bearing, even in Cook County. Scarcely a day passed without a bootlegger's or alky cooker's murder. Then a government informer was murdered, and on December 6, 1928, the police chief of South Chicago Heights—slated to testify before a grand jury against two bootleggers—had the top of his head blown off by shotgun blasts through the living room window of his home.

That did it. The local police and county sheriff's men clearly could not cope, assuming they were not themselves involved. A combined force of Chicago detectives and government agents descended on Chicago Heights in a dawn raid January 6, 1929. Their sweep included the estate of Oliver J. Ellis, a "respectable" friend of bootleggers. In a large outbuilding, the raiders found booze, 433 slot machines and a safe brimming with records.

The raid otherwise came to nothing. Though none of those arrested talked, including Ellis, records in his safe spoke eloquently of some $1.5 million in profits from slots for one year, plus records of checks Ellis had received and payed. The Special Intelligence Unit ran down each check. One, dated the previous summer, June 27, 1928, had suggestively gone through the account of James Carroll in the Pinkert State Bank of Cicero. The SIU focused on that check, determined to identify James Carroll. Bank officials declared they had no idea who or where he was, despite massive activity in his account—suspicious ignorance in so small a bank.

There matters had stood when Al Capone contracted pneumonia in January 1929.

On December 5, 1928, in a Cleveland hotel, twenty-seven gangsters from crime centers such as Chicago, New York, St. Louis and Newark had met to work out differences and arrange greater cooperation. Patsy Lolordo had attended from Chicago, just over a month before his murder. Joseph Guinta accompanied Lolordo. Four years earlier Tony Lombardo had brought Guinta from Brooklyn to help run the Unione. Guinta was then twenty-two. Now he acted as Lolordo's principal aide.

The Cleveland conferees had all been Sicilian, with Unione ties, but the rationality of cooperation and negotiation championed by Johnny Torrio had started to permeate all organized "business" gangs. The Unione, for instance, had long since admitted non-Sicilians. The conclave's existence came to light when an alert room clerk tipped off detectives about his guests.

The nationwide reaction to the St. Valentine's Day explosion required a wider conference. It may have been called by Frank Costello, rising crime star in New York. Enoch J. Johnson would play host in May—Nucky Johnson's Atlantic City, New Jersey, being both an agreeable locale and so corrupt, with Johnson in such firm control, that there could be no question of the police annoyance the Cleveland conferees had suffered.

Before Capone left for Atlantic City he had to dispose of some nasty internal business.

Perhaps Capone's seeming detachment and frequent lengthy absences from Chicago inspired the trouble: when the cat's away the rats forget their place. Perhaps it was just that success seduces and breeds manic thoughts.

Joseph Guinta succeeded Patsy Lolordo after the Unione president's January 9, 1929, murder. A dandified hyperactive, Guinta loved power and dancing, his gyrations in aid of both pursuits so frequent and frenzied many called him "Hop Toad." His elaborate dress, his cocky demeanor and exhibitionism all spoke of a vanity that could make a twenty-six-year-old listen to whisperings of unlikely ambition.

John Scalise should have known better. He should have been able to assess the vanishingly thin chance of successfully betraying someone who commanded such loyalty from the outfit and who had so extensive an intelligence net throughout the city. But Scalise had become puffed up by his repeated role in so many of Capone's famous successes. Soon after the massacre, Scalise was heard to brag, "I am the most powerful man in Chicago."

Two reports—not mutually exclusive—explained how Capone discovered the planned betrayal. One was that Scalise had been seen and heard huddled with Joe Aiello in a Waukegan restaurant, a waiter sending the tale to Capone. Scalise and older, dimmer Albert Anselmi would accept Aiello's offer of $50,000 for Capone's murder. Guinta had elevated Scalise to vice-presidency of the Unione. With that base, those two and Anselmi would usurp Capone's empire, the North Side ceded to a resurgent Aiello.

The more vivid version credited faithful Frankie Rio with first nosing out the plot, unable to convince Capone of such perfidy from the two whose lives he had refused to barter for much needed peace with Hymie Weiss in 1926. Rio confected a charade to convince Capone. They feigned a furious falling out at a dinner where Scalise and Anselmi were guests, Rio slapping Capone and bolting from the room. Scalise and Anselmi looked up Rio the next day and proposed an alliance, detailing their plans with Aiello and Guinta.

With confirmation of treachery, Capone's hurt fury demanded more than instant vengeance. "It was Nitti's idea," says George Meyer. "I was in an office and Capone came in with Nitti and Joe Fischetti." They started to talk and Meyer got up to leave, thinking the big shots wanted privacy. "Stick around," Capone told Meyer. "You're gonna know about it anyway." Nitti suggested a banquet for the outfit's top people, the three plotters guests of honor. The preceding gaiety and sense of camaraderie and security would make the subsequent terror all the more exquisite.

Invitations were issued for Tuesday night, May 7, at The Plantation, a roadhouse and casino that dripped Old South magnolia charm near Hammond, Indiana, just over the line from Burnham, Johnny Torrio's first suburban colony. The banquet would be in a private back room. "We frisked everyone going in as usual," says Meyer.

An autopsy showed that Capone lulled Scalise, Anselmi and Guinta with a heavy meal, washed down with streams of wine, brandy and coffee. When they were sated, Capone struck. He personally beat each one with a cut-down baseball bat, Scalise the most severely, Guinta the least. "They said," recalls Meyer, who remained in the anteroom on guard, "Capone got so worked up they thought he had a heart attack."

Barely alive after the beating, each was shot, Scalise's little finger torn away as he raised his hand to ward off the bullet headed for his blackened eye. Their wounds suggested more shots by three or four different assailants, slugs pumped into their backs after they fell prone on the floor, dying or already dead. A pathologist who examined them said he'd never seen such badly beaten bodies.

About one-thirty next morning two Hammond policemen found the trio in a lonely spot called Spooners' Nook. Scalise and Guinta were kneeling on the floor in the back of a stolen car, their torsos on the back seat; Anselmi lay twenty-two feet away in the road.

Capone left almost immediately for the summit at the President Hotel in Atlantic City. Nearly every important crime figure from the East and Midwest attended, with a glorious American disregard for creed or national origin: Costello, Meyer Lansky, Charles Luciano and Dutch Schultz from New York; Charles Solomon from Boston; Abner Zwillman, overlord of New Jersey; Abe Bernstein from Detroit's Purple Gang; Max Hoff, kingpin of Philadelphia; and others from Florida and New Orleans. All factions—except Moran—were represented from Chicago, the hot spot whose killings were giving bootlegging, gambling and prostitution a bad name, threatening a national cleanup. Capone brought with him Jack Guzik, Frank Nitti and the inevitable Frank Rio.

"I told them," Capone remembered later, "there was business enough to make us all rich and it was time to stop all the killings and look on our business as other men look on theirs, as something to work at and forget when we go home at night. It wasn't an easy matter for men who had been fighting for years to agree on a peaceful business program. But we finally decided to forget the past and begin all over again and we drew up a written agreement and each man signed on the dotted line.''

Johnny Torrio, an important presence at the meeting, must have been proud of his disciple. The agreements followed the pattern that Torrio had negotiated in Chicago before 1923 and that Capone had tried to enforce ever since: rational pooling of corruption efforts; fixed territories; a united front against outsiders, the incorruptible and reform; and arbitration by an umbrella commission, chaired by Torrio. Above all, they agreed to end the public violence that had quickened public opinion against them.

Moran, still in hiding, might present a problem. Some Chicago Sicilians might resent the deaths of Scalise, Anselmi and Guinta. Whether it was the newfound commission's idea, Torrio's or his own, Capone decided to remove himself for a season as target of revenge and focus of public outcry. He set up his own arrest in Philadelphia.

According to Capone, he, Frankie Rio and two other bodyguards left Atlantic City by car early the afternoon of May 16 in plenty of time to catch the Chicago-bound *Broadway Limited* from New York when it pulled into North Philadelphia Station at 4:40. A cylinder head blew some fifteen miles south of Camden, however, and they missed their train, limping into the City of Brotherly Love about six-thirty. Leaving luggage at North Philadelphia, they returned downtown to the Broad Street Station to make reservations for the next Chicago train, leaving at 9:05. Capone thought it best not to be recognized. Strolling along Market Street, the four spotted the marquee of the Stanley Theater at Nineteenth Street: Willard Mack in *Voice of the City*—a ''thrilling Detective Story''—plus a stage show, Fred Waring's Pennsylvanians.

Two detectives, Lieutenant John J. Creeden and James H. Malone, were driving east on Market when Shooey Malone spotted Capone entering the theater. Malone had seen Capone in Florida at the Sharkey-Stribling fight when there (accounts varied) either on vacation or medical leave. That was about seven-fifteen. Creeden and Malone called for backup police, then waited. About eight-fifteen, Capone and Rio stepped into the Stanley's foyer, the other two well behind them.

At that point, accounts sharply differed. In the most vivid version, Malone said, ''You're 'Scarface Al' Capone.''

''Right,'' Capone was supposed to have said, ''Who are you?'' (Or he replied, ''My name is Al Brown—call me Capone if you want to'' before he asked who this stranger was.)

When the detectives flashed their shields, Capone said, ''Oh, you're bulls; here's my gun,'' surrendering a snub-nosed .38 and whispering ''Bulls,'' to Rio, who was thrusting himself protectively forward. ''Oh, that so?'' said Rio. ''Here's my gun too.'' The other two bodyguards merged with the crowd, never seen again.

Other first reports gave variations of the same essential difference: at first glimpse of Malone, Capone caroled, ''Hello, Shooey!'' then volunteered ''I've got a rod on me, Shooey,'' and immediately produced the snub-nosed .38 from his topcoat pocket.

Malone quickly denounced all ''Hello, Shooey'' versions and denied rumors that he and Capone knew each other. He'd only *seen* him at the fight; they hadn't even met. Lieutenant Creeden confirmed Malone, adding that no words had passed between the sides until after the disarming, and then only Capone's wondering why they were being picked up, since they were just passing through town.

By the time Capone and Rio were booked, fingerprinted and photographed, the hour was late; police had to roust Magistrate Edward P. Carney from his Hotel Sylvania bed for arraignment at the Twenty-second and Buttonwood precinct station. At city hall, attorneys Bernard L. Lemisch and Cornelius Haggarty, who somehow magically appeared to represent the gun-toters, had received a promise that no arraignment would take place without their presence. But by 11:35, before the lawyers arrived, Magistrate Carney had acted. He ordered Capone and Rio held on $35,000 bail, each.

Capone had less than $30 in his pockets, not enough for decent tipping money; Rio had $12. They were hustled over to the detective bureau in city hall. When the lawyers charged in—moments too late—they were furious. ''I've seen you hold men for carrying guns in less than four hundred dollars bail,'' said Haggarty.

''I would have held these men in one hundred thousand dollars bail if I had the authority,'' the magistrate said.

Director of Public Safety Lemuel B. Schofield wanted his turn. From midnight to two A.M., in front of a few officials—and a stenographer—he chatted with Capone, whom he found in a ''reminiscent mood,'' as Major Schofield later put it. ''He seems to be at a point where he is anxious to be at peace not only with the gangsters, but with the more serious elements in law.'' Schofield was mightily impressed. ''He talked throughout as a serious man talking of a responsible matter.''

Schofield's impression of Rio was different. Asked for a statement, Rio snarled, ''Give me a mouthpiece and I'll talk.'' Capone waved him down. ''Listen, boy,'' he said, ''you are my friend and have been a faithful bodyguard, but I'll do the talking. We are in a bad jam here and the only way out is to tell the truth and in return expect to be treated as leniently as possible under the law.'' The truth Capone told was of course highly selective and self-serving, but it rings no less true as an index to his state of mind.

''I went into the racket in Chicago four and a half years ago,'' he said, fudging the years of his apprenticeship to Torrio. ''During the last two years I've been trying to get out. But once in the racket you're always in it, it seems. The parasites trail you, begging for favors and for money, and you can never get away from them, no matter where you go.

"I have a wife and an eleven-year-old boy I idolize, and a beautiful home at Palm Island, Florida. If I could go there and forget it all I would be the happiest man in the world. I want peace and I'm willing to live and let live. I'm tired of gang murders and gang shootings.

"With the idea in mind of making peace among the gangsters in Chicago I spent the week in Atlantic City, and I have the word of each of the men participating that there shall be no more shootings."

When Schofield asked him to elaborate on the peace conference, Capone said, "We stopped at the President Hotel, where I registered under an assumed name." Then he went on, exercising more hope than memory, that "Bugs Moran . . . and three or four other Chicago gang leaders whose names I don't care to mention participated. We talked over our troubles for three days. We all agreed at the end of that time to sign on the dotted line to bury the past and forget warfare in the future for the general good of all concerned."

Director Schofield asked the statesman how he had done in the now-past gang wars and whether the odds had been in his favor. "I'm satisfied," came the reply, "but it's an awful life to live. You fear death every moment, and, worse than death, you fear the rats of the game, who would run around and tell the police if you didn't constantly satisfy them with money and favors. I never was able to leave my home without my bodyguard, Rio. He lives with me and has gone with me constantly during the last two years."

Capone did not include in the category of "rats" the three who had sought to betray him, but they were plainly on his mind, since he brought them up when Schofield asked how he ever achieved peace of mind harried so by the constant threat of death.

"Well," said Capone. "I'm like any other man. I've been in this racket long enough to realize that a man in my game must take the breaks, the fortunes of war.

"Three of my friends were killed in the last two weeks in Chicago. That certainly is not conducive to peace of mind. I haven't had peace of mind in years. . . ."

"What are you doing now?" asked Schofield.

"I'm retired," said Capone, straight-faced, "and living on my money."

Schofield couldn't quite swallow that. "You should get out of the racket and forget it," he said.

"I can't," said Capone, "because of the parasites in the game. They follow you no matter where you go. I fear the parasites more than death."

Capone and Rio spent the rest of the night, two to eight A.M., getting what sleep they could on benches at the detective bureau. Philadelphia's detectives were in for a treat. They held a two-man lineup (then called a "stand up"), the entire squad invited. At command, the two turned profile and front, put on hats and doffed them, standing against the lines that measured height, under the glaring floodlights. Rio owned to his record: a 1915 arrest and six months for stealing bonds, many other arrests on charges including gun-toting—all dismissed.

Captain Andrew Emanuel turned to Capone. "You are charged," he said, "with being a suspicious character and with carrying concealed deadly weapons. What have you to say?"

Capone laughed, "Oh, nothing, nothing."

After Capone recited his arrest record, noting the dismissals, Captain Emanuel said, "You've never done any time, anywhere."

"No," Capone confirmed with quiet pride, "not a minute."

In another part of city hall, detectives Malone and Creeden were huddled with the district attorney. At 10:15 they addressed a grand jury, and at 10:25 Philadelphia had an indictment. At 11:30 the trial began before Judge John E. Walsh. Detectives surveyed the courtroom, weapons in their hands; uniformed officers patrolled the corridors.

Capone seemed unperturbed, gassing pleasantly with bystanders before the first gavel. "What's the weight of that ring?" asked one, eyeing Capone's Jagersfontein pinkie sparkler.

"It's exactly eleven and a half carats."

"Worth about fifty thousand, eh?"

"Well, you made a good guess."

At about twelve-fifteen, after a sidebar conference with the judge, Lemisch engaged his clients in a few minutes' earnest discussion. Observers saw Capone's face turn a "dull reddish hue"; but he nodded agreement as his lawyer spoke. Lemisch turned to the judge: his clients would change their pleas to "guilty."

"All right," said Judge Walsh. Then, without pause, "Each of the prisoners is sentenced to one year's imprisonment."

It was the maximum penalty for an offense that usually drew only a fine, ninety days at most. Capone was going to jail for a lot longer than he had expected.

"It's the breaks, kid, it's the breaks," he commented to a bailiff as he was led away.

The story circulated almost immediately that the arrest had been set up. In Chicago, Crime Commission head Frank Loesch figured it was to "escape vengeance of rival gangsters"; investigator Pat Roche called it a "desperate measure . . . to escape death." Mayor of Philadelphia Harry A. Mackey thought Capone was "glad to get into jail to evade the killers who were on his trail"; Capone's own lawyer, Cornelius Haggarty, allowed there might be something to such opinions. *The New York Times* decided "that if he is in prison the reason is that he wants to be." The speed—just over sixteen hours from arrest to jail—seemed unprecedented and suspect. The Philadelphia *Record* stated flatly, " 'Scarface Al' Capone went deliberately to jail."

Not everyone believed that. Capone's sister Mafalda at first refused to believe in the *possibility* of her adored brother's finding himself in such a pickle: "He never gets in jail," she wailed, hearing of newspaper dispatches that at 12:50 Capone had entered Moyamensing, the Philadelphia County workhouse. In Florida, Mae Capone also doubted the setup theory. "What would he want to go to jail for?" she asked. "That's one place he never had a desire to go. He liked to talk of Europe, of Palm Beach, of the famous race tracks, of the scenes at the big fights—but jail—oh, no, not for Al."

Then why was he carrying a gun? Since his Joliet arrest, Capone had sensibly decided to leave gun-toting to his bodyguard. Why no money for bail? Why none of the usual delays? Why a guilty plea?

Naturally, the principals of the arrest denied a setup. "It's enough to make anybody want to quit being a copper," complained Shooey Malone. "Here we take a chance and pinch two of the most dangerous characters in the racket. They have guns on them and know how to use them. Then everyone hollers 'Frame up.' " He could scarcely say anything else.

Neither could Capone. "I didn't give myself up," he said. "I'm here because I'm here"; if his car hadn't blown that cylinder head, he'd never have been there. What else could he say? Admitting he had given himself up could excite both friends and enemies to foolishness in the delusion that he had turned yellow. As he put it some months later, "You don't get a reputation like mine hiding in jails."

The jail he soon inhabited could only have augmented his reputation for toughness. Moyamensing housed only short-term prisoners and those awaiting trial. After one night there, a Black Maria transported Capone and Rio to Holmesburg, the regular county prison in the northeast outskirts of Philadelphia, a grim, dank fortress backed up against the Delaware River, with a fetid creek meandering through the yard. Holmesburg was, a Chicago paper exulted, "known throughout crookdom as the toughest place between the two oceans in which to take a rap." Its guards gloried in their reputation as "manbeaters," and a prisoner was "lucky" to "come out with nothing worse than rheumatism." The food had sparked riots not long before.

Capone's hair was cropped, his $135 custom suit swapped for a white cotton shirt and grayish blue baggy trousers with a broad black stripe down each seam, his white *borsalino* for a prison-gray cloth cap. "It's not much like home, here," Capone remarked to the warden.

Why, some argued, would Capone choose a city with so tough a jail? Why not Chicago or a neighboring county, where, as Druggan and Lake and Torrio had shown, jail terms for the connected were not onerous. Because, as John Stege pointed out, throughout Illinois, police could not search people without a warrant or probable cause; warrantless concealed-weapons arrests were thrown out of court. Why not Miami, where the lockup sat atop the high-rise county building with a view of Biscayne Bay? Because Capone wanted to live in Miami and did not dare further fuel the opposition he had already encountered.

Besides, he probably expected to stay in easygoing Moyamensing with the usual trifling sentence of ninety days or so, just long enough for the world to cool down. Unhappily the fix was not *thoroughly* in. Philadelphia had just been wracked by a grand jury exposé of police corruption, and (as a Chicago paper pointed out) had "seized eagerly upon this gaudy opportunity of showing Chicago some fancy justice."

The boys in the outfit got word from Nucky Johnson that it had indeed been a setup.

Holmesburg was not the *only* prison available. But for now, Capone was in jail for a year—ten months with good behavior—and that, it seemed, was where he was going to stay.

CHAPTER 22

Get Capone

THE 1928 DEMOCRATIC NOMINATION of Governor Alfred E. Smith of New York turned out only marginally less gaga than a Republican nomination of Big Bill Thompson would have been. Secretary of Commerce Herbert C. Hoover trounced Smith by a landslide, a plurality of over six million votes, carrying all but eight states. Hoover took office March 4, 1929, between the St. Valentine's Day massacre and Capone's delayed federal grand jury appearance.

The portly president got in the habit each morning of gathering his closest advisors for mild exercise. They would toss a medicine ball among them as they arranged the nation's affairs. Soon, each session of Hoover's "Medicine Ball Cabinet" started with Hoover's asking, to the thump of ball against belly, "Have you got that fellow Capone, yet?" and ended with Hoover's demand, "Remember, now; I want that man Capone in jail."

Rumor had it that Hoover's implacable enmity originated the previous January in Miami when he stayed at J. C. Penney's Palm Island estate. Entering the lobby of a hotel, Hoover had been receiving plaudits due a president-elect when suddenly the crowd deserted him, rushing to Capone as he came in. A subsidiary version had Hoover furious because the all-hours roistering at number 93 had disturbed his sleep. Hoover emphatically denied personal animus. He told Elmer Irey, head of the Treasury's Special Intelligence Unit, that he had never seen or heard Capone. While he had always thought something *should* be done about so infamous a gangster, the spur came when *Chicago Daily News* publisher Frank Knox led a delegation of equally prominent Chicagoans to demand federal action, since local authorities could or would do nothing.

Until then, Irey had believed the miffed-feelings canard, which seemed plausible, neatly encapsulating the reasons anyone of Hoover's massive dignity and sense of propriety would detest Capone. ''Probably no private citizen in American life,'' wrote *The New York Times* about the gangster, ''has ever had so much publicity in so short a period.'' He had become, the Chicago *Daily Times* complained, our country's ''trademark . . . known in the jungles of Java or the wastes of Lapland,'' indeed better known throughout the world than any other American, even Charles Lindbergh or Henry Ford! Personal feelings aside, such a situation would inevitably scandalize a fine Quaker gentleman like Hoover. At the annual Washington Press Club Gridiron dinner, a newsman spoofing a notably Dry senator tried to steer the good ship Prohibition to safe harbor, calling for volunteers. ''Capone,'' a stowaway, stepped forward, explaining that he owed to Prohibition ''all that I am and all that I hope to be,'' and then sang parody lyrics to ''The Curse of an Aching Heart.'' Hoover's tight polite smile did not reflect unrestrained merriment at the thought that the doings of a gangster could be thrown up in mockery to the President of the United States.

The Cabinet member directly charged with getting Capone was Secretary of the Treasury Andrew Mellon, since J. Edgar Hoover's FBI had refused so unpromising an assignment. When George Johnson became U.S. attorney in Chicago at the end of 1927, he had requested special Prohibition agents. One of them shot a relatively innocent bystander on a booze raid and was almost lynched. Johnson gave up on Prohibition agents. But the man who was determined to make his name, and realized that getting Capone would be the surest way, would not give up on that. Johnson asked for Elmer Irey's SIU.

Born 1888 in Kansas City, raised in Washington, Irey had no politics and had never voted. He had become a stenographer in the Post Office Department and worked his way up to respectable rank in the well-reputed corps of postal inspectors. He transferred to Treasury as head of the SIU through the intercession of a friend.

Irey had sent a dozen agents to Johnson, led by Arthur P. Madden. After the January 1929 South Chicago Heights raids that uncovered the records of Oliver Ellis, two of Madden's agents began tracing that intriguing $2,130 check. At the time, Ralph Capone's case still pended.

Agent Archie Martin, a dynamo, teamed with small, dogged Nels Tessem, whom Irey described as a ''human comptometer.'' Having traced the check to the account of James Carroll at the Pinkert State Bank in Cicero, and with help from four other agents back at headquarters, Martin and Tessem scoured some six million Pinkert bank entries. After six months, they noticed that the Carroll account had opened simultaneously with the closing of an account in the name of James Carter, the Carroll opening balance matching Carter's closing balance. The Carter account, in turn, had opened with the closing balance of James Costello, Jr. Pinkert officials knew nothing of those substantial depositors, either. Nor of Harry Roberts, whose balance had opened Costello's account, nor of Harry White, whose closed account funded Roberts's new one.

The trail ended with White. On October 27, 1925, Ralph Capone had closed an account in his own name, the proceeds opening White's. Confronted with these facts, the bankers protested that they had never seen Ralph since he closed his account, nor of course had they ever seen Ralph's aliases, but they admitted that messengers had brought the deposits to those accounts, always accompanied by the same obviously hoodlum guard whose name they did not know. Based on their reluctant description, SIU agents produced a picture of Antonio Arresso, one of Ralph's bodyguards. Bank employees agreed: that was the man who guarded the deposits.

Just for the hell of it, the SIU checked the balance of James Carter, the alias current on October 4, 1927, the day Ralph swore that he was too broke to pay his tax debt but would borrow $1,000 if the IRS agreed to settle. The account contained $25,236.15.

It wasn't the one they wanted most, but the government had their first Capone.

Meanwhile, the one that Philadelphia had wanted out.

Mae came to visit, along with Theresa and Mafalda, but abandoned plans to move to Philadelphia when she learned that Holmesburg allowed only one visit each month. Prisoners could receive any number of letters, censored, but could write only two a month, heavily censored.

Ninety days, even in Holmesburg, might have been supportable as a useful cooling-off period; a year would be intolerable. Capone offered $50,000 to any lawyer or group who could finagle him out of jail; some of his associates let Philadelphia D.A. John Monaghan know money would reward his helpful attitude. More formally, Bernard Lemisch's law partner, Congressman Benjamin M. Golder, undertook Capone's appeal, claiming Judge Walsh had been prejudiced. Poor Capone had been "coerced" by the "hostile and theatrical display," the crowds, and all those armed detectives.

For the appeal, Capone turned circumspect. "You see," he told a reporter, explaining an unwonted disinclination to grant interviews, "I want to get out of here. If the public is constantly reminded that I still am in jail it will be that much worse for me. The less I say the quicker the public will forget me." That did not include good works that were anonymous—until somehow made public. But in spite of Capone's $1,000 donation to a children's hospital, the court turned down his first appeal. Golder made another.

Meanwhile, Capone's friends knew how to induce other Philadelphians to show more brotherly love than the D.A. and courts had. Authorities transferred Capone and Rio on August 8 from Holmesburg to the Eastern State Penitentiary. Later, those responsible could not agree whether they had acted because of threats against Capone's life or because of overcrowding. Unlike Holmesburg, which assigned numbers only to felons, not for gun-toting miscreants, the state awarded each of its prisoners a number. Capone was now C-5527.

To the eye, Eastern seemed no less grim. The hulking Norman castle, in service since 1829, filled the block from Twenty-first to Twenty-second Streets along Fairmount, just north of downtown Philadelphia. Although popularly still called by its original name, "Cherry Hill," as Eastern's official name implied, the state ran it, not the county, and it was paradise compared with Holmesburg. Prisoners called the warden, Herbert B. Smith, "Hard Boiled." As far as Capone could see, the nickname derived only from his initials.

The warden did not consider grim surroundings sound rehabilitative penology; he let Capone furnish his own commodious cell—located in what everyone called the "Park Avenue" block—with a rug on the floor, paintings on the walls, a polished desk on which a lamp suffused the cell with a soft glow, a French dresser, a smoking stand (shaped like a butler holding an ashtray), a radio console, a vase with cut flowers, and two comfortable cots, the other for his embezzler cellmate. Officials announced that "Capone was much pleased with his new surroundings." Smith was also lenient about letting Capone openly employ fellow prisoners as servants. Most important, Capone had plenty of visitors, got and sent lots of mail, the warden thoughtfully discarding the routine junk mail and begging letters without troubling his guest. Capone stayed in touch with Ralph and Jack Guzik, continuing to conduct business.

When Capone's second appeal fared no better than the first, Golder set about making more, including an appeal for parole on grounds that the sentencing alone had exerted a "salutary effect" and that the prisoner "has been sufficiently punished." But Capone settled in, had some minor sinus surgery on his nose. Two weeks later, in early September 1929, Herbert M. Goddard, vice-president of the board of prison trustees and a doctor, pulled Capone's tonsils. Goddard would get to know his patient well.

The night of October 8, a deputy U.S. marshal and SIU agent Clarence Converse—Pat Roche's old sidekick—arrested Ralph, in tuxedo, at Chicago Stadium, making him miss a scorching draw between featherweights Earl Mastro and Bud Taylor.

The government wanted Ralph held for preliminary hearing on tax charges, which featured a novelty. Resurrecting a Civil War statute drawn to attack war profiteers on grounds they had tried to "cheat, swindle or defraud" the government, the government construed Ralph's lying about inability to pay as that sort of fraud, which carried penalties more severe than tax evasion. Questioned, Ralph claimed that the cash in all those bank accounts was simply betting money "entrusted" to him, not his to pay taxes on. He spent the night in the detective bureau, making $35,000 bail next day in time to catch the last two innings of the World Series' second game, for which he had a pocketful of tickets when arrested (Philadelphia swamped Chicago, 9–3). He would be indicted in November (with Druggan and Lake—plus McGurn and Rolfe on the Mann Act charge) and tried the following April.

When Al Capone's appeal process had started, he had told other prisoners he expected to be out in time to see that World Series. By the time the Athletics had won in five games, he realized the fix was not going to be easy, maybe not possible. A lot of people blamed *him* for all the fuss of Prohibition. Overlooking the point that much of what he was blamed for he had indeed ordered, he complained, "Every time a boy falls off a tricycle, every time a black cat has gray kittens, every time someone stubs a toe, every time there's a murder or a fire or the marines land in Nicaragua, the police and the newspapers holler 'get Capone.' " He said, "They've hung everything on me except the Chicago fire." Universal blame had turned him peevish, but had not quite eradicated his humor. "Tell them," he instructed his lawyer, when the stock market crashed on October 24, 1929, "I deny absolutely that I am responsible."

By then he had made five appeals, all rejected. The state supreme court turned down his sixth in December. With no further need to remain low-profile, he started to mingle, to speak to the press, and to make small headlines again: CAPONE GAINS ELEVEN POUNDS and CAPONE LOLLS IN LUXURIOUSLY FURNISHED CELL and such like. He sent Christmas baskets to the poor families of seventy-five fellow prisoners and bought about $14,000 worth of prisoner handicraft to send as gifts.

Treasury's SIU agents kept poring over the leads afforded by records uncovered in raids. They concentrated on two targets whose loss Capone would find most devastating after his brother, Guzik and Frank Nitti.

Francesco Raffele Nitto had been born exactly ten years before Capone, on January 27, 1889, in Augori, Sicily, and brought to the United States at age two and a half. He started the process of naturalization twenty years later, on March 9, 1921.

As Nitti, he had become known as Capone's "enforcer," maintainer of the outfit's internal discipline, supervisor of necessary violence against outsiders, rival gangsters, fractious customers or extortees. Nitti did not look or act tough. He certainly in no way resembled the hopped-up, dead-eyed gunsel in the 1987 film *The Untouchables* (Nitti chain-smoked cigarettes, but would sip only the occasional glass of wine). Like Torrio, though, Nitti somehow instilled more terror precisely because of his flaccid smallness.

He had started as a barber, fencing stolen goods on the side, before he joined the outfit. With slicked-down hair parted slightly right of middle and large, liquid dark eyes, he looked like a bank teller worried about what the head cashier meant by that remark a week ago last Tuesday. A government agent claimed that a moustache Nitti grew as a disguise when trying to evade the SIU made him look like Hitler. It did not; it make him look even more rabbity.

Looks deceived. Although the misinformed or overly enthusiastic sometimes pegged Nitti as the "brains" of the outfit, which drastically overstated it, he was smart and tough-minded, and Capone had learned to rely on him. Soon Nitti became recognized as the outfit's number two. Especially with Capone away in Florida—and now Philadelphia—the men had come to look to Nitti so readily that he took it as his due after Capone's return. "He always acted kind of snotty,"

George Meyer remembers, "like he was in charge." He could be remarkably offhanded even with Capone. "We were in the elevator at the Lexington, once," says Meyer, "Capone and Nitti and some others, and Nitti says to Capone about some problem, 'I'll handle it, you just keep out of it,' and the rest of us just *looked* at each other."

With Capone in jail, everything cleared through Nitti. As one contemporary newsman put it, nothing important happened "until Nitti has given the word." He daily presided over the outfit's executive meeting at the Lexington, as the same newsman wrote,

> . . . and to them come the hoodlums of the mob to make their complaints—that Danny Stanton . . . is selling his bottled beer too cheaply; that one of the Loop's saloon keepers is "cheating" by buying some of his whiskey elsewhere; that some "copper" or bureau squad is getting too hungry . . . .
>
> . . . The "boys" are sent out with pistols or "pineapples"; the hungry policeman is warned or transferred; the friendly beer or booze distributor is cautioned to maintain prices . . . .

Part of Nitti's duties involved oversight of the gambling operation, raking off his share of the profits. As with Ralph, this involvement gave the SIU its wedge. From seized records and squeezed informers, agents filled in a detailed portrait of how one Cicero operation worked, presumably the pattern for most. Every morning, from the previous day's take, the casino kept $10,000 to work with. One of the managers took the excess to the friendly Pinkert bank and bought cashier's checks made out to "J. C. Dunbar," a fiction. The SIU found Nitti's signature endorsing one of those checks. Made out for $1,000, it had been presented for collection through the Schiff Trust and Savings Bank in Cicero.

Schiff Bank officials huffily affected total ignorance. They had never heard of Nitti and had no record of any such depositor. Nels Tessem began his usual exhaustive search, examining all the bank's deposit slips for the day that check had cleared. Nitti's name never appeared and no slip failed to match the name and account of a bona fide depositor; unlike Ralph, Nitti used no alias. Tessem constructed his own set of bank books for that day. They showed an encouraging discrepancy: the bank had taken in exactly $1,000 more that day than it had paid out or credited to accounts. Armed with this evidence, Tessem demanded the bank's general ledger and all general ledger tickets itemizing each transaction.

He hit pay dirt. The ledger and tickets revealed a string of checks collected on Nitti's behalf. With this screw to twist, the SIU soon got the bank's president to admit he had arranged with Nitti to clear checks and deliver their proceeds without having Nitti's name show on the books, something then legal if neither usual nor ethical. On March 14, 1930, just before Capone came due for release, a federal grand jury indicted Nitti on five counts of evading $158,823.21 in taxes on a provable income of $742,887.81 for the years 1925 through 1927.

The government kept the indictment secret, hoping to grab Nitti before the news leaked, but Nitti had already gotten word and disappeared. The need for secrecy gone, the government announced the indictment on March 22, 1930.

The case against Jack Guzik took digging of a different sort, but led more directly to Capone. The man most directly responsible for it was Frank J. Wilson. Irey had assigned Wilson, from the SIU's Baltimore office, to focus on Capone. By then forty-two years old, balding, with cold eyes flinty behind wire-rim glasses, a square cleft chin, mouth a lipless slash, Wilson feared "nothing that walks," as Irey once put it. He matched Tessem for doggedness; he would pore over a set of books, one of his endless chain of noisome nickel stogies clenched in that mouth, "eighteen hours a day, seven days a week, forever" to find what he wanted in them. Associates said that "he sweats ice water." Wilson had also shown in the past that he didn't mind treading "thin legal ice" to nail someone he thought guilty. "His methods," an admiring biographer wrote, "were often brutal."

Guzik had been much less circumspect than Nitti about endorsing those J. C. Dunbar checks; his signature adorned about half of them. It would hugely enhance the case to locate whoever had turned gambling receipts into checks at the Pinkert bank. Perhaps he had also distributed some of those checks Guzik so blithely endorsed.

No one currently at the bank would tell them a thing about "Dunbar." A former teller, though, remembered him well and described him accurately. They all called him "Fred," and knew he worked as head cashier at the Ship, which had become Capone's principal Cicero casino. Fred had been fussy, always insisting on crisp new bills whenever he got cash; they discovered why one day when, despite his tough look, he blanched at the sight of a cockroach on a stack of bills: Fred had a pathological fear of bugs. The teller remembered the incident especially because Fred had gotten so upset he had forgotten to fork over his usual $5 tip.

With the teller's lead, the SIU had little trouble identifying Fred Ries. Wilson traded a minor hoodlum freedom for information to learn that Ries was holed up in St. Louis. His flight had been mere precaution; not dreaming he'd be identified, much less traced, Ries had not bothered to adopt an alias. In fact, the St. Louis postal inspector told Wilson they were about to dispatch a special delivery addressed to a Fred Ries at a local hotel. Wilson followed the letter through the door; it had been sent by Louis Lipschultz, Guzik's brother-in-law, and contained money and instructions to lam for California.

At first, Ries didn't know anything about any Dunbar and had never seen the Pinkert bank. Wilson committed his prisoner as a material witness, choosing an ancient lockup in Danville, mid-state Illinois. It featured what he later described as "a specially designed cell," alive with vermin. "This ain't fit for a dog," he remembered Ries crying. He'd be shifted to a hotel when he talked—and by the way, what would Capone do when word got out that Ries had spent time with the government? Ries might brush off bugs, but after four days' immersion in the reality of what Wilson's question had suggested, on September

18, 1930, he unloaded. He would testify against Guzik two months later, Guzik drawing a sentence of five years and one day.

Ries now had no choice: for continued protection, he'd have to testify against Capone—as soon as the government could build an indictable case. The government naturally wanted to keep him whole and available. The investigators hit on the idea of sending him on a cruise to South America. But in those days, the budget would not stand such extravagance. A private group came to the rescue. Robert Isham Randolph, chairman of the Chicago Association of Commerce, had formed a special group of six men (including Samuel Insull, who was stealing millions with pyramided utility companies) called the Citizen's Committee for the Prevention and Punishment of Crime. When he refused to give newspapers the names of the other five members, the group was called "The Secret Six." They gave the SIU enough money to send Ries to safety until needed.

That would be a while, because it would take much more digging to tie income to Capone and to discover a starting point for assets. As Wilson had earlier remarked to Irey, "Nitti and Guzik got careless. Al hasn't." At the time of Capone's release from jail in the spring of 1930, the SIU would still lack proof against him.

The manner of Capone's release, the posturing and the fuss surrounding it, would have told a visitor from outer space everything worth knowing about Capone's celebrity and the general state of corruption.

With two months off for good behavior, Capone's (and Rio's) sentence would expire at midnight, March 16; in theory, he could walk through the little door that opened in Eastern's massive iron gate at 12:01 A.M., Monday, March 17, 1930, St. Patrick's Day.

How would he get out of town? Some reports made reservations for him, ten months late, on the same *Broadway Limited* he missed in May. Others chartered a fourteen-passenger Ford trimotor, standing by at $200 an hour on the tarmac of Camden's Central Airport.

How would he reach train or plane? The least problem might be federal agents and Chicago police rumored eager to pick him up as he stepped to freedom. More sinister, the streets of Philadelphia reportedly teemed with gunmen waiting for a shot as soon as Capone appeared in the gate. The Moran remnant supposedly lurked nearby. Even some Klondike O'Donnell troops, thirsting to avenge Duffy, Doherty and McSwiggin, were thought to be in town.

Capone's prison surgeon and newfound friend, Herbert Goddard, professed concern. When that door in the front gate opened at night, light silhouetted anyone stepping through, a perfect target. "We certainly owe it to the man to see he gets out with no bullet blemishes," said Dr. Goddard.

Goddard had come to respect his patient and prisoner enormously. "It's hard to believe," chirped the beguiled doctor, "the things that are told of him. In seven years of prison supervision, I have never seen a prisoner so cheery, so kind, so unassuming. He never asks a favor. He does his work—lately that of a file clerk—faithfully and correctly.

"Why, that man would make good at anything. Power? he's full of it! He's more than a model prisoner. He's perfect!"

Goddard fretted needlessly over snipers' having a back-lit target. The moment of Capone's release was not automatic. John S. Fisher, governor of Pennsylvania, had to sign the Pardons Board's remission of the two months good time. Fisher had been on a West Indies cruise and was due into New York on Sunday, the 16th. Some expected him to dash to Harrisburg to sign the commutation so Capone could leave at the first possible second. But the governor's secretary announced that Capone would be treated like any other prisoner, his papers would be signed in the ordinary discharge of business. Fisher confirmed that he would sign only "in the routine way."

Few believed it. Several hundred sensation-seekers and reporters (and, for all anyone knew, gunmen) congregated outside the Fairmount Street gate by midnight, hoping for a glimpse of—what? A fast car taking Capone and Rio in a mad dash to safety? Maybe a specially bulletproofed sedan? Prison officials derided those ideas, too; Capone would get no special treatment *or* transport—except maybe a police escort to the city limits. By two A.M. most of the disappointed mob had wandered off; no one had come through the gate. Maybe the authorities *weren't* lying.

The watch resumed on announcement that the governor had signed Capone's papers at 11:10 that morning. The secretary of the Pardons Board said he'd deliver them in his own sweet time. "I don't know whether I'll take them down myself or put them in the mail, special delivery," said Francis H. Hoy, Jr. "After I get some of my work done I'll make up my mind. There will be no special consideration in the Capone case." A Philadelphia afternoon newspaper calculated that the soonest Capone could expect release was between eight and nine o'clock that night. Even so, the watchers at the gate steadily numbered near three hundred through the day, topping out at five hundred.

"What the hell's the use of you people standing around here?" crowed Warden Smith to those still huddled about the gate at eight o'clock that night. "We certainly stuck one in your eye that time." Exactly twenty-four hours before, at eight o'clock, Sunday evening, they had whisked Capone (whom Smith called "the big guy") and Rio out in an ordinary prison car, past the gathering crowd, none of whom recognized their quarry. The car deposited Capone and Rio thirty miles west in the new state pen at Graterford. Monday morning, Smith himself drove to Harrisburg for the release papers once Governor Fisher had signed them. Smith got them from Hoy and phoned Graterford. Capone and Rio were released—with no hullabaloo and no crowds—about two o'clock that March 17 afternoon. Wasn't this unprecedented treatment? "Yes," admitted Smith. "But my action was proper and justifiable, as I was ordered to do so by Dr. Goddard."

In his official statement on the release, Goddard explained that they had been worried about those reports of lurking enemies.

Members of the Capone outfit believed Smith had an autographed photo of Capone hanging in his office. Outside Eastern's gate that Monday night, a voice from the crowd called, ''How much did you get for this, warden?''

''You get the hell out of here and stay out,'' argued Hard Boiled Smith, face flushed.

Al Capone was free.

CHAPTER 23

No Place Like Home

CAPONE WAS FREE—but where? ''Al Capone is back in town and sitting at an elaborate banquet table,'' one Chicago reporter wrote the day after Capone's release. ''Al Capone is not back in town,'' he continued, parodying the confusion. ''Al Capone is on a train somewhere between here and Philadelphia. He's in an airplane flying back to Cicero. He's no such thing!'' The plane in Camden never took off. Reporters searched the *Broadway Limited* and found no trace of Capone.

''Boo!'' another paper started its front-page story two days later. ''I'm Al Capone. You're Al Capone. . . . Here he comes! There he goes! He's in that plane. He's on that bicycle! Sh-h! It's Al.'' The police of Charlotte, North Carolina, ''expressed the belief'' he was headed their way. Miami police had the approaches to Palm Island covered. Hundreds of Chicago police had every terminal staked out and the Prairie Avenue house under siege.

Capone's twelve-year-old nephew, Ralph junior, gathered *his* gang and engaged the police in a snowball fight. A reporter tried to pump him, using a bag of candy for bait. ''Where's grandma?''

''Out.''

''Is grandma going to have a special kind of spaghetti for Uncle Al's dinner?''

''Yeah, walnut flavored, probably.''

That was all the hard news Ralphie would vouchsafe. To further questions he replied, ''Say, I won't tell you anything. Another paper sent some people out here to play marbles with me. I won ninety cents from them and didn't tell them a thing.''

Tough talk abounded. In Chicago, John Stege announced that Capone would be arrested on sight as part of Chicago's campaign to rid the city of hoodlums. The governor of Florida, Doyle E. Carlton—whom Capone had counted a pal when Carlton was still a Tampa lawyer—added a new twist. Carlton sent identical telegrams to the sheriffs of his state's sixty-seven counties: IT IS REPORTED THAT AL CAPONE IS ON HIS WAY TO FLORIDA. ARREST IF HE COMES YOUR WAY AND ESCORT TO STATE BORDER WITH INSTRUCTIONS NOT TO RETURN. HE CANNOT REMAIN IN FLORIDA. IF YOU NEED ADDITIONAL ASSISTANCE CALL ME.

In fact, Capone and Rio had driven to Chicago. As he later explained, Capone had donned his glasses, introduced himself and signed hotel registers "as Doctor So-and-So." No one had recognized him.

Federal agents knew where he was. As soon as Capone reached Chicago, he got wildly drunk, a fact discovered by a federal wiretap on Ralph's Montmartre Café.

Prohibition enforcers had not noticeably improved, but their new Chicago supervisor, Alexander Jamie, was both effective and honest. The brother of Jamie's wife was a twenty-six-year-old Prohibition agent named Eliot Ness, a University of Chicago graduate, tennis player and Ph.D. Toward the end of September 1929, Ness prevailed on Jamie to have their superiors institute a special group of raiders composed of men known to be as yet uncorrupted, the unit small enough so its supervisor could see they stayed that way. Ness was appointed leader of the group of ten, all under thirty years of age. Newspapers would soon call them "The Untouchables." They had installed the wiretap on the nightclub Ralph used as headquarters.

"Is Ralph there?" the eavesdroppers heard when Capone's whereabouts still puzzled the world.

"On the phone," came Ralph's gruff voice.

"Listen, Ralph, we're up in Room Seven One Eight in the Western," said the caller (the Hawthorne Hotel had adopted that name to escape its taint), "and Al is really getting out of hand. He's in terrible shape. Will you come up, please? You're the only one who can handle him when he gets like this. We've sent for a lot of towels."

As soon as he sobered up, Capone decided to clear up his status with the police and prosecutors. He had his lawyer, Tommy Nash, phone to ask John Stege if he had any charges against Capone. The police had no warrant, Stege answered, but were arresting hoodlums on sight. "I'll bring him over," Nash had said. "He wants to get it over with." Tony Berardi, Capone's favorite news photographer, says that he and Chicago *American* editor Harry Read convinced Capone to make his inquiries in person, Berardi tagging along to photograph the rounds.

They stopped first at police headquarters, where Stege waited with Harry Ditchburne from the state's attorney's office. Capone was his usual sartorial vision, marred only by a hand swathed in bandages. "I burned the palm of it," he told the inquisitive, "taking a roast out of the oven." During the interview, detectives kept popping in and out of Stege's office. "I want the men to see him," Stege explained, "because I have issued orders that he is to be thrown into a cell every time a policeman sees him on the street." On what grounds? "You're not a good citizen," Ditchburne charged. "If you were walking along the street with your brother and he was killed you wouldn't come in here and tell us who killed him."

"Well," said Capone—who in fact *hadn't* said much when police had killed brother Frank—"put yourself in my place and see what you'd do."

Ditchburne said they had to protect the public; the murder rate had been getting out of hand. "That's why we are going to drive you out of town," Stege added.

"But," protested Capone's lawyer, "you haven't any right to arrest him unless you have evidence of some crime against him." He turned to Ditchburne. "You as a lawyer know the police can't do that."

"I'm not here to tell the police what not to do. I'm here to advise them what to do. I'm not interested in protecting Capone. If Capone feels that he is being arrested wrongfully he has his remedy. He can sue for false arrest."

The idea tickled Stege. "Go as far as you like in suing me. . . ."

"I don't want to sue anybody." said Capone. "All I want is not to be arrested if I come downtown."

"You're out of luck," Stege snapped. "Your day is done. How soon are you going to get out of town?"

Capone hoped to leave for Florida the following week, Governor Carlton's telegrams notwithstanding. Stege said he was free to leave headquarters; but next time they'd put him in the lockup. "You'd better advise him to get out of Chicago," he said to Nash.

"Lenin and Trotsky and others rebelled against that kind of treatment," said the lawyer.

"I hope Capone goes to Russia," said Chicago's deputy police chief.

Instead, Stege sent Capone on a tour of the federal building and the state's attorney's office. No one wanted Capone for anything. As State's Attorney John Swanson put it, later, "If I want to see anybody, I'll phone the chief of police and tell him to arrest the man."

That night, in his Lexington hotel suite, Capone turned reflective for the benefit of the *Tribune*'s Genevieve Forbes Herrick.

He'd gotten that year in Philadelphia "not for carrying a gun, but because my name is Capone. . . . Yes, sir, there's a lot of grief attached to the limelight." What in his record deserved a year's sentence?

"All I ever did," he said, "was to sell beer and whiskey to our best people. All I ever did was to supply a demand that was pretty popular. Why, the very guys that make my trade good are the ones that yell the loudest about me. Some of the leading judges use the stuff. . . .

"The funny part of the whole thing is that a man in this line of business has so much company. I mean his customers. If people did not want beer and wouldn't drink it, a fellow would be crazy for going around trying to sell it.

"I've seen gambling houses, too, in my travels, you understand, and I never saw anyone put a gun at a man and make him go in. I never heard of anyone being forced to go to a place to have some fun. I have read in the newspapers, though, of bank cashiers being put in cars, with pistols stuck in their slats, and taken to the bank, where they had to open the vault for the fellow with the gun.

"It really looks like taking a drink was worse than robbing a bank. Maybe I'm wrong. Maybe it is."

He was tired of it all, and wanted only to be let alone. "Say, if I was just plain Izzy Polatski, living in Chicago, I'd not stand out in the gutter trying to get a peep at Al Capone. I'd attend to my business and let him attend to his; no use making a laughing stock of the city."

He pressed a buzzer on his desk and one of his young gentlemen unobtrusively slid in. "Please ask my wife and sister to come here," Capone asked. When Mae and Mafalda arrived, Capone made introductions. After small talk, the two left "in a swirl of blue chiffon" to Herrick's puzzlement: What was that all about? Capone had wanted to illustrate his point. "Did you notice my wife's hair?" he asked. Sure, lustrous and—"No," said Capone, "I mean the streak of gray. She's only twenty-eight"—gallantly knocking five off Mae's thirty-three—"and she's got gray hair worrying over things here in Chicago."

Well, he said, he would "wind up my affairs here" and leave for Miami in about three weeks. (Since telling Stege it would be a week, Capone had learned of an appointment scheduled for mid-April.) All that fuss about keeping him out was all politics—and a newspaper trying to sell papers.

But how about the Florida raid? asked Herrick. The day before, police had stormed the Palm Island house, confiscated a bunch of booze and arrested six, including John and Albert Capone, plus Jack McGurn, his identity temporarily cloaked in a variant of his Sherman Hotel alias, "James Vincent."

"Ask me anything you want," Capone answered, deftly sidestepping. "Only ask me something I can answer. I haven't been down at my Miami home for ten months so I don't know a thing about the raid."

Soon that raid would occasion trials that raised important constitutional issues—and considerable hilarity. For now, only Governor Carlton's arrest-on-sight telegrams to the state's sheriffs mattered. Capone hired two shrewd Miami lawyers whose own telegram to Governor Carlton asked, "By what authority of law you or the sheriffs of the state may seize and banish from the state a citizen of the United States who is not charged with any crime"—and was a Florida property owner and taxpayer to boot!

Several days later, a federal judge in Florida wondered the same thing. "We live under a reign of law in these United States," wrote Halstead L. Ritter. He noted that his opinion "must be rendered without reference to popular opinion or consequences." A newspaper headlined the story,

FEDERAL JUDGE WEEPS
FOR ABUSED AL CAPONE

Maybe Capone would find no warmer official welcome in Miami than he had in Chicago, but there were those who did want him. An open letter from the Rapid City, South Dakota, chamber of commerce extended "the glad hand of welcome into a community practically free from crime." In those parts, Capone would find that "the stranger is not judged by reports of his past record. . . . Let him who is without sin cast the first stone."

"I'll cast the first stone," snapped South Dakota's governor. "We don't want men like Capone in this state and we won't have him here." Deadwood chimed in that Capone wasn't welcome there, either; they were still trying to forget Wild Bill and Calamity Jane.

Capone thanked Rapid City, but the city boy really didn't want to live in the Black Hills.

Another community tried. The incumbent mayor of Monticello, Iowa, faced no opponent for reelection, but Capone almost beat him anyway with fifty write-in votes; sixty-five more, and he would have been elected. "It was jealousy," a town official explained. Capone's travails with Chicago, Florida and South Dakota had all been front-page stuff. "Monticello has never been on big city maps." This might be her chance. "Monticello will become known as the biggest little town in eastern Iowa."

The mayor manqué had other problems. The "affairs" he had to wind up before leaving for Florida consisted principally of his increasingly parlous tax situation.

A later writer professed astonishment that Capone had ignored the 1927 *Sullivan* ruling, which pronounced illegal earnings taxable, and that Capone had not tried to settle with the government earlier. The writer quoted Capone's overheard dinner table statement, "The income-tax law is a lot of bunk. The government can't collect legal taxes from illegal money" as earnest of either inexcusable incomprehension or overmastering greed. But it wasn't that simple; Capone's situation was labyrinthine.

Sullivan's meaning remained equivocal, and no one could be certain how far it went. The following October, 1930, the SIU would track down Frank Nitti, hiding under an alias in Berwyn, just west of Cicero. Pleading guilty on December 20, 1930, he would say—self-servingly, to be sure, but with little to gain from lying, since he was already sentenced and would leave for Leavenworth next day—"I didn't pay income taxes because the laws were not clear. I talked with half a dozen attorneys and they didn't know any more than I did. In Nineteen

Twenty-six the Circuit Court of Appeals held that income from illicit sources could not be taxed. The next year, the Supreme Court ruled differently.'' Sixty years later, in Chicago, the American Bar Association staged a mock retrial of Capone's tax trial. Defense counsel included Terence F. MacCarthy, director of the Federal Defender Program in Chicago. ''There was still, absolutely,'' says MacCarthy, ''some question in those days.''

Sullivan's clarity to one side, until Capone returned from Philadelphia, filing returns or offering to settle might reasonably not have appeared the smart move anyway. Capone knew he had not left any easy trail of assets or income for the IRS. Filing a return would shatter that shield, as would an offer to settle. Owning to *any* level of income would have reverberations both for the past and for every year in the future, a starting point for both specific item and net worth prosecutions.

That's exactly what was about to happen with Jack Guzik. Presumably alarmed by *Sullivan*, he had filed returns for 1927 through 1929, admitting to $18,000, $24,000 and $18,000 income for those years. On trial in November 1930, Guzik's defense would be that gambling losses had offset most wins and that those checks he had endorsed represented gross, not net, income. Guzik's conviction and sentence showed the danger of admitting income.

Nevertheless, while Capone still sat in the Eastern Pen, Torrio referred him to his own tax lawyer, Lawrence P. Mattingly of Washington. The tax indictments of Ralph, Druggan and Lake in late 1929 had rendered the government's intentions transparent; so did Nitti's indictment three days before Capone's release from Philadelphia. How could it have looked to Capone? Despite all caution, who knew how much the government had been able to discover? A settlement, even with its future risks, would get him off the hook for the past. Any tax lawyer would tell him that government policy had always been to forbear prosecution of citizens who came forward before formal legal action. Congress had fashioned the tax laws to encourage citizens to seek settlement.

On March 23, 1930, six days after Capone left jail, Mattingly wrote to the head of the IRS Chicago office, C. W. Herrick:

> Mr. Alphonse Capone, residing at 2135 South Michigan Avenue, Chicago [the Lexington], has authorized me to make an exact computation of his income tax liability for the year 1929 and prior years, the amount of which he will pay as soon as determined. Mr. Capone has never filed income tax returns. On account of inadequate or absence of records, determining his liability will be difficult . . .

so Mattingly wanted them to ''examine Mr. Capone and his records the same as you would any other delinquent taxpayer with the view of satisfying your office as to his liability.'' Mattingly attached Capone's power of attorney for tax matters. Since Mattingly would be tied up on other cases until mid-April, the

IRS invited him to bring his client in for a discussion April 17, 1930—as it happened, in the middle of Ralph's tax trial.

Herrick had a number of colleagues at the conference, plus a stenographer. Mattingly presented Capone in gleaming duotone: dark blue double-breasted suit, four points of a dazzling white linen hankie poking up from breast pocket, white-polka-dot blue tie, blue silk socks with white clocks, and white-tipped sport shoes. When Capone pulled out a silk for-blow handkerchief, lily of the valley scented the air.

Capone should not have been there. Thomas R. Mulroy, Jr.—once a federal prosecutor, now defending clients as a trial attorney with Chicago's prestigious Jenner and Block—was one of the prosecutors at the ABA's mock retrial of Capone. "You *never*," says Mulroy, "bring your client to a conference like that."

It got worse. "Now, Mr. Capone," said IRS agent-in-charge Herrick, stenographer scribbling away, "just so we all understand the situation, you and Mr. Mattingly are here in an effort to clean up your income tax liability. I want to say this, in order that there may be no misunderstanding, that any statement you make here will naturally be the subject of such investigation and verification as we can make; that is, in the nature of income or anything of that sort. . . ." So far, okay; checking and finding were different things, and Capone had *not* been careless. But Herrick went on, "And I think it is only fair to say that any statements which are made here, which could be used against you, would probably be used. I want you to know your rights. . . ."

Capone's tax-expert attorney replied, "Insofar as Mr. Capone can answer any questions without admitting his liability to criminal action, he is here to cooperate with you and work with you. . . . I don't feel I can . . . permit him to make any statement or admission that might subject him to criminal prosecution." Then Mattingly said what he thought the magic words: "Our action is without prejudice . . ."

"*That*," says the widely admired senior federal judge Prentice H. Marshall, who presided over the ABA mock trial, "was the vernacular in those days with regard to settlement conferences." Today, Marshall says, a lawyer would spell it out, getting IRS agreement that his client was "making statements *solely* for the purpose of attempting to negotiate a settlement and not for any other purpose." The agent-in-charge had already announced another possible purpose, yet Mattingly let the conference proceed.

"It makes no sense," says Tom Mulroy. "You never, under any circumstances, if there's a transcription going on, let your client continue when somebody says, 'Do you waive your rights?' The *Miranda* decision didn't come until the sixties, but here they were, giving him his *Miranda* warnings! With his lawyer there! Who let him go on, anyway! It was goofy."

The goofiness conceded, defense counsel Terry MacCarthy gives Capone high marks for the way he handled the interview, dodging or refusing to answer whenever a question cut too near the bone, deferring tricky ones to his lawyer.

For instance, Herrick once tried to get cute. ''How long, Mr. Capone,'' he asked, ''have you enjoyed a large income?''

''I never had much of an income, a large income.''

''I will state it a little differently—an income that might be taxable.'' That meant anything over $5,000 a year.

''I would rather,'' said Capone, ''let my lawyer answer that question.''

Another time, after Mattingly admitted that Capone had furnished the money to buy the Palm Island estate in Mae's name, an aide shot a question to Capone. ''What was the source of the money that you used to make your cash payment?'' Capone passed to Mattingly, who objected. Herrick quickly followed with the key question: ''Was that source of a nature that it would constitute taxable income?''

''I would rather not answer that question,'' said Capone. His lawyer would give, as one of the questioners put it, any statement of ''assets and liabilities at the present time.''

The agents asked about recordkeeping. Capone changed the answer he had given in Florida at the February 14, 1929, inquiry; now he said he didn't keep records. They asked about checking accounts (none); property (none); income (minimal); stocks and bonds (none); horses and dog racing (Capone was only a bettor). The feds did get Capone to agree that after 1926 he was no longer an employee of Torrio. He handled everything in cash, had no canceled checks or safe deposit boxes. ''What did you do with your money?'' one of the frustrated questioners growled. ''Carry it on your person?''

''Carried it on my person,'' Capone agreed blandly.

At the end of the questions, the IRS still had no specifics about size or sources of income, but they did have an admission that Capone did have some income, if Herrick's warning would serve to defeat Mattingly's ''without prejudice'' before a court.

The stenographer stopped writing and departed, the record of proceedings at an end. At that point, Frank Wilson claimed years later,

> Scarface stuck his big paw in his pocket, pulled out six Corona cigars, and shoved them at me. ''Smoke?''
>
> ''No,'' I said shortly.
>
> He scowled. ''Somebody's trying to push me around. You better take good care of yourself, Wilson.''
>
> ''You bet I will,'' I told him.
>
> There were no handshakes as is usual when we close an interview with a taxpayer. I didn't intend to soil my hands with his bloodstained paw.

The trouble with that account is that the stenographic record listed all those at the meeting and specified that Louis H. Wilson—the IRS man in charge of fraud in Chicago—attended. *Not* Frank Wilson. As we'll see, Louis Wilson later testified about what was said. Frank Wilson was never mentioned, his presence recorded only by himself.

Ralph Capone's jury would hear fifteen days of financial testimony. Although much was complex, much was straightforward and damning. Jewelry salesmen had received from Ralph checks he signed with the aliases of those accounts the SIU had unearthed. Barkeeps had paid for beer deliveries with checks that ended up in those accounts, the sums often divisible by the $55-a-barrel price of beer.

On April 25, 1930, nine days after his brother's sparring match with the IRS, Ralph heard the jury find him guilty. The judge gave him three years and a $10,000 fine. Ralph's only formal statement was, "The verdict speaks for itself"—but he may have muttered, "I don't understand this at all," as he strode from court, on bail pending appeal.

Lawrence Mattingly had a few more sessions with the IRS—alone. On September 20, 1930, he would sum up his client's position in what became known as The Mattingly Letter, addressed to agent-in-charge Herrick "Re: Alphonse Capone."

"The following statement," the letter started, again invoking what the lawyer supposed an infallible incantation,

> is made without prejudice to the rights of the above named taxpayer in any proceedings that may be instituted against him. The facts stated are upon information and belief only.

He gave some vital statistics, getting Sonny's age right "now nearly 12," and pointed out that Capone since 1922 had been sole support of his mother, sister (Mafalda), and brother (Matthew).

> Prior to the latter part of the year 1925 he was employed at salary which at no time exceeded $75.00 per week. During the years 1925 to 1929 inclusive, he was the recipient of considerable sums of money, title to which vested in him by right of possession only. [In other words, it was not income.]

He had entered some unspecified enterprises with three unnamed "associates at about the end of . . . 1925"—that is, after Torrio's shooting and abdication. But since he had no capital, his profit participation was limited until 1928 and 1929 when he got one sixth, like his partners, one third going to "a group of regular employees."

The furniture in his Palm Island home cost no more than $20,000, and the house still sat under a $30,000 mortgage. "His indebtedness to his associates has rarely ever been less than $75,000 since 1927. It has frequently been much more.''

Mattingly concluded,

> Notwithstanding that two of the taxpayer's associates from whom I have sought information with respect of the amount of the taxpayer's income insist that his income never exceeded $50,000 in any one year, I am of the opinion that his taxable income for the years 1926 and 1927 might be fairly fixed at not to exceed $26,000 and $40,000 respectively and for the years 1928 and 1929 not to exceed $100,000 per year.

The Mattingly Letter gave the government the starting point it needed to establish evasion in a net worth prosecution.

Why? For all Mattingly's reputation, was it ineptitude? ''No,'' says Tom Mulroy. ''I think what Mattingly missed was that there wasn't ever going to be a civil settlement in this case—that this was a criminal case from the first day.'' The lawyer and Capone thought they had drawn the government's teeth with their voluntary offer to settle, nonprosecution under those circumstances so firmly established as policy. But policy is not law, and the government in no way bound by it. President Hoover was determined to get Capone, to see him in jail, no matter what.

So was Florida.

CHAPTER 24

Decency Strikes Back

"THE DECENCY OF the community is now asserting itself," one in a series of Miami *Daily News* front-page editorials had shouted a week before Capone's release in Philadelphia. That decency took the "form of a movement to drive Capone out of here, if he should come. . . . [H]e should be escorted to the county line and told plainly that the compass points north." Another in the series lauded the governor's telegram to the sixty-seven sheriffs. That represented "an expression of the decency of the state of Florida," confounding those critics who, because Capone had wintered *and* intermittently summered there unmolested since 1927, had made "a wrong appraisal of this great southern country." The paper reminded readers that a "generation or two ago a man guilty of crime and regarded as guilty by the appraising sense of the community—even though he might have been acquitted—would have been hanged to the first tree." Today, of course, "These measures . . . are to be condemned. We cannot countenance them now, and yet . . ."

This was not a case of rednecks espousing lynch law. The Miami *Daily News* was owned by James M. Cox, once governor of Ohio and Democratic candidate for president in 1920. Most of Cox's newspaper chain still remained in Ohio.

The local leader of the crusade against Capone was Carl Fisher, born January 12, 1874, in Indiana. A sixth-grade dropout (like Capone), Fisher had made his fortune in Indianapolis, where he built the Indianapolis Speedway. In 1911 he had sold his company, which made carbide-gas headlights, and moved to Florida, where he lent Miami Beach pioneer John Collins $50,000 to finish the first causeway from Miami. Ceded enormous tracts of land as part of the deal,

Fisher immediately began building hotels and homes and quickly became Miami Beach's prime developer. He was also a renowned yachtsman; his friends called him "Skipper" and "Skip." Fisher first brought James Cox, an old friend, to Miami.

Another carpetbag soldier in the anti-Capone fight was Daniel Mahoney, Cox's son-in-law and general manager of Cox's Ohio newspaper group.

At Mahoney's instigation, Fisher orchestrated Governor Carlton's telegrams to the sixty-seven sheriffs. After they were sent, Mahoney wrote to "Dear Skip" about his "great deal of pleasure" in reading what the governor had done. "If the community as a whole responds as it should to the efforts of our paper, we should . . . be able to clear Miami and Miami Beach of this undesirable element."

Fisher sent Carlton a telegram telling him that "following several important meetings here among our best people," Carlton could "expect a wire the next day saying the best element of Miami Beach is thoroughly back of you in your stand against gang rule." Later, when a judge sentenced Capone's brother Albert to jail for vagrancy, Cox himself sent Fisher a telegram:

> THE EXPRESSED DECENCY AND COURAGE OF MIAMIBEACH THROUGH YOUR POLICE JUDGE THIS MORNING . . . IS THE MOST EPOCHAL EVENT SINCE YOU DECLARED WAR ON THE MOSQUITOES AND THE SWAMPS SEVENTEEN YEARS AGO STOP IT MAY TAKE TIME TO GET THE OLD SKIPPER GOING BUT HE IS HELL WHEN HE GETS A GOING

Capone was the target, not sin; this was no crusade by blue noses to "clean up" Miami. Those with a taste for dissipation had no trouble spending money in southern Florida, and a bargain it was. Bootleggers operated in such profusion that during the season competition had pushed prices down to absurd levels: $45 a *case* for uncut, smuggled scotch, compared with $90 in Chicago. Gambling ran so wide open that action could be found, one newspaper promised, "by anyone with a bankroll and the urge."

No one objected, including Miami's better element. As a Miami *Daily News* editorial put it, "our people have favored a liberal policy for the winter months." But they reserved such tolerance for "our boys," not outsiders. "Gangsters," the editorial continued, ". . . recognize no middle ground as between regulation and wide-open license." It was "a case now of gangsters ruining Miami as they have ruined Chicago, or a triumph of righteousness."

As for the crusaders, Fisher was a lush, grown gross on drink; he would die nine years later of a gastric hemorrhage. He womanized tirelessly even during his marriages. Gambling? As a historian of Miami Beach put it, he would bet on "anything that moved," though he did uninsistently oppose regular casinos, fearing they would draw gangsters. He forbade formal gambling in his hotels.

Dan Mahoney was equally unsaintly. Known as "Big Dan" and "Big Irish," he was a wheeler-dealer of lusty appetites, an adventurer who had chased

Pancho Villa with General Pershing. As the campaign against bootlegger Capone gathered speed, Mahoney wrote Fisher from Dayton, Ohio, that he planned to be down for a visit "April 1st at which time I hope to drink a tall, cold one with you. I have been so good up here it hurts. I am looking forward with a great deal of pleasure to our trip to Havana."

Capone was partly right in what he had told reporter Herrick about the business motive for his persecution by Mahoney and Cox—whom a longtime Miami newsman remembers as "an imperious bastard." The Miami *Daily News* saw in him a useful stick with which to beat the front-running *Herald*. Local politics were Byzantine in their inbred corruption. Cox's paper supported the incumbents and charged the *Herald* with support for some outs—like former police chief Leslie Quigg and Fred Pine. Quigg had been removed for brutality and corruption in 1928 (but would be reinstated in 1937); Pine had been fired by the governor as county solicitor in 1923 for his "liberal liquor policies."

The *Daily News*'s favorites were no better. Director of Public Safety Samuel D. McCreary had prepared for his career in law-enforcement supervision as manager of El Comodoro Hotel, working for Robert Knight, brother of a Miami commissioner. In 1935, McCreary and Knight would be tried and acquitted in the face of remarkably persuasive evidence that they had sold protection.

But unlike them, Pine had ties to Capone. The Reverend William H. Sledge had once visited 93 Palm Island, not in his pastoral role, but in his sideline of renting cars. He had overheard Capone arrange to send a case of champagne and two of scotch to Pine. And Pine, Cox's paper rightly charged, was the *Herald*'s candidate!

Furthermore, the *Herald*'s editorial bullyragging of Capone lacked appropriate heat. Publisher Frank Shutts once insisted that the Community Chest return a donation of $1,000 from Capone. But his paper kept making finicky demands that all anti-Capone activity be "clearly within the law"—an unbecoming nicety when Shutts openly had his horse-racing interests and, the *Daily News* hinted, interests in other gambling, too.

By his own lights, Frank Shutts took an entirely consistent stand. He exemplified the old-line Miami oligarchy. Whatever their personal morality or tastes, they blinked the "liberal policy" as vital for Miami's lifeblood tourism. In any case, the bootlegging, the gambling, the corruption was *theirs*. If not run by them, it was run by people they knew, hometown fellows who knew their place and acknowledged the oligarchy's ascendency.

A Joseph P. Widener might buy half the Tallahassee legislature to get horse racing legalized, but in the Miami oligarchs' eyes that materially differed from Capone's putting 70 percent of Chicago's police on the pad. D. W. Shannon might get shot and killed one night, but that differed from all those "rides" in Chicago. Red Shannon was their own beloved local bootlegger, respectfully making his deliveries to customers' yachts. He died manfully and honorably, killed by the Coast Guard, not even shooting it out, but simply trying to outdistance them; an outraged Miami wanted the Coast Guardsmen tried for murder.

In all of Miami's local vice and corruption there lurked no hint of racketeering or extortion, no bombs or wholesale beatings (except by the police). That was it: homegrown corruption was consensual and lacked all sense of menace. Its practitioners operated without evident capacity for, or easy recourse to, violence.

In Chicago the standard apology held that gangsters killed only each other. Certainly Capone never personally killed or ordered killed anyone whose absence impoverished the world. But one need not mourn the departed to deplore murder. More to the point, once let outside gangsters in on the action—a Capone most of all—and with their guns and muscle it might be impossible to dislodge them; the oligarchs would surrender the control they exercised, with no further say as to limits. The Miami *Daily News* had a point about gangsters recognizing "no middle ground" between measured toleration and whoopee.

This rational apprehension granted, the force of the Miami oligarchs' reaction to Capone also had more discreditably visceral roots. After all, he truly was not trying to take over. He had an interest in some gambling places—but only a few, including an Everglades mansion in the next county, forty miles out on the Tamiami Trail. He expedited shipments of smuggled booze to Chicago and had an interest in some local speakeasies; they still call part of what's now a west-side country club "Capone's Tower." He also maintained an office in what is now the U-Haul Corporation's main Miami building—which featured a freight elevator that could accommodate a seven-passenger sedan and what modern police think were cells. But Capone did not even dabble in local bootlegging. He came close to the truth when he endlessly reiterated that all he wanted to do in Miami was relax and enjoy himself.

That was the real sticking point for the Miami better element. "We give a Negro porter on the train a tip," one Miami *Daily News* front-page editorial ran, "and he blacks our boots and renders servility. . . ." That's what Capone expected of them with his openhanded patronage of their stores and workmen and his $1,000 charity checks. He was treating Miami the way Miami treated blacks! An Italian gangster from a northern city just could not come down and expect to live in their swankest neighborhood—and not humbly, not quietly; Capone did not cringe and pull his forelock as local bootleggers did. He kept rubbing Miami's nose in his presence, giving parties, making appearances, and attracting attention.

Worst of all, among the greedy and raffish he mesmerized were the literal scions of oligarchy, like Parker Henderson, son of a former mayor. Jack Sewell was another. His father, John, had been Miami's mayor and was a perennial county commissioner. His uncle, E. G. Sewell, had been several times mayor and was currently a commissioner. John and Ev were partners in Sewell Brothers, one of Miami's finest stores. Young Jack dined out on stories about Capone: how Jack had given him—after $1,000 or so worth of purchases at Sewell Brothers—a new belt and a Panama hat, and Capone had asked to shake Jack's hand, saying, "This is the first time anybody ever gave me anything." Jack also liked to tell of the time he walked into 93 Palm Island in the middle of a poker game, stacks of $1,000 bills covering the table. Capone, just getting up, said to Jack, "I'm quitting. These scoundrels have taken me for two hundred fifty thousand."

Ev Sewell was not amused by his nephew's tales. When Ev later voted for a vagrancy ordinance specifically directed at Capone, he stressed that he acted in full knowledge of Capone's lavish Sewell Brothers patronage.

The raid on Palm Island that reporter Genevieve Herrick had asked Capone about was entirely righteous, valid warrant and all. Police in Toledo, Ohio, had identified Ray Nugent—the man Alvin Karpis placed among the St. Valentine's Day massacrers—as one of five who had killed a policeman after an armored car holdup in April 1928. In Cincinnati the police suspected him of a dozen killings, but had evidence only for a second-degree-murder charge involving a bootleg deal, from which Crane Neck had flown, circulars out for his arrest. Several months before Capone's release in Philadelphia, the Miami police sergeant in charge of the identification bureau had spotted Nugent speeding in a car with Ralph Capone and two women, all drunk. Nugent said he was staying at the Palm Island estate. The Capones notoriously helped and harbored ex-cons and those on the lam.

On inquiry, Toledo police said they could not immediately produce witnesses, which left only the Cincinnati charge. Since Florida law made second-degree murder a bailable offense, Nugent walked on a $10,000 bond. By early March, when Toledo was ready to have Nugent picked up, he had jumped bail. The sheriff's department dispatched a deputy, with warrant, to look for him at 93 Palm Island. The deputy did not find Crane Neck, but he could scarcely miss the scads of liquor round about.

Sheriff M. P. Lehman conceived a masterstroke. He would raid 93 Palm Island, probable cause provided by the liquor his deputy had seen there a couple weeks before, while there on legitimate business. The Miami Beach police gleefully joined, but Miami said no, Palm Island lay outside their jurisdiction. City manager Frank H. Wharton added that it smacked of "a grandstand play"—whereupon Lehman, miffed, lifted twenty-seven deputy badges he had issued to Miami police and officials.

On March 20, raiders found the liquor they knew they would. Only the caretaker, longtime Miami resident Frank Newton, was at the estate. Five others had been swimming and they soon wandered into the net: brothers Albert and John; Jack McGurn (alias James Vincent); tiny "Diamond Lou" Cowan from Cicero; and Leo J. Brennan, an unlucky casual visitor.

Newton and John were charged with possession and vagrancy, bottles having been found in the closet of John's room. Some trials and appeals later, it all came to nothing—except a $500 fine for Frankie Newton, who took the heat and claimed to own *all* the liquor, including the twelve bottles in John's closet.

Capone couldn't take a hint. He obtained that federal injunction against Florida's sixty-seven sheriffs. "Of course Capone wouldn't know what you meant if you mentioned the Constitution,'' the Miami *Daily News* sniffed; ''to him the term would probably imply some special brand of spaghetti. . . .'' The Miami authorities did not give up. While Capone stayed in Chicago through the rest of March and more than halfway through April, hoping to clear up the tax situation at the meeting on April 17, Miami kept hinting he was not welcome. On April 1, 1930, they picked up Jack McGurn with Louise Rolfe on the fifth hole of a Miami Beach golf course. Twelve days later came brother Albert's turn, arrested for vagrancy, also on the links.

Capone ignored the hints. Right after his tax meeting ended, Capone left for Florida anyway. His train arrived early on a rainy Easter Sunday, April 20, 1930; with Capone rode his nephew, Ralphie, and a Chicago alderman. His local lawyers, Vincent C. Giblin and J. Fritz Gordon, met him at the Hollywood station, and they all drove in with Chicago *American* editor Harry Read, who happened to be on the same train. Capone told a reporter, ''I am here for a rest which I think I deserve.''

There is no rest for the wicked. Two days later, the Dade County state's attorney, N. Vernon Hawthorne, filed suit to have 93 Palm Island padlocked as a nuisance. With all that liquor around, all those vagrants staying there, it stood as ''a place frequented by common gamblers . . . habitual loafers, idle and disorderly persons . . .''

The process would take a while; meantime, Capone relaxed. Like Fisher and Mahoney, he sampled the joys of Havana—after a short quiz by the chief of the national secret police, for whom he posed as a ''retired businessman from Chicago.'' In his party were Sylvie Agoglia, whose meat cleaver struck the first blow against Pegleg Lonergan's group in Brooklyn, and editor Harry Read—though Read later protested he had paid his own way.

Back on Palm Island, Capone loafed and fished from his yacht, *Arrow*, reeling in one of the biggest sailfish caught that season in the Gulf Stream: seventy pounds, almost eight feet long.

On the morning of May 8, the manager of the Olympia Theater in downtown Miami called to invite Capone to the matinee of his new show, *The New Adventure of Fu Manchu*, with Warner Oland and Jean Arthur. Capone traveled light, driving toward the Olympia with brother John and Chicago alderman Albert J. Prignano, plus bodyguard Nick Circella, none of them armed.

Miami's director of public safety, Sam McCreary, had ordered the police chief, Guy C. Reeve, to station detectives at the Miami end of the causeway to arrest Capone anytime he showed up. When the two detectives stopped him, Capone asked what he was being arrested *for*. ''Investigation.'' Could he call home and tell someone, or call his lawyers? No. Capone shrugged and invited a detective to ride in his car, sending one of his passengers to ride hostage in the squad car. They reached headquarters at exactly two-thirty P.M.

Sam McCreary was there. ''Mr. Director,'' Capone asked him, polite as you please, ''what am I going to do to get out?''

''I don't know exactly the procedure,'' said McCreary. ''I will say this to you; we put you in here on suspicion; you are not going to be abused and you are not going to be mistreated.''

''Well,'' said Capone, ''I don't mind taking a rap myself, but I want you to release my brother.''

''He is with you and we are treating him just like we are going to treat everybody else that we are suspicious of.''

On that note, the desk officer told Capone to turn out his pockets and surrender his ''valuables,'' an apt description in this case: the usual jewelry and $1,160 for an afternoon at the movies. Capone demurred. As he later told a judge, ''I would not put my money up and valuables in his care without a receipt.''

According to Capone, McCreary then instructed Chief Reeve to issue no receipt. ''He said,'' Capone told the judge, ''. . . 'Take it off him and throw them back in the shit house.' '' The officials present gave a less colorful account. When Capone asked for a receipt, the booking officer said that he didn't need one and that the city would be responsible. Unreassured, Capone insisted. Chief Reeve intervened: Capone could have his receipt if he wanted. McCreary cut in to ask if receipts were customary, and on being told no, said, ''Use him like any other prisoner brought in.'' The circling officers were ready to *take* the stuff. Capone gave up, snarling ''That dirty son of a bitch, Cox, is the cause of all this'' as he unloaded.

He later claimed to have asked again to phone his lawyers; the issue remained in dispute. More likely, he didn't bother, knowing the answer. They plainly meant to keep him incommunicado. The jailer conducted his prisoners to cells, with orders to put Capone apart, upstairs. When Reeve saw that the cell was an ''outside'' one, its six-inch barred window opening onto the street, he shifted Capone to an inner cell ''to avoid a crowd gathering since they could not see him in the inside cell.'' Did that also mean, he was asked later, no possibility of communication with the outside? ''I might have had that in mind,'' answered Reeve. ''I don't deny that.''

It was, Capone said, ''a cell where there was no window and no air, and he told them not to give me anything to eat or any water . . .'' Of course, he'd had his usual Gargantuan lunch, and it *was*, after all, only mid-afternoon, a point Capone conceded. ''It was not a question of food,'' he later explained, ''but to try to be dirty and try to make it miserable for someone.'' Also, the ban on water was temporary if not imaginary: a trustee soon fetched a pitcher of ice water.

After a while, Director McCreary drifted back to the cells to deliver the oligarchy's final hint. Capone would be arrested anytime, anywhere, in any company, every time he set foot within the jurisdiction of the city of Miami.

''You mean if my mother and wife and kid are with me?'' Capone asked.

''Yes; I will arrest them too.''

''If you was me what would you?''

''I would not want to be you.''

''No,'' said Capone; ''you are no such of a man.''

About two hours after the arrest, lawyers Giblin and Gordon charged into the station. Someone had seen the arrest and had finally dropped by the office to mention what had been done to their client. At the station, excitable Fritz Gordon tussled with McCreary, who had commanded that everyone going back to see Capone must be searched for weapons, even his lawyers. It ended in a shoving match, Gordon running from the station, pursued by police who forcibly searched him—inspiring the Miami Bar Association to vote a committee of investigation despite the cogent question of one member, "Who is going to keep the committee from being searched?"

Soon, the party moved to a courtroom, where Judge Uly O. Thompson ordered the prisoners' release. Capone never did get fed.

Five nights later, about eight-thirty, Capone sat in his box at the fights in the American Legion hall when a detective tapped him on the shoulder. Chief Reeve wanted to see him in the back of the hall. He was under arrest; so were his companions, Agoglia, Prignano and Circella. This time no one had to alert Capone's lawyers; Vincent Giblin was also in his box. Still, the four spent the night in jail because at that hour Giblin could not raise a judge to issue habeas corpus writs. Capone didn't miss much: Joe "Kid" Peck languidly outpointed favorite Jimmy Spivey for the lightweight "Dixie Championship." A sportswriter admitted that "there were slow spots in the bout."

Next day, Judge Thompson asked Sam McCreary why the pickup order still stood. Because McCreary had received so many complaints that Capone menaced society and reduced Miami's citizens to fear. The defense quizzed Miami city manager Wharton, who had once scorned the Palm Island liquor raid as grandstanding, but had since gotten religion. Exactly who had complained?

"Well, Roddy Burdine for one." R. B. Burdine owned Burdine's, which shared honors with Sewell Brothers as the area's poshest department stores; and it was he who had gone to Palm Island to solicit Capone's $1,000 Community Chest donation.

"Did Roddy Burdine tell you that?" asked Capone's lawyer.

"Yes sir! in my office, two or three days ago."

"Did he ever tell you that he had visited the Capone home?"

"No, sir."

"Who else made complaints that you remember?"

"Well, the president of the First National Bank."

"Mr. Romfh?" Edward C. Romfh was another former mayor.

"Yes, he said Capone was a menace to the community."

"Who else complained?"

"Those particularly, I don't recall others at this time."

"Did you have a thousand complaints?"

"No, I wouldn't say that."

"Did you have a hundred?"

"Yes."

"And out of the hundred you can't recall but two men?"

Laughter rocked the standing-room-only audience. On reflection, Wharton recollected that Mayor Reeder, plus city commissioners John Knight and Ev Sewell had also complained.

"Did Commissioner Sewell say that Jack Sewell ever visited the Capone residence?"

"He didn't talk about Jack."

Judge Thompson released all four prisoners. McCreary's pickup order was amended to allow instant bail. The American Legion sent Capone back the $24 he had paid for his box.

To give the color of law to the pickups, the Miami city council pushed through a new vagrancy ordinance that included "persons having visible means of support acquired by unlawful or illegal means or methods" as well as those "dangerous to public safety or peace of the city of Miami" and those "known or reputed to be crooks, gangsters or hijackers." They thoughtfully added the proviso that if any part were to be found unconstitutional, the rest would stay in effect.

At the same time, the hearing to padlock Capone's Palm Island home came before Circuit Judge Paul D. Barns. Only Carl Fisher had testified when Judge Barns adjourned for weeks to consider various motions. While Capone fixed him with the Look, Fisher, unfazed, swore that Capone's presence had shattered property values and spawned scores of protest meetings.

Was it, Vincent Giblin wanted to know, Fisher who had directed the Miami Beach force to station policemen outside Capone's compound ever since he arrived?

"No sir," the oligarch replied, "it wasn't my order or request; it was my suggestion."

Did *he* have any gambling interests? No, Fisher could answer honestly: Giblin hadn't asked about personal betting. "Are you interested financially or are you the owner of any place wherein intoxicating liquors are illegally possessed?"

"Not that I know of."

Fisher had stayed in Miami to testify as his civic duty, but was anxious to leave for Montauk, on Long Island, New York, which he hoped to develop into the Miami Beach of the north (it and he would quickly go bust). He left soon after testifying, pausing only to write to Michael J. Glenn, the private detective who supervised security at his hotels. A month before he left he had "ordered some beer from Tom Harbin." But the beer hadn't arrived,

> . . . and I commenced to feel like there was some sort of set up that might cause trouble, and as we were not there to use the beer we don't want to have it landed there. At the same time, we don't like to see Harbin get stuck.
>
> Dan Mahoney said he would take a few cases, Ed Romfh would

> probably like a few, and I would like to have you help Harbin out on this beer situation. But I don't want any of it left at the house. We have not a suitable place for a large amount of it, still I would like to be able to get some of it this fall when we get back.
>
> You will have to handle it rather carefully so we will not get in a jam. I am afraid with the Capone situation as it is, we might get into a tough hole. . . .

Noblesse oblige meant seeing that *our* bootleggers didn't get stuck. As for bank president Ed Romfh, complaining about that lawbreaker, Capone, surely was thirsty work.

The Miami police arrested Capone again the night of May 19, again at the fights; but he was out on $100 bail in time to see the boxing—which again "wasn't so hot." Boogie Saab, the "Syrian Slammer," lost to George Harmon, the "St. Augustine Surprise"; Snooks Campbell demolished Red Hooks.

Just before noon on May 24, when Miami police stopped Capone a fourth time—on his way into his lawyers' office—Giblin advised him not to go along except under compulsion, and the police left.

This was enough. Capone charged the mayor, Commissioner John Knight, Sam McCreary and Jim Cox with conspiracy, and McCreary with false imprisonment. Justice of the Peace Warren L. Newcomb—whom private detective Mike Glenn suspected of being on the take from Capone—held a preliminary hearing to determine if enough evidence existed for a trial. Capone told his side, and McCreary and the police told theirs.

Capone displayed the subtle tenacity that makes a fine witness, never allowing opposing counsel to frame the context of questioning. Capone would always turn the questions back to his point—that Sam McCreary was behind the arrest and imprisonment. "And when you arrived at the police station," counsel asked, ". . . you were not locked up by Mister S. D. McCreary?"

"His orders," Capone answered.

"He didn't lock you up himself, did he?"

"No, but his orders were."

"Someone else locked you up?"

"Under his order, yes."

"Wasn't the Chief of Police there at that time?"

"Yes; he told the Chief of Police to do it."

"He told the Chief of Police to do it?"

"Exactly."

"But he didn't do it himself?"

"No; but it was his orders."

"And at the time he told the Chief of Police to do that you were already under arrest?"

"Well, I presume so," Capone said, "from the way they arrest people from orders from him. I guess that I was under arrest."

Asked about his occupation, Capone responded that he dealt in real estate. "Do you buy a real estate license?" counsel asked. Before his own lawyer could object, Capone snapped, "What should I answer a question like that for?"

Capone's hireling or not, Squire Newcomb almost immediately dismissed the conspiracy charges against all, then adjourned to ponder the false-imprisonment charge against McCreary. The threat was enough—especially after Newcomb told police that if they bothered Capone while under subpoena as a prosecution witness, "I'll arrest *you*." The city manager announced that Capone would no longer be arrested on sight.

Capone continued wooing Miami. A teacher at Sonny's school relayed the request of some students to swim in the Capone pool. Capone agreed—provided their parents sent written consent. A gang showed up, perhaps as many as seventy-five, including girls. While they swam, servants spread tables with chicken, cake and soda, balloons and noisemakers. As the guests departed, Sonny gave each a box of candy. About fifty adults showed up several days later for a banquet and "musicale," trading their engraved invitations at the door for an American flag lapel pin. Some were friends, but thirty-two were "approachable" Miamians, the Miami *Daily News* harrumphed. His guests hailed Capone as a "new businessman of the community," and presented him a fountain pen.

"In my opinion," Mike Glenn wrote Carl Fisher, "he is trying to build up his case here by attempting to show that reputable citizens here do not consider him undesirable."

Judge Barns resumed the padlock hearing on June 10, 1930. Capone glistened each day at his snorkiest, one day a billowy cloud of white linen, the next coordinated hues of blue, followed by brown, then gray, one day's diamond stickpin "the size of a machine gun bullet" giving way to a comparably sized pearl.

A parade told of seeing a deluge of liquor served at his home. Some testified with evident satisfaction, others with the last degree of reluctance (State's Attorney Hawthorne introduced one of his witnesses by swearing he wouldn't believe the man under oath. He accused another of being drunk on the stand; "Now listen, Vernon . . . ," the tippler said).

When asked if *he* had been offered a drink, Roddy Burdine decently said, "I was there as Mr. Capone's guest, and it is not my intention to discuss such things unless I am forced to do it; I think it is against the etiquette of society." It was certainly against the etiquette of Capone's society, the difference being that when Hawthorne and the judge insisted, Burdine squealed.

He had arrived at ten o'clock on a Friday morning to find a champagne party in progress, maybe two dozen bottles cooling in an ice-filled sink—a very attractive arrangement, Burdine thought, with Capone graciously helping his "colored man" (Brownie, surely) serve. After Capone tendered a check for that $1,000 Community Chest donation, he asked if Burdine, president of the country club, might not arrange a party and invite him as guest to meet some of the better element in those congenial surroundings.

Burdine was afraid, he testified, that "it would not go over so big. . . . I knew that I could not carry Mr. Capone among my friends. I would not do it. If I did it would be severely criticized and I do not believe they would have stood for it; I know they would not have stood for it. . . . I told him I would see what could be done, but of course when I mentioned it out there they did not even discuss the thing because they knew it would not do."

Why not? "Well, Mr. Capone is what we call a gangster . . . and he just does not fit into what I call society, what I call my friends; he just would not mix. I know I am going to be severely criticized for going over there and mixing for the purpose I went, but I could not take him to my friends; I would not be seen with him. I could not."

Other witnesses testified to Capone's undesirability, the terror his presence wrought in the community, the scandal of his known criminal occupations. "Every time an automobile tire explodes on the causeway," said Palm Island resident Edward Robinson, "people wonder if a gang war hasn't started." The president of the Miami Beach Chamber of Commerce, Thomas J. Pancost, called Capone's house "a harbor for undesirables." He hotly denied that the Pancost Hotel had a bar in the cellar, though he had to admit it did harbor a few slot machines.

Judge Barns did not doubt any of the testimony. Clearly, Capone kept and served liquor; plainly a lot of people didn't like his being on Palm Island. "It is apparent," the judge's opinion ran, "that the popular action to take in this matter would be to enjoin Al Capone from further occupancy of the premises . . . but if I am to abide by my oath, I can not do it, believing that . . . the only cause of annoyance is the mere presence of Al Capone upon the premises. . . . [T]he law does not provide for the expulsion of undesirables, as such."

The judge then cut through the nonsense. "If a community is embarrassed by the mere presence of any particular individual, it certainly does not have to deal with him either socially or in business, either of which would of course encourage his residence being continued. However, to some, the smell of money is good, regardless from whence it cometh."

Two hours after Judge Barns's verdict, County Solicitor George E. McCaskill filed four charges of perjury against Capone in the false-imprisonment hearing: he had lied, claimed the county, about asking at the station to phone his lawyers; about McCreary's threatening to have his valuables thrown "in the shit house," about being denied food and water; and about being told his family would be arrested, too.

They might have had him, but McCaskill had blown it from the start, overeager because of his precarious political situation. Despite Fred Pine's 1923 dismissal from the post, he had been reelected in 1929 to replace Robert Taylor as county solicitor, only to be denied the office by Governor Carlton, who had appointed McCaskill instead. Now McCaskill and Pine were in a runoff following the June 3 primary. McCaskill figured a Capone conviction would help him win if he could contrive it before the June 24 election, so he hurriedly made his charges while Squire Newcomb still dithered over his decision. Had McCaskill waited for the decision, the perjury would have been perfected; now, technically, the case could be reopened for further testimony. On Giblin's application, Newcomb allowed Capone back on the stand.

"Mr. Capone," Giblin asked, "do you remember your previous testimony?"

"I do."

"Do you remember testifying that at the police station you made a request to use the telephone?"

"That's the part I want to correct, with permission of the court. I was excited about my valuables and do not remember if I asked or not. I know that I asked the officer in the police car on the way to the police station if I would be allowed to use the telephone. . . ."

"Was there any similar request made at the police station?"

"Not to the best of my knowledge."

"Is the rest of your testimony as transcribed true?"

"To the best of my knowledge, I wouldn't swear to it."

It is hard to nail someone for perjury over statements to which, even if retroactively, he will not swear. Capone's retraction of testimony worked. Nevertheless, McCaskill persisted. The first of the hopeless cases would come to trial in July, each charge tried separately. Capone asked that it be delayed: he had urgent business in Chicago.

On June 9, 1930, Alfred J. Lingle had been publicly murdered at midday in the Loop. Jake Lingle wore one of Capone's belt buckles, "AJL" in diamonds, and flattered himself they were friends, although sentiment did not impel Capone north. Lingle had been a reporter for the *Tribune*, and the heat was on. All the gangs' business would be devastated until someone delivered to Colonel McCormick the killer of his employee.

Miami refused to delay the perjury trial. But by the time the "telephone" charge came to trial, on July 12, "shit house" had already been thrown out. Judge E. C. Collins directed a not guilty verdict, and McCaskill dropped the remaining charges.

Miami would have to put up with Capone, and a somewhat less munificent Capone at that. As one frequent guest commented, Capone had become "more careful" in whom he invited to Palm Island. Dark rumors circulated that Capone had his eye on an estate farther north, near Palm Beach, a peninsula still called, locally, "Capone's Island." He never did move; but shopkeepers of Miami could hear the rumors only with dismay in this first off-season of the Depression.

CHAPTER 25

Public Enemies

BACK IN APRIL 1930, the Chicago Crime Commission had issued a list of the twenty-eight Chicago criminals who most urgently belonged in jail. Frank Loesch gave them the catchy title "Public Enemies."

Capone, just out of Pennsylvania's jail, of course led. Others included obvious choices like brother Ralph, Jack Guzik, Jack McGurn, Frank McErlane, Danny Stanton (who had taken over from tubercular Ralph Sheldon), Jack Zuta, George Moran and a moderately resurgent Joe Aiello. But some entries must have puzzled the knowledgeable. Many were has-beens like Druggan and Lake, Joe Saltis and Spike O'Donnell. Still more—like William Niemoth, Joseph Genaro and Leo Mongoven (who was Moran's bodyguard)—were relative nonentities compared with such heavyweights as Frank Nitti and Claude Maddox, who did not make the list.

Other names were missing because not many people yet knew they belonged on such a list.

After the fact, everyone wondered how colleagues could *not* have known about Jake Lingle. The casual observer would have reckoned him a most unusual newspaperman.

Born July 26, 1891, he ended his formal education with the Calhoun grammar school. Until age twenty, his only job was as office boy and stock clerk for Schoeling, a surgical supply company. In 1912 he became a $12-a-week copyboy for the *Tribune* and soon graduated to cub reporter—but of a limited kind. Lingle never got the hang of writing news stories. He would always be a "legman," gathering facts which he fed to rewrite. June of 1930 found him a thirty-eight-year-old crime-news legman, salary $65 a week, whose byline had never appeared.

Other newspapers employed legmen who could not write, but this was a legman with a difference. He would only infrequently pop into police headquarters at Eleventh and State, which most crime reporters haunted. Instead, Lingle wandered the city, extracting stories—many of them exclusives—from his staggeringly large collection of great friends, who included the current governor of Illinois, its attorney general and its first assistant attorney general. For years, he had been closest of pals with Chicago's current chief of police, Bill Russell, who claimed to be "fonder of him than I could be of my own son." The night of June 9, 1930, the Board of Trade planned a banquet to celebrate its new building; the printed program listed Jake Lingle among the leading Chicago figures invited as guests.

This legman also spent a lot of time at the track, would sometimes bet $1,000 on a race. He often rode in a chauffeured limousine at his own expense. For some months he had occupied suite 2706 at the Stevens Hotel while his wife and two children still lived in their home in the West Side's Valley, the neighborhood where Lingle and Helen Sullivan had grown up together, marrying when Lingle was thirty.

The suite did not signal a marital rift. He had just bought his family what investigators would later describe as "a rather pretentious" lakeshore summer home in Long Beach, Indiana, paying $10,000 of its $16,000 price in cash. He would take Helen for winter stays in Cuba, once accompanied by John Stege and his wife.

Lingle lived well for a $65-a-week legman. Immediately after his death, the *Tribune* explained that he had kept on with newspaper work "not because it paid him what his efforts and ability deserved, but because he loved it." He had talked about inheritances from his father—$50,000 was the sum everyone heard—and an uncle. Colleagues believed that Lingle's assets had then mushroomed to around $100,000, partly because of hot tips from another great friend, stock operator Arthur Cutten. Yet the 1929 Crash had not affected Lingle's style of living.

Everyone knew that Lingle listed among his great friends all the top criminals, connections of obvious value for someone his paper at first termed "one of the cleverest police reporters of his time." Capone had granted him interviews in Philadelphia when seeing few reporters. But it was an assignment Lingle *failed* on that led to evidence of how strong a position he had with gangland. Lingle had gone to cover Capone's release from Philadelphia, but had been left at the gate with all the others. A day or so later, Eliot Ness's wiretappers heard a furious Lingle demand of Ralph Capone, "Where's Al? I've been looking all over for him and nobody seems to know where he is."

"I don't know where is, either, Jake," Ralph lied, knowing his brother was thrashing about, drunk in the Western. "I haven't heard a word from him since he got out."

"Jesus, Ralph, this makes it very bad for me. I'm supposed to have my finger on these things, you know. It makes it very embarrassing with my paper. Now get this, I want you to call me the minute you hear from him. Tell him I want to see him right away."

Ralph meekly agreed. Within the hour Lingle called again, and when Ralph professed continued ignorance, Lingle snapped, "Listen, you guys ain't giving me the runaround, are you? Just remember, I wouldn't do that if I was you."

"Now, Jake, you know I wouldn't do that. It's just that I haven't heard from Al. What else can I tell you?"

"Okay, okay. Just remember to tell him that I want to talk to him right away."

What manner of legman was this, the wiretappers wondered, to take so high-handed a line—and have crude, mean, crazy son of a bitch Ralph placate him!

Lingle's clout had always seemed unaccountably robust. Five years before, he and two other *Tribune* reporters had been caught with booze by Prohibition agents. The case—assigned to Bill McSwiggin—simply evaporated. That was before Bill Russell became police chief. Now, knowing circles called Lingle Chicago's "unofficial chief of police"; they whispered that Lingle set the price of beer in Chicago and that no major gambling joint opened without an okay routed through him.

That meant Jake Lingle had also made enemies. News photographer Tony Berardi remembers his wariness. "He would walk like this," Berardi says, taking a couple of cautious steps, then casting a furtive glance over one shoulder; two steps further, then the same wary glance to the other side. "I said to myself," Berardi says, "Why is he doing it? He's scared half to death."

Lingle had many sources of apprehension. John J. McLaughlin, for instance, had been a state senator and was known as "Boss" McLaughlin. He wanted to open a casino and went for permission to the state's attorney, with whom he had served in the legislature. Swanson hadn't the heart to give his old colleague what was later demurely termed "a definite refusal." In late May 1930, McLaughlin checked with Lingle, who told him to talk with Chief Russell.

Instead, McLaughlin opened on schedule at 606 West Madison, and a squad from the police chief's office promptly raided. That night McLaughlin called Lingle at the *Tribune*.

"Swanson told me it was all right to go ahead and I don't see why Russell is butting in," said McLaughlin.

"I don't believe Swanson told you any such thing," said Lingle; "but if it's true, you get Swanson to write a letter to Russell, notifying him that it is all right for you to go ahead."

"Do you think Swanson is crazy? He wouldn't write such a letter."

"Well, Russell can't let you run, that's final," said the $65 legman.

With a curse, McLaughlin said, "I'll catch up with you, and it won't be long, either," then slammed down the receiver.

At about the same time, the state's attorney's office raided the Biltmore Athletic Club. They dismantled the gambling joint, eliciting an anguished cry from Lingle to Pat Roche, by then Swanson's chief investigator. "You have put me in a terrible jam. I told that outfit they could run, but I didn't know they were going to go with such a bang."

It was noised around that Lingle may also have broken his pick with Capone. The outfit's dog track had operated under an injunction forbidding police raids. Recently, when the state supreme court had ruled dog racing illegal, dissolving the injunction and closing the track, newspapers reported Capone as having told Lingle, "Well, the racket is through, and as far as I'm concerned, so are you."

Jake Lingle would not make it to the Board of Trade banquet on the night of June 9.

He set out from the Stevens, on Michigan Avenue, sometime before noon of that bright, sunshiny Monday, temperature a mild sixty-one degrees heading toward the day's high of sixty-eight. At middle height, running to fat, dewlaps becoming a double chin, he nonetheless looked dapper in blue serge with muted gray stripe, dark blue silk tie, black and white sport shoes and straw skimmer.

He stopped at his bank to make a $1,200 cash deposit, checked in at the *Tribune*, then lunched at the Sherman Hotel. He planned a day at the races, and would catch the 1:30 Illinois Central train to Washington Park in suburban Homewood, south of the city. He left the Stevens about one-fifteen for the short walk east on Randolph, pausing on the southwest corner of Randolph and Michigan to buy the *Racing Form*. The newsstand stood next to the west entrance of a pedestrian underpass, called a "subway," that tunneled under Michigan. On the east side of the avenue, stairs led up to the street, but the subway continued for another eighty-five feet toward the station.

As Lingle started toward the subway's west mouth, a man seated in a car nearby called, "Hey, Jake," and when Lingle looked up, said, "Don't forget to play Hy Schneider in the third." Lingle laughed and waved; he already had that horse. The morning line put Hy Schneider at 12–1 in the third, a mile race for three-year-olds and up. Lingle was about to save some money: long shot Paraphrase would win, paying $23.46; Hy Schneider would run out of the money.

Lingle disappeared down the subway, his usual cigar—which he smoked despite an ulcer—clenched in his teeth. He held the *Racing Form* open, rapt in it as he walked. He should have been glancing over his shoulder.

Witnesses later described a man nearly six feet tall, athletic and young, "like a college senior," natty in a gray suit and straw skimmer over light brown or blond hair. He was rushing to overtake Lingle.

Someone put a snub-nosed .38 to within inches of the back of Lingle's head and pulled the trigger once.

Lingle pitched forward, cigar still in his teeth, the end glowing, the open *Racing Form* still in his grasp. He was dead as he hit the pavement, his hat inches away, blood pooling around it. He could never have known he was in danger.

The killer dropped the .38. The blond man was seen doubling back through the tunnel at a quick walk. He turned into the stairwell leading up to the east side of Michigan. He may not have been alone. First reports spoke of a second man—shorter, with dark hair, wearing a blue suit. One witness thought the two and Lingle might have been walking together, with the blond man lagging behind. After the shot, instead of doubling back, the dark-haired man continued east, out of the subway. He leaped the railing and scrambled up the embankment to Michigan Avenue, where he disappeared.

Although it was presumably the killer (and presumably the blond man) who appeared in the intersection of Randolph and Michigan, that point would later be in dispute. In any case, he dodged through traffic to the northwest corner, then sprinted west on Randolph as the cry "Stop that man" arose from onlookers who had followed him up from the subway.

A traffic policeman, Anthony L. Ruthy, heard the cry and took off after the runaway, who darted into an alley, took several more turns, outdistancing his pursuers, and lost himself in the crowd. The chase ended.

The killer had dropped a left-hand silk glove, which perhaps accounted for the absence of usable fingerprints on the .38, whose serial number had been filed off.

Jake Lingle had $1,469 in his pocket. John Boettiger, the first *Tribune* reporter on the scene, quietly appropriated fourteen $100 bills—and duly turned them into the paper—thinking $1,469 an unseemly sum to be found on a $65 legman. The money began the revelations about Lingle and set the pattern for the *Tribune*'s overweening role in the investigation.

Although Colonel McCormick had not known Lingle's name, he saw this gangland-style killing as an attack on the *Tribune*, and he vowed *this* gang killer would be caught and punished. He could not rely on the police and the state's attorney. Throughout Prohibition they had won only one conviction of a gang-connected murderer, and that only because Sam Vinci had shot his brother's killer *at* the coroner's inquest—and even so did not receive the death penalty. McCormick demanded a special prosecutor. Charles F. Rathbun, a member of the *Tribune*'s outside law firm, was appointed, with Pat Roche as chief investigator. The *Tribune* posted a $25,000 reward, matched by Hearst's *Herald and Examiner*, another $5,000 coming from the *Post*. In all, rewards would total $55,825. McCormick would pay all extraordinary expenses of the investigation.

As the state's attorney would later put it, the *Tribune* soon discovered "that Lingle's money operations were entangled." Probate records revealed that his father had left him $500, not $50,000; his uncle's bequest had been $1,150. If Lingle had cashed out his two stock market accounts on September 20, 1929, he would have realized $85,980.66. Instead, the Crash caught him, and he lost at least $39,500, cash, and owed brokerages another $25,000 or so. Yet he hadn't stopped. One of his first two accounts had been jointly held with Bill Russell—who, with all the revelations, was forced to resign as police chief, replaced by John H. Alcock.

After the Crash, Lingle pumped another $28,000 into new stock accounts despite continuing losses. Also, in nearly two and a half years from the start of 1928 up to his death he had deposited $63,900, cash, to his bank account, most of it, judging from the checks he drew, lost at the track. Encomiums to Jake the Press Martyr and "cleverest police reporter of his time" abruptly ended.

"Alfred Lingle now takes a different character," McCormick's personally written editorial ran, "one in which he was unknown to the management of the *Tribune*. . . . He was not, and he could not have been, a great reporter. His ability did not contain these possibilities. . . ." The indication certainly seemed "that Alfred Lingle was killed because he was using his *Tribune* position to profit from criminal operations and not because he was serving the *Tribune* as it thought he was. Events will prove that this newspaper has nothing to cover in this connection."

Lingle's immediate superior, the day city editor, explained how a newspaper could miss for years such a story under its own nose. It had something to do with the old adage about the cobbler's children going shoeless.

So Lingle's murder had not been a "threat to the press" or an attempt to muzzle "exposure in the *Tribune* of facts about gangland." But then who had killed Lingle, and for what reason?

Calvin Goddard furnished the first solid lead. He restored the filed serial number on the murder weapon, and the Colt company identified it as part of a June 1928 shipment of six guns to Peter von Frantzius.

"Now, I want you to get this straight, von Frantzius," growled Coroner Bundesen as he questioned the arms seller. They had gone easy on him last time, but Lingle's murder bulked even larger than the massacre, and "if you don't do your part you'll find your way to the penitentiary." Von Frantzius stalled all he dared, but when investigators found a sale in his books with no name recorded, they sweated him until he admitted that North Side hoodlum Frank Foster had bought all six guns. Actually, "Foster" was only his most usual alias; he also used "Frost" and "Citro," his real name being Ferdinand Bruna, whose Romanian parents had settled in San Francisco. Foster had been accompanied by Ted Newberry. A search began for both men, especially Foster, who had skipped out of Chicago right after the killing, telling friends, "This town is getting too hot for me."

On July 1, 1930, Frank Foster was arrested in Los Angeles. Special prosecutor Rathbun faced a dilemma. He secured Foster's murder indictment in order to facilitate extradition, but despite a report that Officer Ruthy had identified Foster from a photograph as the man he had chased, Rathbun and the police denied that Foster had pulled the trigger. Otherwise, their case against the six-foot blond disintegrated, and he was the prosecutors' favorite murder suspect, although no one claimed to have seen him pull the trigger. Foster matched the early description of the second man—shorter, barely five feet eight inches tall, weighing less, more swarthy, with distinctly dark hair. Yet convicting Foster would be next to impossible, given what all witnesses except Ruthy said, and Rathbun *had* to deliver a conviction to Colonel McCormick. The second man was never again mentioned.

The favorite theories about why Lingle was killed involved his extortion of money from North Side operators, specifically Jack Zuta. The Sheridan Wave Tournament Club, a posh gambling joint at 621 Waveland Avenue, near Sheridan, had been padlocked in the aftermath of the massacre. Engraved invitations had gone out for its reopening, scheduled for the night of June 9, 1930. One version had Jake Lingle demanding 50 percent of the take for permission to open; a variation had him demanding $15,000, but up front. The managers turned him down flat, and Lingle reportedly told them, "If this joint is opened up, you'll see more squad cars in front ready to raid it than you ever saw in your life before." The managers appealed to Zuta, and he reportedly arranged to have this impediment eliminated.

A more direct story said that Lingle had received $50,000 from Zuta to buy permission to open dog tracks despite the supreme court ruling. Lingle could not deliver, yet refused to return the money. Zuta felt he had to avenge himself or face crippling ridicule. Frank Foster was known as Zuta's man.

On the night of July 1, 1930, not long after Foster was apprehended in Los Angeles, someone in Chicago gave a big boost to the theories that put Zuta behind Lingle's murder. The day before, police had picked up Zuta for questioning, along with a henchman, Albert Bratz. Held overnight on a disorderly conduct charge, on July 1 Zuta was quizzed about the murder by Roche. He swore he didn't know a thing. By 10:25 that night, back at Eleventh and State, Zuta's lawyers had obtained the usual writ requiring bail. Zuta was free to leave with Bratz, plus Zuta's chief lieutenant, Solly Vision, who had been with a woman, Leona Bernstein, when both of them were arrested.

Zuta was worried. Lieutenant George Barker, whose squad had collared him, was on his way out, going off duty. Zuta stopped him. "Lieutenant," he pleaded. "I don't want to go outside. I get it that a mob is gunning for me. I'll never get home alive. You took me from a safe spot. Now get me back there, will you?"

"What's the matter?" Barker laughed with disdain at the cringing pimp. "You're not scared are you?"

"Well, some of my friends don't like me very well." Spurred by Colonel McCormick, police had been cracking down and Pat Roche was savaging all the gangs, stressing that the heat would continue until he had the killer. The gangs blamed Zuta.

Barker was headed north. He'd take Zuta and the others as far as the Loop, where they could grab other transport.

Barker's Pontiac had passed Jackson, approaching Quincy, which dead-ended at State from the left, when Zuta—slumped in the rear seat between Bratz and Bernstein—cried, "My God, we're followed!" A dark blue Chrysler gunned up between Barker and the curb. One of its passengers, in a tan suit and Panama hat, swung out the rear door onto the running board, yanking an automatic from his shoulder holster. He pumped seven rounds into the detective's car. Barker jammed on his brakes and leaped out, pulling his service pistol. At thirty-three he was the youngest lieutenant on the force. As a World War I Marine he had been wounded twice, and was one of eight from a 250-man company who had survived the heavy fighting of Château-Thierry, Soissons and St-Mihiel. Now he stood in the middle of the street and returned fire as the assault car stopped, its other rear-seat passenger joining in, shooting through a window.

With the assailants intent on Barker, Zuta and his three companions slithered out of the Pontiac and across the street, disappearing into the dumbstruck crowd. Uniformed patrolman William Smith rushed onto the scene, pistol out, and the Chrysler zoomed away. Smith aimed at Barker, a gunman in civilian clothes, next to a civilian car. Just in time, Barker flashed his badge. Smith joined him as Barker dove behind the wheel and screeched off in pursuit of the Chrysler.

The fugitive car belched a dense cloud from the exhaust; the assailants had rigged the Chrysler to emit a smoke screen. Barker floored the accelerator through the cloud, emerging in time to spot his quarry turning east into Madison. Barker careened around the corner, then again as the gangsters swerved north on Wabash. Barker had closed the distance when his engine died; a bullet had ruptured the gas tank.

The killers sped away. They had fired about twenty rounds at Barker; Barker had emptied his pistol. Neither side suffered casualties. But Barker's car had blocked a streetcar. Its motorman, Elbert Lusader, thirty-eight, father of three, took a bullet in the neck as he stood at the controls, boggling at the action. He collapsed, dying in the hospital an hour later. Across the street, sixty-nine-year-old Olaf Svenste had been plodding to his job as a watchman when another bullet winged him in the arm.

Next day, Bratz showed up in court on the disorderly charge; he announced that Zuta feared for his life and would stay in hiding. Chief Alcock suspended Lieutenant Barker briefly for having offered Zuta protection.

None of this directly involved Capone, still in Florida getting ready for his perjury trial. No one seriously thought he had ordered Lingle's death. Notwithstanding reports that both Foster and Newberry had defected to Capone some months before—with Newberry identified as the gunman on the running board of the Chrysler—neither the Lingle killing nor the try at Zuta faintly resembled a Capone murder operation, both being public, chancy and haphazard.

Even so, the outfit suffered more than other gangs; it had more to raid and more to lose. And Capone was about to be propelled into the hubbub through his inveterate gabbiness.

John T. Rogers of the St. Louis *Post Dispatch* had been among the first to print revelations about Lingle, acting on a tip from Treasury agent Frank Wilson, who had just arranged to interview Lingle about Capone when the legman was murdered. Frankie Pope, one of the outfit's casino managers, had told Wilson about Lingle's ties to the gangs. After the attack on Zuta, Harry T. Brundidge of the rival St. Louis *Star* started digging into other possible press entanglements with Chicago gangland.

His series (gleefully reprinted by the *Tribune*) revealed that Julius Rosenheim, who had been murdered in February, had been a paid informer for Leland Reese, crime reporter for the *Chicago Daily News*; Rosenheim had used his connection with Reese to shake down gangsters, and Reese himself had been threatened with death. Then there was the trip to Havana with Capone that *American*'s city editor Harry Read had made. Ted Tod, crime reporter for the *Herald and Examiner*, took money as a public relations man for Moran's dog track while the state's attorney's office, which Tod covered, tried to close it. Matt Foley, the same paper's assistant circulation manager, had run a fake lottery based on the Kentucky Derby and was now a fugitive. If Lingle had been "unofficial police chief," Bill Stuart of the *American* was "unofficial mayor," so tight was he with Thompson. James Murphy of the *Daily Times* had part interest in a speakeasy. Another reporter was said to receive a nickel for each bag of cement sold in the city; another had a price schedule for divorce lawyers who wanted their names in the paper. "Only the dumb wits in the newspaper game in Chicago are without a racket," one of the smart ones told Brundidge, explaining that when a stranger had tried to muscle into *his*, he'd had the interloper's arms and legs broken.

Without introduction or invitation, Brundidge caught a train for Miami to find out what Capone knew. At about ten o'clock the night of July 11 (the day before Capone beat the perjury charges), Brundidge was introducing himself.

"This is a surprise," said Capone. "Come on in."

The reporter found Capone "intelligent, happy go lucky, and affable" with a "dark, kindly face, big sparkling eyes." His "whole demeanor" was "that of an overgrown boy." Capone's personality struck Brundidge as so "exceptionally pleasing," it took "no vivid imagination to understand the reasons for his huge success in his chosen field." In fact, mused Brundidge, anyone who knew nothing about his past would think Capone "a playful, lovable chap, as harmless as a big St. Bernard dog." Brundidge had been similarly impressed by Colonel McCormick. They settled down to a four-hour interview, including Capone's usual prideful tour of the estate.

"You seem to have raised merry hell in Chicago," said the big, playful fellow as they seated themselves on the sun porch. "What brings you here?"

Lingle, of course. "Why ask me?" Capone said, but added, "The Chicago police know who killed him." Jake was his friend, "up to the very day he died." And any rumors about a falling out between them were bunk.

"What about Jake's diamond belt buckle?" Brundidge asked.

"I gave it to him." Why? "He was my friend." What was his trouble? That addiction to horses.

"How many other 'Lingles' are there in Chicago in the newspaper racket?" asked Brundidge.

"Phooey," said Capone, " 'dun't esk.' " Maybe he was not the one to say it, but "newspapers and newspaper men should be busy suppressing rackets and not supporting them."

"How many newspaper men have you had on your pay roll?"

Capone paused, considered, shrugged. "Plenty."

He leaned over, put his left arm around Brundidge's shoulder, and gave it an avuncular *abbraccio*. "Listen, Harry," he said, "I like your face. Let me give you a hot tip. Lay off Chicago and the money hungry reporters. You're right; because you're right you're wrong. You can't buck it, not even with the backing of your newspaper, because it is too big a proposition. No one man will ever realize just how big it is, so lay off."

"You mean?"

"I mean they'll make a monkey out of you before you get through. No matter what dope you have to give that grand jury, the boys will prove you're a liar and a faker. You'll get a trimming."

"I'm going to quote you as saying that."

"If you do, I'll deny it."

Capone proved right on both predictions. After Brundidge's series appeared, Capone indeed denied having said all that, claiming they had spoken for only ten minutes or so. Brundidge pointed out that other papers had stories telling of his four-hour interview before his first piece ran. When the grand jury called the St. Louis reporter, his Chicago newsmen sources flocked to testify it had all been a put-on. Harry Reutlinger, assistant to Harry Read on the *American*, swore that when Brundidge kept hectoring him for instances of Chicago newsmen gone wrong, "finally, in jest, I told him some wild tales which no reasonable man would have taken seriously." As Capone predicted, the grand jury held Brundidge's evidence mere "hearsay" which "could not be substantiated."

The Lingle case would drag on into the fall, then winter. Rathbun kept probing, while Roche kept the pressure on all the gangs with unfixable raids.

Something had to be done. His perjury trial past, Capone returned to Chicago toward the end of July 1930 to take charge.

Someone identified only as a "wealthy businessman" approached Rathbun and Roche. Capone wanted a conference. Afraid of being compromised, the investigators appointed a proxy—known to history only as "Operative Number 1"—to represent them at the first meeting, held in October at a borrowed mansion.

"Here's what I want to tell you," Capone said, according to the operative's recollection, "and I won't be long about it. I can't stand the gaff of these raids and pinches. If it's going to keep up, I'll have to pack up and get out of Chicago."

"So far as I can tell you, the gaff is on for keeps. This town has been burning up since Jake Lingle was murdered."

"Well, I didn't kill Jake Lingle, did I?"

"We don't know who killed him."

"Why didn't you ask me? Maybe I can find out for you."

"Maybe you can."

"I don't know what the fellow that killed Jake looks like. I know none of my fellows did it. I liked Lingle, and certainly I didn't have any reason to kill him."

Operative Number 1 gave Capone the official description of the killer. In turn, Capone told what he'd heard: essentially the dog track theory, that Lingle was paid $30,000 (not $50,000), but could not deliver. "When the gang saw that they could not go they blamed Jake Lingle," Capone summed up, "and I think that's why he was pushed. But I don't know who they used to do the job; it must have been some fellow from out of town. I'll try to find out."

"You can do this if you want to, Capone, but I don't think it'll help you with Pat Roche."

Capone knew better. If Colonel McCormick was mollified with a convicted killer, he would withdraw both his funding and his pressure. Rathbun would return to his law firm, Roche to the state's attorney, Chicago to normal. All Capone apparently had to do was deliver a serviceable six-foot blond.

Only one hitch. The defense called Officer Tony Ruthy at a bail hearing for Foster, and to the surprise and consternation of both defense and prosecution, Ruthy positively identified *Foster* as the man he chased, insisting that the fellow had dark hair, not blond.

After a while, Capone's representative called on Operative Number 1. "Al wants to know if you will take the killer of Lingle dead?" Unfortunately, a dead, untryable body would scarcely satisfy McCormick, let alone Roche. The search went on.

Tribune reporter John Boettiger insisted it was Rathbun's and Roche's idea to "set a gangster to catch a gangster," using ex–Pinkerton detective and ex-con John Hagen to ferret out the killer from inside the underworld. Others supposed John Hagen a Capone plant, working on his instructions. If Rathbun and Roche did recruit Hagen, it counted as an amazing coincidence that he struck on exactly the right connections so quickly. Through a St. Louis gangster named Pat Hogan, he identified a lamster from a St. Louis labor-racket murder as prime suspect. At first, Hagen only got the name "Buster," then Lou Bader. The real name was Leo V. Brothers. He stood the requisite height, had light brown or blond hair, and had been in Chicago on June 9, 1930.

Captured in December, Brothers went on trial March 16, 1931. Witnesses divided about evenly between those who said he was and was not the blond everyone but Ruthy saw running away. Oddly, the most damaging witness turned out to be Ruthy, whom the defense called to repeat his identification of Foster as the man he had chased. An accident had fractured the policeman's skull long before the Lingle murder, but he had been kept on duty despite episodes of hallucination. Perfectly lucid at Foster's bail hearing, at Brothers's trial Ruthy almost parodied a stage lunatic. When asked if he had trouble with his head, he replied, "I had plenty of visions. I saw everybody from the good Lord down. I saw Abe Lincoln. I never saw Lincoln in my life. Gee, it must be terrible to think that this world will some day be ruled by the yellow race." So much for the awkward fact of a policeman having identified someone who could not possibly have been Brothers. In Howard Browne's novel *Pork City*, Roche says, "So they got to him. . . . Ruthy's either been paid off or he's had the fear of God put in him." Maybe not, but Browne also noted that Brothers had *very* high-priced counsel, four of them, yet had been visibly broke when picked up.

The verdict came after twenty-seven hours of wrangling: guilty, with the sentence set at fourteen years, the minimum. One holdout for "not guilty" had agreed to that as a compromise.

Browne's novel asserted that Capone had engineered the entire comedy, including the laughable sentence and the sequel for Brothers. Capone did remark, with the assurance of someone who knew, that Brothers's conviction was "the biggest frame-up I ever saw." After painstaking research—including a poll of many police and newsmen involved, Browne is positive that Foster killed Lingle but of course cannot prove any of it.

Three weeks after the trial, Officer Ruthy returned to duty in spite of his visions. A bandit shot and killed him, along with another officer, in circumstances untouched by suspicion. Brothers emerged from prison in 1940, after only eight years, and returned to St. Louis, where he beat the 1929 murder indictment that he had fled. He immediately became partner in mob-dominated taxi and loan companies, and was murdered by an unidentified gunman in 1950.

Capone had not been entirely forthcoming with Operative Number 1. On his return from Florida, he had set in train the only action he could hope might effect an accommodation with the *Tribune*, short of producing Lingle's killer. He would do, right, what Ted Newberry had botched: taking care of Jack Zuta. As August approached, Zuta could be charged with nothing, yet everyone agreed he'd ordered Lingle's murder. Colonel McCormick certainly knew that. Perhaps Zuta's own killing would be seen as a good-faith offering. In any case it would serve as punishment for having so cavalierly loosed such retribution on everyone else's racket.

Capone's network soon found Zuta holed up in Wisconsin. He had registered at several resorts in the last month as J. H. Goodman from Aurora, Illinois. Other guests remembered him as a short dumpy big-spender. Lately he and a companion had been at the Homestead. When the companion left, Zuta had moved nearby to the Lake View Hotel on Upper Nemahbin Lake near Delafield, about twenty-five miles west of Milwaukee. During the last week of July, a group of men, somewhere between nine and fifteen of them, took a cottage about three miles from the Lake View.

The evening of August 1, 1930, a Friday, Zuta stood by the nickelodeon in the Lake View's dance hall, feeding it money. Six men drove from the cottage to the Lake View. Three guarded the exits while three others, led by a large, solidly built man, strode into the dance hall. All carried guns. One covered the bartender; another kept watch on the couples dancing. Their leader went up to Zuta, whose nickel had just started a current show tune, "It May Be Good for You but It's So Bad for Me."

"Turn around."

When Zuta did, slugs ripped into him, the first taking him right in the mouth. He fell, and the other two gunmen stepped over to fire into his body, sixteen bullets in all.

No one doubted who had organized the killing. One report commented on its "businesslike fashion" and "machinelike precision." Nothing unusual, like a conviction, resulted, not even when Calvin Goddard matched bullets taken from Zuta's body with a bullet from a gun found on Danny Stanton, the massive leader of the murder party. The night of the murder, Capone conspicuously hosted a party for about one hundred close pals at the Western Hotel in Cicero.

In Zuta's car, police found another of the guns bought by Foster from von Frantzius. Pat Roche tracked down four safe deposit boxes belonging to Zuta, one empty, three packed with records and memorabilia of corruption that Zuta had squirreled away over the years, reaching back to 1921. Some five hundred canceled checks, purposely uncollected notes and IOUs, memos, letters and records compromised police, politicians, newspapermen and judges. One notation showed a weekly $3,500 payment to the East Chicago Avenue precinct house. The police chief of Evanston had sent a handwritten note to "Dear Jack," begging the loan of "four 'C's' for a couple of months," signed, "Your old pal, Bill Freeman."

When asked if these finds would go to a grand jury, Swanson said the "connections . . . are being traced." But wouldn't that attitude look to the public like a cover-up? "They will just have to take it that way then," said the public's servant.

"A crook is a crook," said the public's enemy, Capone, "and there's something healthy about his frankness in the matter. But any guy who pretends he is enforcing the law and steals on his authority is a swell snake. The worst type of these punks is the big politician. You can only get a little of his time because he spends so much time covering up that no one will know that he is a thief. A hard-working crook will—and can—get those birds by the dozen, but right down in his heart he won't depend on them—hates the sight of them."

CHAPTER 26

The Beginning of the End

THE SUMMER OF 1930 marked ten full years of constitutional Prohibition. Doggerel in Franklin P. Adams's column *The Conning Tower* summed up the country's feelings:

> Prohibition is an awful flop,
>   We like it.
> It can't stop what it's meant to stop.
>   We like it.
> It's filled our land with vice and crime,
> It's left a trail of graft and slime,
> It don't prohibit worth a dime,
>   Nevertheless we're for it.

No one thought it other than a flop, but the deepening Depression made circumventing Prohibition an ever more important source of income on both sides of the law.

One man, twenty years old in 1930, part of an enormous Italian family, fondly remembers his moonlighting job. Three nights a week he would carry fifty 100-pound sacks of sugar to the third floor of his family's home, where some of Capone's North Side allies maintained a still. He would whistle on his way up so the bootleggers knew it was he coming. Once a week, his employers left an envelope on the stairs, usually with $25 or $30, sometimes as much as $70, "depending on what they felt like paying me," he says. "I was tickled to death to get it because somebody had to feed the family, and my dad was making I think something like fifty cents an hour, with all those children.''

On the other side of the law, money might come informally, like the five-dollar tips that Officer Edwin F. McNichols used to get every time Capone passed his way when McNichols was on traffic duty in the theater district; or it could be a regular payoff, like Zuta's $3,500 a week. Or police could search out targets of opportunity. The man who hauled sugar sacks once rode along to help unload the five-gallon cans when the bootleggers carted alky to Cicero for processing. The ball-shift Dodge truck had modified suspension so passing police would not spot the load's inordinate weight, investigate and cut themselves in. That night, alert police stopped them anyhow and received an on-the-spot payoff.

Joseph Refke remembers from his days on the force literally sniffing out the pungent odor of stills with his partner. ''We were looking for some dough,'' he says, especially when Chicago, nearly bankrupt, paid city salaries with scrip and tax warrants.

The lucky could count on practically regular supplementary incomes. When Arthur T. Ristig transferred to a new district, his captain took him on a walking tour. ''We went into a gambling joint,'' Ristig remembers. ''The captain called the owner over and says, 'Take care of this man, here,' he says. 'He's going to be on the post, and I want you to take care of him every week.' 'Okay; yessir, yessir, we'll take care of him.' We go to another joint—and before I got through, I got about thirty bucks from them. And I said, 'What's all this for?' And the captain says, 'Ahhh—just do what you always do, mind your own business. You just made some good friends.' I said, 'Fine, that's good.' ''

A little later, when Ristig was assigned to the state's attorney's office, he was in Turner's, a steak house with gambling in the cellar. He spotted two of George Moran's boys at the bar, .38s practically tumbling out of their pockets. Moran was downstairs. Ristig had Turner summon Moran—whom Ristig knew, as he says, ''very well.'' He steered Moran to one side and said, ''They your two men?''

''Yeah.''

''Well, Jesus! tell them to stick those revolvers—somewhere. Before somebody takes them away. You know me. What the hell, I'm from the state's attorney's office—and a friend of yours—and I never make any problems.''

Moran motioned to his gunmen. ''C'mere,'' he said. Then, with a jerk of his thumb, ''*Out*'' to one and ''*Out*'' to the other. ''Bury those pistols,'' he called after them. ''We don't want them around.'' Moran turned to Ristig with a grin and said, ''That's the way it should be done—instead of running in here with twenty policemen and pinching them.''

''That's right,'' said Ristig, ''everybody's got to make a living, George.''

Intramural relations often tended to be less cozy. Lingle's murder, for instance, marked the eleventh killing in ten days, most of them barely mentioned in the press.

Santa Baldwin still remembers the shock when the man who pushed a cart

At the Roman Baths, Miami Beach, with future heavyweight champ Jim Braddock
HISTORICAL ASSOCIATION OF SOUTHERN FLORIDA

Sonny in the Biltmore tournament, with his bodyguard as caddy
HISTORICAL ASSOCIATION OF SOUTHERN FLORIDA

Line of fire from an Aiello gun nest in the Atlantic Hotel. Gunmen wait for Capone opposite Hinky Dink Kenna's cigar store on Clark Street.
CHICAGO SUN-TIMES

ANTHONY BERARDI

Six of the seven victims of the St. Valentine's Day massacre: (from top) *Pete Gusenberg, Weinshank, Heyer, May, Schwimmer and Clark* (against wall). *Frank Gusenberg died in the hospital.*

Frank Gusenberg
GODDARD COLLECTION

Pete Gusenberg
GODDARD COLLECTION

Albert Weinshank
GODDARD COLLECTION

Adam Heyer
GODDARD COLLECTION

James Clark
GODDARD COLLECTION

John May
GODDARD COLLECTION

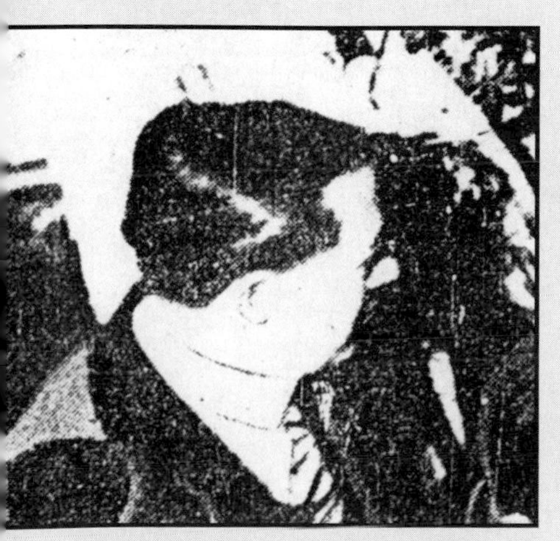

Reinhart Schwimmer at O'Banion's funeral
DAVID SCHOENBERG/DICK DUFFIN

ANTHONY BERARDI

Highball, the only survivor, still chained to a truck in the garage

Fred Burke, massacre suspect, 1931
CHICAGO SUN-TIMES

John Stege during Capone's round o
authorities after his release from
Philadelphia
ANTHONY BERARDI

CHICAGO TRIBUNE

Jake Lingle in the Tribune *newsroom*

Lingle dead in the "subway"

CHICAGO TRIBUNE

Frank Foster, probable killer
CHICAGO TRIBUNE

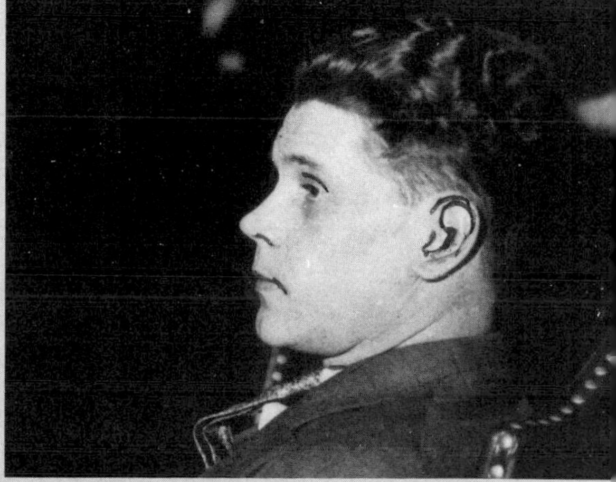

Leo Brothers, convicted killer
CHICAGO TRIBUNE

Coroner Herman Bundesen (left) *questions gun dealer von Frantzius* (right) *after Lingle's killing.*

THE UNITED STATES
TREASURY DEPARTMENT — BUREAU OF PROHIBITION
Washington, D.C. MAY 20, 1927
THESE PRESENTS WITNESS THAT
Eliot Ness
(whose photograph and signature under official seal are annexed), is regularly appointed and commissioned by the Bureau of Prohibition of the United States Treasury Department to enforce the laws and regulations relating to the manufacture, sale, transportation, control and taxation of alcohol and of intoxicating liquors, under the authority vested in the Bureau of Prohibition by the Congress of the United States, and the Secretary of the Treasury.
J M Doran
Commissioner of Prohibition.
COUNTERSIGNED:
Assistant Commissioner--Prohibition.
This commission is void if autographed photograph of appointee does not bear the seal of the Treasury Department.
No. 557
Eliot Ness

CHICAGO SUN-TIMES

Eliot Ness's Prohibition agent ID

Investigator Pat Roche in 1930
CHICAGO SUN-TIMES

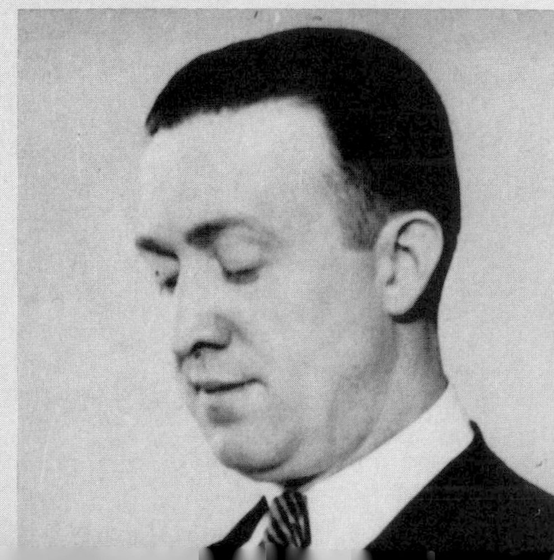

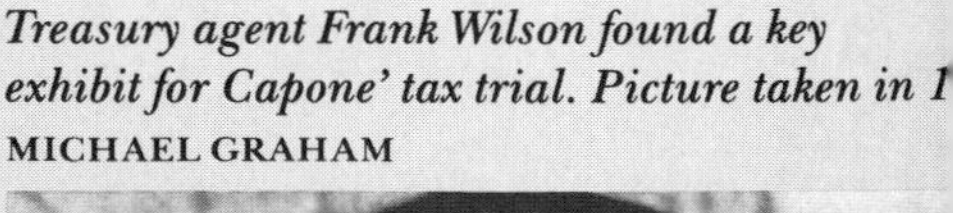

Treasury agent Frank Wilson found a key exhibit for Capone' tax trial. Picture taken in 1
MICHAEL GRAHAM

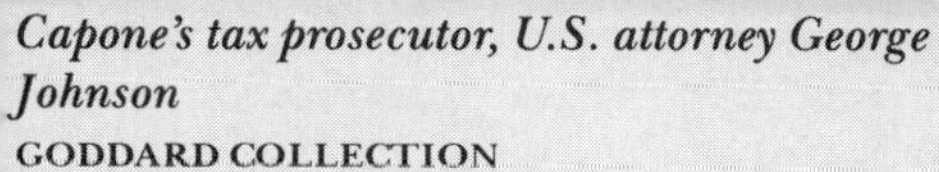

Capone's tax prosecutor, U.S. attorney George Johnson
GODDARD COLLECTION

CHICAGO TRIBUNE

Judge James Wilkerson, just after he rejected Capone's plea bargain, 1931

Capone gave him ten dollars every day of the trial.

ANTHONY BERARDI

CHICAGO TRIBUNE

Capone between lawyers Ahern (left) *and Fink* (right)

Capone in handcuffs during his appeal, while in Atlanta Penitentiary

CHICAGO SUN-TIMES

Capone in an FBI car on the day he was finally released
CHICAGO SUN-TIMES

A dapper Capone (right) *with his last lawyer, Abe Teitelbaum, in the early 1940s*
CHICAGO SUN-TIMES

The Capone family monument, Mount Carmel Cemetery
MARK LEVELL

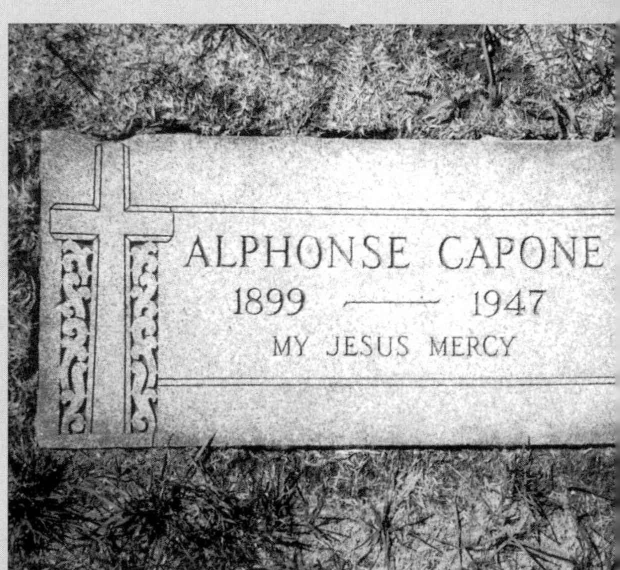

around her Little Sicily neighborhood, sharpening knives and scissors, wound up shot in the street. He had opened a small still in his basement. The Provenzanos, local Capone allies, had objected in the usual manner.

One day, the man who hauled sugar came across an object lesson to his neighborhood—a husky specimen lying on the sidewalk, half his head blown away, though the face was intact. "The blood was running into the gutter, and—I'll never forget—out of his mouth was a cough drop: you know, laying there, out of his mouth."

Mayhem also continued as usual among the more infamous. Frank McErlane, for instance, stayed his accustomed awful self. He once distinguished himself by being held on five simultaneous charges: drunk and disorderly, carrying a concealed weapon, firing a shotgun indiscriminately around the neighborhood, driving with forged license plates, and biting his sister on the cheek.

On January 28, 1930, McErlane was rushed to the German Deaconess Hospital, his left leg shattered by a bullet from an automatic that medics found lying on the floor. McErlane claimed it was an accident. Actually, his wife probably shot him, but it may have been John Oberta, with whom McErlane had been feuding since Joe Saltis removed himself to Wisconsin. Either way, Oberta tried to capitalize on the shooting. The night of February 24 McErlane lay in his hospital bed, almost healed, but his leg still in traction, when two or three gunmen invaded his room and started blazing away. McErlane snatched an automatic from under his pillow (the police later contemplated a concealed weapon charge) and blazed back, five of his shots shattering the doorjamb. His fire, while inaccurate, unnerved the assailants enough to rout them. They had winged him three times, nothing serious. One of the attackers dropped a .45 automatic.

"I'll take care of this matter myself," was all McErlane replied to questions about who had shot him. Leg healed, the new wounds superficial, McErlane was out of the hospital very soon after the attack. Nine days after it, March 5, John Oberta was found in his own Lincoln, shot to death, having fired three from his belly gun; a .45, unfired, was in his pocket. Some yards away lay his driver, Sam Malaga, also dead. The gun found abandoned in McErlane's hospital room was traced to Malaga. McErlane had taken care of it.

Despite the disaster inexorably closing in, the summer of 1930 saw Capone at the top of his form, his supremacy unchallengeable. He never bothered to reach any formal accommodation with Moran, because real trouble from the North Side had become unthinkable.

Capone continued to position the outfit in other fields, anticipating the end of Prohibition. Of more immediate importance, expansion appeared his only hope for blunting tax prosecution. If he could inextricably enmesh himself in the city's legitimate economic and political apparatus, he might be reckoned too powerful or valuable for harsh punishment. Racketeering might become a bargaining chip.

He had moved in on labor unions a long time before; now he stepped up the pace. He selected targets to maximize both return and clout, for instance going after the teamsters and the chauffeurs, since transport was central to much business and industry. He targeted the building trades, of supercharged importance with Chicago projecting a World's Fair for 1933. Capone moved on the plumbers, backing them in their fight with the steamfitters to see who would get the city's monopoly to inspect refrigerators. He also tried to maneuver through ordinances that would solidify his grasp on street maintenance.

More directly, he moved on city hall, sending Danny Stanton to organize the municipal workers. He already had in place Danny Serritella—as city sealer, part of Thompson's cabinet—plus Roland V. Libonati, identified as Capone's man in the state legislature (and later the U.S. Congress), and William V. Pacelli, a former state representative who had become Twentieth Ward alderman.

Along with expansion, Capone also kept increasing his booze business, shipping as far east as New York, south to Tulsa and Hot Springs, and west to Omaha. He could make up in volume for booze's low yield. Much of his supply, supplementing the Detroit shipments, came over routes Capone had developed. For example, one $100,000 load of G&W, Old Crow and Indian Hill bourbon, all distilled in the United States, was exported to Canada and on to Bimini in the Bahamas, then smuggled back into Florida and shipped from Jacksonville over various rail lines, billed to fake consignees as "lumber," ending in an Illinois Central station. The feds intercepted this load, thanks to papers seized in raids on Ralph's Cotton Club and Greyhound Inn.

Capone's beer trade, because of its bulk nature, had to remain local, but he could still expand in two ways. First, although the major gang territories remained inviolate, his muscle told independents they henceforth took from Capone or got out of the business; in some cases, as with Matt Kolb in the northwest, the outfit just muscled in. Second, Capone began horizontal expansion. Speakeasies had to take his pretzels and soda and buy his towels and linen, which had to be washed at a Capone laundry.

He also edged into new fields, like his move, through Murray Humphreys and Frank Maritote, to establish his own milk distribution company, Meadowmoor Dairy. He learned with astonished delight that milk incorporated a larger markup than booze or beer, and he realized that demand for it was steadier and nearly universal. "Honest to God," he'd tell his people, "we've been in the wrong racket right along."

Capone's fame expanded with his empire; he had become a civic landmark. When the Daughters of the Nile convened at Chicago's Medinah Temple, one complained, "Why, I haven't seen Al Capone since I've been in Chicago. And I've been here three days. I thought he would be on the reception committee." He was "America's trademark" around the world. Five Spanish actors, a stage director and two French scriptwriters stopped in Chicago on their way west to

MGM and asked to see one sight. ''Where's Capone?'' their spokesman demanded. The answer could never be Hollywood. A producer had offered Capone a million dollars for a cameo at the end of a gangster film, but Los Angeles district attorney Buron R. Fitts threatened to ''file every charge I could think of'' if Capone turned up in the jurisdiction.

In Russia, head commissar Vyacheslav Molotov derogated Capone as the logical culmination of capitalist rapacity, making Capone thereafter a standby in *Krokodil*, the humor magazine. A correspondent for the French weekly *Voilà* wrote that his compatriots saw Capone as Robin Hood, battling ''the sumptuary laws of puritanical America.'' In England, London's *Evening Standard* saw the French as ''The Al Capones of Europe,'' a cartoon showing Premier Pierre Laval and Foreign Minister Aristide Briand, feet up on the table, Briand barking into a phone, ''Wassat? Some guys muscling in on our territory? Bump 'em off, kid, bump 'em off.''

Cornel Capovici tacked a picture of Capone to the front of his house in Oradea, Romania, and insisted that this was his long-lost son. Foreign correspondent John Gunther reported that Viennese papers called Capone the ''real mayor of Chicago.''

Capone's fame had penetrated even Oklahoma—sort of. A witness before a state senate committee admitted about certain shady tactics, ''I'd say it's doing like Al would do.''

''Al who?'' asked a senator.

''Why, Al Pecan, the big Chicago gangster.''

At home his popularity soared, its nurturing not uncalculated. Once, at a Northwestern football game, packs of Boy Scouts made Dyche Stadium ring with ovation upon Capone's usual flashy entrance. He had bought the tickets so they could attend. But adulation became an unbought habit. At a Cicero high school game the cry went up, ''There's Al!'' followed by huzzahs, though he had bought tickets only for himself and six bodyguards. At another high school, in the more elegant suburb of Elgin, guest speaker William A. Rahn got in Dutch with parents for admiring Capone's ability and exhorting the kids to emulate his drive for success, if not his career path. Rahn had been on the federal grand jury that indicted Ralph for tax fraud.

Partly because Capone adored the popularity, he abhorred the implication of ''Scarface.'' He confronted newspaper publisher Merrill C. Meigs about it: Was it fair to harp on a disfigurement, styling him ''Scarface Al'' at every mention? After pondering, Meigs promised to use the sobriquet only in direct quotes. ''The man was right,'' he told colleagues. ''I just had not thought of it before.''

Of course Capone's central image problem remained his activities. He endlessly tried to talk the public out of belief that they merited what ''Scarface'' evoked. One technique was to absurdly overstate matters. ''There's a lot of people in Chicago,'' he said, ''that have got me pegged for one of these bloodthirsty monsters you read about in the storybooks—the kind that tortures his victims, cuts off their ears, puts out their eyes with a red-hot poker, and grins while he's doing it. Yeah—that's me.''

Then he talked sense. He cherished no illusions about himself. ''I'm no angel,'' he said. ''I'm not posing as a model for youth. I've had to do a lot of things I didn't like to do. But I'm not as black as I'm painted. I'm human. I've got a heart in me. I'll go as deep in my pocket as any man to help any guy that needs help. I can't stand to see anybody hungry or cold or helpless. Many a poor family in Chicago thinks I'm Santa Claus.'' Like his soup kitchens serving three thousand a day. ''I don't take any credit to myself for being charitable,'' he said. Just to show ''I'm not the worst man in the world.''

Nor was he the carouser everyone supposed; just ask his wife. ''She ought to know about it,'' he said, ''and I'll take my chances on what she says.'' Actually, they were apart so frequently Mae might not have known about most of the lapses (although syphilis surely provided a clue). For instance, Mae hadn't known Capone was in Atlantic City or Philadelphia until his arrest. But none of that mattered. Standing by her man, Mae would sum up her feelings after his death: ''The public has one idea of my husband. I have another. I will treasure my memory and I always will love him.'' And aside from the growingly rare binges, these days a thirty-one-year-old Capone indeed more usually played the homebody in the evening. Robed and slippered, he would linger over board games with Sonny or listen to the gramophone.

''I'm nuts about music,'' he said. ''Music makes me forget I'm Al Capone and lifts me up until I think I'm only a block or two from heaven.'' He preferred Italian opera, but also was fond of jazz and pop music. For one night of his week-long gala before the second Tunney-Dempsey fight, he had hired the band of Jule Styne—writer of such songs as ''Three Coins in the Fountain,'' ''It Seems to Me I've Heard That Song Before,'' and ''It's Magic.'' As Capone and Styne arranged the engagement, their talk turned to Gershwin's ''Rhapsody in Blue,'' which both agreed was splendid. Capone shyly asked if maybe he could conduct that one number. ''I did not feel I should say no,'' Styne remembered many years later—another vivid memory being of Snorky's silk shirt, itself of rhapsodic splendor. At the party, Capone conducted. ''He wasn't in time,'' said Styne, ''but all his life it must have been something he had wanted to do and by the end of the number he had tears in his eyes. He gave everyone in the band a hundred dollars and me a thousand.''

That was how Capone saw himself; why couldn't everyone else? Even the killings: people just didn't *understand*. Asked what a man thought as he killed another in gang wars, Capone replied, ''Well, maybe he thinks that the law of self-defense, the way God looks at it, is a little broader than the law books have it.'' It included what we would today call ''preemptive strikes''—like the O'Banion handshake and the Weiss ambush. ''Maybe it means killing a man who'd kill you if he saw you first. Maybe it means killing a man in defense of your business—the way you make the money to take care of your wife and child.

I think it does. You can't blame me for thinking there's worse fellows in the world than me.''

Such had become Capone's éclat that everyone wanted to get in on the act.

While Capone was still in Philadelphia, some of the boys had started shaking down visiting show business stars. Mae West had been forced to hand over $3,000, lesser sums demanded from actors Wallace Ford and William Gaxton, from singing cowboy Roy Rogers, comedian Lou Holtz, and from entertainers Rudy Vallee and Harry Richman.

Much later, Richman told a different version. He asserted that he had been nervous because his first wife was then married to Frankie Lake, who Richman seemed to think was part of Capone's outfit. He professed to believe that Capone might therefore want to harm him. He never specified why he believed that, but it could only have proceeded from the ludicrous theory that gangsters reflexively slaughter anyone who enjoyed even *prior* knowledge of their women (or, in this case, presumed subordinates' women). In other words, the lead-in to Richman's story was idiotic.

But as he recounted it, when the manager of ''George White's Scandals'' announced that Capone intended to visit Richman's dressing room, the entertainer panicked. Came a knock on the door, Richman later wrote, and,

> There stood the ruler of the underworld in the entire United States, a man whose name was respected even by the worst gangland leaders in Paris, London, and Rome.
>
> Before I had time to faint, he grabbed me and put his arms around me. ''Richman, you're the greatest!''

Afterward, when street hoodlums mugged Richman, Capone ordered return of his money and jewelry, then armed him with a letter that threatened any who dared molest Richman that they'd answer to ''Yours truly, Al Capone.''

George Jessel recounted a more likely story with the same payoff. Capone invited him to dinner one night. ''Dining with Capone,'' Jessel wrote later, ''was very important to me at that time in the city of Chicago because in those days actors were being shaken down for almost any reason.'' Such a mark of intimate favor told everyone to lay off of another entertainer Capone thought the greatest. Jessel generously included his pal Eddie Cantor in this unexclusive group, telling about the time Capone supposedly invited the pop-eyed singer-comedian to Palm Island. Assuming it was a command performance, Cantor broke into his routines on arrival—to the amazement of his host, who had merely invited him to dinner as token of his vast admiration and esteem.

Ethel Barrymore got in on the act, too. When her 1928 *Kingdom of God* reached Chicago on tour, she spotted Capone in the opening night audience. Barrymore played a nun who suffered and transcended three acts' worth of unwed mothers, orphans and the dying. She could see Capone, holding Mae's hand, tears bathing his face throughout the performance. ''Other actors were threatened in Chicago,'' she wrote in her autobiography, but not she. ''Word had evidently gone out to leave me alone.'' Capone thought her the greatest.

In 1927, comedian Joe E. Lewis was warned by Jack McGurn not to refuse a new contract at the Green Mill, a cabaret in which McGurn held an interest unconnected with the outfit. Lewis took a better offer anyway. Three men invaded his room, two bashing his head open with gun butts, the third ripping his jaw with a knife, slashing him twelve times. They left him for dead, but he slowly and painfully recovered. There is reason to question Lewis's assertion that the men were sent by McGurn. But that question aside, Lewis proudly underlined that later Capone told him, ''Why the hell didn't you come to me when you had your trouble? I'd have straightened things out.'' Capone wanted to back Lewis in opening his own club. Capone thought Lewis the *greatest.*

In his autobiography, newspaperman and screenwriter Ben Hecht told of a visit from two of Capone's gentlemen holding Hecht's script for *Scarface* and wanting to know if it was about Capone. ''God, no,'' Hecht answered, ''I don't even know Al.''

Then why ''*Scarface*'' they reasonably wondered; ''Everybody'll think it's him.''

''That's the reason,'' Hecht said. ''Al is one of the most famous and fascinating men of our time. If we call the movie *Scarface*, everybody will want to see it, figuring it's about Al. That's part of the racket we call showmanship.''

Hecht wrote that they went away satisfied.

In fact, the movie derived from W. R. Burnett's best-selling roman à clef about Capone. In the movie, Paul Muni played Tony Camonte, who had come from Five Points in 1920, using Joe Black as an alias, to work for Johnny Lovo, from whom he took over. From his headquarters on Twenty-second Street in the First Ward, Camonte struggled against his rival O'Hara on the North Side. Not everything was an exact match: the scar, which Camonte claimed he got in the war, was an X on his cheek.

Even if Hecht's visitors didn't read books, Hecht specified that they had read the script. Yet they bought his assurances? And if they were really annoyed, why wasn't Hecht dead after the movie appeared?

Whatever the various permutations of truth in those accounts, one incident certainly took place only in the imagination of Clara Bow's press agent. It was an attempt, in November 1930, to garner some favorable press for Bow when she was embroiled in squalid embezzlement charges back and forth with her onetime secretary and pal Daisy De Voe. A story appeared in the Los Angeles *Examiner* that Capone had slipped into Los Angeles to watch the It girl on the set at Paramount, leading the entire crew in frenzied applause after a take. Paramount immediately denied Capone had been on the lot; the Los Angeles police denied that he was anywhere near California.

Whether or not any of these stories is true in whole or part (could Ethel Barrymore really spot tears on a cheek across the footlights?), it is testament to

Capone's unique position that people who were themselves celebrities would boast of any connection with him and obviously consider his admiration validation of their talent. To whatever extent they invented the connection, their eagerness to parade it becomes that much more astounding.

Even the good guys in the fight against Capone wanted in on the act. Elmer Irey, Frank Wilson and Eliot Ness all told their stories later, in book form. All had done laudable work, but they could not resist making it seem perilous as well as meritorious. Since the world hailed or damned their target as the most deadly gangster of all, the observed fact that he "went quietly" somehow didn't sit right. So they capitalized on Capone's reputation as the most dangerous adversary imaginable (an entirely deserved reputation when he faced other criminals) to efface the fact that the desperado never did anything desperate against federal agents.

Consider what Irey and Wilson wrote about the informer and the undercover agent who helped them.

The informer was Edward J. O'Hare, who had been a lawyer in St. Louis, where his most significant client was Owen P. Smith, inventor of the mechanical rabbit that made dog racing possible. Between them, Smith and O'Hare exercised national control of dog tracks. As a sideline, in 1923, O'Hare helped steal some $200,000 of whiskey from a bonded warehouse. The whiskey belonged to George Remus, known in the early twenties as "king of the bootleggers." O'Hare was convicted and given a year's sentence, but won an appeal. Remus refused to testify at the retrial because O'Hare started to make restitution—which included giving Remus a share of his dog track interests.

After Smith died in 1927, O'Hare was *the* man in dog racing. He tied in with Capone, managing his Hawthorne Kennel Club and leveraging the connection to branch out further across the country. Nevertheless, O'Hare expressed the most profound contempt for gangsters, feeling he could use them without becoming one. By 1930, he wanted to change sides. The overture came through St. Louis reporter John Rogers. According to Rogers, the dream of O'Hare's adored boy Edward H., called Butch, was to attend Annapolis, and O'Hare hoped his cooperation with the government would smooth the way. O'Hare was once called "the best stool pigeon the government ever had"; Frank Wilson nominated him "the most important single factor" in getting Capone.

The undercover agent was Michael F. Malone, one of the SIU agents already working on the case. He was "black Irish" from Jersey City, five-foot-eight and two hundred pounds, "the greatest natural undercover worker the Service ever had," Wilson called him. (Incidentally, both Irey and Wilson claimed credit for initiating Malone's undercover role.)

Bankrolled by Chicago's Secret Six, Malone bought a wardrobe from Wanamaker's in Philadelphia and hung around there and in Brooklyn long enough to pick up current gangland gossip. Then he checked into the Lexington Hotel, using the alias "Michael Lepito." He became a fixture in the Lexington lobby, minding his own business, waiting for the outfit to make the first moves. They engaged him in enough conversation to get his hints that he was on the lam, a grifter from Philadelphia and Brooklyn; they opened the mail that he had written to himself, sent from Philadelphia by friends. They soon included him in their card games, parties and talk—his chance to gather inside information, though at constant risk of exposure and presumed death.

That was the story as Irey and Wilson told it. Neither lied about Malone's ploy; they merely exaggerated the danger beyond recognition. Obviously, a few searching calls by the outfit to Philadelphia and Brooklyn would have blown Malone's cover. So why weren't the calls made? Perhaps they were—as we'll see shortly. But if not, it's because there was no need to check on "Lepito" with care; he was not joining the outfit. And if the outfit had spotted Malone as a federal agent, his death was anything but sure; it was not even likely. Police certainly got killed, and federal agents, too, but not by the organized gangs, who knew better. Besides, why kill him? He sat at no inner councils, could have seen and heard nothing Capone didn't want him to learn.

Plainly, Malone lounged about the Lexington, ingratiated himself with the boys, picked up what loose talk he could. But that patent truth lacks drama. And if Malone, in the den, was not in mortal peril, how could federal agents at large possibly be? So when Malone went to a party upstairs at the New Florence Restaurant, across Michigan from the Lexington, Irey and Wilson professed to tremble for his safety. "We all knew," Irey wrote, "of Capone's penchant for throwing a party prior to batting out the brains of unfaithful followers." Wilson wrote merely that he "was sweating bullets" until Malone called him after the dinner.

Why settle for vicarious danger? According to Wilson, he got a tip from Ed O'Hare that Capone had imported four "Mafia Sicilian killers from Brooklyn" to kill Wilson, Pat Roche, U.S. Attorney George Johnson, and local SIU agent-in-charge Arthur Madden. In Irey's version, Wilson learned of the plot not from O'Hare but from Malone, who warned that five killers were gunning for only three targets: Wilson, Madden and . . . Irey.

Tribune reporter John Boettiger added some circumstantial corroboration. He wrote that word had come from "a government undercover man" (presumably Malone), who had learned of a highest-level meeting at which Capone had outlined his assassination plan—and for the first time told henchmen of his negotiations with Operative Number 1, a detail that supposedly authenticated the story and showed that it was no idle rumor, since Capone was the only one in the outfit who knew about Operative Number 1.

To his credit, Pat Roche never believed the assassination story. Sure enough, when scrutiny disclosed no trace of the imported killers in Chicago, word filtered out of the Lexington that cooler heads had dissuaded Capone.

One explanation makes sense. Maybe the outfit did check up on Malone. He had arrested Willie Heeney six years before; perhaps Heeney recognized him. Maybe someone had seen him around town before he became "Lepito." However unlikely it is that Capone would have projected the murder of a federal prosecutor and federal agents, it's even more improbable that anyone in the outfit's top echelon (a notably discreet lot) would have breathed a word of such top secret plans (not to mention those delicate negotiations with Operative Number 1) in earshot of lesser outfit members—let alone an outsider—without orders. But if the outfit had indeed unmasked Malone, why not feed him tales of imported killers and get the opposition chasing phantoms and their tails? That charade could have inadvertently snared Ed O'Hare, who might well have heard some of the talk being circulated for "Lepito's" benefit.

No one had to help Eliot Ness get in the act. George Meyer claims that Ness would sometimes be given addresses of minor warehouses to raid—with expendables standing by to take a pinch—presumably in exchange for Ness's passing up, at least temporarily, more vital locations and personnel. Another observer saw him lead such obvious show raids. Even so, the head Untouchable and his unit unarguably destroyed and seized a lot of Capone's assets.

On the other hand, in his 1957 book, *The Untouchables,* Ness boasted that "Chicago was drying up fast" as result of his raids and that finally "Chicago became dry as the proverbial bone." He had "put Capone out of the beer and alky business."

He did not. Nor did Ness and his group, as he averred, contribute directly and materially to Capone's "undoing." In the event, the government never tried Capone on the only indictment for which the Untouchables gathered evidence—conspiracy to violate the Prohibition laws.

Ness and his men racked up an entirely praiseworthy record of effectiveness and incorruptibility. They cost Capone a lot of money during a crucial period. Naturally Capone tried to counter in the usual way, only to find that the Untouchables were just that. If Ness did sometime trade one target for another, neither he nor his men were seduced by the handsomest bribe offers. When some of Ness's $2,800-a-year men turned down $1,000 a *week*, Capone's emissary said plaintively, "Everyone here in Chicago gets along good with each other. Why not you?"

Virtue and a job well done seemed not enough for Ness. His book kept invoking The Menace. Ness knew that Capone "had to crush us to survive." Yet by Ness's own account, The Menace never made a convincing move toward bruising, let alone crushing, any federal officer. No setups, none of Capone's minutely planned ambushes. (Frank Wilson did write about a warning *he* had received that Capone had him and his wife staked out at their hotel, the Sheridan Plaza, their every move charted; Wilson countered that peril by hurriedly switching residence to the Palmer House, a maneuver that evidently so confounded Capone he never got onto them again.) The only time Ness faced "a Maffia gunman," it was one lonely punk trailing him in a car that Ness easily forced to the curb only to find the punk sitting dazed, his gun still nestled in its holster. Another time, Ness caught a lone terrorist skulking outside his fiancée's house.

In the course of pummeling the unresisting intruder, Ness discovered that the fellow's shoulder holster was empty; he had left his piece in the car. The only member of the group to be killed was not really an Untouchable, but Ness's driver, an ex-con. It was no gang murder; police almost instantly caught the killer, who hanged himself in his cell.

In the furor over Lingle's murder—which might have indicated that no one's life was safe from The Menace—Ness sought to allay his fiancée's fears. "There's nothing to worry about," he told her. "Why, they wouldn't think of shooting a federal man." That was the truth. Paul W. Robsky, an original Untouchable, remained a federal agent until he retired in 1951 and became an airline guard. "We all carried guns, naturally," he reminisced, "but we never had to use them in Nineteen Thirty to Thirty-one because Capone gave the word to his boys at the breweries not to oppose federal officers. If there was a raid, he told them to take it or to try to get away—but no rough stuff." Capone had enough trouble without killing federal officers, who would anyway be replaced at once by more.

This hankering after the appearance of dangers braved was silly. The government people did their jobs exceptionally well, and the job descriptions did not include trying armed conclusions with gangsters. That applied particularly to Frank Wilson, essentially an investigative accountant, whose forte was dogged, exhaustive, *sitzfleisch* examination of records. That—not anyone's high-hearted daring—led to the single most important revelation in the getting of Capone.

By the summer of 1930, Wilson and his people had been diligently pursuing a link between Capone and provable income for two years. As Wilson had told Irey, Capone had not seemed to make a mistake. Of approximately one million checks the SIU examined, two bore Capone's endorsement, and those for unimpressive sums.

One night, everyone else gone home, Wilson decided to scan the exhibits of seized records once again; maybe he and his colleagues had missed something. The SIU had been crammed into a cubbyhole in the Old Post Office building with barely room for four filing cabinets and two desks. About one A.M., muzzy with weariness, Wilson quit. As he went to stow the documents he had been examining, he bumped into the filing cabinet drawer, slamming it shut, the lock automatically engaging. He could not find the key.

A less exacting man might have let the fruitless papers lie out till morning. Wilson poked around in an adjoining storeroom, where he spied a rickety unlocked filing cabinet. To make room, he removed an armful of dusty envelopes from a drawer, and discovered in the back "a heavy package covered in brown paper." Without knowing quite why he bothered, he cut the package's string, revealing three notebooks of the kind distinctively used as ledgers.

The first he opened seemed innocuous. After one glance at the second ledger, Wilson's weariness vanished. Columns headed "Bird cage," "21," "Craps," "Faro," "Roulette," and "Horse Bets"—and the figures in the columns—proclaimed this the cash receipts ledger of a large-scale gambling operation. Odds were it had to do with Capone.

The entries covered a period from 1924 to 1926 and showed Wilson exciting sums. In twenty-two months, profits from this one joint totaled $587,721.95. Every few pages the ledger's bookkeeper struck a total and detailed the disposition of funds. On one page, for December 1924, Wilson saw that the ending total cash in the house bank had been $31,443 before adding a $17,500 item and subtracting a $6,256 loss in ''Races.'' Reserving a new $10,000 bank, that left $32,687 to be divided according to formula. ''Town''—obviously the local payoff—received the most, 20 percent or $6,537.42; ''Ralph'' (Capone, Wilson could reasonably hope) and ''Pete'' (who turned out to be Capone gambling boss Pete Penovich) each got 5 percent or $1,634.35. Four payments of $5,720.22—17 percent each—went to ''Frank,'' ''Lou,'' ''D,'' and most suggestively, ''J&A''—which could mean Jack (Guzik) and *Al*.

A large entry at the top of the page read, ''Frank Paid 17500 for Al''—which matched the $17,500 additional house profit (Pete Penovich would later testify it represented a horse bet Capone had lost). It connected an ''Al'' with the operation. And The Mattingly Letter pegged Capone's share of operations at one sixth, 17 percent!

Next morning Wilson learned that the ledger had been seized in a raid on Capone's Hawthorne Smoke Shop right after the McSwiggin killing. Pat Roche had it in his possession, but neither he nor anyone at that time discerned possibilities in such records. As retired policeman Arthur Ristig puts it, ''Nobody ever gave that stuff a thought. To us it was just a bunch of garbage we'd throw to the side. If somebody ever knew we were sitting on the red hot tomato all the time they'd be shaking those guys down for money forever!'' When Roche had left for the state's attorney's office, he had turned everything over—and there it sat, unconsidered, gathering dust.

Now Wilson had to find someone who could tie Capone directly to the ledger. Three hands had written in it, one especially distinctive, neat and precise. The SIU collected handwriting samples from every hoodlum and ancillary connected with the outfit, checking voting registers, bail bonds, bank accounts and every other imaginable source. On a deposit slip from a Cicero bank they finally found a match. It had been signed by one Leslie Adelbert Shumway—Lew, they called him. An informant told Wilson that ''Shumway is a perfect little gentleman, refined, slight, harmless—not like a racketeer at all.'' The SIU learned Shumway had left Chicago, but otherwise had no idea where he could be found.

The following February, 1931, Ed O'Hare tipped Wilson that Shumway had long since moved to the Miami area, where he worked as a cashier at either a horse or dog track. When Wilson found him at the Biscayne Bay Kennel Club, Shumway pretended ignorance. ''Oh, you're mistaken, Mr. Wilson,'' he squeaked, shaking. ''I don't know Al Capone.'' The man who broke tough Fred Ries would have little trouble with a perfect little gentleman.

If Shumway refused to talk, Wilson promised to send a deputy marshal loudly looking for him at the track, waving a subpoena while shouting his name. How long would he last once Capone heard about that? On the other hand, if Shumway agreed to testify, the government would protect him. "Play ball, Lew," Wilson told the little wreck, "and I'll guarantee that Mrs. Shumway will not become a widow."

Shumway would talk. What he said also led Wilson to the Reverend Henry Hoover and his raiders, who would testify to Capone's admission of ownership of the gambling joint across from the Hawthorne Hotel.

Some thought *Tribune* lawyer Charles Rathbun gave Judge Lyle the idea. Newspaperman Walter Spirko says no, "he was put up to it by the newspapers; he was a publicity hound." No one imagined that Judge John H. Lyle himself conceived the idea of issuing vagrancy warrants against all on the public enemies list. Lyle was a lean, long-faced grump who had come to Chicago from Tennessee in 1900 at age eighteen. He had worked at many jobs—including some professional boxing—until he became a lawyer. He served in the Illinois legislature and on the Chicago City Council, where he once engaged in a brawl on the council floor. Lyle won election to the Municipal Court in 1924.

According to his colleagues he lacked judicial temperament. To be sure, many of his fellow judges were crooked, and he was not; still, they had a point. He habitually set unconstitutionally high bail, routinely reduced on appeal; his $100,000 would become $5,000. He bombinated from the bench against criminals, particularly Capone, calling for his speedy electrocution, forgetting to specify any charges. Not only would he issue arrest and search warrants to any officers who asked for them, regardless of probable cause, he also tagged along on the raids. Four times, the chief justice transferred Lyle from criminal to civil matters where he could torture propriety a little less. But the public adored his windy toughness against known criminals (though it resulted in no more action than his colleagues' sinuous crookedness did). A combination of public and newspaper pressure always restored him to the criminal bench.

The notion of charging millionaires as vagrants was not entirely demented. Prosecutors could argue that the targets had, as the statutes read, no *visible* means of support. It seemed worth a try.

Capone's name, naturally, appeared on one of the first warrants that Lyle issued in September 1930, and Capone flirted with the idea of turning himself in. But Lyle would not guarantee to emissaries Capone's immediate bail. Capone had two reasons for staying out of Lyle's toils, even for a short while.

First, he was planning the next murder. When Capone was still in Philadelphia, Joe Aiello had returned to follow Hop Toad Guinta as head of the Unione, which automatically gave him a renewed power base, inspiring him to resurrect dreams of getting Capone. Gossip put him again plotting with Moran—who wisely stayed clear of Chicago, first in Minnesota, then Lake County, north of the city.

Aiello might no longer represent a mortal threat or serious competition, but he was a chronic annoyance, the only one left.

The operation was classic Capone. Aiello thought himself safe, holed up at the apartment of his partner in the Italo-American Importing Company. Pasquale Prestogiacoma—mercifully known as Patsy Presto—lived at 205 Kolmar Avenue. It took Capone's network little time to locate Aiello. About ten days before October 23, 1930—night of the killing—two polite, nice-looking young gentlemen, one of whom called himself Morris Friend, rented a second-story apartment across the street, at 202 Kolmar, its window facing Patsy Presto's front door. A walkway ran along the side of Presto's building. So a polite, nice-looking young gentleman who called himself Henry Jacobson rented a third floor apartment at 4518 West End Avenue, this one in the back of the building, its window overlooking that walkway.

Joe Aiello had not stirred out of Presto's apartment since "Friend" had taken the apartment across Kolmar. Either the watchers' vigil ran 86,400 seconds a day or their communications system was a miracle, because at 8:20 that Thursday night a cab was summoned to 405 Kolmar. At 8:30 driver James Ruane stood by his cab outside. Aiello and Presto opened the front door and stepped out; Aiello's pockets held an address in Brownsville, Texas, and a train ticket. A machine gun opened up from across the street.

Ruane crouched behind his cab, unscratched. Presto ducked back inside, also unscratched, though .45 slugs tore through windows and walls, tattering a thousand-dollar couch in his living room. Aiello, hit, staggered, then hobbled around the corner of the building—smack into a stream of slugs from the West End machine gun nest. Aiello died with over thirty holes in him. Although some residents had seen gunmen fleeing the West End apartment, police could identify no one, arrested no one.

Politics ruled the second reason Capone would not let Judge Lyle put him behind bars even for the length of time it might take to find a complaisant judge and a writ. As the fall declined to winter, Lyle planned to contest the Republican nomination for mayor with Thompson, who, incredibly, announced for a fourth term. Capone would do nothing that might help Lyle. Even before Lyle formally announced his candidacy, in late December, Capone took off for Florida, beyond the reach of Lyle's vagrancy warrant.

Toward the beginning of November, he made his first move to test the bargaining powers of his racketeering. He sent Michael J. Galvin of the Teamsters to John P. McGoorty, then chief justice of Chicago's Criminal Court. With what McGoorty termed "cool effrontery," Capone offered to get out of labor racketeering if he could have a free hand—no more raids—with beer. McGoorty, who anyway could not commit the federal agents, turned him down.

Capone's determination to frustrate Lyle meant that he had to miss the December 14, 1930, wedding of his sister, Mafalda, to John J. Maritote, Frank Maritote's twenty-three-year-old kid brother.

People whispered that the happy couple barely knew each other, and were in fact not very happy about the match, supposedly arranged to more firmly cement the two families. They said the groom had been about to marry someone else. Nineteen-year-old Mafalda denied it all, naming John her childhood sweetheart.

After the ceremony, hundreds moved to Ralph's Cotton Club for an all-night celebration while the families and the favored repaired to the nearby apartment of a friend for a more intimate reception, featuring a $2,100 wedding cake in the shape of a yacht, nine feet long, four high and three wide, with "Honolulu" in icing on its prow. That was the presumptive honeymoon destination, although the couple ended up in Cuba, as would appear when Italy-born John later had naturalization problems over false declarations he made as they reentered the country.

Capone was right to stay away. Police blanketed the ceremony and reception looking for him in order to serve Lyle's warrant; they hauled off five guests they found armed. Capone later gave the couple a house and, to get started in life together, $50,000; or maybe it was $75,000. Reports varied, and neither Capone nor anyone in either family discussed the matter publicly.

CHAPTER 27

Gotten . . .

ELEANOR MEDILL PATTERSON—cousin of Colonel McCormick and sister of Joseph Patterson, who owned the New York *Daily News*—ran the Washington *Herald* for Hearst, and would end up buying it, merged with the Washington *Times*. On impulse, early in January 1931, she stopped her car outside 93 Palm Island. "Come in," said Capone, "let me show you around." Cissie Patterson found his manner "like that of any kindly, hospitable man, proud of his estate. . . . He has," she wrote,

> the neck and shoulders of a wrestler. One of those prodigious Italians, thick-chested, close to six feet tall. The muscles of his arms stretched the sleeves of his light brown suit, so that it seemed to be cut too small for him.
>
> Once I looked at his eyes. Ice-gray, ice-cold eyes. You can't anymore look into the eyes of Capone than you can look into the eyes of a tiger.
>
> Now to get to the root of the whole interview. . . .

He complained of a raw deal. "I don't interfere with big business. None of the big business guys can say I ever took a dollar from 'em," he told her. (Perhaps he had forgotten, among others, the Balaban and Katz theater chain from whom the outfit collected $1,000 a week per theater on pain of stink bombs being loosed.) In fact, Capone continued, he had broken a strike for a *big* newspaper. (Danny Serritella, as president of Capone's newsboys union, wrote that Capone had settled a threatened drivers' strike against the *Tribune*; Colonel McCormick denied it. But such a projected strike had indeed been inexplicably called off.)

"And what," Capone asked, "do I get for doing 'em a favor? Here they've been ever since clamped on my back. Why can't they let me alone?"

Maybe he should leave the country, Patterson suggested. How about living in Italy?

"Why should I? My family, my wife, my kid, my racket—they're all in Chicago. I'm no Eyetalian. I was born in New York thirty-one years ago."

As they talked, Capone's hands mesmerized Patterson. "Enormous," they seemed to her,

> powerful enough to tackle—well, almost anything, although superficially soft from lack of exposure, and highly manicured.

But she kept coming back to those eyes. Even absent the Look, they transfixed their object. "Capone's eyes," Patterson wrote,

> are "dime novel" gangster's eyes. Ice-gray. Ice-cold. I could feel their menace. The stirring of the tiger. For just a second I went a little sick. I had to fight the impulse to jump up and run blindly away.

When he called for a servant,

> a man in a white apron simply tore into the room. "My goodness," I half whispered, "I wish I could get service like that at home."

Home for Cissie Patterson was a Dupont Circle mansion she had lent the Coolidges when the White House needed repairs.

> "Tell me, Mr. Capone," I ventured presently, feeling somewhat refreshed after the lemonade we had consumed, "what do you think about Prohibition? Do you think it will ever be done away with," I asked.
>
> "Yes, I do, and I'm all for that time to come. Prohibition has made nothing but trouble—trouble for all of us. Worst thing ever hit the country."

Capone walked her to the front gate, which he unbolted himself, waving off the "seven or eight neatly dressed, athletic young men" who, Patterson had noted, stood discreetly, unobtrusively, yet constantly and watchfully present.

> "Well, good-bye, Mr. Capone." I shook hands with him. "Good luck to you," I called as the motor started, and I meant it sincerely.
>
> It has been said many times, with truth, that women have a special kind of sympathy for gangsters. If you don't understand why, consult Dr. Freud.

Capone's trial on the long-pending federal contempt charge was set for February 25, 1931, in Chicago. Though he left for Chicago four days early to direct the usual election arts on behalf of Big Bill Thompson in the February 24 primary against Judge Lyle, Capone still took care to dodge that vagrancy warrant.

During the campaign, Thompson had called Lyle the "nutty judge." Lyle had called Thompson a "blubbering jungle hippopotamus." With Capone's help, the blubbering hippo won, 296,242 votes to the nutty judge's 228,401. Capone said, "Lyle tried to make me an issue and the public has spoken." Now it didn't matter if police served Lyle's warrant.

The usual mob scene attended Capone's court appearance. The star wore a sumptuous blue suit, accented by white silk hankie, pearl gray spats and diamond-studded platinum watch chain. One newspaper snorted that the police and court personnel, and especially the "crowds of the worshipful laity," treated him "with the ostentation and consideration due Chicago's biggest home town boy made good."

Capone gazed about, visibly impressed by the courtroom's white marble, its dull gold ceiling and trim, the Constitution's preamble in fading letters over the entrance. Judge Wilkerson would hear the case without a jury. Before the judge appeared, Capone chatted with the press.

No, he would not return to Florida after the trial, the weather had been foul. No, he would not write his autobiography. "The last bid I had was two million dollars," he said. It did not tempt him. "I'm not going into the literary business. That would be cutting in on the work of the boys who are writing about me." He had read one of those books, and hadn't thought much of it; he hadn't recognized himself. "I guess maybe I could write a better one, but that sort of stuff isn't my line." Neither was making movies. He flashed his grin. "I'm no Mary Pickford."

Nor—it quickly appeared when the trial started—had he been an invalid when he was due to appear before the grand jury on March 12, 1929. Police at Hialeah told of seeing him at the track almost from the start of the meeting on January 17; records and testimony told of his trips to the Bahamas in early February. Stenographer Ruth Gaskin noted his rude good health at that interview the day of the massacre.

When the noon break came, police finally got to serve Judge Lyle's vagrancy warrant. By now it was a frayed, crinkled document that had passed through many hands while police had searched for Capone. Two sergeants trotted Capone down to the detective bureau, where the chief of detectives entertained him with coffee and sandwiches, and Capone entertained the chief with assertions that he made his living in real estate. Judge Frank M. Padden set $10,000 bail and continued the case for a week. Capone's lawyer had already arranged a bond, secured by an $80,000 property, a neat trick for most vagrants. The police considerately rushed Capone back in time for the one-thirty resumption of federal court.

All afternoon Capone heard more testaments to his fine fettle that February and March. After court, police escorted Capone back to the Lexington. ''We don't want Capone killed in Chicago,'' explained Police Chief Alcock. When Capone's cab stopped at a light and a swarthy man approached it, a police lieutenant jumped from a following squad car to grapple with the possible assailant. ''I'm only a bondsman trying to drum up some trade,'' whined the frightened hustler.

''You can't be a vagrant,'' someone called to Capone, who showed up for the second day's hearing in a misty-morning gray ensemble replacing yesterday's rich midnight blue. ''Two suits.''

''I'm buying this one on time,'' Capone called back.

The defense started. Young, chubby, ineffectual-looking Dr. Phillips turned jelly under cross-examination by the government's lead prosecutor, Jacob I. Grossman. Phillips had to admit that statements in the affidavit he had sent the court to excuse Capone's nonappearance were false. He hadn't *really* practiced four years in Chicago as he had sworn; it was one year, and only as an intern. Capone hadn't *actually* spent six weeks in bed as the doctor's affidavit had insisted, only two or three weeks. He had only meant such statements, he now explained, to be ''approximate.''

When he had first seen Capone on January 13, 1929, his wire to the family doctor, David Omens, stressed that the illness was not too serious. How did that square with his affidavit? ''I didn't want Doctor Omens telephoning the family,'' Phillips replied, ''because the family was already excited.''

Phillips had not composed the affidavit. Miami lawyer John Stokes had—on Capone's orders. Phillips had been in the office while Stokes dictated it, but had not been listening closely and hadn't really read the thing before signing. ''I scanned it rapidly,'' he admitted. ''I am sorry I didn't read it carefully.''

Ann Fagan, the day nurse who attended Capone, and Nora Hawkins, the night nurse, did much better. Both exuded professionalism; cross-examination could not bend them. Capone had been a sick man the middle two weeks of January. Dr. Phillips was partial to what Fagan called ''home remedies''; Capone had been layered with mustard plasters until there wasn't enough room left on his hide for a ''post no bills'' sign. Nights, Hawkins had found him uncomfortable, with chest pains, running a 104-degree temperature. ''He was coughing hard enough to shake your head off.'' He certainly hadn't been in any shape to gallivant about at racetracks while they were on duty.

Their testimony countered that element of the prosecution case, but the defense could not otherwise impeach the rest of it. Capone had told the press that he'd ''probably go on as the last witness.'' That was nonsense bravado; he dared not expose himself to cross-examination. The defense rested.

During prosecutor Grossman's summation, Judge Wilkerson pointed out that little turned on the exact date when Capone's Hialeah jaunts began. He was unarguably well enough to fly to Bimini by February 2, so he surely had been well enough by March 5—when Phillips signed his affidavit—to have returned the grand jury's subpoena in Chicago a week later.

Capone's lawyers did what they could. Ben Epstein argued that the grand jury, after all, had evinced no urgent need for him when he had shown up; the prosecutors had barely questioned him at first, held him over a week.

That wasn't the point, said the judge, nor was the fact that Capone had stalled his appearance for only eight days. The court summoned all before it only one way, by subpoena. When issued, Wilkerson intoned, "it is to be respected, it is to be obeyed, it is not to be trifled with, it is not to be flouted."

Judge Wilkerson's verdict was guilty. "Yeah," Capone muttered, "I felt it coming." The judge at once sentenced Capone to six months in the county jail, where federal prisoners went for sentences under a year. Capone had not felt that coming. Chewing gum furiously as an aid to composure, he said as he left the court, "If the judge thinks it's correct, he ought to know. You can't overrule the judge." He remained on $5,000 bail, pending appeal, but appellate courts seldom overturned contempt convictions. On however scantling a charge, they had him.

Now came Chicago's turn with Lyle's vagrancy warrant. In prosecution theory, no matter how many suits Capone possessed, he'd had "no legitimate occupation for years," and that made him a vagrant. In practice, on March 4 the state needed a continuance to March 20; on March 20 they needed another to April 3.

That day the alleged vagrant showed up in another gray medley, an oxford single-breasted suit, gray stripes on his dark blue tie, gray topcoat, and the expected fedora. The state showed up in disarray. It seemed that the complaining policeman had signed the vagrancy complaint as a favor to a brother officer and had nothing on Capone, in fact didn't even know him. "We have been unable," prosecutor Harry Ditchburne admitted, "to find anyone who knows anything about Mr. Capone."

Judge Padden glared down from the bench. "Couldn't you find a single copper?"

"The state moves," said Ditchburne, "that the charge be dismissed."

On April 7, despite a rousing campaign song to the tune of "The Sidewalks of New York" that started,

> Tony, Tony, where is your pushcart at?
> Can you imagine a World's Fair mayor
> With a name like that?

Big Bill Thompson lost to the Democrat, Anton Cermak, 475,613 to 667,529, the largest margin then recorded in a Chicago mayoral election. Cermak brought to municipal corruption a creative intelligence that fashioned a machine for the decades.

Meanwhile—though supposedly no outsider knew it—U.S. Attorney George Johnson had secured the first tax evasion indictment against Capone. SIU agent Frank Wilson had broken bookkeeper Leslie Shumway on February 21, 1931. The government whisked Shumway in front of a federal grand jury, which, on March 13, voted a tax evasion indictment against Capone for the year 1924. They had to rush, because on March 15, 1930, the six-year statute of limitations would have expired. (Until 1954, tax returns were due in March.) The government hoped to keep the indictment secret until ready for further indictments. In fact, Capone's network informed him almost as soon as the grand jury returned it. His lawyers immediately set about trying to arrange the outcome.

Capone's life and business routine continued. For reasons they never explained, newspapers always called the old pimp, Mike Heitler, "Mike de Pike." The night of April 29, 1931, a Wednesday, Heitler played cards in a cigar store west of the Loop with Lawrence Mangano and Frankie Pope, two of Capone's stalwarts, plus two minor hoods, identified only as "Fritz" and "Hank." Capone had steadily pushed Heitler out of operations until he had become, as Pat Roche said, "a down and out, dope wrecked ex-convict." At nine o'clock, Mike de Pike drove them all to a restaurant in the nearby Chicago and Northwestern railway terminal. The car belonged to Emily Mulcher, his live-in companion of twenty-three years in spite of his still undissolved marriage. She got a call from Heitler a couple of hours later. He wanted a phone number from his address book. When she got back on the line and said "Hello," a strange voice growled, "What the hell do you want?" and the stranger hung up.

Early Thursday, out in Barrington, a town about thirty miles northwest of Cicero, Hattie Ganusch saw flames on the estate of Otis Spencer and called the police. In the ruins of an icehouse they found a charred body, unidentifiable save for dental work and a scrap of distinctive underwear that had somehow survived the inferno.

About fifteen miles closer to town, in Itasca, police found Mulcher's car, also burned, in the back seat a revolver with all six rounds discharged by the heat. Another eight miles closer, in the Des Plaines River at River Grove, police found the car's license plates.

The police held Mangano for three days before concluding they had no case. Lawyer Mike Ahern relayed Capone's offer to visit Pat Roche on the understanding that if they could not prove complicity he'd be free to go. Roche talked tough enough to run for mayor or police chief. "When we want Capone," he announced, "we'll go after him. When we arrest him he'll be tossed into jail like any other hoodlum. If terms must be made we'll dictate the terms, not Capone."

No one would be charged with anything. But rumors circulated: embittered, Heitler had supposedly written a letter to State's Attorney Swanson, outlining Capone's operations. Capone had gotten hold of a copy and had told Heitler he was through.

When such a letter surfaced, months later, it asserted that Capone had told Jack Zuta that Jake Lingle was a double-crosser who "is going to get his." It accused Mangano and Frankie Rio of killing Zuta—with no mention of Danny Stanton. The letter supposedly had been dictated by Heitler, who was illiterate, to Emily Mulcher, who was nearly so. She said he had dictated such a letter, about eight months before his murder, but had destroyed it. As published, the letter ended with a warning that if Heitler was killed, eight men would be responsible—most of all Billy Skidmore, who had hated Heitler for years, but also Capone, Jack Guzik and Lawrence Mangano.

"No such letter ever came into my office," Swanson said, though he'd heard talk at the time that Pat Roche had received it. Roche said he hadn't. Nor did he think much of the contents. "No one would believe Heitler in his life or trust him."

Capone said, "I know nothing whatever about the whole damn business." On grounds of fastidiousness he had for years refused to have any truck with Heitler—"you see," he explained, "that fellow had a reputation of being a professional stool pigeon."

Though absorbed by such daily routine, Capone knew what was coming. Friday afternoon, June 5, 1931, the federal grand jury indicted him on twenty-two counts, four each for attempting to "evade and defeat" the income tax laws for each of the years 1925 through 1929. These were felony counts. The jury also brought in two misdemeanor counts for failing to file returns in 1928 and 1929 (the statute of limitations for misdemeanors ran only three years).

Including the earlier indictment, and figuring from records seized in raids over the years, the government charged that Capone had enjoyed an income of $1,038,660.84 from 1924 through 1929, which meant he'd cheated the government out of $215,080.48 in taxes:

	Income	Tax due
1924	$123,102.89	$32,489.24
1925	257,286.98	55,365.25
1926	195,677.00	39,962.75
1927	218,057.04	45,557.76
1928	140,536.93	25,887.72
1929	104,000.00	15,817.76

Prosecutors did not suggest that these figures—which *The New York Times* joked were "astonishing in their meagerness" for "the very pinnacle of crookdom"—constituted Capone's total income, only the portion they expected to prove to a jury. Moreover, that income included only gambling profits, not a penny from beer, booze, brothels or racketeering.

One week later, the grand jury added a third indictment, charging Capone and sixty-eight others with conspiracy to violate the Prohibition laws. This indictment stretched from 1922 to 1931, and part of it reflected the work of Eliot Ness and his men, whose digging and wiretaps discovered, among more damning items, the fact that Murray Humphreys had a dog he called "Snorky."

Capone made bail of $50,000 for all three indictments. The government talked of calling Johnny Torrio as a witness as well as Louis La Cava, who had once shared a safe deposit box with Capone, but who had fallen out with him and fled to New York. Capone's situation looked bad; the government had *never* lost a major tax case.

On the other hand, the prosecutors could make only a circumstantial case. Without the Mattingly admissions—which might not be admitted in evidence—they had no starting point for figuring Capone's income. And this was Chicago. Even federal juries had assailable greeds and fears, with Capone and his people nonpareils at getting to jurors. What's more, George Johnson labored under tremendous pressure to put Capone away before the World's Fair would open in 1933, if possible before Hoover would run again for the presidency in November 1932. Such a brisk result appeared unlikely if the government went to trial.

So both sides stood ready to deal. As soon as Capone learned of the first indictment, his lawyers made overtures to plea-bargain, being careful to ask if the court would buy a deal. Johnson assured them that in the cases where he had made a sentence recommendation on a guilty plea, "the courts . . . had followed my recommendation."

Of course, the appearance of a deal wouldn't do; everyone had pretenses to uphold. For public consumption both sides continued to posture right up to what they imagined would be the end. The day before Capone was to enter his plea, one of his lawyers insisted that they intended to fight the case. Johnson declared, "We don't expect him to try to plead guilty and secure a lighter sentence. We want no compromise with Capone." Even in court, waiting for the judge on June 16, 1931, lawyer Mike Ahern said, ". . . And as to reports that we have bargained with the government for leniency, I'll say that isn't true. We have made no overtures whatsoever." Capone played along. Waiting to plead, confident the fix was in, he said, "I'm going to throw myself completely on the mercy of the court, for I know that I'll get a fair deal from Judge Wilkerson."

It took about three minutes. Standing between two lawyers, dressed in a camel's hair suit of a hue that struck observers as somewhere between "bilious sulphur" and "shrieking banana," Capone murmured "guilty" three times as the clerk called the indictment numbers. Wilkerson set sentencing for June 30, two weeks off. Capone appeared unworried.

President Hoover applauded the result, but he had little company. If Capone received the maximum possible sentences, all running consecutively, he could serve thirty-two years on taxes and another two for conspiracy, paying a total of $90,000 in fines. But no guilty plea to tax evasion had ever drawn more than eighteen months. From the first, estimates of the time Capone would do ranged between two and four years. The United Press heard it would be exactly two and a half years; soon Capone was confirming that as the arranged sentence and everyone took it *for a fact*.

The sentence's brevity rankled; the knowledge of what it was for infuriated. ''He was arrested,'' one magazine commented, ''on charges of violating the income tax laws—yes, it has its funny side. . . .'' The St. Paul *News* called it ''a devastating criticism of our legal machinery.'' The Louisville *Courier-Journal* argued it was ''not conducive to American pride that gangsters, guilty of every abomination . . . should be found guilty only of failing to pay taxes on their ill-gotten gains.'' At best, mused the *News*, ''better in jail for three years or so than not in jail at all.'' *The New Republic* disagreed. Such a sentence could ''be described only as a victory'' for Capone. His only mistake, they observed, had been failure to fix the federal government with quite the same efficiency he had shown locally. ''In the meantime, let's hear no more about Capone's defeat. The defeat is Chicago's.''

It also counted as a defeat for Judge Wilkerson, and he did not lightly suffer defeats. Later, prosecutor Johnson guessed that it was Capone's boasting of getting off with only two and a half years that Wilkerson could not stomach. More likely it was the general assurance that Capone *and* the prosecution had the judge in the bag. Wilkerson was a proud man. Solid, stolid, stocky, with a seamed face topped by an iron-gray cowlick, the judge had been on the federal bench since 1922 and was considered one of the best qualified ever appointed up to that time. Previously, he had chaired the Illinois Commerce Commission, racked up an enviable record as a U. S. Attorney, and had been a very successful lawyer with an extensive civil and criminal practice. No one dictated to James H. Wilkerson.

Mike Ahern appeared without Capone on June 29, the day before the scheduled sentencing, to ask for a continuance; Capone was tied up with civil litigation in Florida. On condition that Capone start jail immediately when sentenced, Wilkerson put it off to July 30. Ahern mentioned that, after all, the conspiracy charge was minor and guilty pleas for tax violation had never drawn heavy sentences. Wilkerson replied, ominously, ''There are conspiracies and conspiracies and tax violations and tax violations, but I'll hear you fully on July thirty.''

The litigation that kept Capone in Florida smacked of charade. The day after Capone's June indictment in Chicago, his Miami lawyer, Vincent Giblin, obtained a writ of attachment on the Palm Island estate and its contents in payment of $50,000 he claimed Capone owed in legal fees. That night, Giblin had the sheriff back up a truck and start carting off the wicker sun-porch furniture, an exercise interrupted at midnight by the coming of the sabbath. Some, observed the Miami *Daily News*, ''were unkind enough to express the belief that the suit was 'friendly' and brought to prevent the government from levying on the property'' to collect back taxes.

To counter such speculation, word went around that Giblin, ''an ex–football player of demonic energy,'' had charged into 93 Palm Island, ''grabbed [Capone] by the shirtfront and threatened to knock his teeth down his throat.''

Lawyer Giblin would have known that Capone, in his own home, had to retreat no further, legally, before using whatever force necessary to defend himself from assault. Forget the bodyguards ever-present even with Cissie Patterson there; Capone, the ex-bouncer, needed no help handling the ex–football player. Jack Sewell, the husky young Miami clothing store heir, had once sparred with Gene Tunney. When Capone learned that, he had asked Sewell to spar. Sewell figured the pudgy cigar smoker would be no match. ''I found out right away,'' Sewell recalled, ''that he wasn't easy. He was as hard as nails, strong and quick, and he had a very good defense.'' Four of Capone's hard-eyed guards monitored the friendly bout.

The tip-off came when Capone ''settled'' with Giblin for $10,000, out of which Giblin agreed to pay a $1,000 bond and to represent all members of the household, at no further charge, in matters pending. So much for the claim of $50,000.

Just before Wilkerson had granted the continuance, Capone attended the American Derby at Washington Park. He occupied a box with his guest, the widow of heavyweight champion Bob Fitzsimmons. His bodyguards had the adjoining box. In the crowd also sat prosecutor George Johnson, who was treated to the spectacle of the band striking up the song, ''This Is a Lonesome Town When You're Not Around'' in Capone's honor.

The press could score it as his victory, and his fans could serenade him, but as Capone had said when in Philadelphia, ''Jail's a bad place under any circumstances, and don't let them kid you.'' And that's where he was headed.

CHAPTER 28

. . . for Good

CAPONE SPENT THE day before he thought he was going to jail making a public case for leniency. In slippers and black silk pajamas piped with white, he received the press at the Lexington.

"When I leave this hotel," he said, "it's my goodbye to the racket. When I say I'm through, I mean I'm—" his fist crashed on a tabletop "*—through*." He thought there was no money in it anymore, and said that most of the outfit agreed. "That's why I think they'll do the same as I'm going to do when I come back from Leavenworth—go into honest business."

How about rumors that Johnny Torrio would take over while he was gone? Capone guffawed. "Torrio feels like I do. Why, he wouldn't get back into the racket for a million dollars. Torrio got out, and I'm getting out"—teeth gleamed in a smile—"with the help of the United States government." He had a wife, son "and as fine a mother as any man on earth." His parting would pain them, but then they'd have "a husband, father and son who never again can be looked upon as a law violator." That's why he pleaded guilty—plus consideration for his about-to-be-fellow taxpayers. He intended to "accept whatever punishment is given without contesting the government's charges or causing the government to spend a lot of money in expensive trials and in fighting appeals."

He felt "wonderful, but nervous" about tomorrow's date. "I've been made an issue, I guess, and I'm not complaining. But why don't they go after all these bankers who took the savings of thousands of poor people and lost them in bank failures? How about that? Isn't it lots worse to take the last few dollars some small family has saved . . . than to sell a little beer, a little alky?"

He nursed no hard feelings. "If the United States government thinks it can clean up Chicago by sending me to jail, well, it's all right with me. I guess maybe I owe the government this stretch in jail, anyway.'' He'd regret missing the World's Fair; he'd never seen one. He hoped Chicago's would be a great success.

''I'll see you in court tomorrow!''

''The consensus,'' wrote a reporter the next morning, expected a sentence of ''two or three years—perhaps two and one-half—with a fine of about $10,000.''

Reporters gathered about Capone in the courtroom, waiting for Judge Wilkerson to appear at ten o'clock. ''There's nothing to all this,'' Capone said. ''It won't take long. But, of course, I'm a little nervous. I'm only human.'' No, his family wouldn't be in court. ''Now let me alone for a few minutes, boys.''

The court's opening statement changed everything. Wilkerson acknowledged that Capone's guilty pleas had been predicated on the U.S. attorney's recommendation as to sentencing. ''It is always understood, however,'' Wilkerson hurled the bombshell,

> that . . . the court does not bind itself to adopt them or to enter judgement in conformity therewith. . . . The parties to a criminal case may not stipulate as to the judgement to be entered. . . .
>
> This defendant must understand that he cannot have an agreement as to the judgement to be entered in the case.

An agreement was exactly what Capone thought he did have, and now the deal was off. It got worse. ''If the defendant,'' said the judge, ''asks leniency of the court he should be ready to answer all proper questions put by the court touching the matters which he has confessed by his plea.''

This would take longer than Capone had thought. Wilkerson had to hear a civil matter for the rest of the morning, so he continued Capone's case until two o'clock that afternoon.

When they regathered, U.S. Attorney George Johnson appeared only marginally less distraught than the defendant and his lawyer, Mike Ahern. Johnson's hands shook and his voice rose in an excited tremolo as he related the genesis of the deal. The government had wanted to ''avoid the hazards of a trial'' and to jail Capone ''at an early date.'' When Capone's lawyers approached them after the first indictment, the government entered negotiations. Treasury had signed off on the deal, as had the attorney general. The government had agreed to recommend a sentence for each indictment and recommend that they all run concurrently. The reporters' consensus had been on the money. ''I think I owe this further duty'' to the defense, Johnson said. He wanted to make it clear ''that if they had not understood that I would make the recommendation, I do not believe they would have pleaded guilty.''

Ahern emphatically agreed. As for Capone's testifying, Ahern told the judge that ''it was expressly stipulated by me as a condition of the plea that the defendant should be asked no questions.''

Wilkerson replied that the court would listen to recommendations, "but the thing that the defendant cannot think—must not think—is that in the end . . . the court is bound to enter judgement according to those recommendations." The judge had been stewing. "There have been some unfortunate things in connection with this case," he said. "There have been some publications which were contemptuous in character, and tending to bring the administration of justice in the federal court into disrepute. They have even gone so far as to announce in advance what the period of punishment would be." Wilkerson would not abide such talk. "It is time," he said, "for somebody to impress upon this defendant that it is utterly impossible to bargain with a federal court."

Next day, Wilkerson granted Capone's motion to withdraw the guilty pleas on taxes, but held up on the Prohibition conspiracy plea. He ordered the grand jury to examine that indictment to see if they couldn't, instead, indict for Prohibition *violations* under the Jones Law, not mere conspiracy. Such a conviction would carry a much longer sentence.

When the federal grand jury in Chicago tried to report why it could not indict Capone under the Jones Law, Wilkerson refused to hear them. He wanted indictments, he said, not reports. Capone withdrew his guilty plea, never to be tried on Prohibition charges.

Wilkerson set October 6, 1931, for Capone's tax trial.

That summer, Cornelius Vanderbilt, Jr., arranged an interview with Capone in the Lexington office. Capone said he envied bankers—"your dad's friends"—because compared with their depredations he was "a piker." He then harangued Vanderbilt for an hour about the economic, political and moral decline of America, impressing his interviewer with his speech, manner and cogency. In the end, Capone said, the only cure for the country's ills would be to "kick out state governments, jail all mayors and have the whole show run by the Federal people."

That seemed brave, Vanderbilt said, considering who was on the verge of getting him. "Oh, they are only trying to scare me," said Capone with a stagy yawn. "They know very well there'd be hell in this city if they put me away. Who else can keep the small-time racketeers from annoying decent folks?"

Capone sounded more confident than he could have felt. When Vanderbilt's piece appeared in the middle of the trial, Capone would deny having had any such conversation, calling it merely a business meeting.

Perhaps sensing his downfall, some of his fans switched sides. Three days before his trial he attended the Northwestern-Nebraska football game and heard himself booed. Northwestern's president had asked the police if Capone couldn't be ejected, but Chief Freeman, Zuta's "old pal," replied that the man and his entourage all had tickets and had not caused any trouble.

After three torpid quarters, Capone, easy to spot in his purple suit, moved for the exit. A group of children trailed after him and started booing. Jack McGurn turned, scowled and shook his fist, but the crowd swelled the chorus, drowning out a small troop of friendly, loyal Boy Scouts who tried to defend their benefactor with cries of ''Yea-a-a Al.''

Just before his trial started, Capone tried to improve the odds in two tested ways. A representative visited the man who had recruited Elmer Irey for his Treasury post and offered $1.5 million if Capone escaped with no jail sentence. More directly, Capone went after the jury. Ed O'Hare called SIU's Frank Wilson for a clandestine meeting. He showed Wilson a list of ten names. They were, he said, numbers 30 through 39 on the jury list that even Judge Wilkerson had not yet received. Capone's people were offering the veniremen the usual mix of money, political jobs, prize fight tickets and threats. Wilson, Elmer Irey and George Johnson sped with the ten names to Wilkerson, who called next day—when he received his own copy of the complete list—to say that those ten names were indeed jurors numbers 30 through 39. ''Bring your case into court as planned, gentlemen,'' Wilson recalled the judge's saying. ''Leave the rest to me.''

Wilkerson gave no notice to the defense, but swapped jury panels with Judge John P. Barnes, which meant sixty new veniremen. Many of the new ones evidenced nervousness. Fifty of them wanted to be excused, though only eighteen made it. But none had been fixed, and those chosen would be sequestered, out of reach.

The judge conducted the voir dire, asking questions posed by both sides. He would exclude no jurors merely because they knew about Capone and the charges against him—a category that in 1931 America included everyone except hermits and the catatonic—only those who swore they had already decided the case. He dismissed a scant six prospects for cause, and both sides had to exhaust their peremptory challenges. Capone faced exactly what he least wanted: an untampered-with jury, only one member from Chicago, one from Waukegan, and the rest from the sticks, a venue notoriously less understanding about Capone's line of work.

When not in court, the jury would be locked in seven connecting rooms at a nearby hotel, where they would be lodged and fed at government expense, $4.60 a day. They were allowed no phone calls, but could send and receive all the censored letters they wanted. Bailiffs would scan their newspapers and magazines, snipping out all references to Capone; that also meant those media could not be used for smuggling in offers or threats.

''If I told you I wasn't worried,'' Capone confessed to reporters, ''I wouldn't be telling the truth.''

The next day, Wednesday, October 7, the government opened. George Johnson attended every day, but allowed his associates to present most of the case. Dwight H. Green, who would parlay this experience into the governorship of Illinois, acted as Johnson's principal assistant. The others were Capone's contempt prosecutor, Jacob Grossman, Samuel G. Clawson and William J. Froelich. Their first witness, local IRS man Charles W. Arndt, testified that his records contained no income tax returns from Alphonse Capone for the years 1924 through 1929, not under that name or any of his aliases, not even "Snorky."

The Reverend Mr. Hoover and his two raiders, Morgan and Bragg, told of their May 1925 foray against the gambling joint across from the Hawthorne Hotel, when Capone said he was owner as he forced his way in the door.

Leslie Shumway took the stand. To *New York Times* reporter Meyer Berger he looked like "a nervous, fussy deacon in some little country church, a drab, baldheaded man of uncertain age." A handkerchief or his hand covered his mouth through most of his testimony and he spoke almost inaudibly, averting his gaze from Capone and from Phil D'Andrea, the crack-shot bodyguard who sat just behind Capone, both impaling him on the Look.

The effect of the ex-bookkeeper's testimony was more subtle than at first appeared. The government had the overwhelming problem of establishing Capone's ownership. Backed by seized records, especially the cash ledger Frank Wilson had stumbled upon, Shumway's testimony could firmly establish gambling profits. But could the government convince the jury that Capone got any? The testimony of Hoover and his raiders that Capone had claimed ownership of the gambling hell they invaded did not prove ownership. If you walk into a bank and say you own it, you cannot then help yourself to its assets. But if you walk in, say you own it, tell everyone to take the afternoon off, and they *do*, onlookers might believe you.

Although Shumway could not document Capone's ownership, he testified to behavior that strongly suggested it: Capone's air of proprietorship and access to the back rooms where customers were not permitted. Most of all, Capone issued orders that sounded very bosslike. For instance, one of Shumway's duties had been to carry profits from the Subway to a safe in the Hawthorne. Originally he had been escorted by two armed goons, but had objected: if anyone started shooting in a stickup, he'd be in the middle. Thereafter they let him carry the money unescorted. Soon, Capone asked him what he would do if someone did stick him up.

"I would just let them take it," the nervous deacon replied.

"That is right," said Capone. Who but a boss would dare give permission to so insouciantly surrender money that belonged to gangsters?

On the third day, Thursday, October 8, came what one observer called Capone's "sunken road at Waterloo." The government introduced the subject of Mattingly.

Defense counsel objected forcefully. But was their objection as effective as it should have been? "What I see, reading the transcript of the trial," says Terry MacCarthy, one of two defense lawyers at the ABA's 1990 retrial of Capone, "is that these people were really inept criminal defense lawyers. Certainly by today's standards they did not know what they were doing. Their trial advocacy skills were terrible." Judge Marshall, who presided at the ABA retrial, agrees: "The practice of law has become truly more sophisticated," he says. "Those were simpler days."

To start with, it was the wrong team. Mike Ahern, tall, sharp-featured, was half of Chicago's preeminent criminal defense partnership, Nash and Ahern. Tommy Nash had represented Capone in court before, but for reasons never alluded to in contemporary news reports, and not known even to Nash's son—today still a practicing Chicago lawyer—Nash was not present for this trial. Yet he was the litigator of the team, the courtroom wizard; Ahern was the back-room, library lawyer, the researcher and writer of briefs, possessing only indifferent courtroom skills. For this trial he teamed with Albert Fink—slightly shorter and portly—who laughed easily and whose eyes habitually twinkled behind gold-rimmed glasses, giving him a jolly appearance. He played the character, often prefacing his more impassioned statements with "Oh, my conscience." While his reputation as a trial lawyer stood high, he lacked Nash's touch both as legal strategist and court tactician.

Defense counsel barely brushed by the right objection to the introduction of the Mattingly material. Ahern correctly pointed out that for purposes of collecting taxes, Congress encouraged citizens "to come in and settle with their government before criminal or civil action is started so it won't have to be started." But the defense quickly strayed off into nonsense, and did it maladroitly.

"It is human nature to evade taxes," Ahern proclaimed. "We had our Boston Tea Party—"

"And what is this?" Judge Wilkerson interrupted. "Is this another Boston Tea Party?"

"No," admitted Ahern. "I don't know what it is."

Then Fink argued that Mattingly had exceeded his authority, and didn't know what he was doing. "He was to get Capone out of a criminal liability," Fink said. "And after the Treasury officials had told him the statements he made would be used against him, he went ahead and made statements which, if admitted into evidence, are most damning. I want to ask this lawyer why he did that. I want to cross examine him."

What Fink said undercut his case. First, if Mattingly was indeed trying "to get Capone out of a criminal liability," it was not a bona fide offer to compromise, which cannot be made in order to avert or short-circuit criminal prosecution. More important, no amount of ineptitude or wrongheadedness in his tax lawyer's tactics would let Capone off the hook. The defense had to insist that Mattingly was right, that his repeated "without prejudice" exempted from other use all statements made to negotiate the amount of tax Capone would have to pay. True, as the IRS agents stressed at the time, they could not grant immunity; statements Capone or his lawyer made about gambling or bootlegging or prostitution might well have been used for prosecution of those offenses. But the defense had to distinguish admissions of income as different, holding that when offered to reach a compromise they could not be used in a later prosecution for tax evasion.

The issue naturally arose at the ABA's 1990 retrial. "I believe," says Judge Marshall, "it was a bona fide offer in compromise. And under the current federal rules of evidence—and, indeed, under the rules of evidence as they existed back then—bona fide offers to compromise are not received in evidence.'' The rule makes sense. After all, in such settlement negotiations citizens might even admit to untruths if they consider doing so less trouble and less costly than fighting the issue. ''Pay the two dollars,'' as the old joke goes. Judge Marshall at first ruled that he would exclude all of the Mattingly evidence from the ABA retrial, reversing himself only when the prosecution protested that then they couldn't mount a case and the ABA's program would have to be scrapped. At the real trial, Judge Wilkerson unhesitatingly allowed all of the Mattingly testimony.

''This is the last toe,'' moaned Albert Fink as Samuel Clawson started to introduce The Mattingly Letter. It may have been a wrestling term—or reference to completion of his next image: ''They have got him nailed to the cross now. This is just putting the last toe on him.'' He did not exactly want to waive his objection, Fink pouted, but he kind of hoped the judge would overrule it and let *all* the material in ''because there are some things in this letter that at least indicate the lawyer is crazy.''

That was precisely the wrong tack. If the material had to go in, the defense's only sensible line was to argue not, as Fink did, that Mattingly was crazy ''to make statements that may get [Capone] into the penitentiary,'' but that the content of Mattingly's statements was unreliable, stressing that taxpayers and their tax lawyers might say anything in order to settle. Calling Mattingly's admissions damning and talking penitentiary played the government's game.

The IRS agent in charge of fraud in Chicago—Louis H. Wilson, *not* Frank J. Wilson—testified about Mattingly's approaches offering to settle, about preliminary meetings on April 10 and 16, 1930, and about the April 17 meeting with Capone. The transcript of that meeting was read into the record. Finally, Louis Wilson told of his September 20 meeting at which Mattingly produced his letter. Altogether, those admissions were indeed damning.

The Mattingly matter culminated the next morning, Friday, October 9. The government read to the jury The Mattingly Letter with its admissions of income that provided something like a starting point to measure Capone's assets. After that, all expenditures and receipts of money the government could prove made a case.

The jury heard what Capone had said in Florida about his gambling interests. They started to hear about his expenditures—the rooms he had at the Metropole and the Lexington, paying weekly in $100 and $500 bills peeled off a huge roll; the $1,633 extra rent he paid for two days of his Dempsey-Tunney fight party. When Ahern objected to the recital as not being germane to the issue of income, the judge replied, ''I presume that what he paid out he must have taken in.'' Ahern did not object to that, yet the judge's words were decidedly presumptuous, since proving what Capone took in was the trial's sole point.

Parker Henderson told of his dealings with Capone and of fetching money for him at Western Union. A parade of Western Union employees detailed $77,550 in money transfers. The parade stopped with John Fotre, Western Union's manager in the Lexington. Small, slight, looking professorial in horn-rims, Fotre demonstrated an extremity of reluctant nervousness, eyes darting about the court as he niggled with prosecutor Green over whether he *knew* it was indeed another brother of Jack Guzik, Sam Guzik, who had sent a money order signed "Sam Guzik." The judge took over the questioning, further discomfiting Fotre. "You think it over," Wilkerson charged as he rose for the day.

The manager might not have wanted to offend such good customers. Or he might have been scared—although neither he nor any other witness applied for protection. According to one account, Frank Wilson reproached him after the session, and Fotre squeaked, "What can you expect when they let one of Capone's hoodlums sit there with his hand on his gun?" Obviously, he meant Phil D'Andrea.

Working half days on Saturdays was routine in those times, so the court sat the next morning to hear more about Capone's prodigality. In Miami his meat bill ran $200 to $250 a week at a time when sirloin steak cost 33 cents a pound; he bought $3 to $4 a day worth of bread and cake when a three-layer strawberry cake cost 38 cents. He spent $2,100 on landscaping in 1928 alone, $859.07 for Chinese rugs and other floor covering—down from $1,896.97 the year before and $2,775 in 1925, when he also paid $21,550 for furniture. In 1927 he had spent $2,835 for custom suits, down from his high of $6,180 in 1925; he spent another $869 for shirts in '28. His annual Columbus Day party cost $3,160 in 1926, his Kentucky Derby party that year, $4,925.

Of course Capone spent on others besides himself and family. His church contributions hit a high of $15,600 in 1926, contribution to the police widows' and orphans' fund $58,000 in 1925. His Miami telephone bill in 1929 (when he sat in a Philadelphia jail from May on!) ran $3,141.50—an astounding figure to jurors whose bills, if they had a telephone, might average $36 a year. Miami contractor Curt Otto Koernitzer testified that he had built Capone a garage and bathhouse and was paid $6,000 in cash by Mae. A second garage, plus other outbuildings, cost over $10,000.

When the court rose at noon, Capone left with his lawyers and Phil D'Andrea. Two SIU agents edged up to the group, seized D'Andrea, hustled him into an antechamber and took from his belt on his right hip a loaded pistol. Judge Wilkerson ordered him held without bail despite D'Andrea's plea that as a municipal court bailiff he had a gun permit. That would not have excused carrying a weapon into federal court anyway, but in fact the permit had expired with the advent of Cermak's administration. The judge would continue the case until Capone's trial ended, heeding defense objections that action during the trial might prejudice the outcome.

Monday, October 12, the bemused jury heard more about snorky high life, Capone's taste for canary and green, his monogrammed shirts, the ties and hankies bought in wholesale lots, sterling dinner service and jewelry, the diamond-studded belt buckles. At mention of glove silk underwear, the "rural gentlemen of simple and rather careless habits of dress," wrote Meyer Berger, ''pursed lips.'' Capone joined in the laughter that greeted this testimony, but his scars flashed livid against his blushes. Berger continued,

> It was too much for the rural gentlemen in the box. They look like men who would have no use for Chinese rugs and wouldn't feel right in canary chairs and green rockers. One of them spanned his forehead with a tired hand, as if dazed by it all.

The prosecution decided to plug one possible out. Jacob Grossman led Miami contractor Henry E. Keller through the time he had asked Capone what his first job was and Capone had replied, ''Tendin' bar on Coney Island.''

Albert Fink sprang up with an objection. ''What's the idea of introducing this?'' he asked.

''I wish to show,'' Grossman informed the court, ''that his wealth was not inherited. I want to show something of his background.''

''I see,'' said Fink. ''You wish to show that he was self-made, like Andrew Jackson and Abraham Lincoln and Herbert Hoover?''

''Not exactly.''

On Tuesday, October 13, to everyone's surprise, the prosecution rested its case with John Torrio and Louis La Cava still sitting outside, uncalled. The most important witness of the day, Fred Ries, gave enough of an inside view to convince any jury of gambling profits—with the strong implication that Capone had an interest in them. Ralph Capone had once announced a reorganization—Jimmy Mondi out as gambling manager and Pete Penovich in. The *world* knew Ralph acted for his brother. Guzik once told Ries that for bookkeeping purposes he was not to give money to ''anybody besides himself and the man he would send for it, not even to Al. . . .'' That ''not even to Al'' spoke volumes. Like Shumway, Ries saw Capone deferred to and saw him acting with proprietorial assurance. Plainly, had Capone told everyone to take the day off, the joint would have closed.

A judge must decide whether the prosecution has made enough of a case to be worth submitting to a jury for a verdict. Thirty years later, reading the transcripts, Judge Marshall thinks that Shumway and Ries—unimpeached as they were by the defense—had made a submissible case against Capone. At the time, Judge Wilkerson had no doubts.

The defense wanted a few days to prepare. ''There was no work done on this case by the defense,'' argued Albert Fink, ''until the plea of guilty was withdrawn.'' The judge gave them until ten o'clock the next day, Wednesday, October 14.

They labored under a tremendous difficulty. They hadn't worked out a defense because, after all, they had never intended defending Capone. ''The case was a guilty plea from the beginning,'' says Tom Mulroy, a prosecutor in the

ABA's 1990 retrial. "He tried to plead guilty once and they wouldn't let him. It was a laydown. He pleaded guilty, and when that didn't work they thought they had fixed the jury."

The defense had no witnesses who could rebut Capone's expenditures, or receipt of money by wire, or the strong inference of his participation in proven gambling profits. What witnesses they did introduce were exactly what Mulroy calls them, "laughable." Ahern and Fink trotted out a succession of bookies who swore that Capone had been a monumentally bad handicapper; "He lost most of the time," said Oscar Gutter. Samuel Rothschild said that while Capone might win individual bets, he never registered an entire winning day against Rothschild's book.

"That," says Terry MacCarthy, one of Capone's ABA retrial defenders, "is no defense at all." Then as now, the tax rule was that gambling losings can be deducted only from winnings; all the defense did was reinforce in the jurors' minds how much money Capone had at his disposal.

The case really climaxed on October 16 with the testimony of Pete Penovich, manager of the Subway when Ries was cashier. Big, burly Penovich had dark slick hair and small dark eyes in a beefy face. "He was a better witness for the government," Terry MacCarthy states, "than any the government put on." Under cross-examination, Penovich explained how he had thrown in with the Capones because the outfit could provide a better location and better protection than he had for his independent Cicero operation. He had gone along when his share was slashed from 25 percent to 5 percent, and had "condescended" to operate under Frankie Pope when Ralph placed Pope in charge.

What was the point of the defense strategy? Tom Mulroy thinks they were stalling, "maybe to try to make another run at that jury." Only something like that could explain the defense's horrible trial tactic in not simply stipulating to all those expenditures rather than suffer the cumulative effect of twenty-three witnesses pounding the jury with details of Capone's many rich purchases over the years. "What are you going to cross-examine those guys about?" Mulroy asks. "And in fact they didn't cross-examine most of them. So you stipulate to it. But I don't think they were in any hurry."

That day Edward G. Robinson, star of the current hit *Little Caesar*, showed up as a spectator, "to give Snorky a few pointers," one reporter remarked, on "how a real gangster should act." At 12:12 A.M. next morning, Thursday, October 15, Warden David Moneypenny gave a signal at the Cook County jail, and Frank Jordan, thirty years old, the robber who had killed Officer Anthony Ruthy—chaser of Jake Lingle's killer—died in the electric chair.

The defense rested at two o'clock that afternoon. Among the usual defense motions for a directed verdict of not guilty, they briefly argued—as we'll see, the wrong way—that all charges before 1928 should be struck on grounds that a three-year statute of limitations applied to tax matters.

Jacob Grossman gave the first of the prosecution's closing arguments, summarizing the evidence. Then he made the real case for the jurors. "You have the high privilege," he exhorted them, "of putting the stamp of disapproval on the whole Capone organization and the conduct of the defendant."

Ahern and Fink summed up. Mostly they mouthed the blather of lawyers who have no substantive defense. The government's crusade against Capone reminded Ahern of Cato the Elder's insistent "Carthage must be destroyed." When they addressed reality, they made some sense.

"If the defendant's name were not Al Capone," Fink said, "there would be no case and we would die laughing at the evidence." There wasn't, he maintained, "a scintilla of evidence that he made a dollar" in 1924. In 1928, the government had shown expenditures galore, but again no firm income. All right, he paid for furniture with checks signed by Guzik; maybe it was a loan. As for 1929, Capone had started trying to pay his tax two days after release from Philadelphia. "How in the world," Fink asked, "can you find him guilty of willful intent to defraud the government" in 1929? The government had prosecuted "merely because his name is Al Capone."

Of course, that was the trouble. The jurors knew who he was, much of what he had done, and what he stood for. "Do not be swayed," Fink begged, "by the argument that the defendant is a bad man. He may be everything he is said to be, but do not for that reason find him guilty of something of which he has not been proven guilty." But the jurors, like everyone else, knew Capone had made a lot of money and had not paid tax on it.

Another celebrity visited that day: Bea Lillie, in town with a show, contracted stage fright when introduced to Capone, could stammer only "Well . . . well . . ."

Saturday, October 17, 1931, the trial ended. George Johnson spoke for the first time, closing the government's summation. He urged common sense on the jurors. Where did those Western Union money orders come from? What was the money? gifts? inheritances? life insurance? The jurors, he told them, should "draw your own conclusions."

As for charges that the government was prosecuting only because of who the defendant was, Johnson said he didn't want the jury to think of him as Al Capone. "Future generations," he said, "will not remember this case because of the name Alphonse Capone, but because it will establish whether or not a man can go so far beyond the law as to be able to escape the law." In other words, can a man get away with murder and not answer at least for *tax evasion*? In other words, remember this was Al Capone.

The case went to the jury at 2:40 that afternoon, and, considering the complex daftness of the verdict, it is a wonder that they took only eight hours to reach it. But one of them had been a holdout, and the verdict represented compromise.

When Capone rushed back from the Lexington and stood in court at a little before eleven o'clock that night, fate teased him. The first verdict he heard was "not guilty" on the single count of the first indictment, covering 1924. The next thing he heard dashed hope: "guilty" of tax evasion in 1925 on felony count 1 of the second indictment. But then, incredibly, hope sprang again: Capone heard "not guilty" for the next three counts in that same year. It was the same for 1926 and 1927: guilty on the first count covering each year, not guilty on the other three more or less identical counts. Amazingly, the jury found Capone not guilty of tax evasion for the years 1928 and 1929—but guilty of the misdemeanor counts for those years that charged willful failure to file tax returns.

The verdict left the prosecution scratching their heads. How could Capone be guilty of failing to file when he wasn't guilty of evasion? Guilty of one felony count in a year and not of three others that merely rang the legal changes on the first count, relying on the same evidence? They needed a moment to huddle over it.

Fifteen minutes later, they returned to court. Johnson announced that the government would accept the verdict as rendered. Of the twenty-three counts, Capone stood guilty on 1, 5 and 9 of the second indictment, felonies carrying a possible five-year sentence with a $10,000 fine, each, guilty on 13 and 18, each a misdemeanor with a possible year in jail and a $10,000 fine. The jury found him not guilty on eighteen of the twenty-three counts.

Thirty years later, at the ABA mock retrial in Chicago, a panel of fourteen federal jurors voted Capone not guilty. What was the difference? For one, Capone's "new" lawyers stuck to the effective assertions that the government could document no income, that Capone honestly did not know he had to file, relying on mistaken tax information, and that the government had unfairly targeted him, weakest of the three.

Most of all, the modern defense team's cross-examination could expose the methods used to compel the testimony of Ries and Shumway—essentially, coercion. The jurors Judge Marshall talked to after the ABA trial "said they had no real confidence in the credibility" of witnesses so coerced. In 1931, the defense did not know about that compulsion; besides, as Tom Mulroy says, "Back then nobody cared whether anybody's testimony was coerced; that's the way the police did it."

Mulroy thinks the 1990 not guilty verdict reasonable, as does Judge Marshall. Mulroy adds a telling point. If the real trial were held today, with a real Capone in the dock, he thinks a modern federal judge might not permit the case to go to the jury, since even with Mattingly, the government never could establish an unequivocal starting point for a net worth prosecution. "But," he says, "if a judge did not dismiss it, if you could overcome that expenditures rule and go to a jury, I think common sense would take over." A modern jury would know what the real jury knew: Capone *was* guilty, however technically insufficient the government's case.

* * *

A week later, Saturday, October 24, 1931, Capone stood in court for sentencing. As he had throughout the trial, he wore what for him was a suit of muted hue, in this case only slightly purple.

Under today's federal sentencing guidelines, for the same offense he could receive from seven years three months to nine years. In reality, Judge Wilkerson read the first sentence for count 1: the maximum permissible five years and $10,000 fine. That was not unexpected; Capone knew he would surely get what Jack Guzik had gotten. Then the judge gave him the same on count 5—and Capone waited to hear the word "concurrently," which means prisoners serve time on similar counts all at the same time. That would mean a total of five years in prison.

The judge did not mention how the time must be served. He continued with the same sentence on count 9, then the maximum one-year sentence for each misdemeanor count, with the maximum fine. Then Capone heard the bad news. Wilkerson allowed the first two felony counts, 1 and 5, to run concurrently, but Capone had to serve count 9 consecutively; that meant ten years in a federal prison. One of the misdemeanor counts would also run concurrently with the first felony count, but Capone had to serve the other year consecutively to the first and the six-month contempt sentence after that. Judge Wilkerson had contrived the sentence so that even if appeal overturned the felony convictions, Capone would still do two and a half years.

Absent successful appeal, Capone would do ten years in federal prison, followed by one year in county jail. The fines were cumulative, $50,000, and Capone would pay all costs of prosecution, finally reckoned at $7,692.29.

Judge Wilkerson refused bail pending appeal. Capone would go straight to county jail, to be held until shipped to a federal penitentiary.

The day after the guilty verdict, in Morton Grove, Matt Kolb was gunned down in a roadhouse. He hadn't listened when the outfit declared him out. Capone might be going to jail, but business continued. At first it looked as though Capone would continue to run it, jail or no jail.

CHAPTER 29

To Jail and Prison . . .

''I GUESS IT's all over,'' said Al Capone to his lawyers after he heard the judge. ''You've done all you could''—an opinion that would not endure.

''Goodbye, Al, old man,'' said Albert Fink, clutching his client's hand, a catch in his voice.

U.S. Marshal Henry C. W. Laubenheimer, stout and grim, led Capone away, surrounded by deputies, while his lawyers argued that he should be kept in Chicago pending their motions on appeal.

Outside the courtroom, a slight gray functionary approached brandishing a paper. ''Mr. Capone, I want to serve this—''

Capone bellowed a curse and pulled back a leg as though to kick the fellow. The surrounding deputies grabbed Capone, letting the assistant tax collector serve him with a demand for $137,324—a figure that would change and be negotiated.

Downstairs, Capone managed a wan smile for photographers. ''Get enough, boys. You won't see me for a long, long time,'' he said before he scrambled into a taxi, flanked by deputies, his taxi trailed by another filled with the rest of the escort. ''It was my own fault,'' he said on the ride to the county jail. ''Publicity—that's what got me.''

When he reached the jail, more reporters waited. What did he think of the sentence? ''It was a little below the belt,'' Capone answered. ''But I guess if I have to do it I can.'' He intended to keep his nose clean and build maximum time off for good behavior.

Put first in a receiving cell, he retreated to a far corner when news photographers clustered about. ''Please don't take my picture here, fellows,'' Capone pleaded. ''Think of my family.'' As a deputy warden led him to a less accessible cell, one photographer strained in for a close shot. Capone exploded, snatched up a bucket and started for the man. "I'll knock your block off!" he raged. Guards grappled him back.

By evening, Capone was settled into a commodious cell in the hospital section, D block. He shoved aside a tin plate of corned beef and cabbage, untouched; he had nibbled a bit of rice pudding, sipped a bit of coffee. Otherwise, he acted the model prisoner. "He knows," one guard said, "how to do time." A jail radio played "The World Is Waiting for the Sunrise."

Capone's lawyers obtained a stay easily enough; that kept the government from hauling Capone to federal prison until his appeals had been heard. But Judge Wilkerson refused to permit the prisoner to serve his year's misdemeanor sentence first, and also refused bail. That meant Capone had to stay in the Cook County jail although no time he spent there would count toward serving his sentence, which would not start, technically, until the government turned the key on him in a federal prison.

Capone's trial over, Judge Wilkerson gave Phil D'Andrea six months in county jail for bringing a gun into the court. D'Andrea also stayed in D block, eventually in a large private cell, crammed with file cabinets to help him discharge his duties as president of the Unione. He had succeeded Joe Aiello.

For Capone's control over the outfit, Cook County jail was almost as if he had never left home. About six weeks after he arrived, anonymous telegrams to Judge Wilkerson, George Johnson and Warden David Moneypenny charged that Capone received visitors ad lib, day and night, their limos, guarded by hoods, parked outside the jail. The telegrams also charged that Capone made phone calls, sent telegrams, dispatched messengers, and even had secretarial help. In short, Capone ran the outfit virtually unimpeded.

Warden Moneypenny immediately denied any special treatment. Capone wasn't *really* in the hospital ward: D-5 was a dormitory cell for those "recuperating." Moneypenny led reporters on a tour to demonstrate the accommodations' suitable bleakness. Capone did his part, asking the visitors, "I'm in jail; aren't they satisfied?" He waved his hand to encompass the dreary setting. "Anybody who wants this place is welcome to it. I don't see anything swell about this."

The government exonerated Moneypenny of pampering, but henceforth Marshal Laubenheimer would have to approve any but family visits. That set the pattern for the rest of Capone's six-and-a-half-month stay in county jail. Not surprisingly, it turned out that "Mr. Smith" and "Mr. Jones," coming to see Capone or as special visitors to others in D-5, were really outfit members. In answer, someone devised a plan to have "trusted guards" accompany each visitor, access only by written pass. Politicians then applied for passes, handing them over to the likes of Joe Fusco, Murray Humphreys and Jack Guzik. So Laubenheimer ordained a twenty-four-hour guard on Capone.

The visits never stopped. At one point, the story went, Lucky Luciano and

Dutch Schultz showed up shepherded by Johnny Torrio; Capone hoped to mediate differences between the two. For privacy, Warden Moneypenny let them use the death chamber as a conference room, where it pleased Capone to take the chair in The Chair. The conference came to nothing, Schultz displaying the lone-hand stubbornness that would precipitate his murder three years later.

Mike Ahern pursued Capone's appeal—a performance that led Judge Prentice Marshall to say that if they were both practicing law today, "I would not refer a client of mine to him." Observers assumed that the Mattingly material would figure in the appeal. Another element would surely be Judge Wilkerson's ruling that a six-year, not a three-year, statute of limitations applied to tax prosecutions, a labyrinthine issue. The law stated that only tax *fraud* ran for six years, other tax offenses three; but circuit courts had held that evasion constituted fraud. Another factor was how much time the accused had spent outside the jurisdiction in which they had committed their offense, that time not counting toward the statute of limitations. Albert Fink, who had raised the issue merely in passing during the trial, never mentioned the time Capone had spent in Florida and Philadelphia, so no record was established for review on appeal. That turned out not to matter, because Ahern's appeal never mentioned the statute of limitations or Mattingly. It attacked only the government's failure to provide a bill of particulars detailing how Capone had evaded taxes. At worst, that counted as harmless error.

On February 27, Assistant Warden Edward Nettles called Capone out of a card game to tell him that the Circuit Court of Appeals had rejected his plea. Capone shrugged. He could still hope for Supreme Court review.

Three days later a kidnapper stole Charles Augustus Lindbergh, Jr., from his crib in Hopewell, New Jersey. "It's the most outrageous thing I have ever heard of," Capone said, offering $10,000 for information.

Frankie Rio showed up in Hopewell with a proposal: if the government would let Capone out temporarily, Capone thought he could effect the baby's return within forty-eight hours. "I know a lot of people who might be valuable in finding the child," Capone's message ran. "There's nothing I can do here behind the bars, but I'm pretty sure there would be if I could get out for a while." He explained to Hearst editor Arthur Brisbane that he didn't expect freedom, just ad hoc parole. He didn't mention public relations benefits. He told the editor that he'd "give any bond they require," and leave his brother, John, as hostage against his return to jail. "You don't suppose I would doublecross my own brother?" he asked.

Heading the Lindbergh investigation was founder and superintendent of the New Jersey State Police, Colonel H. Norman Schwarzkopf, whose own baby boy would one day join the army and make general. After Capone's offer, the Lindberghs called in SIU agents Arthur Madden and Frank Wilson, who presumably knew how to deal with him. On their advice, the family turned down the offer. Capone renewed it six weeks later, after the distraught parents paid $50,000 ransom yet did not get back their son (who in fact had been murdered). Although they would give "credit where it's due" if Capone got their baby back, the Lindberghs still refused to call for Capone's release. After working so long and hard to get him, the government probably would not have let him go anyway.

On May 2, 1932, the Supreme Court told the government it could keep Capone. A certified copy of the Court's refusal to grant certiorari arrived in Chicago next day.

At nearly the last minute, the Justice Department changed its mind about where to send Capone. Sam Guzik had been convicted of some crime and sent to Leavenworth. Nitti had been there with him until Nitti's release on March 24, 1932. Jack Guzik had been sent there on April 8. There was too much outfit at Leavenworth and too little discipline: both Nitti and Sam Guzik had wangled scandalously soft jobs. Nitti, dressed in civilian clothes, drove a car for the superintendent of the prison farm; Sam pottered in the truck garden. Ralph Capone had started his sentence at Leavenworth on November 7, 1931, but Justice had transferred him to McNeil Island, Washington, on December 10. They now decided that Al Capone would serve his time in Atlanta, toughest of the federal prisons. Capone—who had said good-bye to his family that afternoon—did not get word of the change until he heard it on the radio shortly before leaving county jail for the Dearborn Station, where the *Dixie Flyer* would depart at eleven-thirty.

At nine-thirty the night of May 3, 1932, Marshal Laubenheimer arrived with a document instructing the sheriff of Cook County to deliver up "the bodies of one Alphonse Capone and one Vito Morici," a scrawny twenty-six-year-old car thief on his way to Florida for trial.

The jail's central yard was hedged with photographers and newsmen, chockablock with Prohibition agents, deputy marshals and Chicago detectives. The barred windows looking down megaphoned cheers as camera flashes flared. "You'd think Mussolini was passing through," said Capone, waving.

"I'll take what they give me for two reasons," he told reporters on the train, next day. "One is, the only way you can expect to get a pardon is to go along and be a good prisoner. The other is, I won't have any say about it." Maybe they'd make him a cook; he was great with Italian food. "I hope the warden likes spaghetti." That night, May 4, the *Dixie Flyer* pulled into Atlanta's Union Station four minutes early, at 7:46.

The white stone facade of the U.S. Penitentiary at Atlanta rose sixty feet, ran six hundred feet, becoming thirty-foot-high stone walls on the sides and back. A massive outer gate stood between tall Doric pillars. At 9:10, a small group trooped up the stone steps to the gate as the keeper within called, "Who are you?"

"The United States marshal from Chicago," cried Laubenheimer, "bringing Alphonse Capone, a prisoner.'' The marshal, the escort and prisoner—subdued, tense, seen trembling with emotion—passed through to an inner gate, where Warden A. C. Aderhold met them.

The warden shook hands with the marshal, then turned. ''What's your name?'' he asked.

''Alphonse Capone,'' said a barely audible voice.

''What's your sentence?''

''Eleven years.''

''The sentence here, warden,'' Laubenheimer interjected, ''is ten years.'' The other year would be served back in county jail.

''Well,'' continued the warden, ''your number will be four-oh-eight-eight-six.'' Number 40886 handed over the $231 he had in his pocket; the only other items of value he had were sixteen religious medals, a rosary, a scapular, a fountain pen, one key, a wallet and a nail clipper. In return he received a set of blue denim overalls, ''40886'' stitched on the jacket and trouser leg.

A photographer poised for a shot. ''Get on your way,'' a guard snarled; photographs of federal prisoners were generally forbidden. Next day, when Marshal Laubenheimer returned to pick up some papers, even he heard nothing about the body he had delivered. Number 40886 was a federal prisoner, cut off from the world. In theory.

Settled in, Capone followed the routine as he had known he must. Up at the 6:00 A.M. bell, breakfast 6:30 to 7:00, work from 7:00 to 11:30, back to the cell he shared with seven others for the noon count, then lunch; back to the cell at 12:20, work again from 1:30 to 4:30, dinner, then back to the cell for the night at 5:30. The radio played from 7:00 to 10:00, then lights out.

Capone was assigned to the shoe shop, where he became a competent cobbler. The administration called him a model prisoner. ''He obeys every order the second it's given,'' said one official. Even so, Capone took prison hard. In county jail, on appeal, he had expressed active delight in sleep. ''Sleeping is like escaping,'' he had told a guard who remarked on the ease with which he conked out for eight hours at a stretch. ''I'm nobody's prisoner when I'm asleep. So that knocks off a third of my term, see?'' Now, in for good, he would thrash about in nightmare, grunting and crying, ''No, no!'' Once a cellmate had to slap him awake.

Capone naturally wanted out. By fall 1932, it looked like he might stand a chance. On April 11, 1932, less than a month before Capone arrived in Atlanta, the Supreme Court ruled in the *Schwarton* case that tax evasion did not constitute fraud, which meant the three-year, not the six-year, statute of limitations should have ruled Capone's case. Applied retroactively, that would outlaw prosecution on counts covering 1925 through 1927, his only felony convictions. Since he was acquitted of felony counts for 1928 and 1929, a favorable ruling would mean he had to serve only two and a half years on his '28 and '29 misdemeanor counts and six-month contempt conviction.

In August, two sharp appeals lawyers from Washington were hired—ostensibly by Theresa. William E. Leahy had been named by the government as special prosecutor against Nick Arnstein, Fanny Brice's husband, and had defended former secretary of the interior Albert B. Fall in the Teapot Dome trials. On September 21, 1932, Leahy and co-counsel William J. Hughes filed for a writ of habeas corpus in the federal district that included Atlanta.

After several hearings and delays, Capone's lawyers argued the case in New Orleans on April 26, 1933, before federal judge E. Marvin Underwood. Dwight Green, one of the tax-trial team, appeared for the government. The issue of timeliness proved paramount. In *Schwarton* the defense had raised the statute question at the beginning of the trial, had objected at once to the judge's six-year ruling and made it part of their appeal. Albert Fink had brought it up as an afterthought, never really an issue. He agreed with Judge Wilkerson that circuit courts had been following the six-year rule. Moreover, Ahern had made it no part of Capone's original appeal. For Judge Underwood to review the issue now, he held, would be like giving another district court appeal jurisdiction over the trial district. Writ denied.

A cellmate later quoted Capone about his lawyers: "They're overpaid dumb bastards who couldn't spring a pickpocket." Investigators heard that when Capone realized how ill-served he had been by Ahern and Fink, he sent collectors around to get back the fee he had paid them. They were lucky that was all he collected.

Prohibition started to die even before Franklin Roosevelt took office on March 4, 1933. On February 16, the Senate voted 63 to 23 to submit the Twenty-first Amendment to the state legislatures, undoing the Eighteenth. Four days later, the House voted 289 to 121 for submission. Meanwhile, the new administration put through a bill allowing the sale of 3.2 percent beer. On December 5, 1933, Utah became the thirty-sixth state to ratify repeal. The government quietly dropped all further Prohibition prosecutions, including the one against Capone and others of the outfit.

All that "model prisoner" talk aside, word circulated almost as soon as he had arrived in Atlanta that Capone had taken over. A cellmate claimed that he enjoyed such special privileges as smoking $2 cigars. Another prisoner charged that Capone wore specially constructed $25 shoes, moved through the yard with a bodyguard and, most of all, still ran the outfit by specially permitted, uncensored "business correspondence."

The warden instantly denied the reports. Atlanta's most famous prisoner was just plain old 40886, wearing regulation shoes and smoking only what he could buy from the commissary on the $10 a month he was allowed, just like everyone else. True, Capone received more fan mail than anyone else, almost more mail than the warden received, much of it panhandling letters. But no matter how much mail Capone received, he could reply with only the regulation two letters a week. In fact, because "of reports," the warden wrote, "that his friends would probably try to smuggle money and guns to him," he received visitors under more stringent monitoring than other prisoners. Except for that *lack* of privilege, Capone was "being accorded the same treatment as any other prisoner in the institution, no better and no worse."

That was nonsense. Capone "lived like a king," another ex-con claimed later. Though an exaggeration—he was a prisoner and lived in a cell—events and the prison's own records bear out the cons and the rumors rather than the warden.

Capone got the prison doctor, William F. Ossenfort, to prescribe a special pair of shoes, incorporating arch supports, that were built for him at Leavenworth. Several hired cons did indeed surround him as a bodyguard when he walked the yard.

Even the prison's disciplinary reports suggested privilege: they were always catching the model prisoner in some wretched excess. One guard wrote him up for having too much underwear (seven pair of drawers) and too many socks (ten pair); he had eight sets of sheets and two homemade feather pillows in addition to the government-issue one. His "punishment" was to turn in the extra. Another guard found the wooden slats from Capone's bed in a keg; in their place Capone had rigged a more comfortable bed cradle out of plumber's chain attached to the frame by springs. He could even get away with what a guard called "insolence." E. W. Yates complained that Capone had sassed him when ordered to wash a window. "This is the second time this prisoner has become insolent to me in front of the other members of the detail," Yates wrote to Deputy Warden Julian A. Schoen. The punishment was the lightest possible: the model prisoner was "reprimanded and warned."

In Capone's eight-man cell—in addition to occasional excess underwear—he kept his picture album, two rugs, shaving gear, mirror, bathrobe, typewriter, tennis racket, tennis shoes, alarm clock and a twenty-four-volume set of the *Encyclopaedia Britannica*. When he wanted to use the tennis racket and shoes, he never waited. For doubles, a player would leave, making room for him in the game he wanted; for singles, he'd point to his opponent of choice and the other player would retire.

A clue to his power could be found in a recess carved in his tennis racket's handle. He might have a couple of thousand dollars secreted there at any given time. One cellmate was Morris Rudensky—generally called "Red," though Capone called him "Rusty"—a safecracker Capone had known slightly in Chicago. Rudensky later claimed he had arranged for the money to be smuggled in by a trusty, and that he kept it for Capone in a hollowed-out broom handle. Perhaps, though neither Capone nor the outfit needed lessons in subversion and bribery. As Warden Aderhold had said, Capone, like other prisoners, could buy only one $10 book of commissary scrip a month. Many prisoners could not afford one, so Capone would give them the money, letting them use some of it, keeping the rest himself. For guards and officials, straight cash transactions were probable.

The FBI investigated, trying to put together an indictable case against Capone for smuggling money into a federal prison. Thousands of memos and nearly five years later, on February 2, 1937, the special agent in charge wrote to J. Edgar Hoover that since there had been no action in six months, he would—saving the director's displeasure—declare the case closed.

Something had to be done. James V. Bennett, who became director of the Bureau of Prisons, later admitted that Capone had become "too big a problem for our officers at Atlanta to handle. He behaved well enough at Atlanta, but he still comported himself as the king of crime," even continuing to have a say, by messages, in running the outfit. The government had a solution.

CHAPTER 30

. . . to Hell and Back

On August 1, 1933, President Roosevelt's attorney general, Homer S. Cummings, asked an aide to consider "would it not be well to think of having a special prison . . . It would be in a remote place—on an island, or in Alaska, so that the persons incarcerated would not be in constant communication with friends outside." A week later, Justice had picked a site.

It squatted lonely in the bay, one and one-quarter miles off the north shore of San Francisco, ripped by six-to-nine-knot, fifty-one-degree tidal currents. In 1775, Spanish sailors who first saw it covered with roosting birds named it "Isla de Alcatraces"—Island of Pelicans. It looked so unpromising, with precipitous 130-foot cliffs rising to a craggy surface covered only with guano, that Lieutenant Juan Manuel de Avala and his men did not deign to explore, and the island remained uninhabited for some seventy years.

Alcatraz Island was everything Cummings wanted. It had served as a fort, then a military prison. In 1933 a disciplinary barracks still sat on the island, which the U.S. Army planned to abandon soon. Justice took over and spent $260,000 to make the jail facilities as escapeproof as the island itself.

The Depression budget allowed only half the cells to be made of toolproof steel; but gun galleries and five watch towers that provided interlocking lines of fire were constructed. Gates to and from work areas featured electric locks operable only from the towers. The main gate required two officers to open, one of them seated behind steel, electronically controlling a steel panel that shielded the lock. Prisoners and visitors had to pass through metal detectors so sensitive that Capone's arch supports would set them jangling (he had the supports replaced by plastic). On his mother's first visit, the metal stays of her corset triggered the alarm.

The Army cleared out on July 19, 1934. On August 1 the Federal Bureau of Prisons took over, and next day Warden James A. Johnston took office. After a year as warden at Folsom in 1913 followed by twelve years at San Quentin, he had become a banker. He was noted as a prison reformer and believer in rehabilitation. The government installed him as head of a prison that did not even pretend to rehabilitate, only to warehouse.

Saturday night, August 18, 1934, while Homer Cummings made a final inspection of Alcatraz, guards in Atlanta removed fifty-three prisoners from their cells. "It was after chow," recalled Capone's cellmate Rudensky, "when Swede, a bland faceless guard, came up to our cell and rapped his stick across the bars." Four other guards marched up as Swede unlocked the cell, and three more came running when Capone raged that he wasn't going anywhere. The prison grapevine had vibrated for weeks about Alcatraz. Capone hurled himself on the nearest guard in a fury, spewing obscenities. The guards subdued him. "You dirty sons of bitches," was the last howl Rudensky heard as they dragged Capone away, "you'll never take me out of here!"

With the others slated for Alcatraz, Capone was stripped, searched, and given a fresh set of clothes. The government would take no chance on escape attempts from within or rescues from outside. A special train had been shunted into the prison yard. The barred windows of its steel cars were covered with heavy wire mesh. The guns of guards in screened cages could sweep the length of the aisles. At five A.M. next day, the train chugged out of the prison yard, its fifty-three prisoners shackled to their seats. The train stopped at none of the usual stations, its progress uncharted except by phoned reports to Warden Johnston.

Early Wednesday morning, August 22, the train skirted the turmoil of Oakland and glided into the tiny, almost unused yard at Tiburon, where the cars could be run right onto a barge for the short trip across the bay to the Alcatraz dock.

There, guards struck off leg-irons and the prisoners trudged up the hill to the rear gate, handcuffed in a column of twos. At the sight of the main prison building looming over them, one of the prisoners broke, sobbing.

When all were counted and processed, stripped, searched, bathed, fed, reclothed, and locked in, Warden Johnston wired Cummings the agreed code: FIFTY THREE CRATES OF FURNITURE FROM ATLANTA RECEIVED IN GOOD CONDITION INSTALLED NO BREAKAGE. A little later 103 crates arrived from Leavenworth, none of them outfit members, plus smaller consignments from McNeil Island and Lewisburg, and eight ringleaders of a hunger strike in the Washington, D.C., lockup.

Capone became number 85, cell 181.

* * *

The Justice Department and Warden Johnston had designed incorruptibility into the Rock, as Alcatraz became known. Capone would run *nothing* on or from Alcatraz; he wouldn't even know what was happening outside. There would be no smuggled letters or messages. All incoming letters were censored, then retyped by guards with prohibited subjects omitted, which included the faintest whiff of business or the doings of former associates. Censors excised even mention of current events. No newspapers were allowed; magazines had to be more than seven months old. The only source of news was new arrivals. At best, prisoners could write one letter a week, rigorously censored, and only to their immediate family members. With special permission, they could write to their attorneys, those letters censored like the rest.

Alcatraz scheduled no regular visiting days. Only immediate family could visit, only two of them each month, and they had to write the warden for permission each time. Convicted felons were excluded, which barred Ralph. Visitors and prisoners made no physical contact. They sat on opposite sides of plate glass, the "communication system" a perforated metal strip so poor a sound conductor that voices had to be loud enough to be heard by the guards, who would chop off the visit at the hint of a proscribed topic.

"I have a lot of friends," Capone told Warden Johnston at their first interview, "and I expect to have lots of visitors. . . ."

"Capone," the warden told him, "your friends and associates will not be permitted to come here as visitors." Blood relatives only, two a month, no exceptions.

No one would smuggle money into Capone, and he could not have spent it anyway. Johnston purposively eliminated all currencies that one prisoner might use to gain influence over others. Alcatraz offered no commissary. There were no cigars. Management provided everything convicts were allowed to ingest. On the other hand, Johnston made sure they got enough, forestalling the food riots common at other prisons. Bureau regulations specified at least 2,100 calories a day for each prisoner; at Alcatraz they averaged 3,100 to 3,600. For one typical dinner, they could take any or all of soup, Beefaroni, beans, cabbage, onions, chili peppers, biscuits, pudding *and* ice cream, ice tea and coffee. Lunch one day was milanaise soup, baked meat pie, spiced beets, Swiss chard, bread and tea.

Smokers could draw three packs of cigarettes a week, another usual medium of exchange in prisons. But the cigarettes were government issue: among the forbidden luxuries Capone had carted along from Atlanta, all of which had to be shipped home, were a carton of cigarettes and four packs of gum. At Alcatraz, the value of tobacco suffered further deflation because those who ran out of ready-mades could help themselves from dispensers in each cell block to all the roll-your-own makings they craved.

Single cells and around-the-clock scrutiny by one guard to each three inmates devalued to furtive gropings the other traditional currency of prison influence, sex.

The system worked against Capone. He could not buy popularity or influence. When he offered to equip the prison band, pay for a tennis court, lend money or have it sent to other prisoners' families, Johnston turned him down.

Like all the others with *full* privileges, Capone lived in a nine-by-five-foot cell, so narrow he could stand in the middle and press the palm of each hand against the opposite wall. He had a drop-down cot, table and chair, a washstand and a toilet.

From 4:50 P.M. lockup (with lights snapped out at 9:30) to the 6:30 A.M. wake-up bell, he saw no one but passing guards. Cells had no radios. Lights came back on with the wake-up bell. Three times a week each prisoner could shave with a razor and blade passed through the bars, then reclaimed by a guard. Prisoners marched in silence to breakfast at 6:55 and ate in silence. Ten canisters in the mess-hall walls would spray tear gas in case of any disturbance. With the others, Capone lined up at 7:20 in silence, worked in silence from 7:30 to 11:30. If he had a question about work, he had to direct it to a guard or one of the civilian foremen. Prisoners could smoke and talk, but not "congregate," during the eight-minute rest break at 9:30. They ate lunch from 11:40 to 12:00, returned to cells for a "lockup count," worked from 12:20 to 4:15 with another break at 2:30, ate dinner at 4:25, then were locked in for the night. Guards made thirteen head counts day and night, shop foremen six more at work.

Weekends, prisoners could mingle and talk two hours each afternoon while exercising or at their chosen hobbies. Capone took up music, becoming proficient on the tenor banjo and a lutelike instrument called the mandola.

After about four years, Johnston relaxed the rule of silence and some strictures on awareness of the outside world. Prisoners then got to see four movies a year, on holidays, Shirley Temple their favorite star. They could subscribe to approved current magazines.

Warden Johnston had been alert from the first to the danger of Capone's trying to dominate. "It was apparent," Johnston remembered from their initial meeting, "that he wanted to impress other prisoners by asking me questions as if he were their leader. I wanted to make sure that they didn't get any such idea." Johnston assigned Capone his number, issued standard instructions, "and told him to move along."

Even so, Capone retained star quality. The director of the Bureau of Prisons, James Bennett, called him "the most prominent gangster of all time." While Johnston might on principle put Capone in his place, the warden acknowledged that of all the prisoners, "the name that topped the list was Capone." At Rotary, at San Francisco clubs like the Commonwealth and Olympic, from reporters, all he heard was, "How is Al Capone? What does he do there? Is he boss of the other prisoners?" Even other federal officers would start conversations with "How is your star boarder?" Johnston's patience expired the night Sanford Bates, Bennett's predecessor as director, called to ask if radio reports were true that Capone was allowed to order silk underwear from London.

The bureau did its best to stifle outside talk of Capone and all others, proud, one memo put it, "that Alcatraz Penitentiary is as inaccessible to the outside as any prison could be." The Alcatraz prisoner "should lose the place in the public notice which attended his capture and trial." That would never quite happen with Capone. His doings would continue to leak, rating at least small-headline treatment.

And Capone never stopped trying to beglimmer the authorities, in which he partly succeeded. Despite put-downs, Warden Johnston found him fascinating, and was always ready to grant him interviews.

"Maybe you don't know it, Warden," Capone told him at one, "and maybe you won't believe it, but a lot of big businessmen used to be glad to be friends with me when I was on top and they wanted me to do things for them."

Shocked, the ex-banker wanted to know what kind of legitimate businessmen would need a gangster's help. Capone regaled the warden with tales of settling the newspaper circulation wars and strikes, dealing directly with "the highest top guys that own the newspapers."

"That is very interesting," Johnston said, "you may want to tell me some more sometime."

Gladly. But Capone received no extra privileges, nonetheless. And since he could not buy favor with his fellow inmates, his prestige diminished because he would not engage in the only activity at Alcatraz that could enhance it: defiance.

When Director Bennett first visited to interview inmates, the warden explained why he had stationed a guard with a rifle aimed at each convict who faced Bennett. These desperate men "might regard it as an accomplishment to assault the prison director." It didn't matter that they would be instantly caught and punished. The typical Alcatraz inmate, age thirty-five, had to serve a twenty-five-year sentence—knowing that some state had filed detainers on him to serve another sentence when released from federal prison. The first 178 arrivals faced 117 detainers among them. One inmate said, "Life gets so monotonous you feel like bucking the rules" just for something different. With some regularity they staged strikes and mini-mutinies, squelched with absolute regularity and speed. Yet each new ringleader found many willing collaborators. Most figured they had nothing to lose, and as Johnston once wrote, most "were in a mood to risk anything not caring much whether they lived or died."

By design, Alcatraz offered none of the usual prison incentives to good behavior. There were no trusties, special privileges, eased rules or the like. Instead, inmates could anticipate only punishment for the least infraction. Unfortunately, the main disincentive was loss of "good time," the ten days that could be knocked off a sentence for each month's good behavior. What did that mean to a man looking at twenty-five years, with detainers waiting? The other sanction involved worse treatment than usual, immersion in isolation. But that was only marginally worse, while on the plus side, the defiant won their fellows' respect and admiration as indomitable tough guys.

Capone did have something to lose. When he first trudged through the

Alcatraz gate, he had already served over two years of his ten-year federal prison sentence. With 1,200 days potential good time cutting the total to just over six years and eight months, he might have to spend only four years and four months on the Rock, with nothing but that single year in county jail waiting at the other end—and that, too, was reducible by good time. He would remain the "model prisoner" he had been at Atlanta, a rational decision that cost him prestige and popularity.

Officials first assigned Capone to work in the laundry, which handled washing for the area's military bases. Everyday toil in that fetid, steamy, airless din was drudgery enough. In January 1935, when an army transport docked and buried them in soiled bedding and clothes, the extra push drove thirty-six laundry-room convicts to strike. Capone refused to join, earning him their instant hatred. A month later, one of the strikers, William Collier, an ex-soldier doing life for killing an officer, exploded when he thought Capone was shoving clothes at him through the mangle too fast. He hurled a sopping mess of wash at Capone. The two mixed it up for the few moments it took guards to descend on them.

This may have been the first convict fight since federal prisoners had arrived the previous August; it was certainly the first inkling the outside world had of how the Rock disciplined its troublesome.

To "cool off," the brawlers spent the night in the dungeon that the Spanish had hewn out of rock (a punishment soon abolished). Then they went into isolation cells for eight days. These were in D block, most of them regular barred cells; others, called collectively "the Hole," featured steel doors that kept the cell in darkness. Isolation prisoners had no cots, but slept on mattresses at night; those in the Hole slept on concrete. They got all the water they wanted—to wash down their daily ration of four slices of bread, punctuated twice a week by reduced portions of regular meals, minus any such treats as dessert. The reduced meals would become standard D-block fare after the courts ordered an end to bread-and-water diets.

Almost all Alcatraz inmates pulled some time in isolation or the Hole, prescribed punishment for the insubordination and rule breaking that constituted most prisoner defiance. Fighting among convicts did not mean automatic loss of good time, which required a special in-house trial. Capone's laundry set-to had not hurt his chances for early release, and of course did nothing to lower his status. But he still would not join any activity that might impact his release. That doomed him to obloquy.

A year later, January 20, 1936, John Paul Chase—sidekick of Lester Gillis, better known as Baby Face Nelson—instigated a serious revolt. A prisoner named Jack Allen had died, officially from pneumonia, but in audible agony from a perforated ulcer—inmates maintained—when the duty medical officer refused hospital care for someone he thought a malingerer. Chase started the ruckus in the laundry, leading a chant: "Who's a son of a bitch?" "The doctor!" "Who killed Allen?" "The doctor!" Capone was not present, having been transferred after his fight from the laundry to duty in the shower room, but the strike metastasized quickly throughout the prison, and Capone, along with some others, refused to join.

He heard "dirty yellow rat" hurled at him, which left him unmoved. "Those guys are crazy. They can't get anything out of this," he said, asking to remain in his cell until it was over. "I have to protect my skin if I'm going to get out of here alive." He also had to protect his good time—even at the cost of what passed in that society for a good name.

Johnston's invariable strikebreaking technique, invariably successful, was to keep dissidents locked in their cells on short rations. This time the strikers anticipated the warden and declared a hunger strike. That soon ended, except for a couple of diehards. One of these was James C. Lucas, a Texas bank robber in for thirty years. The guards finally force-fed him.

Jimmy Lucas gloried in his reputation as a real hard case. Later, he claimed Capone had squealed on him; Capone claimed he had merely ignored a shakedown request for money. Another prisoner said Lucas had asked Capone for $15,000 to finance a scheme to have machine guns smuggled into them, with a speedboat waiting after they shot their way out. Capone supposedly laughed and said they were crazy; he wanted no part of it or of Lucas and his friends. He had made a point of shunning all prison cliques.

The morning of June 23, 1936, found Capone at work in the basement, swabbing between the shower room and the clothing room, his back to the barber shop about ten feet away. Lucas, waiting for his monthly haircut, snatched up a pair of scissors, tore the four-inch blades apart (a single blade having more penetrating power), and hurled himself at Capone's back. Capone whirled about and clubbed Lucas sprawling with a fist. A guard, fifteen feet from the fracas, rushed between them. Capone walked to the prison hospital, where the doctor pulled a few stitches; Lucas had managed only a half-inch-long slash in the lower left side of Capone's back, penetrating no more than a quarter inch. Guards dragooned Lucas to the Hole, his only punishment, since risk of escape during a mainland trial outweighed any conceivable sentence he might draw.

The press reported other fights and other attacks. One prisoner had tried to choke Capone on his way to a dentist's appointment and Capone had decked his assailant. A Chicago mail robber once described simply as "Charles (Limpy) Cleaver, desperado" had punched Capone. Someone had tried to poison Capone's coffee with lye. Someone had heaved a solid iron sash weight at him, Capone saved by the warning of Roy Gardner, a mail robber whom Capone also fought later. He had brawled with kidnapper Harmon Whaley. The warden denied all such reports.

Some former prisoners asserted that Capone couldn't take it: he was yellow; he could be scathelessly pushed around by any punk and was loathed by all. It's always gratifying to note how the mighty are fallen. When Capone had manned a mangle, washing the Army's dirty linen, one soldier wrote home: Guess who was doing his laundry! When Capone next swabbed the shower room, some inmates delighted in calling him "the wop with the mop." (His later assignments included sweeping the yard and delivering books and magazines from the library to cells.)

An opposite and even more improbable view came from an ex-con who gushed, "The boys . . . why, they love him."

Capone undoubtedly lost status and popularity in Alcatraz, but suffered few practical consequences because of it. The warden gave a balanced, accurate picture of how Capone stood. "As a matter of fact," Johnston wrote, "he had friends and enemies in the prison." Another observer wrote that "Capone got along as well as most others and better than some." He still bulked too large, too much the hard-hitting ex-bouncer, for many to risk hassling even if they despised him. "Most of the time he scrupulously kept his nose clean and stayed out of trouble."

Of course, no prisoner could avoid the inherent trouble of life in that brutal, crazed, suicidal atmosphere. Even a relative short-timer like Capone had no immunity from the daily surrounding madness. He had been attacked, could be again at any time, maybe next time fatally. He always had to walk warily. Who knew when another inmate would go berserk? One, chopping old tires in the shop where they made rubber mats for the Navy, suddenly used his hatchet to hack all the fingers off his left hand, then asked the nearest fellow inmate to take the hatchet and do the same for his right. Everyone scattered: at any moment he might discover a different target for the hatchet. After the hospital and the Hole, he rejoined the prison population, not reckoned quite insane enough for shipment elsewhere.

There were suicides. Escape attempts were so monotonously thwarted, some seemed indistinguishable from suicide—like that of the inmate who calmly scaled a fence under the eye of an armed guard, disregarded shouted warnings, ignored two warning shots, and died with the third shot. Johnston permitted no overt mistreatment. But guards who each minute had to face the ferociously dangerous were not gentle if given the least provocation. Those who moved among prisoners carried only metal "gas billies," which shot tear gas and doubled as bone-breaking truncheons. When one inmate attacked Johnston, injuring him, the guards' gas billies left the inmate a "drooling wreck."

Asked what life on the Rock was like, someone who had spent a mere sixteen months there replied, "It's hell."

The man who had planned to avoid a third of his sentence in sleep faced what inmates called "hell nights." As a released prisoner said, "A fellow like Capone just lays there on his side staring at the cell wall reminiscing"—how sweet life was outside, wondering what the world does without him, and if he'll survive to join it again. And that was before Capone cracked.

Just before, in late January 1938, Capone gave a deposition to Seymour M. Klein, an assistant U.S. Attorney from New York, where for two years the government had been preparing a tax case against Johnny Torrio. With Klein came SIU agent James N. Sullivan, who had worked on Capone's case with

Frank Wilson. Delighted at the chance to talk with anyone from the outside, Capone rambled on for two afternoons. He certainly talked about Torrio. "I carried a gun for him," Capone said, "I'd go the limit for him." Klein left with a fifty-page, ten-thousand-word deposition that imputed no usable specifics and was worth almost nothing. Over a year later, in March 1939, when the case finally came to trial, the government did not call Capone. Midway through, Torrio changed his plea to guilty and drew two and a half years in Leavenworth.

Fortunately, the government had a persuasive case without Capone. Even had he been willing to turn on his old mentor, by then he would not have made a convincing witness.

An enduring myth had it that Capone took his understandable dread of being punctured by machine gun bullets to the extreme of also refusing to be punctured by a hypodermic needle for diagnosis and treatment of syphilis until he already displayed signs of syphilitic dementia. In fact, he had a Wassermann test almost as soon as he reached Atlanta—a procedure that required his permission. When it proved "2 plus positive," consulting urologist Dr. Steven T. Brown immediately placed Capone on a course of bismuth injections. Heavy metals and arsenicals, which could attenuate the disease and render it asymptomatic, constituted the only efficacious treatment in those pre-penicillin days. By September 7, 1932, his blood test read negative, though "as a matter of course," the doctor placed Capone "on mixed treatment as a follow-up." The doctor also noted Capone's "history of having had partial treatment" before he entered Atlanta—though that is less reliable, since it probably came from the prisoner rather than prison records.

Without penicillin, it would already have been too late for curative treatment a month after he contracted syphilis. Despite the treatments then available, the disease would march its terrible course—though not necessarily toward madness. A modern expert, Lydia Bayne, MD, of San Francisco General, cites studies showing that under 5 percent of all those who contract syphilis progress to the paretic or demented tertiary stage called neurosyphilis. Robert Rolfs, MD, of the government's Centers for Disease Control in Atlanta, adds that other debilities can afflict another 20 to 25 percent. Still, the odds against paresis rate no worse than nineteen to one.

Early in 1938 Capone showed that his luck with spirochetes matched his luck with horses; the odds beat him. On January 21, Mae wrote for permission to make her monthly visit, this time with Mafalda, asking for a date toward the end of February. January 31, Johnston wrote back, instructing her to catch the ten o'clock boat from San Francisco, Monday morning, February 28.

Saturday, February 5, 1938, Capone donned his blue uniform. But those were worn only Sundays and holidays; a guard made him change into his workday gray coveralls. He seemed confused, not realizing breakfast time had come. Afterward, when the inmates returned to their cells before the work-detail shape-up, Capone turned into the wrong tier. A guard later found him in his cell, wrote Warden Johnston later, "looking bilious and trying to vomit."

The prison doctor summoned a specialist from San Francisco, psychiatrist Dr. Edward W. Twitchell. A spinal tap showed a positive Wassermann of "4 plus." Washington issued a muted press release, but newspapers already told of a Capone raving, battling guards, spitting at inmates, straitjacketed or lashed to a bed—when he wasn't mooncalfing into space, compulsively making and re-making his cot, or bursting into Italian arias. They also stated that he had fallen down in a coma. Speculation insisted that Capone would be transferred to some place like the government's medical facility at Springfield, Missouri. Portions of the reports were true.

"What happened to you this morning?" the warden asked Capone in the hospital. Capone grinned. "I dunno, Warden, they tell me I acted like I was a little wacky."

A telegram arrived:

FEBRUARY 9, 1938 MIAMI BEACH FLORIDA

JAMES A JOHNSTON

DEAR SIR DUE TO THE RUMORS WOULD LIKE TO LEAVE AT ONCE SO I COULD BE NEAR MY HUSBAND IF ANYTHING SHOULD HAPPEN THAT HE WOULD NEED ME BUT WOULD NOT LIKE TO MAKE THE TRIP AND FIND THAT HE HAS ALREADY BEEN TRANSFERRED KINDLY PLEASE ADVISE ME BY RETURN WIRE COLLECT WESTERN UNION RESPECTFULLY YOURS

MRS ALPHONSE CAPONE
93 PALM ISLAND

RESPONSE WESTERN UNION COLLECT 2-9-38

RETEL NINTH HAVE JUST HAD REPORT FROM PHYSICIANS ADVISING ME THAT YOUR HUSBAND IS QUIET, COMMUNICATIVE, COOPERATIVE, APPARENTLY COMPREHENDS HIS CONDITION AND NECESSITY OF FOLLOWING DOCTORS ORDERS AND THAT IT IS NOT NOW NECESSARY TO RESTRAIN HIM THEREFORE DISCOUNT RUMORS STOP HOWEVER, THEY CANNOT DEFINITELY DETERMINE WHAT CHANGES MAY OCCUR AND THEY DO NOT WANT TO PREDICT STOP IN THE CIRCUMSTANCES I SUGGEST YOU AWAIT FURTHER ADVICE AND IN THE MEANTIME KEEP IN TOUCH WITH DIRECTOR OF BUREAU, WASHINGTON, TO WHOM REPORTS WILL BE FORWARDED

J A JOHNSTON

A couple of months later, Johnston wrote to Director Bennett that Capone's recent letter to his mother betrayed no "deviation from his regular and normal use of language or any tremor or nervousness in penmanship."

Capone acted normally—for a neurosyphilitic under treatment. Ups and downs would continue, including periods of seeming remission, but on a downward slope. That summer Capone gave permission for a lumbar puncture. His spinal fluid confirmed the diagnosis of tertiary syphilis. On examination, Capone presented classic symptoms of paresis, with its distorted speech, plus tabes, another syphilitic affliction characterized by shuffling gait. He stumbled over such test phrases as "Methodist Episcopal," "truly rural," and "Roman Catholicism."

The examining psychiatrist found him cooperative and lucid, speech and memory good except during occasional convulsive attacks, though his reasoning, judgment and powers of concentration had noticeably fragmented. "He is," Dr. Romney M. Ritchey's report concluded,

> definitely expansive in mood and has developed extensive plans as to his activities after his release. He is going extensively into Charity work and will build factories and industries, etc, to furnish employment to everyone needing it. He takes a great deal of pleasure in perfecting these plans and relating them.
>
> His mood is happy and he has no enemies whom he cannot excuse readily for their mistaken views etc etc. He still has some disturbances of consciousness at times as his mind wanders and he hears God and the Angels verbally reply to his prayers etc. He however retains partial insight into these and says that he probably imagines some of the things he hears. These experiences are pleasant and he enjoys them and feels friendly toward those about him although he is very unstable and easily aroused by any excitement or confusion taking place on the Ward.

Capone would spend the rest of his time on Alcatraz Island under treatment in the hospital.

That time did not amount to much. Allowed full reduction for good behavior, Capone's federal prison term would expire January 6, 1939, leaving just the year's misdemeanor sentence.

First, though, he had to pay the fine and costs associated with the felony counts, $37,617.51. The government had pressed him for $322,842 in back taxes but later compromised to $157,416, since the jury had, after all, found him not guilty of tax evasion for 1924, 1928 and 1929. Besides, the government could locate no major attachable assets beyond the Palm Island estate. Although that rested in Mae's name, the government had served an assessment on her. Following Mike Ahern's bad advice, she ignored the assessment instead of protesting it, giving the government a right to go after the estate for Capone's tax debt. Thereafter, the IRS had twice threatened to sell the estate, both times blocked at the last minute. In November of 1936, Ralph had obtained a $35,000 mortgage and had, he wrote Capone, "managed to borrow enough to make up the difference'' to satisfy a lien for $52,103.30. In September of 1937, he ponied up another $17,194, barely beating foreclosure.

Two days before the Alcatraz release date, brother John bought a cashier's check for $35,000, cash. Capone's new lawyer, Abraham Teitelbaum, handed that and $2,962.29 in cash to the federal clerk in Chicago, the additional $74.78 over the fine and court costs to cover miscellaneous processing fees.

Government officials did not want Capone back in Chicago; Cook County jail was not their idea of secure imprisonment. At the same time, they wanted to continue treatment. They effected a change in Wilkerson's order so they could allow service of the misdemeanor sentence at the Federal Correctional Institution at Terminal Island, California, between San Pedro and Long Beach, south of downtown Los Angeles.

The night of January 6, 1939, an associate warden ferried Capone to Oakland. He stayed handcuffed to guards, six weights shackled to each leg. His guards, flourishing machine guns, barricaded a drawing room on *The Lark*, an express train to Los Angeles. The government had heard talk of a vigilante attack. Just in case, the guards hustled Capone off *The Lark* at Glendale, a suburban stop north of the city. A sedan met the train and drove the guarded, manacled, and woefully disoriented prisoner to the Terminal Island facility, which he entered at ten o'clock on the morning of January 7. Alcatraz number 85 became Terminal Island number 397.

George Hess, MD, had known Capone at Atlanta and at Alcatraz, where for a time Hess had been chief medical officer. Now he held that post at Terminal Island. He immediately put Capone in the jail hospital. ''During the first few days,'' the doctor soon reported to the bureau's medical director,

> this patient was definitely confused, indifferent and somewhat depressed. The depression was punctuated with periods of irritability but he was at all times the same cooperative person as always. His stream of thought was superficial and the speech was somewhat slurring in character. His reflexes were still unchanged, i.e. fixed pupils, knee jerks etc.
>
> I wish to advise, at this time, that Capone shows signs of improvement and I am of the opinion that the new environment is proving of definite value.

Not being on Alcatraz was like a month in the country. Hess continued Capone's regime of bismuth and Tryparsamide, a shot of each once a week.

''The guy's as screwy as a bedbug,'' one guard told a reporter shortly after Capone arrived, glassy-eyed and babbling. A week later, Mae visited and came away saying her husband was fine, entirely lucid. A month and a half after that, the head of the Secret Service detail in Chicago visited Terminal Island, chatted with Capone a while and pronounced him perfectly sane and well, the ''picture of health.''

Perhaps. When a visiting minister asked who among his seventy-five congregants felt the need of prayer, which among them would stand and witness, Capone got to his feet. Warden Edwin J. Lloyd promptly banned the Reverend Silas A. Thweat from the jail for his ecstatic recital to the press of the epiphany.

Capone could earn enough additional good time so that his final release would be on November 19, 1939. The bureau had rejected lawyer Teitelbaum's claim that Capone's time should have been computed from October 24, 1931, when he entered Cook County jail, not May 4, 1932, when he entered Atlanta. Wilkerson's sentence had been pellucid on the point. But the bureau did agree to remit three days representing that time in jail between October 24 and October 27, when the stay of execution became effective. Capone could be released November 16, 1939.

First, though, some arrangements had to be made. Capone still owed the $20,000 fine associated with the two misdemeanor counts. November 3, John paid with checks. Next, the family had to be convinced that Capone must not just be ''freed,'' but must go directly to some medical facility for continued treatment. That was easy. Good time was not automatic; the family had to agree, because the government could keep Capone until his full sentence expired on May 3, 1942. Capone would be sent east for release from Lewisburg penitentiary, near Harrisburg, Pennsylvania, then rushed to nearby Baltimore, where Joseph E. Moore, MD, a premier expert in neurosyphilis, would treat Capone in Union Memorial Hospital.

A week before the release, the boys in Chicago gave Capone a coming-out present. For a year, Ed O'Hare's associates had thought him strangely disturbed, for the past ten days downright agitated. While theories abounded as to why, his role in helping convict Capone had become common enough knowledge to find its way into newspapers, including the fact that his contact had been Frank Wilson. A letter warned that Capone wanted revenge.

The afternoon of November 8, O'Hare left his office at Sportsman's Park—closed as Capone's Hawthorne Kennel Club in 1930, reborn for horse racing in 1932. While O'Hare was driving to the Loop, a .32 Spanish automatic lay on the seat next to him. He normally packed a gun only when he was loaded with cash; today he had $53 in his pocket. Driving up Ogden, O'Hare had reached Rockwell, about halfway to the Loop, when a sedan drew next to his coupe. O'Hare evidently saw it because he floored the accelerator. As he did, two shotgun blasts ripped his head and neck. His car slewed across the boulevard, over the streetcar tracks, and crashed against a light pole. Ed O'Hare was dead.

In his pocket police found a note telling him ''Mr. Woltz phoned.'' Woltz wondered what O'Hare knew about an ex-bootlegger and bank robber named Clyde N. Nimerick; O'Hare should call Mr. Bennett with any information. George Woltz was an FBI agent in Chicago, as was Bennett. O'Hare had continued to inform, working his passage away from outfit connections. The note was signed "Toni"—Antoinette M. Cavaretta—always described as O'Hare's "confidential secretary." Later she would marry Frank Nitti.

The night of November 13, a boat ferried Capone to San Pedro and a car drove him to San Bernardino, where he boarded a train that would travel east by way of St. Louis, avoiding Chicago. The Bureau of Prisons refused comment on his whereabouts. FBI agents shoved a reporter who tracked them and tried to approach Capone.

On November 16, 1939, the government drove Capone clandestinely to Gettysburg. They turned him over to family and doctors—with the drama and secrecy of a spy swap—at a designated crossroads twelve miles east of town.

CHAPTER 31

Unpunctured End

THE BODY'S DEFENSES include the "blood-brain barrier" shielding spinal cord and brain from chemicals in the bloodstream. For Capone, that meant arsenicals and heavy-metal injections could not effectively counterattack the tertiary syphilis ravaging his mind. Penicillin does penetrate the barrier, but in those pre-penicillin days, only the addition of heat could retard the otherwise implacable progress of mental dilapidation. Doctors might elevate a patient's temperature to 107 degrees. They originally did that by infecting the patient with malaria. By 1940, Dr. Moore commanded less drastic hyperthermia techniques.

Mae, Theresa and Ralph drove Capone to the hospital from Gettysburg. He occupied a two-room suite, costing a lordly $30 each day, the spare room used by visiting family. Dr. Moore had some question about his own fees. When he asked James Bennett about his chances of being paid, the Bureau of Prison's director reminded him that "the Government has been trying to collect some $300,000 in back income tax" and had "not succeeded in finding any property on which levy can be made." Of course, Capone found money for the doctor and hospital. When he left, he planted a large weeping cherry tree, which still graces the front of Union Memorial.

Even if doctors could not then cure syphilis, with continued treatment symptoms would come and go. Capone might enjoy a couple of almost entirely lucid years before he needed more hyperthermia. By January 8, 1940, Dr. Moore proclaimed Capone well enough to become an outpatient. With Mae and Theresa, Capone rented a house in Mount Washington, a Baltimore suburb. Neighbors at first expressed nervousness, but soon took the recluse for granted. John often visited from Villanova, Pennsylvania, where he had a business.

On March 19, 1940, treatment ended. The family set off for Palm Island by car, stopping only to eat and for roadside naps. Thirty hours later, Capone had returned to Florida.

The remainder of his life declined into a routine determinedly bare of the excitement thought bad for paretics. Mae and Sonny of course lived at Palm Island with him, along with Mae's brother, Daniel Coughlin, and his wife, Winnie. Danny's duties as business agent for the Bartenders and Waiters Union did not interfere with his more important function of chauffeuring Capone. Winnie was that family's dynamo. She operated Winnie's Waffle Shop and Winnie's Little Club, a combination bar and restaurant that, according to one patron, "did a tremendous business, day and night." Another of Mae's sisters, Muriel, and her husband, Louis Clark, were the only other permanent residents, except for a usefully yappy fox terrier that sounded the alarm at any stranger's approach. Faithful Brownie and a maid, Rose, lived out.

The days passed, long and lazy. For much of each day Capone lounged in pajamas and robe, fishing from his pier, playing cards. Danny Coughlin would take him to driving ranges to practice his golf swing. In a few months, Capone was leading demure expeditions to local nightclubs, his party at a back table, quietly filing out before midnight. He and Mae would dine together at restaurants, a bodyguard watchful at the bar. Almost no one seemed to notice.

A few nights, Capone threw modest parties on Palm Island: drinks, hors d'oeuvres, and a four-piece band. A guest at one of the parties found him the genial, gracious host he'd always been, though not especially talkative. That night someone told a joke about an old Italian on his deathbed, surrounded by his family. As they pestered him with the same question, "Pa, where'd you hide your money?" his left forefinger tapped weakly on the first joint of his right forefinger, which he repeatedly raised and let fall, his right hand lying palm up on the counterpane. One son leaned close to ask. "Pa, why're you doing that with your fingers?" Summoning his last strength, the old man bolted upright. "Because I'm too weak to go like this," he croaked, his left hand banging the crook of his right arm in the classic *sto cacco* gesture.

Capone dissolved in laughter, perhaps because he had long been making a similar gesture to the IRS, which still couldn't locate assets to satisfy his back-tax debt. That first summer, a federal judge ruled that Capone owed $265,877.71. Next winter, when the government wanted Capone to appear for an inquiry into his assets and ability to pay, he suddenly became "unable to attend" because of poor health. Ralph and lawyer Abe Teitelbaum produced a doctor's certificate from Dr. Phillips—to whose office Danny Coughlin drove Capone for treatment twice a week. Perhaps fearing that Dr. Phillips's certificates lacked credibility after the contempt trial, Capone showed up at the hearing after all, natty and lucid, in a striped suit, white bow tie and straw hat, hiding behind sunglasses, puffing a cigar.

Capone and his household cost about $40,000 a year to maintain, a sum that somehow just appeared. As James Bennett had assured Dr. Moore, Capone had no assets the government could prove. The outfit took care of him.

That connection argued at least some retained power—witnessed by Norman Kassoff. After a career that included service as a Miami homicide detective, Kassoff became director of operations for the Dade County medical examiner. When he was seven or eight years old, sometime around 1940 or 1941, his mother and father wanted to open a funeral home. For months they found themselves bound in red tape; they couldn't shake loose the requisite licenses. Finally Mrs. Kassoff called her brother in New York, Charles "the Bug" Workman, Louis Lepke's premier Murder, Incorporated, killer. About two hours later, Mrs. Kassoff got a call telling her to go immediately to 93 Palm Island. Having no one to baby-sit, she took Norman along. He remembers Capone coming in from the pier, in pajamas, robe and floppy slippers. Capone offered coffee, called Mrs. Kassoff by her first name, though they had never met, and assured her, "I'll see to it personally." The next day another call informed the Kassoffs that it was all set. The licenses came through at once.

A few months later Norman and his mother were walking on Miami Beach's chic Lincoln Road when they saw Capone, very stylish in sports clothes, coming out of Pierre's, a fashionable tailor. Capone recognized Mrs. Kassoff, remembered her name, wanted to know if everything had been taken care of to her satisfaction. He was headed for Hialeah; would they like to accompany him? No? Well, then . . . he bent over to shake hands solemnly with Norman, and was on his way.

At just about this period, the eldest Capone brother—Vincenzo, or Jimmy—surfaced. He had changed his name to Richard James Hart and lived in tiny Homer, Nebraska (population 477), where for a time he had been town marshal. He claimed the nickname "Two-Gun," and told reporters he could uncap a beer bottle at a hundred feet "firing from either hip." When based in Sioux City he had been "sole law authority" for the Indians on the Coeur d'Alene, Kootenai and Kalispell reservations in Iowa. His tale provided an appealingly ironic picture of a Capone gone *right*. Some of it was true.

From Brooklyn he had headed west. He had worked ranches, and after four years, became a circus roustabout. In 1919 he jumped a freight in Homer, met and married Kathleen Winch, daughter of a local grocer (he had saved the Winches during a flash flood), and turned lawman when he failed as a house painter and paperhanger. He gave the Winches and fellow Homerians a different version of his background, including fantasy service in World War I, which won him appointment as local American Legion commander. Three years later, the last as a state sheriff, he started his stint with the Indian Service, interdicting liquor sales among the Winnebagos and Omahas, who loathed him.

After transfer to Sioux City, he killed an Indian in a barroom brawl but escaped charges because his victim was a bootlegger. He could not, though, escape the man's relatives; their attack cost him one of his eyes. (He later claimed it was lost fighting gangsters.) Another murder charge followed his transfer to Coeur d'Alene, Idaho, but he never faced trial for it.

Back in Homer—town marshal once again—he started pilfering from shops on his night rounds, including canned goods from his father-in-law's grocery. The town took away his set of keys and his badge. After the American Legion finally asked for proof of his war service, the vets took away his membership.

Jimmy and Kathleen had four sons, one killed in World War II in the Philippines. Broke, a cataract dimming his good eye, Jimmy finally got in touch with the family. Only Theresa, of course, even vaguely recognized him. He visited Ralph in Wisconsin, then Al in Miami. Only after that did Kathleen learn his true identity. Ralph sent him regular checks—a boon he would return, as we'll see, when Ralph needed connivance from the ''lawman'' Capone.

Sonny remained the only entirely blameless male in the family. He attended St. Patrick's School, run by the diocesan cathedral. In his senior year, 1936–1937, one classmate was the son of Cubans who had fled when their patron, dictator Gerardo Machado, was overthrown. Sonny and young Desi Arnaz became friends.

Portentously, if perhaps naturally, Sonny's slightest misdemeanors—a speeding ticket, once—made headlines. Another time he scraped his car against four trees that bordered a golf course when he tried to pass another car, and the son of Capone had to post a $150 reckless driving bond. He could not escape his name, though he would try.

Since he stood no chance of anonymity anywhere near Miami, he went off to Notre Dame, using his father's alias, ''Brown.'' His identity became known anyhow, and resigned to being pointed out, he returned to study business administration at the University of Miami. His father's release from prison found Sonny twenty-one years old, partly deaf, but remarkably well-adjusted and sweet-tempered considering what must have been, despite doting parents, a *most* peculiar childhood, the strangeness persisting into adulthood. For instance, when Sonny played in a golf tournament at the Biltmore in Coral Gables, his caddy was one of his father's bodyguards. Once, the bookies challenged the local gangsters to a softball game, the prize a keg of beer. The bookies won 3–2 when Sonny dropped a fly in center field.

On December 30, 1941, at St. Patrick's on Miami Beach, Sonny followed his father's example and married an Irish girl, Diana Ruth Casey, a schoolmate from an old Miami Beach family, proprietors of the popular bar, Casey's Oasis. Her brother, Jim, would become a Miami Beach detective sergeant and would remain fond and protective of Sonny long after the couple divorced and after Diana's death. Ralph Capone, Jr., stood as his cousin's best man.

Sonny and Diana set up house off the beach in the northeast section of Miami, on Tenth Avenue, and started to produce the first of four daughters.

Exempt from war service because of his deafness, Sonny patriotically gave up the flower shop he had opened the September before and asked for war work, assigned to the Miami Air Depot as a mechanic's apprentice.

Capone idolized the granddaughters as he had his little boy, lavishing attention and toys on them.

At last, in 1942, Dr. Moore laid hands on enough penicillin to treat Capone, one of the first neurosyphilitics to receive the war-scarce new miracle drug. It came too late to perform the miracle of curing Capone, but it prolonged remission of the disease's more debilitating symptoms.

To one man who met him then, he seemed almost normal. Pat Purdue joined the Miami Beach police at the beginning of 1942. While a rookie, he ran into Capone strolling along Lincoln Road. As in the old days, Capone still made friendly overtures to beat cops. "How are you today, pal?" he said, stopping Purdue.

"I says, 'Fine,' " Purdue remembers. Only one thing seemed odd about Capone. "He says, 'How you doing, pal?' two or three times—like a broken record, you know? Then I mentioned I had caddied for him back in nineteen twenty-eight. He couldn't get over it. I said, 'Well, you remember watching Sharkey up at Hollywood, don't you?' He said, 'Ooooh, that was you?' I says, 'That's right.' And he said, 'Here, here's a stick of gum.' He always chewed Dentyne—everybody he met, he gave a stick of Dentyne."

Summers, the household moved to Wisconsin, though not to Couderay, which had long since been sold, but near Ralph's place, Racap. Capone and his entourage would take two or three bungalows at Beaver Lodge. The owner considered them the least troublesome guests he'd ever entertained, especially since they paid their large bills so promptly. "What I liked about Al," said the owner, "was the smile he always had." Capone would fish, play pinochle, and pore over baseball scores—always, though, with a companion, usually Ralph or Matt. He seldom went into town, and never alone, and never to nearby Hurley, fabled for the number and raunchiness of its saloons. Capone no longer either drank or smoked.

The citizens of Mercer had started out apprehensive, but soon learned to anticipate their spring greening. Tradespeople would scowl when nosy strangers talked of gangsters. "Mercer," wrote one investigator, "likes to think of him as Mr. Capone, the man with the folding money." He and his followers had modified the ecology of Mercer, introducing to store shelves such big-ticket exotica as caviar, anchovies and petit fours, not to mention $100 and $1,000 bills, which residents were happy to receive once they satisfied themselves that those hitherto unknown denominations were actually government issue.

Sometimes reports would circulate that Capone had been stricken and was deathly ill. Yet he always seemed to recover. Such episodes were normal for his progressive dishevelment. He would indeed recover, but his mental deterioration proceeded relentlessly. Since he was taboparetic, his gait became shambling and lurching, his speech increasingly slurred. ''In time,'' explains Dr. Bayne, ''he would be rendered pitiable.''

In the summer of 1946, in Chicago, two shotgun blasts tore into James M. Ragen, Sr.—an old hand from the newspaper circulation wars who controlled much of the country's horse-race wire services after Moe Annenberg went to jail. Earlier, he had charged that ''the Capone organization'' had been trying to muscle into his business. Before he died of his wounds, Ragen charged that Capone still ran the outfit, and the outfit ran Chicago.

He was right about the outfit. But to those who knew Capone, the notion that by 1946 he could run anything mixed absurdity with sadness. Capone, said Dr. Moore, now had the mind of a twelve-year-old. ''He hasn't sufficient intelligence to run his own life, much less the affairs of a vast crime syndicate.'' Dr. Phillips, who saw him regularly in Miami, said he remained ''nervous and excitable'' and had to be closely attended. ''Al sometimes plays tennis,'' Phillips said, ''he swims and even goes out and mows the lawn occasionally. His wife, Mae, has assumed his entire care. Most of his old associates have abandoned him.''

One of them, Jack Guzik, in scoffing at any Capone role in the Ragen shooting or in the outfit's affairs, said with his usual delicacy, ''He is as nutty as a cuckoo. He can't walk a mile without someone leading him.'' The ''playing tennis'' Phillips mentioned was really batting the ball against a wall for endless hours. He would go for rides, let out of the car alone only to lurch into a drugstore for Dentyne and Sen-Sen. ''He likes to chew Sen-Sen,'' said Phillips. He could still fish and play cards, but would make up rules as he went along, behaving ''like a spoiled brat,'' as one of his few faithful associates said, whenever he lost. ''We always let him win.''

Everyone admired the way Mae stuck by him. ''Mrs. Capone,'' said Phillips, ''has borne a cross greater than that which comes to most people in their lifetime.'' She attended Mass at St. Patrick's several times a week. In all likelihood, so did Capone.

On Monday, January 20, 1947, former U.S. representative Andrew Volstead, age eighty-seven, died at his home in Granite Falls, Minnesota. He had lost his seat in 1922 to a clergyman who heralded himself as ''drier than Volstead.'' Nevertheless, the icon of enforcement, as an obituary editorial stressed, ''continued to believe in the beneficence of Prohibition to his last day.''

About four A.M., the next morning, in Florida, four days after he had turned forty-eight, Capone suffered what his doctors called an apoplectic stroke, probably unconnected with syphilis. A call to St. Patrick's found Father Cloonan, just released from the Chaplains' Corps and on casual duty, awaiting reassignment. He rushed over to Palm Island. At six A. M., he administered extreme unction.

Yet after fourteen hours in a coma Capone rallied, regained consciousness and wanted to talk with Mae and Sonny, who had not left his bedside. Next day, Dr. Phillips told reporters gathered outside the front gate that his patient might survive after all, though he was "definitely not out of danger." The family had been summoned, Theresa, Ralph and Matt on their way.

Capone continued to improve; the talk was of being "out of the woods." Then, late Friday, pneumonia struck. Phillips called in a specialist, but chest man Arthur J. Logie could do nothing; he found both lungs congested and Capone's heart weakened. They kept him alive on oxygen.

Ralph would emerge regularly through the large gate with bulletins and beer for the reporters. He never had anyone to help him and never brought the bottled beer in a basket, just clutched in his stubby arms. That reminded the press of their nickname for him, "Bottles." Capone was back in the headlines—except for the Miami *Daily News*, whose publisher James Cox decreed that the imminent death would be treated like any obituary. "I don't want that son of a bitch on my front page," he said.

Capone had said he always feared dying alone in the street, "punctured with machine gun bullets." He wanted his family about him. His death came from cardiac arrest at 7:25 Saturday night, January 25, 1947, the family at his bedside.

Monsignor William Barry, rector of St. Patrick's, brooked no nonsense. He would shush any talkers, gangsters or not, during his services. He conspired with the family to fool the press. An empty hearse drove over the Seventy-ninth Street causeway from the Philbrick funeral home to the cathedral, where a memorial service was to be held. Requiem Mass was forbidden. Meanwhile another hearse bearing Capone's sheet-bronze coffin drove to Chicago, the press thrown further off the scent by rumors that it was on the same train that bore the family.

Mount Olivet Cemetery lies in Chicago's far South Side at 111th Street. Forty to fifty mourners began gathering at plot 48 at two o'clock, the afternoon of Tuesday, February 4, 1947, shivering in four-degree temperature. It had taken sextons three hours to hack a grave in the frozen, snow-covered ground between the graves of Gabriel on the right and Frank. The sparse turnout betrayed no lessened gangland or political respect. Anthony Accardo had emerged as the outfit's power along with Joseph Aiuppa. He had ordained that, besides the family, only those associates who had become family friends would be welcome. Otherwise, he said, almost everyone in Chicago would want to swarm in. "Al had no enemies," he said.

Capone's cousins, the Fischettis, crowded under the tentlike canopy erected over the grave. Murray Humphreys, Sam Hunt, Willie Heeney, prominent outfit members Tony Capezio and Nick De Grazio—and the otherwise obscure, like Joey Korngold and Robert Ansonio—joined them, collars turned up against the freeze and against the news cameras that ringed them. Jack Guzik, limping, arrived last. Ralph guarded access to the site, rumbling, "Why don't you leave us alone?'' to importunate newsmen. Charley Fischetti was more direct. ''I'll kill any son of a bitch who makes any pictures,'' he snarled.

Monsignor William J. Gorman, chaplain of the Chicago Fire Department and pastor of Resurrection Cathedral, took time to tell the press why he was there. ''The Church,'' he explained, ''never condones evil nor the evil in any man's life. This very brief ceremony is to recognize his penitence and the fact that he died with the sacraments of the church.'' He had not known Capone. But he had been pastor of St. Columbanus', around the corner from the Capone home on Prairie, where Theresa had demonstrated ''unfailing piety,'' never missing a daily Mass or Sunday communion. ''She asked me to conduct this service today.''

It was indeed brief. At three-thirty, the Rago Brothers hearse brought the bronze casket, decked modestly with a blanket of gardenias and topped by fifty orchids. Monsignor Gorman read a few passages from his missal, led the mourners in Our Fathers, Hail Marys and Acts of Contrition. It was over in five minutes. The casket descended.

Police had milled about, saying they wanted to question some gambling suspects, but made no arrests. On the way out, Matt echoed Charley Fischetti's threat to a photographer who tried to snap his mother's picture.

A black granite stone rose eight feet. QUI RIPOSA was carved over the Italian names of Gabriel (Gabriele), then Frank (Salvatore), with the day, month (also in the Italian spelling) and year of their births and deaths, NATO and MORTO. When stonecutters added Capone's name, they called him only AL CAPONE and engraved only his years, no NATO or MORTO, no days or months. Compared with his father's and brother's inscriptions, his looked slapdash and impermanent—which, as we'll see, it was.

CHAPTER 32

The More It Changes

THE WAGES BOTH of crime and civic virtue were paid unevenly.

A few of Capone's associates died with edifying dispatch and messiness. Capone's labor plunderer, Red Barker, was gunned down in June 1932; police suspected the Roger Touhy gang—with the usual results. Barker's partner, Three-Finger Jack White, got his, probably from the inheritors of the outfit when he proved unmanageable. Tiny "Diamond Lou" Cowan died by the usual unknown hands, shot in 1933. That probably does not qualify as a cautionary tale, since he had never been much of a menace and had certainly lived a fuller, more interesting life after association with Capone than he had as a reasonably honest and unregarded newsstand operator.

On the other hand, James Genna of the "Terrible Gennas"—each less deserving than the next—died in bed of heart disease barely a month after Capone's tax conviction, while brothers Sam and Pete survived at least into the forties, though visibly down-and-out. Myles O'Donnell, the frail, less objectionable brother, died young, in February of 1932, but at home and of illness, while Klondike flourished as part of the outfit.

During Capone's trial, Frank McErlane shot his wife, Elfrieda, four times, leaving her in the back seat of her car next to her German shepherd and fox terrier; he had killed them too. No end could have been too severe for him. He died in a hospital bed of pneumonia in October of 1932, the only gratifying detail being that four attendants had to hold him down in his delirium.

Frankie Rio, whose only visible virtue was loyalty, died of natural causes in the mid-thirties.

Jack McGurn was murdered while Capone sat in Alcatraz. Not everyone had liked him; he had been blustery and inclined toward elite laziness. When Capone went to prison, McGurn at first just played a lot of golf. With Repeal and the Depression creating hard times, he tried to muscle his way back into the outfit's action. Late St. Valentine's night, 1936, he went bowling. At about one A.M., February 15, two "unknown" men gunned him down, leaving on his body a comic Valentine that mocked his relative poverty:

> You've lost your job,
> You've lost your dough,
> Your jewels and handsome houses.
> But things could be worse, you know.
> You haven't lost your trousers.

Mindful of their victim's family tradition of long memory and relentless vengeance, sixteen days later three outfit gunmen invaded a poolhall, stalked up to the table where Anthony Demory sat playing cards, and shot to death McGurn's worshipful half-brother and bodyguard.

McGurn's widow, Louise, was soon involved in another fatal auto accident. As with the one that inaugurated her public career at age fifteen, she was unhurt and uncharged; someone else sat at the wheel of her car. Indeed, the rest of Louise's life resembled an ongoing collision in which others sustained most of the injuries. She married twice again, spending minimal time with her husbands. In 1940 police arrested her for possession of a gun that her current live-in had used in a stickup (they could not prove she knew of its use). In 1943 police grabbed her for harboring an army deserter. "It's the darnedest thing," she mused, "but every time I get pinched it ain't because I did anything—it's because of the guy I'm with." In the late 1980s she was reported living in northern California.

The 1950s saw an end to many of the old outfit, most of them by natural causes. Charley Fischetti died in 1951. Rocco died in 1964 at age sixty-one of a heart attack. Willie Heeney died of throat cancer in 1951 at age sixty-three. When Louis Campagna—who had threatened Joe Aiello in jail—died in June of 1955, John and Albert Capone showed up at the funeral, as well as Tony Accardo, Sam Hunt, Claude Maddox, Tony Capezio and another outfit power, Rocco De Grazio. Maddox and Hunt would themselves soon be gone. Phil D'Andrea died peacefully in the fifties. Frank Maritote, however, got it from a shotgun.

In 1965, Murray Humphreys, whom one former colleague remembers as "the nicest guy" in the outfit, pitched face down on the costly carpet of his expensive Marina City apartment in Chicago. Cardiac arrest felled him seven hours after the FBI had served a perjury warrant. Still a valued fixer in the outfit, he had lied to a federal grand jury. He was in his mid-sixties.

Frank Galluccio, the man who cut Capone, died of a heart attack in 1960. Joe Fusco—always more bootlegger than racketeer—survived into the 1980s, still in the liquor business. Tony Accardo and Joey Aiuppa were alive in 1992—Aiuppa in federal prison, Accardo untouched by the law.

Of the former Chicago gang chiefs, hulking Danny Stanton lasted, criminal as ever, until shot in a saloon in 1943, victim of an intramural dispute.

George Moran went to the Ohio State Prison in 1946 for the $10,000 robbery of a tavern owner in Dayton. On release, the feds took him on a detainer for robbery of an Ansonia, Ohio, bank. He died of lung cancer in Leavenworth on February 25, 1957.

After a wrenching divorce in which he denied paternity of his oldest son, Joe Saltis—who had been living in flophouses—died broke and alone in the county hospital in August of 1947.

Spike O'Donnell suffered a severe heart attack in 1956 and died of another in August of 1962, age seventy-two.

The inseparable Terry Druggan and Frank Lake quarreled over nothing at a party and split up in the early thirties. Druggan spent the rest of his life in and out of hospitals, suffering from ulcers, asthma and heart disease. He died friendless and in penury in 1954. Lake moved to Detroit, lived in an exclusive suburb as president of a coal and ice company, his wife given to society-page good works. No one seemed to care how he had come by his stake. In 1947 he died young, in his fifties, but of natural causes and in the bosom of his family.

Frank Nitti went earliest of the outfit's heavyweights, in 1943. He'd had a very close call even earlier. With Capone's conviction, Mayor Cermak moved to recapture control of gambling and vice. Ted Newberry would be Cermak's point man in the new order, so the old guard had to go.

About twelve-thirty on the afternoon of December 19, 1932, two detective sergeants, Harry Lang and Harry Miller (brother of Herschie and Davy), invaded Nitti's office at 221 North La Salle; they carried no warrant, but on their way into the building had enlisted two passing uniformed policemen.

Nitti had his hands up when Lang shot him three times—in the back near the spinal cord, then, when he spun around, in the right side of the neck and right chest. Lang gave himself a flesh wound in the arm, which he later charged Nitti had inflicted. Unfortunately for the plot, Nitti lived and Christopher Callaghan, one of the uniformed officers, testified truthfully about what happened. The judge directed a not guilty verdict at Nitti's assault trial. Now Lang and Miller needed a good lawyer. They retained Abraham Lincoln Marovitz, who had started as an assistant state's attorney, under Crowe, become a famed criminal defense counsel, and then an honored federal judge (and still serves as a senior judge). He won them a verdict of simple assault, which carried a $100 fine, promptly paid. That ended the case.

On January 7, 1933, Ted Newberry's body turned up in the ditch alongside an Indiana byroad.

Revisionists claim that when, on February 15, Italian-born Giuseppe Zangara tried to assassinate President-elect Roosevelt in Miami, he was sent by Nitti and was aiming for Mayor Cermak—also in Miami on vacation. Zangara fired five shots. None hit Roosevelt, but one did get Cermak. Though not a mortal wound, it induced peritonitis. Three weeks later Cermak died, and two weeks after that Florida executed Zangara, who had pleaded guilty. Voices in his stomach, which constantly ached, had commanded him to smite authority. He said he was aiming at Roosevelt. Nitti, who studied under a master, would have planned better and chosen a more reliable instrument.

When chance exposed a Hollywood extortion scheme of the outfit's, and it became clear that Nitti would be going back to jail, he could not face that prospect nor the likely decline of his power. On March 19, 1943—nine months after he married Ed O'Hare's former secretary—Nitti took a stroll along the Illinois Central tracks near his Riverside home and put a bullet in his brain.

Jack Guzik collapsed with a heart attack in February of 1956 at age seventy. He was in St. Hubert's English Grill, where he used to hand out payments to city employees. His son, carrying on the family tradition, was then serving time in Arizona on a morals charge.

John Torrio, seventy-five, went from his Brooklyn home to get a haircut on March 16, 1957. He had been active mostly in real estate since release from his late-thirties tax conviction. A coronary seized him in the barber chair. He died in the hospital a few hours later.

After Big Bill Thompson's 1931 defeat by Cermak, he descended into unrelieved buffoonery which turned nasty with the approaching war, when he fronted as an apologist for fascist groups. After his death, in March of 1944, investigators found a total of over $1,800,000 in five safe deposit boxes, naturally inspiring thoughts of peculation—although Thompson had, after all, inherited more than that. He was seventy-seven.

Some of the good guys did very nicely. Had he lived, Ed O'Hare would have been prouder than he deserved. His son, Butch, graduated from Annapolis in 1937. Butch O'Hare became a Navy pilot, won the Medal of Honor for shooting down five Japanese bombers in 1942, but died in combat a year later. Chicago renamed its main airport for him.

Dwight Green served two terms as Illinois governor, 1940 to 1948, when he was defeated by Adlai Stevenson. He died of lung cancer in February of 1958 at age sixty-one. His associate, Jacob Grossman, became partner in a law firm and died in 1975 at age eighty-two. Frank Wilson became head of the Secret Service and died in June of 1970. George Johnson became a judge briefly in 1932, then returned to private practice and died in 1949 at age seventy-five.

Judge Wilkerson did not win confirmation to the court of appeals, but served eighteen honorable years on the federal district bench; he died in September of 1948, at age seventy-eight.

Pat Roche spent the last ten years of his life in semiretirement because of frail health. He died in July of 1955 at age sixty.

Eliot Ness went from hero in Chicago to partial goat as director of public safety in Cleveland from 1935 to 1942. He failed notably in a prominent murder case there. During the war he worked for the Federal Security Agency. Afterward, he became president of the Guarantee Paper Corporation in Coudersport, Pennsylvania.

Fate played him an ironic trick. His greatest disappointment was one he never knew about: missing the ubiquitous fame he thirsted for and would achieve only posthumously. Ness died in 1957 with his ''as-told-to'' book, *The Untouchables*, still in galley proofs. He never saw Robert Stack portray him on television or Kevin Costner in the movie.

Some of the artifacts of Capone's life had their own histories.

In February 1952, Mae sold 93 Palm Island to a Cleveland real estate broker for a little over $64,000, including the furnishings. ''You can just say I stole it,'' said Thomas W. Miller. He planned to resell after his brief occupancy took ''the Capone curse'' off the property; but time exorcised that curse very slowly. Miller offered the house three years later for $72,000 and was glad to swap it for a small airport near Canton, Ohio. The house went through at least two other sets of hands, each short-term owner exasperated by rubberneckers, especially those buzzing the dock in motorboats. Finally, in 1971, airline pilot Henry T. Morrison bought it for $56,000. He lives there still, quite happily.

After Theresa's death, November 29, 1952, Mae also sold the house on Prairie. It would become center of a tiff into the late eighties between those who did and those who did not want the house protected as a landmark in the National Register of Historic Places. The nays had it.

A brisk trade developed in Capone's armored limousines. One was brought by an English entrepreneur, who displayed it at fairs in that country, though a dance hall owner finally got it at auction in 1958 for only $510. Another sold in the U.S. for $37,000 in 1971, with its owner turning down $80,000 as an ''insult'' four years later. Still another, slated for use in a movie about Capone, burned up. (Not counting the early *Scarface*, three such movies appeared, Capone played by Rod Steiger, Ben Gazzara and Jason Robards. Robert De Niro played him briefly in the 1987 movie *The Untouchables*. An updated *Scarface*, eponymously played by Al Pacino, veered entirely away from Capone's story, the protagonist a Cuban cocaine dealer.)

Capone's Cicero headquarters, once the Hawthorne Hotel, then called the Western, finally became the Towne. One of Capone's inheritors, Joey Aiuppa, had control of it. On February 17, 1970, Aiuppa sold the hotel. Within hours—entirely by coincidence; no one suggested arson or fraud—a grease fire started in the kitchen and the Towne burned down. The site today is an empty lot.

Urban blight leveled both the Four Deuces and the Metropole. Indeed, brick-strewn empty lots are the most attractive features of the area that was once the Levee and its environs. Unhappily, the Lexington still stands, a sad, derelict hulk, gutted and vandalized. After its seedy glory in Capone's day and the thirties, it devolved further into a brothel and then a flophouse so pestiferous a court order closed it in 1980. Hype about possible Capone riches in its sealed, 125-foot concrete vault led to a television special during which the vault disgorged some empty whiskey bottles. In the late eighties a developer trumpeted talk of refurbishment and renaissance. No sign of either has yet appeared.

The wall against which the St. Valentine's Day massacre occurred fared better. By the mid-sixties, the garage had become an antique shop run by Charles and Thelma Werner. "Hardly a day goes by," sighed Mrs. Werner, "that somebody doesn't come to the door and ask me to show them the back room where those men were killed." She was sick of it. When the city cleared the site two years later for a housing project dedicated to the elderly, the demolition company carefully preserved the whitewashed, blood-flecked bricks of the north wall. Offers flooded in for single bricks, but the company refused to break up the set. Reconstructed, the wall first entered show business as a shopping-center attraction—until women's groups protested. It next became a curiosity in the washroom of a nightclub whose other delights included banjo sing-alongs and moose heads that had belonged to Bonnie and Clyde. Finally the bricks retired from public view, bought for private amusement in the rumpus room of a Canadian businessman whose bid topped about fifty others.

"It's tough," said Ralph in 1941, "being pointed out as Al Capone's brother. Wherever I go, people point me out not as Ralph Capone, the businessman, but as my brother's brother."

Even so, Ralph remained attentive and loyal throughout Capone's last years. Ralph's importance in the outfit, though, plummeted as soon as his brother went to jail, never to rebound. Capone's death found Ralph operating Dreamland, a scruffy dance hall in Stickney, while he maintained an interest in a bottling company and liquor firm and controlled a company that spotted two hundred or so cigarette machines around Cook County.

In the early fifties, Ralph survived another government try at nailing him on tax matters—partly with help from brother Jimmy. The "lawman" swore that *he* owned Ralph's Wisconsin lodge, Racap, having bought it in 1941 with money Theresa had given him. "Ralph runs it for me," he explained. Ralph also "ran" Billy's Bar in Mercer's Rex Hotel, where two of Jimmy's sons worked, yet the government somehow could never prove Ralph had assets much beyond his incessant poor-mouthing.

Ralph's wife divorced him with charges of cruelty that included physical abuse. He had the sadness of seeing Ralph junior commit suicide at age thirty-three. "Risky," as Ralph junior was nicknamed, had drifted in and out of many jobs and one marriage. By November 9, 1950, he had become a bartender. His girlfriend, twenty-one-year-old nightclub comic Jean Kerin, was playing a date in Duluth. Risky missed her. That night, "flooded with alcohol," as the coroner put it, he gulped down a bottle of a cold remedy whose label warned against mixing it with alcohol and started a note in a barely intelligible scrawl:

> Dear Jeanie. Jeanie my sweetheart. I love you, I love you. Jeanie only you I love. Only you. I'm gone

ending with neither punctuation nor signature. For a long time he had used the alias "Ralph Gabriel." His father said he could not imagine why.

Ralph Capone drifted sleazily through the rest of his life, mostly disdained and ignored by the old crowd and new outfit bosses. He died peacefully in November of 1974.

Jimmy died in Homer, entirely blind, the night of September 30, 1952, two months before his mother's death.

Al Capone had tried to keep Matt, the youngest brother, away from trouble. He sent Matt to Villanova College. It was no use, although Matt did manage to avoid the family's usual portion of notoriety until April 1944, when he became suspect in the murder of a racetrack tout committed in Matt's Cicero tavern, The Hall of Fame. He disappeared for eleven months, by which time there appeared to be no case, so it was dropped. Matt died January 31, 1967, age fifty-nine, under an assumed name. Just twenty-five people attended his funeral, where two newsmen had to be pressed into service as pallbearers.

Albert regularly used the aliases Bert Campbell and Bert Novak. In 1942, he legally changed his last name to Rayola, a version of his mother's maiden name. He bought one of the first gambling stamps issued in Illinois when the federal government required them for professionals in 1953, but he generally avoided police trouble, except for a $25 fine he once paid for wife-beating. He died at age seventy-six in June of 1980.

John—"Mimi"—stayed out of the public eye, perhaps because he changed *his* name to John Martin. He survived at least into the eighties.

"Sonny," says retired Miami detective Pat Purdue, "was straight as a die." That straightness rendered Sonny inappropriate for his first postwar job. He quit selling used cars rather than countenance his employer's sharp practices, which included rolling back odometers. For the rest of his life he would have money problems. His father had left no will and had no money to leave.

Much later, Sonny confided to a fellow employee of the port of Miami that in the late 1940s the outfit offered to cut him in on some action. He discussed it with Mae, who said, "Your father broke my heart. Don't you break it." He stayed straight, but needy.

By 1956, Mae and Sonny owned a restaurant called Ted's Grotto at 6970 Collins, just over a block from a bar-restaurant in which Sonny's uncle John had an interest. Mae handled the cash register and Sonny played maître d'hôtel, not the best role for one so shy.

When their restaurant failed, Sonny took a job with a tire company in the north Miami section called Hollywood. Perhaps exacerbated by the trials of raising four daughters on a tire warehouseman's salary, the marriage of Sonny and Diana came to an unrancorous end, Diana taking the four girls to California.

Sonny won the further admiration of his friends by faithfully providing all he could for their support, at one point working two jobs, sixteen hours a day, both menial and low paid.

When Sonny's old St. Patrick's classmate projected a two-part TV segment of *The Untouchables* that would feature Capone's transfer from Atlanta to Alcatraz, Sonny phoned, asking Desi Arnaz for old times' sake to scrap the idea. They hadn't been *that* close. Mae and Mafalda joined in a suit against Desilu, CBS and the sponsor, Westinghouse Electric, asking a total of $1,000,000 damages. They lost.

Sonny had always been quietly proud of the father who had adored him. He had turned himself into one of the better handgun and rifle shots in the country, and at a time when he desperately needed money, the Springfield Firearms Company offered him $25,000 to tour the country demonstrating their line—*if* he would change his name. They considered "Capone" a negative. Sonny refused.

Then something snapped. On Friday, August 6, 1965, he loaded three bags' worth of groceries in his cart at the Kwik Check in north Miami Beach. He also crammed two bottles of aspirin and a pack of flashlight batteries into his pocket, their total retail value $3.50. Two detectives on the sheriff's shoplifting detail nailed him outside the door. "Everyone has a little larceny in them," Sonny told the police. "He's a real good customer," said the store manager. "All I can say is I hope he comes back."

The judge put Al Capone's son on two years' probation, and newspapers around the country put his story on their pages, in Miami and Chicago under multicolumn headlines.

Nine months later, Sonny petitioned the court to change his name. "I should have done this years ago," he said. The judge unburdened himself of a homily on the sins of the father with an exhortation not to besmirch the new name so that "nothing will be visited upon the heads of your children."

Sonny moved to California for a spell, perhaps to be near the four daughters when Diana died in the late eighties. At last report he was back in Florida, retired.

Mafalda felt differently about the family name. She ran a delicatessen restaurant in Chicago, at 10232 South Western. In 1957 she charged a policeman with assault, saying he hit her when she stopped him from beating a drunk he had hauled out of a bar next door. The policeman was suspended. When reporters covering the incident asked if she was Capone's sister, Mafalda snapped, "Yes, and I'm proud of it." And when she died in 1988, her marker inscription read:

MAFALDA MARITOTE
CAPONE

Mae continued to live modestly in Florida, dying about the same time as Mafalda.

* * *

In the 1940 presidential election, New York recorded write-in votes for Al Capone.

In 1959 a respected Philadelphia lawyer let the University of Pennsylvania know that an anonymous group of clients contemplated endowment of the "Al Capone Chair on Taxation." Penn suspected a hoax, but was willing to talk. Nothing came of it.

In 1973 one of Brazil's biggest pop hits was the song "Al Capone."

In 1986 the opening of Capone's Lexington vault—however empty—propelled the show's host back into network TV. That show's rating still holds the record for a non-network syndication.

In 1991 French radio commissioned a ninety-one-year-old retired Chicago newsman to help write a twenty-five-part series on Al Capone.

Capone was little known outside Chicago until 1926, and five years later he was through. Why does his name endure? Why did he so firmly capture the world's imagination?

In 1931, before Capone pleaded guilty, Katherine Fullerton Gerould pondered the phenomenon in a *Harpers Monthly* piece, "Jessica and Al Capone." Gerould allowed that he was "one of the central figures of our time"; but until recently she had considered him only a "factor in American misgovernment, a strong argument against the evil workings of Prohibition, a living reproach to Chicago." It had never occurred to her that others might find him *personally* interesting, even a figure of romance. Then her daughter's friend, twelve-year-old Jessica, who possessed "all the elements of breeding"—indeed displayed alarmingly Monarchist tastes—declared that she found Capone fascinating and nominated him the only man she'd ever consider marrying.

Gerould soon discovered that "intelligent young people" agreed with Jessica. "They do not think him exactly a good citizen. They merely find him more interesting than most good citizens. On the whole, they are not sure that he is a worse citizen than many million others—not, that is, if you search the human heart."

Most of all, they liked his style. "What 'gets' the public, in Al Capone," Gerould wrote, "is simply his efficiency." He had so ordered life that he could loll in Miami and receive constant tribute, push a figurative button there and have the "interfering" eliminated in Chicago. Acclaim for Capone, Gerould thought, "was one with the glorification of Ford and Rockefeller."

Not quite. Capone did show the same gift for organization and the same "genius for mass-production." And, as Gerould observed, "For some generations, we have tended increasingly not to care how money is made, so long as the money is demonstrably there." But this was a mogul with a difference no less important for being obvious. Capone was Ford or Rockefeller with a shoulder holster, Cissie Patterson's tiger. He didn't simply have lawyers who could bend or confound the law like the great trusts and he didn't simply co-opt officials. He flatly broke laws his public either wanted broken or cared little about, yet hadn't the nerve or resources to break themselves, and he might personally knock an officious lawmaker down the stairs. He killed or had killed only those that the public, had they dared, would have happily dispatched themselves or at whose loss, at best, they felt no pang. And in the process, Capone joyfully documented the public's perception of their rulers' corruption. As he said, he met conditions without backing up. In short, he *lived* the people's more miscreant fantasies.

So did others, but Capone did it on a much grander scale. No bootlegger in New York commanded so great a market share, none enjoyed such wide hegemony. Most important, none courted the same exposure. Perhaps wisely, the others shrank from publicity, covered up in the camera's presence, and when caught in the open, beetled. Capone grinned and posed and amused his public. Publicity may finally have undone Capone, but before that it rendered him glorious.

"He was no hero to me," says photographer Tony Berardi. "He hurt the Italian people." Many Italian-American groups agree. They join with Chicago officialdom's queasiness about its past to oppose any move to commemorate the man or the era. As Katherine Gerould pointed out long ago, it's a bum rap. "It is not," she wrote, "because Capone is different that he takes the imagination: it is because he is so gorgeously and typically American."

Many at the time complained "peevishly of the menace of our foreign population," and charged "irritably that our criminals are overwhelmingly of non-Nordic stock." (The Dillingers, Barkers, Nelsons and Van Meters had yet to grab headlines.) "But," Gerould continued, ". . . if such citizens try to use Al Capone for a shining example, they are in error. For the comedy and the tragedy alike of Al Capone . . . are that both his purpose and his practice are one hundred percent American."

Perhaps the family had planned it all along, as the haphazard treatment of Capone's Mount Olivet inscription suggests. Perhaps they finally grew tired of gawkers trampling the family graves. In March of 1950 they had the remains of all three exhumed and moved to Mount Carmel Cemetery, on the far West Side, west and just north of Cicero.

A white granite family monument rises high, a monolith carved so that two columns flank an elongated cross surrounded by an elaborate relief design. CAPONE is engraved near the base. A planting hedges the monument all around at waist height, deliberately obscuring the name. The Mount Carmel staff does not reveal the plot's location to tourists. The family's individual graves gather about the stone, each with its own, simple one-by-three-foot, fifty-pound granite marker set flush with the ground, each inscription in an identical pattern. Capone's marker was twice removed by vandals, then replaced. On it is engraved, next to a cross

ALPHONSE CAPONE
1899——1947
MY JESUS MERCY

Explanation of Sources and Notes

Al Capone was born on January 17, 1899. Although the New York Department of Vital Records finds no birth certificate for him—many births, especially to immigrant families, went unrecorded in those days—the date is well documented. But ''documentation'' also exists for three other dates. That includes a baptismal record (January 18, 1899), the 1900 U.S. Census (May 1898) and World War I draft registration card (January 17, 1896).

How can we be sure of his birth date? Because it's the only one that makes sense. It's the date he always gave, and he gave it in situations when he could gain no conceivable advantage by lying.

Errors in the other ''documentation'' are easily explicable.

The date on the baptismal record was given weeks later by (probably) his godmother, who probably wasn't at the birth and may not have remembered. Or the person recording it may have erred; the baby's baptismal date was first entered as February 5, 1899, then corrected to February 7.

Census takers faced formidable problems of language barriers, accents, faulty memories, information from people who may not have known details (and certainly didn't care), and, above all, yawning indifference on both sides about precision in recording such details; all that a census taker really had to get straight was the number of people living at each address. The next census did not ask for birth dates, only current age. (Whoever responded thought Al was twelve in April of 1910, though he was eleven, and the census taker, Harry R. Johnson, evidently heard the family name as ''Capollia.'')

The coincidence of the January 17 birth date on the draft card is amazing. But why would Capone have told a *draft board* in mid-1917, when the card was issued, that he was born in 1896? If he had wanted to join the army, he could have enlisted; if he had wanted to avoid service, he did better to wait until his age group was called, late in 1918, rather than register in the first wave of call-ups.

In other words it *makes no sense* for that card to have been Capone's. And, in fact, the card belonged to a butcher who lived in Atlantic City.

That manner of reasoning serves as a test for much of the information about Capone and the people around him: if it doesn't make sense, chances are it isn't true. If it does make sense, and matches other established facts or patterns of behavior and style, chances are it is true—especially if information comes from several generally reliable sources.

The notes to events recounted in this book may quarrel with details given by sources on which the text relies at least in part for an account of those same events. A note for one event may flatly reject a source on which the text will rely for an account of another event. Because a source is wrong about one event does not make it unreliable as a source for others. It depends on what makes sense and how *generally* reliable the sources are. It is a matter of diligently comparing all possible sources, mulling them over, then constructing the most plausible account.

Elmer Davis once wrote about a biography of Alexander the Great that it ruined some fine stories by disproving their plausibility, but that one could rely on what was left. It's hard to think of a better standard for anyone writing about the past.

In the notes, as in the text, I use the present tense for information from those I interviewed, past tense for written sources. Interviews are marked by the person's name, followed by "int."—for example, Balsamo int.; LeVell int. A description of each person interviewed can be found in the text, Notes or Acknowledgments.

For written sources, I cite only those on which the text relies, and list them in order of the degree to which they contributed to the text. The only exception to that rule is when important sources disagreed; I cite them for a discussion of why I accept some accounts rather than others, and to present any versions that responsibly differ from the text. Sources listed at the end of a citation, after the words "and" or "also" contributed only slightly to the text. When several sources—especially newspaper accounts—simply repeated the main source, I do not bother to cite them, citing only the most detailed or, if they are the same, the earliest.

Books are cited by the author's name, as listed in the Bibliography, followed by the relevant page numbers. Where multiple books are listed for the same author, the second will have a short title in the citation: "W&K 82–4" refers to pages 82 through 84 of Wendt and Kogan's *Big Bill of Chicago*, "W&K *Lords* 172" refers to page 172 of Wendt and Kogan's *Lords of the Levee*. The citation listing for three books—*Report of the Senate Vice Committee, X Marks the Spot*, and *The Wickersham Report*—do not refer to authors, but they can be found in the Bibliography at those alphabet positions.

Except for the *Literary Digest*, abbreviated as LD because it is so often cited, names of all cited magazines are spelled out. LD articles gave the editors' views of current stories and topics with a sampling of quoted press comment from around the country. No page numbers are given for magazines because the relevant articles can easily be distinguished by their index titles in the magazines.

Abbreviations for the newspapers cited are listed below. Except as noted, the publication date is given, but no page numbers, which often varied, anyway, among different editions. The citation "T 1/7/24" means that the source was the *Chicago Tribune* of January 7, 1924; "MDN 4/2,3,5–8,11/29, 5/18/30" means that the sources are the Miami *Daily News* of April 2, 3, 5 through the 8, and 11 of 1929 and May 18, 1930. The convention among newspapers in the twenties and thirties was not to capitalize words like street and avenue in addresses, not even the famous Loop. I have changed that without notation. I also arbitrarily capitalize "Prohibition" when referring to the phenomenon of the Eighteenth Amendment and "Wet" and "Dry" when referring to people who held positions on Prohibition.

For brevity, the citations use various shorthand notations and abbreviations:

ABA = documents prepared by the American Bar Association in connection with a one-day mock retrial of Al Capone's tax trial; it was held in Chicago in the summer of 1990

BES = story rewritten by the *Chicago Tribune*, 2/14/88, from one that appeared in the

San Francisco *Call Bulletin* in 1937 recounting the Alcatraz experiences of Albert Besmanoff

BOP = documents in the Federal Bureau of Prisons' Capone file

Browne notes = extremely detailed marginal notes made by Howard Browne in several standard source books that he lent me, especially Burns and Wickersham; Browne lived in Chicago during the era, later became a scriptwriter, doing among many others the Gazzara and Robards Capone pictures; he is a fanatical researcher

Campbell Collection = at the Philadelphia Historical Society

Coroner's test. = testimony at hearings of the Coroner's Jury in the St. Valentine's Day massacre

FBI File = memos and other documents on file at the J. Edgar Hoover Building headquarters of the Federal Bureau of Investigation, Washington, D.C.

Fisher Collection = papers of Carl Fisher at the Historical Association of Southern Florida, Miami

Lightnin' = an eccentric publication by the Rev. Elmer L. Williams that exposed the ties among Chicago's criminals and politicians; he called it "The World's Humblest Newspaper," with the line "Published Every Little While"; in fact, the first issue was dated 5/20/25, but subsequent issues were undated

Nat. Arch., Chicago = National Archives—Great Lakes Region; in addition to those records more fully cited in the notes, this includes all records of cases cited from the 21st District Court

Nat. Arch., DC = National Archives, Washington, D.C.; among other exhibits, this includes the affidavits cited that were collected by the FBI for Capone's contempt trial

NFA = those few interviewees who preferred to remain "not for attribution" or, in even fewer cases, those who spoke on the record but wanted particular remarks to be NFA

NEWSPAPER ABBREVIATIONS

A = Chicago *American*, the Hearst afternoon paper in Capone's time

BUL = Philadelphia *Bulletin*

DN = *Chicago Daily News*, the *Tribune*'s quality competitor in Capone's time

DT = Chicago *Daily Times*, a picture tabloid with good crime coverage

H-A = Chicago *Herald American*, final Hearst paper after mergers

H&E = Chicago *Herald and Examiner*, the Hearst morning paper in Capone's time; often the most complete crime coverage

J = Chicago *Journal*

LED = Philadelphia *Public Ledger*

LD = *Literary Digest* (see above)

MDN = Miami *Daily News*

MH = Miami *Herald*

n.d. = those who clipped stories from the various Chicago papers for the files (popularly called the "morgue") that are now at the Chicago *Sun-Times* usually stamped or marked both the paper each clip came from and its date of publication; those few that remained undated are cited as n.d.

n.p. = those stories in the *Sun-Times* file not marked to show the paper they ran in and that have no distinguishing characteristics (e.g., dateline treatment); the citation n.p. 10/2/30 means that the clip was dated October 2, 1930, but cannot be identified as to paper; some sources are n.p., n.d.

NYT = *New York Times*

REC = Philadelphia *Record*

S-T = Chicago *Sun-Times*, successor to the *Sun* and the *Daily Times* and inheritor of the *Daily News* files

SUN = Chicago *Sun*, started just before World War II as an answer to Colonel McCormick's isolationism
T = *Chicago Tribune*
TAX = testimony at Capone's tax evasion trial, 1931; this explains citation of news stories so long after the events discussed

Notes

Page **CHAPTER 1**

17 Gabriel . . . age twenty-eight — U. S. 12th & 13th Census, 1900 and 1910, agreed with his tombstone at Mount Olivet and marker at Mount Carmel (see Chapters 31 and 32); both censuses gave Theresa's birth date as 1870, while her Mount Carmel marker gave the date as 1867, which would have made her twenty-six when the Capones landed; the only family member birth certificate still in the files (of Ermino, John) gives her age as thirty-three in 1901, which would make her birth year 1868, or possibly '67, the information supplied by Sophia Milo; the text follows the census, as being closer in time to the event, although the answers given to the census taker tended to be haphazard, and of course there might have been a language problem; for instance, the 1900 census rendered the family name "Capollia" and gave son Salvatore's birth in January 1896, though he was born, according to his Mount Olivet tombstone and school records, on July 16, 1895; the same census gave Al Capone's birth in May 1898, though he was born January 17, 1899; Census records from Nat. Arch., Chicago; pictures of Capone tombstone and markers from Mark LeVell

Page	Phrase	Source
17	Castellammare di Stabia	Allsop 295 called it ''Castel Amara''
	another [son] on the way	Kobler 17–18 stated that Ralph was born one month after the Capones landed; the 1900 census, however, gave his birth in February 1893 and birthplace as Italy; Kobler gave the eldest son's age as six, but both censuses gave the length of Gabriel's and Theresa's marriage as, respectively nine and nineteen years, making it 1891, and the 1900 census gave his birth as May 1892
	landed . . . just in time	Census 1900 gave 1894 as date of immigration, 1910 gave Gabriel's as 1891, Theresa's as 1893 as though he came first right after their marriage then sent for her; but that cannot have been the case if 1900's was right about at least when Ralph was born
	97 percent . . . peasants	Amfitheatrof 147
	Emigrated to escape	Nelli 18
	He was a barber	Census; Brooklyn City Directories
	bled . . . teeth yanked	Fitzgerald 26
	family . . . had just $17	Amfitheatrof 160
18	ten to twelve days	Moquin 57 (he cited no dollar amount)
	padroni	Amfitheatrof 161
	Italian later recalled	Fitzgerald 15
	hod carrier	Ibid. frontispiece
	haircuts and shaves . . . nickel	Ibid. 26
	Four dollars . . . apartment	Ibid. 14
	coal at 35 cents	Ibid. 16
	''In winter''	Spirko int.
	He could read and write	His signature on his 1906 naturalization certificate is plainly that of a literate man; the 1900 census listed both Gabriel and Theresa as literate, 1910 gave both as able neither to read nor write—but also gave the same for Ralph whose occupation it then gave as ''Printer'' in a ''Newspaper office''; Theresa, if she couldn't write then, certainly had learned by May 1937, when she signed a deposition in a firm, literate hand
	expected . . . barber to read	Weld 138
	job in a grocery store	Kobler 17; Balsamo. int., from interviews with those who knew the family, says that Gabriel had a job, did not own the store
	barber shop . . . 69 Park	City Directories; Naturalization cert., 1900 & 1910 census; John's 1901 birth cert.
19	[baptism]	Balsamo obtained from St. Michael's a certificate of information from the original baptismal record: according to the current pastor, Fr. Delendick, the godmother usually supplied the information; Milo gave

Page	Reference	Source
		Theresa's name as Teresina—a diminutive, like "Terry"—and her maiden name as Raiola; Theresa herself signed the 1937 deposition that way, with an *h*; Gabriel used the Italian spelling of his name, *Gabriele*, on his naturalization certificate, but it was without the final *e* on little Al's baptismal record; the city directory listings gave it sometimes one way, sometimes the other
19	Capones . . . 95 Navy Street	Baptismal record
	[Sands Street]	Kobler 23–4; Willensky 48–9 for "cheap liquor . . . women"; Irish shops: Weld 117; Sands still dead-ends at the Navy Yard, but the Yard is now closed, and Sands is lined with housing developments
	Gabriel took his brood	1900 census; the addresses given in the various censuses are exact; the Capones plainly moved to 69 Park sometime between Al's birth and 1900
	[boarders, neighbors]	Ibid.
	Greeks, Carthaginians	Amfitheatrof 140–9
20	Michelangelo and Leonardo	Ibid. 6
	statue of Columbus	Ibid. 19
	"Once upon a time"	Balsamo 105; the text read not *figlio puttana* but *skoongeel*; Balsamo's collaborator made some interpolations; Balsamo says that *skoongeel*, presumably a corruption of *scungili*, a kind of shell fish, simply is not an Italian insult; the incident is correct, and Balsamo says that his great-uncle Don Battista Balsamo would most likely have used *figlio puttana*; all other citations from his book have been checked for accuracy with Balsamo, and adjustments made where necessary
	butcher's chart of meat	Amfitheatrof 66
	twice swept down . . . welding	Ibid. 69–71
	changed after 1848	Ibid. 72–6
	Gabriel foreswore	Naturalization certificate, obtained at Kings County Court House
	Theresa . . . two more sons	John's birth cert. gives 8/11/01; 1910 census made John and Albert nine and six respectively; Chicago Crime Commission file of police records gave John's date as 4/4/04; Albert's Mount Carmel marker gives 1906, which would have made him four at the 1910 census; John is not buried at Mount Carmel
21	Jimmy had run off	Kobler 26 gave his age as sixteen, which seems probable; news stories when Jimmy reappeared gave his age as about that, but no date
	half the barbers . . . Italians	Moquin 89

21	21 . . . 38 Garfield	1910 census showed them still at 21; they certainly ended up at 38, as Balsamo discovered from neighborhood people who remembered them
	John Jay . . . 141 York	Grodinski int., from schoolboard records; it still stood, though closed, in 1990
	William A. Butler . . . 355 Butler	Ibid. for name and address; Kobler 26 for his record—which has disappeared from the files, according to school secretary, Shirley Poch, who provides the only record left of brother Frank's time there; Kobler's "P.S. 113" is surely a typo; Butler was and is P.S. 133
	Matthew, born 1908	Crime Commission records and Mount Carmel marker agree
	nearby pool hall	Kobler 26 put it at 20 Garfield; Balsamo says that the building, still standing, could never have accommodated a pool hall, whereas there were many of them around the corner on Fourth Avenue; Sullivan *Surrenders* and *North American Review*, Sept. 1929, called him champ of Greenpoint, which Allsop 290 called "fatuous" without specifying why he thought it so
	pitched sandlot baseball	Balsamo 221
	ballroom dancer	*New Republic* 9/9/31, article by Daniel Fuchs, who knew Capone in the Brooklyn days
	five feet ten and a half inches	FBI File
	assortment of . . . jobs	Kobler 26; $23/wk ammunition: BOP form
	"were so given over"	Weld 138
22	One gang's . . . fun	*New Republic* 9/9/31
	possess a "clubhouse"	Kobler 28, quoting a 1927 study of gangs by sociologist Frederick M. Thrasher
	American Beauty cigarettes	Weld 118
	One Irish kid later	Fitzgerald 17
	A Jewish gang member	*New Republic* 9/9/31
	"The Jewish boys"	Fitzgerald 30
	Sicilians reigned there	Kobler 26
	South Brooklyn Rippers	Balsamo int.; also Forty Thieves Juniors, below
	brother Ralph now operated	Archives, City of New York
23	authentic details . . .*	Kutner, though still alive, donated his papers to the Chicago Historical Society. The Society's biographical note warns that Kutner tells different stories at different times; the following details come from those papers; in December 1968, Kutner projected a memoir of his claimed association with Capone; Neil Elliott offered to help, and ended up interviewing Kutner at length, then editing the sessions into a typescript of an interview with a "Mr. X," who

did not want to be further identified; the proposed publisher, Arlington House, sent a copy to Kutner, who immediately forbade any use of the material he had given Elliott; that naturally scuttled the project; years later, Elliott edited the work, changing "Mr. X" to Jack Woodford and eliminating references that would have made it clear that Woodford could not have been Mr. X; the book was published (*My Years with Capone*; Seattle: Woodford Memorial Editions, 1985) by a man who cherishes Woodford's memory and prints anything he can find by his idol; a copy of the original typescript is among Kutner's papers; the work contains improbabilities; one example: Kutner claimed that Jack McGurn invited him along to witness the St. Valentine's Day massacre—not participate, just witness, and for no reason, just because McGurn thought Kutner might want to see something interesting; in fact, Kutner claimed to be at nearly every important event; it got to be too much even for Elliott, who adored Kutner; at one point he remarked that Kutner was like the Little Man Who Wasn't There: he was *always* there; as noted, checkable assertions in the book are at best equivocal, at worst clearly untrue; two examples, one trivial, one key: Kutner claimed in passing that while at the University of Chicago he had been picked as an all Big 10 baseball player, but copies of *Cap and Gown*—yearbooks while he was a student—show that he was not on the university's team; more important, he told Elliott that Capone visited three Chicago psychiatrists, killed the first two when he found that he had told them too much, and wanted to kill the third until Kutner talked him out of it, but the two "dead" ones were never listed in the phone book or city directories of the time, presumably never existed; perhaps most convincing, NFA, who knew Kutner well, and had extensive opportunity to know who was in Capone's circle, declares that Kutner over the years never bragged about knowing Capone, which this source finds incredible had Kutner actually known him.

Kobler 24–5 recounted two stories about this period that seem unreliable; one cited a woman who years later remembered that,

		as a sixteen-year-old teacher, she had Capone in her first-grade class; trouble was she remembered him as a classmate of Lucky Luciano, who was fourteen months older, grew up in Manhattan's Lower East Side so could scarcely have been Capone's classmate (Feder 34–6 [Popular Library edition, 1956]; Gosch 4–5; they differed on which school he attended, but agreed it was in Manhattan); a former U.S. Marine corporal wrote to the Brooklyn *Eagle* nine months after Capone's death (it is the only exhibit in the Brooklyn Historical Society's files under ''Capone'') telling of an incident when Capone was about ten, and challenged a Marine recruit to a fight; again, the memory of a little boy whom the corporal only ''later'' found out was Capone, and the sureness of the identification seem unreliable; also, by then Capone lived over a mile away; what was he doing near the Navy yard ''after school,'' as the story stated?
23	''something . . . nonentity''	*New Republic* 9/9/31
	[Torrio birth, childhood]	McPhaul 38–40
	he promoted boxing	Ibid. 41–3; Irey 155
24	average . . . five feet three inches	Asbury *Gangs* 272
	He knew how to turn	Observation derives from Rebecca West's *The Meaning of Treason*, writing about William Joyce, ''Lord Haw-Haw''
	Monk Eastman	Asbury 276; gang names supplemented by FBI report, Nat. Arch., DC
	Coney Island of vice	Asbury 5–6
	Ninety-nine places of amusement	Ibid. 26
25	1,500 strong	Ibid. 252
	below average height	McPhaul 46
	Warfare continued	Ibid. 49–50
	''glorious struggle''	Amfitheatrof 74
	French armies rushed in	Ibid. 80
	''Iscariot of liberty''	Ibid. 85
	Even the Know-Nothings	Ibid. 103
26	The 1850 census	Ibid. 89
	Not until 1870	Moquin 37
	''Higher walks of American''	Ibid. 39
	Lorenzo Da Ponte	Amfitheatrof 89
	Corsair raids	Ibid. 140–9
	laggiu, ''down there''	Ibid. 160
	no one tasted meat	Ibid. 155–6
	''prehistoric squalor''	Barzini 150
	post–Civil War	Amfitheatrof 158–9
	peak of 285,731	Moquin 38
	Martinis . . . Little Big Horn	Amfitheatrof 191
	''disadvantaged and humble''	Ibid.137

26	[illiteracy rates]	Moquin 106
27	''struggle of right''	Amfitheatrof 81
	New Orleans mob	Moquin 167–70
	lynched Italians in 1893	Ibid. 254
	rates of pauperism	Amfitheatrof 165
	operated garbage scows	Ibid. 168
	''what most people . . . missed''	Hanhardt int.; although his father was a German immigrant, all other relatives were Italian and he considers himself Italian
	''the Italian immigrant who''	Amfitheatrof 163
	average criminality	Stella 73
	1863 Italian . . . report	Amfitheatrof 154
	patronized Gabriel's shop	Balsamo int.
	John Torrio Association	Kobler 26
28	[Torrio to Chicago]	McPhaul 53–4
	ushered Al . . . Forty Thieves Juniors	Balsamo int.
	name of Ioele	Ibid.; he has a picture of the tombstone, which gives birth date
	[description of Yale]	Pilat 274–5; picture in Kobler, following 260 and Balsamo, following 62
	At age seventeen	Pilat 273
	gun-toting charge	Exhibit, New York City Police Museum
	married Maria Delapia	Balsamo int.
	despite Maria's disapproval	Pilat 274
	[Harvard Inn]	Ibid.

Page **CHAPTER 2**

30	Yale . . . enterprises	Pasley 240; Kobler 33; Balsamo passim discusses them; Balsamo 96–7: Sunrise Café
31	An FBI study	Nat. Arch., DC; in *Behind the Green Lights* (New York: Knopf, 1931) retired New York police captain Cornelius W. Willemse wrote about ''dollar beer 'rackets' . . . A 'racket' then meaning a gathering, not a type of criminal activity.'' (pp. 139–40); this was around 1910
	cigars . . . more shelf space	Kobler 33
	Yale . . . appropriated . . . manager	Balsamo 97
	five mile . . . sixty-odd piers	Balsamo int.
	''I'm with Frankie Yale''	Ibid.
32	[Coney Island]	Pilat, McCullough, Snow, all passim; Willensky 173–87
	''Never again''	McCullough 233
	half their drunk arrests	Pilat 113
	Prostitutes—''blisters''	Ibid. 98, 113
	''Al's room''	Balsamo int.; they showed it to him
	Yale beat his brother	Ibid.
33	[Galluccio episode]	Interview by Balsamo, printed in *Chicago* magazine, March 1990, supplemented by

		Balsamo int.; the brief version at Balsamo 89 gave only the nickname ''Galluch''
34	[description of scars]	FBI File
	Lost Battalion	Allsop 297; Pasley 240; *X Marks* 2
	Three rousts by the law	Olean: BOP form; statement to Philadelphia police (see note, Chapter 21); murder details: n.p., n.d. from New Jersey paper located by Balsamo
	Capone met an Irish girl	Kobler 35–6
	Baptized ''Mary''	Balsamo int.; he saw the record book
35	Mae gave birth	Certificate of records from St. Mary Star of the Sea, obtained by Balsamo: it was Sonny's baptismal record
	[marriage]	Ibid.; ''fudged year'': Kobler 36 (he gave marriage date as 12/18, Sonny's birth the ''following year'')
	Arthur Finnegan	Balsamo int.
	William Lovett	height, DSC: Balsamo 31 amended by Balsamo int.; description: picture in Balsamo following 62; cat's tail: Berger 315
36	*—Al, you can come*	Balsamo int.

Page CHAPTER 3

37	Chickagou	Asbury 3
	Pierre Moreau	Ibid. 29
	du Sable . . . Kinsie	Ibid. 4; Asbury gave the name as Baptiste Point de Saible, without Jean; copy editor Boe alertly caught the error
	appointed a constable	Ibid. 40–1
	recorded no arrests	Ibid. 30–1
	received a town charter	Nelli 6
	botched . . . prosecution	Peterson, citing A. T. Andreas, *History of Chicago*, vol. 1, 421–2
38	first executed a criminal	Asbury 39–40
	gambling . . . prostitution	Ibid. 34–7; Wells . . . martyred: 5
	''never seen a town''	Ibid. 38
	eighty-man police force	Ibid. 42–8
	''The political complexion''	Peterson 20
	Cap Hyman wed	Asbury 77; Nelson title: Peterson 34
	''notorious among criminals''	Asbury 61
	Chicago displaced New York	Ibid. 28
	Civil War growth	Ibid. 61
39	Mayor John Wentworth	Ibid. 54–8
	Medill . . . McDonald	Ibid. 144–5
	Seventy-seven American cities	McPhaul 65
	John J. Coughlin	W&K *Lords* 12–33; ''Bathhouse, you look'': 156
	Michael Kenna	Ibid. 73–81; H. G. Wells: 336
40	John P. Hopkins . . . mayor	Ibid. 158–9; also ''It is surprising,'' below
	''Chicago is unique''	Kobler 64 quoting Charles E. Merriam, a University of Chicago professor and re-

		former; his son used the same line: Liebling 107–8, and when Liebling protested that there were others, the son said, "Yes, but they aren't nearly as big"
40	aldermen openly sold	Asbury 156–7; Peterson 76–9; Allsop 253; Bright 137
	"No YMCA ever growed"	McPhaul 65
	"unending and ceaseless"	Bright xxii, introduction by historian Harry E. Barnes
41	a reformer in trouble	Lyle 26
	"Nobody's on the legit"	T 3/22/30; Pasley 349–50; see note in Chapter 25
	"had to make compromises"	NFA
	"I can't change"	Sullivan *Surrenders* 205
	James Colosimo	Asbury 312–4; Kobler 39, 42–4, 51; McPhaul 69–71; Nelli 52
42	"the Levee"	Asbury 122, 243–4, 246–7, 259
43	clattering their ponies	W&K 18–9
	[Thompson background]	W&K 13–9; Bright 7–10 gave "Eastham"; W&K just "Medorah Gale"
	"perennial boy"	Bright: title of Chapter 1
	trek west	Bright 10–12; W&K 20; not quite 15 seems young to get parents' permission, but opponents surely would have jumped on the story later had it not been true
	substitute sports	W&K 28–33
44	[1900 campaign]	Bright 14–5
	Thompson . . . redistricting	W&K 42–3
	saloon in 1905	Nelli 52
45	lasted in parlor houses	Asbury 266
	Colosimo . . . Van Bever	Ibid. 263; McPhaul 84 made it "green livery"
	"white slave ring"	Ibid. 267
	Chicago sociologist's	Reckless 92–3
	"Now, what would you say"	*Report* 509
	woman newspaper reporter	Ibid. 117–8
46	"Poverty is"	Ibid. 23
	Most gentle Mr. Silvani:	Asbury 232–3; Nelli *Business* 80–2 gave some samples from around the country
	murdered about four hundred	Asbury 231
47	[Colosimo] did what he could	Burns 6–8; McPhaul 74–8 (also covering Torrio's action, below)
	"Little John" . . . "J.T."	McPhaul 24
	Now chief lieutenant	Ibid. 104–5
	In 1912	Asbury 298–303, 304; W&K *Lords* 315 made it the suggestion of the "Committee of 15," a civic group described well by Asbury 295–6; McPhaul 107; full name of Funkhouser from review of Kevin Brownlow's *Behind the Mask of Innocence* in *Los Angeles Times* Sunday Book Review, 12/23/90
48	Captain Michael J. Ryan	W&K *Lords* 302–3

48	Automobile registration	Reckless 134
	steel and mill towns	Pasley 15
	Burnham . . . Patton	Bennett 30–1; McPhaul 116–7
	[murder of Birns]	Report 503–9; Asbury 304–6; W&K *Lords* 316–8; McPhaul 111–4
49	Max Nootbaar	Asbury 307–8; W&K *Lords* 319–22
	"reform might stick"	Asbury 308; it was John Jordan, husband of Levee madame Georgie Spencer
	[Lundin background]	W&K 82–3; LD 3/3/23
50	"just a poor Swede"	W&K 103
	[1915 campaign]	Bright 39–65; W&K 84–115; "Thompson tries": Lewis 374
51	When Thompson took office	Asbury 309
	Colosimo's Cafe	Burns 2–3; Kobler 33–4 gave a particularly vivid description
	raked in . . . $50,000	Asbury 312
	Torrio tended . . . humdrum	Asbury 322; McPhaul 80; Reckless 90
	the Four Deuces	Asbury 319; McPhaul 121
52	"industrious . . . spoilsman"	LD 4/19/19
	"Thompson is"	Bright 2
	"unconscious interpreters"	Barzini 136–7
	$1,670,000 . . . securities	W&K 127
	"lies, all lies"	Ibid. 135
	champion blacks	Ibid. 168–9
	"The truth is"	*Survey* 11/20/20
	"Everybody to his taste"	LD 4/19/19
53	"a dozen times, coat collar"	Kobler 67 quoting *Ten Thousand Public Enemies* by Courtney Ryley Cooper
	[Yale] called Capone	Balsamo 42–43
	Second Hand Furniture	Pasley 19
	"any old thing"	Murray 16

Page CHAPTER 4

54	Colosimo . . . outgrown Coughlin	W&K 328–332
	Thompson's candidates swept	Peterson 111; NYT 9/16/20; "roof's off": W&K 189
	broken with . . . Lowden	Bright 102, 113; W&K 158–9
55	"a model of integrity"	NYT 9/20/20; Bright 165 called him "really a competent man"
	Small . . . favorite . . . Lundin	Stuart 111
	Robert E. Crowe	Burns 172; Kobler 171–4; a fine lawyer: Marovitz int.; Judge Marovitz's first job as a lawyer was on his staff
	Garrity . . . Fitzmorris	T 10/23/20; NYT 11/11/20; W&K 192
	"Clear out the crooks"	W&K 192
	Fitzmorris's raids	W&K 192; NYT 11/22,29/20
	Soon . . . declined	W&K 199
56	murdered over 300	NYT 11/26/19
	"is not due to poverty"	NYT 1/8/20
	"No tendency is quite"	*Ladies Home Journal* May 1919
	Anti-Saloon League	Sinclair 85–6
	They "looked forward"	Ibid. 4

56	League terrorized Congress	Ibid. 107–10
	''What Have They Got on You''	Sheet music on exhibit at the New York Public Library, Spring, 1990
	sanctified the dry cause	Sinclair 20; Cincinnati banned the pretzel for its hated German name; also 116–124
	''wartime'' Prohibition	NYT 7/1/19; Sinclair 158–9, 164; the war was over, but the patriotism lingered on
	did not include owning	LD 12/27/19; Sinclair 159
57	Sophie Tucker suggested	New York Public Library display; also Irving Berlin song title
	nickel . . . dime	Slayton 53
	cost twice . . . ten times	Sinclair 231
	Volstead Act	T 10/4/20
	prostitute to Bridgeport	H&E 1/25/25 [*sic*]
	Americans . . . would inform	Sinclair 180
	Prohibition agents . . . $200,000	Ibid. 183–6
58	judges fined few	NYT 8/26/19
	auction off his trousers?	NYT 2/3/20
	voted 391,260 to 144,032	W&K 191; NYT 4/2/19, 4/4/19
	''All the Boys Love Mary''	New York Public Library exhibit; also ''smart little feller''
	A reform group noted	*Survey* 1/26/20; group was the Juvenile Protective Association of Chicago
	''Prohibition enforcement''	NYT 9/25/21
59	like George Babbitt	Sinclair 238
	''When . . . Prohibition start?''	LD 1/22/21
	[theft of liquor]	Allsop 31; Lyle 39 (Browne note for Herschie)
	selling only . . . neighborhoods	McPhaul 158
	Druggan and . . . Lake	DN 7/2/43, 8/14/47 (retrospective pieces)
	Stenson Brewing . . . in court	NYT 7/26/19; Nat. Arch., Chicago
	Three of the four Stenson	Nat. Arch., Chicago
	five Stenson breweries	Browne note; they were the National, Ruehe, Monarch, Atlas & Independent
60	Dale Winter	Background and relationship with Colosimo: Burns 9–12; McPhaul 131–136; Sullivan 82–86; T 5/12–19/20; About age twenty-five: McPhaul 131; she said in T 5/12/20 they had met three years before
	''He among you . . . without sin''	Burns 11; McPhaul 134 identified him as the Rev. John T. Birmingham; it was misrendered, as usual, ''Let him among you who is without sin'' but the text gives the correct quote from John 8:7
	concluded some liaisons	McPhaul 130, quoting H&E 5/14/20
	[arranged auditions]	Burns 11
61	Woolfson, claimed	McPhaul 134; but Victoria may have attacked her, thinking otherwise
	Love visited Big Jim	Analysis based on all sources
	''Never take anything''	Asbury 277; W&K 32 gave the full quote
	Torrio . . . back Jack Guzik	Lyle 49; McPhaul 121 (citing H&E 1/25/25) gave the $25,000 amount

62	"It's your funeral"	W&K *Lords* 340 gave it as "It's you' funeral," evidently under the impression that Torrio had an Italian accent
	Torrio had married	McPhaul 123–6
	"gracious, intelligent"	Ibid. 125
	"Torrio beat"	Bennett 28
	"best and dearest"	Burns 23 gave the full quote from H&E 1/25/25, after Torrio was shot
	Colosimo divorced Victoria	Kobler 71
	without . . . being notified	McPhaul 146
	married Antonio Villano	McPhaul 149; W&K *Lords* 340–1; It was also spelled "Villaini" (T 5/14/20) and "Villani" (Kobler 71)
	Colosimo and Winter eloped	McPhaul 148–9
	Torrio . . . consulted . . . Capone	Inferential, because Capone was best judge of Yale's abilities
63	[Colosimo murder]	T 5/12–16/20 (especially details 5/12 & 5/14); McPhaul 148–153; Allsop 43; Peterson 108; W&K *Lords* 341; reports that he asked if anyone had asked for him seem unlikely: he was on time, and the "anyone" (O'Leary) was very well known to all his staff
64	"Swan"	Kobler 72
	"Jim is gone"	T 5/14/20
	"Me kill Jim?"	Allsop 299
	theories . . . New York	McPhaul 152
	Joseph Gabrela . . . glimpsed	Ibid.; T 5/12,14/20 gave details; Allsop 43 claimed Chicago police "had reason to believe" Torrio paid Yale $10,000
65	"The fellow I saw"	McPhaul 153
	The Church had . . . certainties	Landesco 191–4
	[Colosimo funeral]	Ibid.; also T 5/14–16/20; these—plus Lyle 47, Kobler 74 and McPhaul 154—disagree about the number of judges, congressmen, aldermen, pallbearers, etc., but are congruent about what happened; the text relies on contemporary accounts
	commented one newspaper	Landesco
	After a prayer . . . De Carlo	All from Landesco, except "two brass bands played": NYT 5/16/20
	$7,500 . . . casket	McPhaul; but NYT 5/16/20 made it bronze
	Capone had gone unshaven	Pasley 91
	Colosimo died unwed	T 5/12/20
66	Everyone expected $250,000	T 5/17,18/20; McPhaul 154
	He gave Dale . . . Victoria	Kobler 76; McPhaul 154
	"wasn't a piker's hair"	Burns 19
	"less and less sinister"	Landesco 192 quoting A 5/15/20

Page **CHAPTER 5**

67	"wettest city in the U.S."	NYT 5/3/22: official was the field superintendent of Prohibition enforcement

67	Luigi Barzini maintained	Barzini 118–9
	"genius for organization"	*X Marks* 3, quoting U.S. Attorney Edwin A. Olson in 1926; quote continued "and a profound business sense."
	disciplined "businessmen"	Murray 19
68	Dinny Meehan's murder	Balsamo 24–34
	Torrio sponsored . . . Sonny	Kobler 36 gave Torrio as Sonny's godfather
	"value of a bland smile"	Pasley 19
	"We don't want any trouble"	Bennett 84
	Capone attend night school	Browne int.
	paid . . . $75 a week	The Mattingly Letter: see Chapter 23 et seq.
	about $25,000 a year	Kobler 102; also subsequent details of father's death, with exact date from picture of tombstone supplied by Mark LeVell
	success . . . made up for . . . flee	McPhaul 167
	Rosalia . . . Mafalda	Balsamo int. gives Rosalia as probable name; age is extrapolated from those of Matthew and Mafalda, both of which are exact, given on grave markers, pictures supplied by LeVell; almost nothing is known about Rose: Kobler 321 stated that she had married Frank Maritote, a gang associate of Capone's and older brother of John Maritote, who married Mafalda; but it seems unlikely, since no contemporary account mentioned it at the time of Mafalda's marriage, when newspapers were explaining that it was a marriage arranged to cement the two families, unnecessary if Rose had already married Frank; the *only* mention of her over the years seems to be by a guard who told a reporter that a sister "not Mafalda, the other one" came to say good-bye to Capone with the family when he went to Atlanta (H&E 5/5/31)
	Older brother Ralph	Kobler 103–4, including details about the rest of family and their move
69	"valley of factories"	*Atlantic* magazine, July, 1919
	Gold Coast . . . suburbs	Landesco introduction passim; Zorbaugh passim.
	thirty-four families that shared the block	Pasley 89–90
	1920 census . . . 11.81 percent	Landesco introduction
70	"crime as usual"	E.g., NYT 1/13/20, 12/25/20, 8/13/21; *Current History* magazine, February 1922 (352 murders in Chicago, up 121 percent in twelve years; T 3/16/22 had headline, "Crowe Says Vigilantes May Be Necessary"
	[Druggan and Lake]	DN 7/22/21, 6/20/22, 8/30,31/22, 7/2/43 (retrospective piece); Chicago *American* 8/31/22; T 6/19/22; Wisniewski murder: DN 7/22/21; T 6/19/22; Allsop 41; *Lightnin'*

		Vol. 1, #3; Lyle 70 "credited" Dion O'Banion and Hymie Weiss
70	[Miller and Morton]	T 8/24,25/20; Burns 87
71	Pekin Inn . . . police station	Lyle 137
	O'Banion . . . jackroller	Murray 27
	safecracker "in a small way"	DN 11/10/24
	total of only $9,919	T 1/22/24
	era's first hijacker	Murray 37–9; Sullivan 10
	Price Flavoring Extract	Case 23893, records: Criminal Court of Cook County
	Susquemac Distilling	Case 28982, ibid.
72	pay about $30,000	Sullivan 10
	Aldermen Coughlin and Kenna	McPhaul 154–5
	pool halls and cigar stores	Irey 156
	"With no conscious effort"	McPhaul 166–7
	the "Look"	Balsamo int.
	Torrio . . . move . . . suburbs	Asbury 322–3; Mezzrow 22; Sullivan 76 cited net on three 50-inmate houses as $15,000 a month each, which seems low, even for net
	Torrio liked the control	Bennett 30; Sullivan 76
	brothels . . . furtively	T 8/16/22, 8/19/22
73	"Come on boys"	T 8/16/22
	bolted down benches	Asbury 322–3
	beer, always 90 percent	McPhaul 141
	[Thompson-Lundin split]	Stuart 136–7, 141–2, 144–5, 158–62, 168; Bright 179–191
	even Hearst's *American*	LD 7/2/21
	"crushing defeat"	NYT 6/7/21, and Editorial: 6/8/21
	"whether Chicago . . . meant"	LD 7/2/21 quoting Cleveland *Plain Dealer*
	Small's . . . pardons	McPhaul 168 gave the *Tribune*'s estimate that 40 percent of Small's pardons joined Torrio-Capone; Pasley 36–7 wrote that the pardons amounted to 950 in two years, ten months
	governor was in trouble	Stuart 147–50; Landesco 87; Bennett 30; for intimidation, Torrio sent as leader of the muscle Walter Stevens, known as the "dean of Chicago gunmen" (T 10/23/23) and "Umbrella Mike" Boyle, a talented bag man, for bribery
74	[D'Andrea-Powers]	The entire story of this interesting episode in Chicago ward and ethnic politics of the era—too convoluted and tangential for full treatment here—can be found in T 1/24/21, 5/11/21; Nelli 92–112; Landesco 121–3, 200; Burns 210; the name of the Unione has been rendered often incorrectly as "Unione Siciliano" or "Siciliane"; *Unione* is feminine and the proper form of its adjective is *Siciliana*; it was formed in New York by Ignazio Saietta, whose nickname was "Lupo" (the Wolf), and who got a

		thirty-year sentence for counterfeiting (McPhaul 140) after a long history of crime (Nelli *Business* 74–5, 78–9)
74	[Labriola et al. murders]	Again too complex for full treatment here, these sequels can be found detailed in T 3/9/21, 5/11, 12/21, 6/27/21, 7/7/21; DN 6/29/43 (retrospective piece); Burns 204–13; Pasley 96–7; Landesco 124–5; Bennett 69
76	"His followers may avenge"	T 5/13/21
	Torrio proposed the right	McPhaul 158–161; Kobler 105–6

Page **CHAPTER 6**

77	an Amazon of beer	U.S. District Court, Case 23256, a multiple indictment in 1931 that specified many overt acts, some discussed below, including World Motor Service Company incorporation and Capone's truck-buying
	called "the outfit"	Meyer int.
	August 30, 1922, Capone	The City News Bureau version: Pasley 20; T version: p. 5 of 8/31/21—except as specifically noted, details are an amalgam of both accounts
78	One policeman never doubted	Trilling int.
79	IRS agent Edward P. Waters	TAX: T 10/14/31
	McCormick . . . spelling reform	Browne int.
80	Most of the North Side	Principal guide map Asbury 239, derived, evidently, from a map in the *Tribune*, although it shows the territories of a slightly later period; Allsop 45 provided a very good discussion of territories; Kobler 74–94 detailed the composition of most gangs, with sketches of principals; also McPhaul 160–1
	dispatched twenty-five . . . sixty-three	Chief was Morgan Collins
	"Oh, no . . . swell fellow"	Bennett 7
	"sunny brutality"	Pasley 45
	Wojciechowski . . . age three	Sometimes spelled W*a*jciechowski: e.g. Allsop 89; brightest subordinate: Lyle 91; age three: Browne int.
	Hymie Weiss	Murray 41 wrote "no one can remember why"—but that was well over fifty years later
	often used a rosary	Pasley 24
	sharp features	Allsop 89 called them "Sinatra-like" and wrote of "big, dark, ominous eyes"; Lyle 118 remembered "keen eyes and a hard, intent expression"
	who had steered . . . bootlegging	Asbury 352
	real name Di Ambrosio	Balsamo int.
	pilfering . . . coin boxes	Pasley 23
	"Schemer" . . . harebrained	Kobler 80
	Alterie . . . California	Murray 69

80	.38s . . . with a spare	Lyle 91 specified two were nickel-plated with maplewood handles, the spare a blue-steel, belly .38; also wrote Alterie liked to be called "Three-Gun," but evidently people stuck to "Two"
	"wild . . . and wooly"	T 8/17/25
	Druggan . . . jewel robbery	T 6/19/22
	formed and ran . . . Union	formed, Murray 69; ran, DN 3/12/24
	Miller . . . Morton	Murray 69
	O'Banion . . . idolized	DN 3/12/24
81	"Bugs"	Lyle 201 fn
	"Morrissey"	Criminal Court, Cook County, cases 24784 and 23893; both were in 1921 and in one Moran was called George Morrissey and in the other "George Morrissey otherwise . . . Moran"; Pasley 246 wrote he first went to prison for robbery at age seventeen (in 1910) under the name of George Miller: Lyle 91 made it for horse theft, but Pasley 246 (presumably the source, since they both cite the same 9/17/10 date) did not mention horses
	slow thinking	Allsop 137: "more muscle than brain"; Berardi, who knew him fairly well, marveled that such a clod could have such a bright, charming wife
	"Dapper Dan" . . . Cragin	Murray 54
	labor racketeer	Tried for murder in 1922: e.g., T 8/8/22
	occasional partner	E.g., T 5/17/22; O'Banion had hijacked a truck of whiskey, which he sold to Morton for $22,500—a coup that he shared with Dan McCarthy, another spur-of-the-moment deal that took twenty minutes: Murray 57–8; Criminal Court Case 28982
	interest in a florist shop	T 5/14/23
	tutored O'Banion	*X Marks* 14
	made dinner jackets	Kobler 83
	old Nineteenth Ward	redistricting had renumbered it to 25
	six Genna brothers	Pasley 49, 92; Burns 130–2;
	Samuel . . . Angelo	Burns 130 gave that as order of age, although Bennett 38 called Angelo youngest, and was right
	reporters were even unsure	Pasley gave the date as 1910; Burns, without a date, insisted that Sam was ten, Mike an infant, plainly impossible if 1910 is correct; Kobler 81 followed Burns and evidently extrapolated an immigration date of 1894 from the ages given in news stories (Mike Genna's death certificate gave his birth date as 1/18/95, his father's name "Tony," his mother "Maria Lucari"); Kobler added Burns's assertion that their father was a railroad section hand and that

		the parents died young and "in squalor"; however, Clem Lane, in his series (DN 6/29/43) claimed the brothers left behind in Marsala their parents, a seventh brother and two sisters; a story (T 11/2/26) occasioned by Jim's arrest in Italy had them showing up in the United States "fifteen or so years ago," i.e., 1910, and had *them*, not papa, working as section hands
82	"Klondike." Thick-bodied	No source even speculated on the origin of his nickname; description (also of Myles, below): DN 5/27/26; no source gave exact ages or background of either
	Ragen's Colts	Pasley 146; additional details: Allsop 126–7, Kobler 99–101; their main territory lay somewhat further east, according to Allsop, from Forty-third to Sixty-third Streets, but from Cottage Grove, east (almost by the lake) to only Halstead on the west; the Colts continued, Sheldon's gang operating as a business opportunity for members
83	Daley . . . a Colt	Murray int.
	superpatriotism . . . racism	DN 5/18/43 for "you hit me"; T 10/4/25 told how they planned a trip to Oklahoma to punish the KKK for anti-Catholic activity; Sullivan 1 for killed and injured
	consumptive . . . eighteen	SUN 6/6/44 . . . DN 11/11/23: age given as twenty-one in 1923
	majority . . . comparably young	LeVell int.
	Daniel Stanton	DN 12/12/28, 5/7,18/43 (retrospective)
	no other . . . source of beer	McPhaul 160
	Joseph Saltis	Pasley 24, esp. 135; Burns 58; also Allsop 122
	six-foot and two-hundred-pound	T 6/4/47
	he kept a saloon	Pasley put it at Fifty-first and Artesian, pre-Prohibition; DN 8/2/47 had him "into beer handling from a poolroom at 47th. St. and Damen Av."; a 1925 raid on the "Soltus brothers" was at 2123 West Fifty-first, which would be about right, Damen being 2000 west
	"Polack Joe's shirttail"	Pasley 146
	Born in 1894 in Hungary	T 6/4/47, 2/4/31—the former made it "Austria-Hungary," the latter specified Budapest; DN 8/2/47 had plain "Hungary"
	spelled his name . . . Soltis	Chicago *Times* 6/4/41; J 8/21/25 reported a raid on the "Soltus" brothers' saloon (2123 West Fifty-first); when his ex-wife claimed his body, she spelled it *Soltis*: DN 8/2/47
	Torrio had little trouble	McPhaul 160 had Saltis and McErlane getting beer from a Joliet brewery; but as the

		text below demonstrates, the timing is wrong for that to be true
84	"butter and egg man"	T 4/23/26; DN 6/28/43; Burns 52
	"most brutal gunman"	DN 11/17/24
	1922 . . . another jail sentence	A 1/22/26
	minor . . . gangs existed	Asbury map showed small enclaves, "De Courseys" and "McGeoghegan" just south and east of Sheldon, "Murray" just north of the Gennas; sources did not otherwise mention these gangs; on the other hand, two people—Santa Baldwin and NFA—speak of the Provenzanos, who ruled their Little Sicily neighborhood in the mid-twenties; Maddox and Guilfoyle had gangs spread along North Street, the former centered around Ashland, the latter west of Kedzie; however, it's unlikely that they existed when Torrio started his plan: for instance, Al Winge, an important partner of Guilfoyle's, still served as a police lieutenant at that time
	South Side O'Donnell's	Asbury 326–7; Burns 54–8; Spike's age: T 8/27/62, given when he died
	$12,000 Stockyards . . . job	Burns specified the figure, Asbury did not give one; McPhaul 170 made it $10,000, but that is the sort of detail about which he was careless; he also specified a ten-year sentence, while Burns called it "indeterminate"

Page CHAPTER 7

85	details . . . fishy land deal	W&K 201—the board passed up a chance to buy a school site for $60,000 until the land title could pass into "friendly" hands, when they bought it for $90,000; William A. Bither's sentence was overturned two years later, so was his confederate's nine-month sentence, but by then Crowe had lost interest; Bright 185 claimed it was the *Tribune*'s idea; for details, see NYT 10/1/22; T 3/23/23: Bright 185–96
	Thompson withdrew	NYT 1/27/22; W&K 206;
	"My friends have crucified"	W&K 208
	[the trial]	Details of the scandal and trial, W&K 207–13; Bright 190–6
	defense team . . . Darrow	Bright 191
	Brennan . . . needed a candidate	Stuart 182–4 claimed Dever was one of several suggestions from William L. O'Connell, a reformer Brennan approached first; Schmidt 62–5 insisted that no one knew for sure whence the suggestion first came

85	Dever, then sixty	Schmidt 42–4
86	let Brennan handle patronage	Perhaps the pejorative way to put it (as did Stuart 185 and Bright 202); Schmidt 64 had it that Brennan simply asked Dever not to use patronage to build a competing machine, to which Mrs. Dever replied, "I'm sure he would never do anything of that sort"; in any case the effect noted in the text was the same
	"a dripping Wet"	Schmidt xvi
	"I have never pretended"	Ibid. 88 (from T 10/8/23)
	Captain Morgan A. Collins	medical student: Schmidt 76; saloon owner: *Lightnin'* Vol. 1. #10
	sixteen bookies operating	Landesco 74
	he'd need at least three hundred	T 3/7/23
	"unabated enthusiasm"	T 5/7/23
	arresting five hundred	NYT 4/23/23
	450 in another	NYT 5/14/23
	four thousand blatant saloons	Schmidt 87
	"If I don't get it"	T 10/3/23
	two weeks later	T 10/16/23
	Collins raided [Deuces]	Landesco 40
	Pershing Hotel	Meyer 12 and Meyer int.; Graham says he's heard that too, though it appears in no other source
	$1,000 . . . $100,000 a month	W&K 237
	$5 a barrel	Sullivan 148
	two hundred bookie joints . . . Tennes	Landesco 78
	"drying up Chicago"	LD 12/15/23
	Spike O'Donnell	Pasley 30–1; Burns 54–8; additional details, Allsop 51–2; McPhaul 170–1
	"I can whip . . . Capone"	Pasley 36
87	"Life with me"	Burns 55
	all the real stuff	Allsop 52
	$45 a barrel . . . Torrio's $50	T 10/19/23 . . . DN 9/11/23
	Captain Thomas C. Wolfe	DN 9/11/23; T 10/9/23; he had made at least one such call in June 1923; at Wolfe's suspension hearing, the complaining witness, saloonkeeper Joseph Jungman recanted, saying that Wolfe had simply warned him that he "shouldn't buy beer from highwaymen"; he said all that other stuff he had told Chief Collins (in detail) was made up by the state's attorney's office (DN 1/25/24)
	cutting his price $10	DN 9/9/23
	bizarre accident . . . slapstick	T 5/14/23; Sullivan, who claimed that Louis Alterie had instituted horseback riding; "particularly nervous": *Tribune*; it was part of a string owned by Cornelius N. Shea, Con Shea being an elderly labor racketeer

88	Grief . . . Louis Alterie	DN 11/19/25
	Harry and Alma Guzik	DN 9/27/23; Bennett 30; Burns 25–6
89	Brice F. Armstrong	Criminal Case 11548, 21 U.S. District Court; T and H&E, both 4/8/24 (Armstrong's Puro Products testimony before a U.S. Senate committee); Allsop 48; McPhaul 174–6
	His mother and stepfather	McPhaul 176–9
	Spike O'Donnell could answer	T 9/8,9,19,23/23, 10/10/23; DN 9/9–11/23, 9/10/23, 9/11/23; Pasley 31–3 gave the address of Kveton's saloon as 2300 West Twenty-first—probably a typo because the news account (Trib. 9/9/23) was clear that Kveton's was on Fifty-first, Twenty-first being *way* out of the territory.
90	Ralph Sheldon led them	DN n.d.; he was identified to the police's satisfaction (but of course, no court's) a couple of months later
	"Stick up your hands"	T 10/23/23
	"give us a square deal"	T 9/8/23; DN 9/10/23 had a slightly different version, with Walter merely saying, "Please give me a chance," with a gun muzzle shoved up against him
91	Torrio . . . carrying a pistol	DN 10/9/23; T 10/10/23: Torrio was in court on a gun-toting charge
	combine hit again, harder	T 9/18/23, 9/19/23; the usually reliable Asbury 327 had a very garbled version, with the killing "in the wilds of Cook County," the bodies, trussed up, in a ditch
	Thomas Hoban drove	An assumption: McErlane and McFall were shooters, Hoban a spearcarrier
	They hit Meeghan	Grand Jury presentment in case 31927 of Criminal Court, Cook County
	"every officer of the law"	Pasley 33–4
	"It's plainly a . . . feud"	T 9/18/23
	Police questioned Capone	Pasley 35
	Cicero, a community	Pasley 39, Murray int.; Pasley gave population as seventy thousand, Asbury 331 made it "more than" fifty thousand; Asbury claimed only slots were permitted in Cicero, but the Ship was already in operation by January 1, 1924, according to the indictment in Criminal Case 14843, 21st U.S. District Court; it surely didn't spring into existence in the couple months after Torrio's move, and it belonged not to Torrio and Capone (until they muscled in) but to Vogel and his partners, who included the La Cava brothers, James V. Mondi, an old gambler from the Levee, and Fred Ries
92	Vogel raked off 60 percent	McPhaul 183
	In October of 1923	The entire "slots" story: Asbury 332–3; Lyle 74–76; Landesco 178–9

92	rocking-chair redhead	An assumption that it was the same one: McPhaul 183 described her as "matronly . . . [with] vivid red hair"
93	"an associate of Capone"	Landesco 240
	one-thirty A.M., on December 1, 1923	T 12/3/23, 12/4/23; DN 6/28/42 (Egan's verbatim account); Burns 50–1 has an expanded version of Egan's statement that reads as if Burns made much of it up; Palos Park Golf Club: *X Marks* 9
94	jury . . . indicted Danny McFall	Pasley 35–6
	bought his mother a villa	H&E 1/25/25; Lyle 87
	Secreted some $1,000,000	Irey 163

Page **CHAPTER 8**

95	evade news photographers	T 9/9/23, for instance, made a point of how Spike aided his brothers "in 'covering up' to foil photographers"; they left the courtroom with hats in front of faces
	"Why . . . want to be mean?"	Berardi did not hear the conversation but knows his editor, Harry Read, did say something like the text
	much later in Florida	Bennett 82
96	Forest View	Kobler 123–6 (story of takeover and all Nosek quotes); T 5/31/26 for foundation by vets, layout and burning of Stockade by locals; the police chief was often confused with Frank Dillon, who was called "Porky," a known criminal—which accounts for "may have been another convict" in the text; Chief Dillon always protested being called "Porky," but he certainly behaved criminally; the vets had no one to ask for help: as H&E 4/8/24 shows, even the relatively honest state attorney general Edward Brundage had a crooked side
	Hawthorne Hotel	Pasley 62–3 specified three stories, borne out by a contemporary news photo (T 4/29/26); the address came from Drury, who put the Hawthorne Restaurant, in the hotel—then called the Western—at 4823; Asbury 335 and Burns 39 also gave it as 4823; sources often called it the Hawthorne *Inn*, but its name was visibly "Hotel" in the news photo—confirmed by Nachtman int., who often went there
97	Hawthorne Smoke Shop	TAX: T 10/8/31: testimony by employees
	Capone . . . did not try to	Landesco 180
	the Ship	Bennett 32–3 (Also, below, Lauterbach's)
	Edward G. Kovalinka	Pasley 38–9
	April 1, 1924, election	T 4/2–4/24; NYT 4/2/24; Landesco 179–80; Asbury 333–4; Burns 40–1; others as noted

97	muscle from O'Banion	McPhaul 190
98	According to . . . witnesses	The *Tribune* account quoted the proprietor of a tailor shop who said all three were in there when the police stopped, and that they simply ran out to see what the fuss was about, which seems unlikely—why, then, did the police stop? St. John 179 claimed to have seen it all, and that Frank's gun was still in his pocket, his hand on the butt, when "we rolled over the corpse": but St. John is so wrong about other details of the day (including a claim [181] that he was there when Eddie Tancl was killed *the same day*, though Tancl was killed, as we'll see below, the following November, and under circumstances much different from St. John's description) that he's a reliable source only for details of incidents either corroborated by others (like his beating, below) or inherently likely
	detective bureau squad cars	St. John 178
	A myth grew	The brief (six-inch including one and a half inches of headline), page 21 NYT 4/2/24 account had it that "Tony Camponi [*sic*], alias 'Scarface,' a brother of the dead man, and owner of the 'Four Deuces' . . . escaped after emptying two guns at half a dozen detectives"; the story explained first that "Frank Camponi" was a "member of the Johnny Terrio [*sic*] beer runners' gang"—demonstrating that transmission gets garbled and that Torrio and Capone were still relatively obscure outside Chicago. However, *no* mention is made of Al Capone as being involved in the shootout in the Chicago papers; first book mention seems to have been by Burns, who got around the subsequent identification of Dave Hedlin by insisting on four, not three gunmen, and others followed him; but as the *Tribune* put it, Frank was "known to the police by sight," and if Frank, then a fortiori Al, since he was much better known, indeed already locally notorious—yet the police did not recognize the other gunman, made no mention of seeing Al Capone; his recognizability at that point is indisputable: the *Tribune*'s next-day account did manage again to mangle his name, this time calling him "Tony 'Scarface' Caponi," but that was only some editor's nodding, because it mentioned *him* to give Frank context as brother

Page	Cue	Source
		of the "owner [no longer alleged] of the notorious 'Four Deuces'," and the picture of him that accompanied the story (though on a different page) gave his name as Alphonse; even so, as late as the inquest (where Al appeared), the police could not identify the third gunman; Landesco, Burns and Asbury confirmed later identification of Hedlin
98	Frank's funeral	Landesco 179–180; Kobler 118
99	Klenha . . . got above himself	T 9/22/26; Pasley 66; Asbury 335; Murray 125–6
	town council seemed bent	Ibid.
	"This way they learn"	Ibid. and Murray int.: James Doheny, a *Tribune* byliner had become something of a confidant, almost a sounding board for Capone; Doheny told Murray about it much later
	a federal indictment	Criminal case 14843, 21st U.S. District Court
	stack "against any city"	T 11/21/24
	The fifth-largest city	Ibid.
	case of Robert St. John	St. John 171–4, 181–196; Kobler 151–7 derived his account from St. John's, with interpolations from an interview he had with St. John; Kobler's summary is more coherent than the original, but I mistrust some of the interpolations, like his assertion that a group of ministers led by Henry C. Hoover (who will appear below) paid one of Hymie Weiss's gang $1,000 to torch the Maple Inn: what St. John evidently told Kobler got the timing and some key details wrong; besides, contemporary accounts make it clear that local vigilantes were responsible; Kobler identified St. John's policeman friend as the chief, Theodore Svaboda, who was named in the indictment with Klenha and Kovalinka, above
101	120 . . . 165, saloons	McPhaul 197 made it "at least" 120; T 4/2/24 made it 143; Bennett 33 made it 161; Asbury 335 made it 165: perhaps it turned on who counted and exactly when
	$5,000 a week . . . Stockade	Kobler 116
	Capone's 1924 share	T 10/8/31
	a customized . . . McFarland	TAX: Guy C. Baxton; T 10/13/31
	Capone a fifty-fifty partner	McPhaul 198
	In newsrooms . . . they said	Dornfeld int.
102	"How much money"	Meyer int.; after thirty-two years in various prisons, he became "born again" and works with Chaplain Ray, a radio evangelist who ministers to prisoners; Meyer's pa-

perback was used as a premium to solicit donations on Chaplain Ray's program

102 Louis Cowan — St. John 187–8; except Cadillac and diamonds, DN 11/23/27

"underworld is like the upper" — Murray int.

Jack Guzik — Kobler 12 called him "Russian-born," Murray 122 made him one of five brothers and 141 pictured him at St. Hubert's grill, handing out envelopes to Chicago officials; NFA knew him in the thirties, despised him as uncommonly crude and offensive; he is often called "Jake" Guzik, though he hated that nickname and called himself "Jack"; he was also called "Greasy Thumb" Guzik by the press, although the nickname actually belonged to his brother Harry

Joseph L. Howard — Details of the murder T 5/9/24, 5/10/24; H&E 5/9/24; also Pasley 25–30 and Burns 28—though neither mentioned Guzik as the cause, attributing the killing to Howard's hijacking activity, which makes no sense: Capone, who "wanted no trouble," could have arranged a much quieter and discreet routine disciplinary killing as he did on other occasions; Luis Kutner guessed that Guzik had been making up to one of Howard's girls, and that seems as good a guess as any, though of course Kutner presented it to Elliot as fact; indeed, Kutner claimed *he* was with Capone at the time: witnesses mentioned that "a smaller man" had come in with Capone; the ensuing dialogue showed it obviously wasn't Guzik

"G'wan back to your whores" — Speculation, since witnesses specified no dialogue after Howard's greeting; all writers on the subject gave some variation of the text, e.g., MH 1/6/49 (retrospective) "Go back to your girls," but the combination magic words of "pimp," "whores" and "Dago" seems most likely to have so enraged Capone; as a switch, the *Tribune* stories called him "Alphonse Capone" while the *Herald and Examiner* called him "Tony 'Scarface' Caponi"

103 "I hear the police" — T 6/12/24; Pasley 28–9

most strategically important — Balsamo int. disagrees, claiming that distinction for the killing of Richard Lonergan (see Chapter 13) because that gave Capone credibility with gang leaders outside Chicago

Page	CHAPTER 9	
104	''There's thirty million dollars''	Murray 62; Landesco 94 observed that ''some gangster chiefs, like O'Banion, chafed under Torrio's generalship''
	[O'Banion description]	Pasley 45; Bennett 7–8; Burns 85 (and Browne notes in Burns); Pasley, who wrote first, and knew O'Banion, had his right leg shorter, Burns made it his left, with no inches specified; T 11/11/24 gave four inches but didn't say which leg
105	home life an idyll	Murray 41; T 11/11/24, 11/13/24 (quote from interview at wake with Viola: she also said they had only one car, a modest sedan, though T 11/11/24 made it a half dozen worth $40,000, including the Locomobile in which he drove to the flower shop his last morning); 6081 Ridge: DN 5/17/22
	''The swellest fellow''	DN 11/10/24 (also Devon shop)
	a marvelous business	T 11/13/24: his father claimed it made $200,000 to $300,000 a year
	''childishly irresponsible''	DN 11/19/24
	''It's too hot out here''	DN 7/28/24
	bedraggled girl	H&E 3/15/24
	''perched on a refuse box''	DN 6/24/22
	A police wiretap	Most sources repeated the story: Bennett 9–10; Pasley 46; Burns 87; Landesco 92; DN 11/19/24 is earliest and inspired their ''childishly irresponsible,'' remark; the actual quote had the man call him ''Dionie,'' which is clearly impossible
106	Charles Dion O'Banion	In Maroa: Decatur *Daily Review* 12/8/60; general & in Chicago: DN 11/12/24 (interview with Charles); Pasley 43–4; Burns 80–2 (and Browne notes); father farmer then plasterer: Murray 40; in Chicago: Burns 81–2; Lyle 89–90; Pasley 44; Murray 41–2; T 1/22/24
	to the Mayo Clinic	Burns 91–2
	four years as an altar boy	T 11/13/24; virtually all sources repeated the story, with some embroidery; only Murray 40 gave Fr. O'Brien's full name; Chicago Archdiocese Archives supplied information about his baptism, communion and policy about altar boys; the rector of Holy Name at the time was the Rev. Michael Fitzsimmons
107	Sibley . . . warehouse	NYT 5/30/24; DN 7/9,17/24, 11/13/24
	Give and Take	Review by Ashton Stevens H&E 1/21/24
	[Miller shooting]	Composite of DN 1/21/24 and H&E 1/22/24; also T 1/21,22/24; DN 1/24/24, plus specifics below; Burns 87–8 gave the ''did you hear what he called you'' version

108	Herschie scandalized . . . hoods	H&E had as the drop to its headline: Gangland Aghast as Miller, Believing Himself Dying, Violates Code to Name Assailant
	Millers . . . cheated . . . $60,000	Or maybe the other way around: H&E 3/14/24 (''story . . . that the Miller brothers had 'cheated' him out of $60,000 in a whisky deal'') vs. DN 11/10/24 (''The Millers, it was said, had accused O'Banion of 'shorting' them $60,000'')
	''If I had wanted to bump''	n.p., n.d.—but about time of Duffy killing, below
	''I'm sorry it happened''	Burns 88, though that may have been his version of T 3/1/24: ''That was a misfortune. It was a hot headed affair, and I had hoped it would blow over. But a fellow can't lose a bad break, I guess.''
	plausibility of *any* version	Sullivan wrote that O'Banion had once shown him a story about a man, Arthur Vadis, who had been mysteriously shot in the leg while crossing a Chicago River bridge: O'Banion said he'd been nervous lately, thinking someone was tailing him, had mistaken a car backfire for a shot, whipped out his gun and potted the only one in sight, poor Vadis; now he proposed to send him an apologetic box of cigars; actually, Sullivan may have invented the story: the diligent search by indefatigable local crime historian Mark LeVell failed to turn up any such newspaper report, and none exists in the *Sun-Times* clips file; Sullivan, writing in 1929 for an audience that *remembered* O'Banion, was sure everyone would believe it of him; Capone and Torrio undoubtedly would have
109	Lieutenant William O'Conner	DN 1/22/24; H&E 1/23/24; T 1/23/24
	John Duffy	T 3/1,2,8/24; Burns 88–91 was overly colorful; Lyle 92–3 had Duffy smothering Exley
	''a respectable businessman''	T 3/12/24
110	''alley cat breed''	Allsop 76
	procured a federal license	T 7/10/25; Pasley 92–3
	Henry Spignola	Pasley 48, 98
	Financed by Torrio	Murray 51: only he claimed it, but it seems likely
	Gennas installed . . . stills	Sources all agree (e.g., Pasley 93, Bennett 63), and all seem to have derived from T 7/10/25, quoted below (''The average family . . .''), even to the pipe smoking
	tap into gas and water	Gas: Bennett 69; NFA whose home harbored a still on the third floor mentions water also

110	''The Police Station''	T 11/3/25
	$15 a month . . . $125	J 10/9/26
	year's merit bonus	Police Department Bulletins, supplied by Charles Trilling; the one for 1/25/26 showed some getting a $240 merit bonus, but many only $120; top was one $300 award; Bulletin for 3/3/28 showed the same mix
	Captains drew $500	J 10/9/26
	rotgut . . . $3 . . . $6 to $9	Burns 93–4; the $3 price applies if they cut the 190-proof gallons themselves; Bennett 64–8 gave a fascinating account of how the moonshine was made, ''aged'' and ''flavored,'' quoting Coroner Herman Bundesen; Murray 52 claimed that cheapskates used horse manure for its straw content to prepare their mash
111	Mike Merlo	Hanhardt int.; *Lightnin'* Vol. 1, #6 (quoting an article, undated, in *Liberty* magazine); Pasley 172–3; Burns 98–9; Murray 64–5; Messick and Goldblatt 61 wrote that he was ''one of the few Mafia leaders . . . more concerned with the humanitarian aspects'' of the organization than with ''blood and terror''
	(one of his sons)	Hanhardt credits John Merlo with turning him toward the police force rather than one of the gangs
	''He absolutely rejected''	*Liberty* in *Lightnin'*
	O'Banion hijacked . . . booze	Asbury 47
	Sieben Brewery	Sources for O'Banion's offer (and, partly, subsequent events): Burns 94–5; Asbury 48; McPhaul 200–205 (he made the selling price $250,000, citing Allsop 73–4, but Allsop followed Burns with $500,000); for the main action: DN 5/19,27/24; T 5/20,21/24 and especially, cases 12475 (the federal indictment), 4006, and 12433 (affidavit for search warrant), 21st U.S. District Court; where there is conflict in accounts the text follows the federal records, which include many affidavits of the federal agents and arresting Chicago officers

Page **CHAPTER 10**

114	refused to return the money	Lyle 96 claimed O'Banion openly boasted, ''Guess I rubbed that pimp's nose in the mud, all right''; such pointless vainglory would have been uncharacteristic; certainly the boasting wasn't open enough to make any of the contemporary news reports

114	''Oh, to hell''	Bennett 9, in 1929, gave it as ''O, to hell,'' etc. [''O'' *sic*]; Pasley 50, a year later, dropped the ''O'' and specified ''a five-word sentence''
	they weren't Sicilian	Bennett tried to get around that obvious flaw with the phrase, ''Grouping the big shots under the general term, 'Sicilians' ''
	''You can't trust''	Meyer 13–4, confirmed by Meyer int.
115	''Dion was all right''	Sullivan 48–9; Bennett 9 gave a version of the dispute over how much bribe money to pay
	''Say,'' said O'Banion	T 7/8/24
	[trial]	DN 7/9–11,17/24
116	''Chicago amnesia''	Kobler 96
	''They warned me''	T 7/11/24
	vacation in Colorado	DN 11/12/24; T 11/13/24; Murray 62; he returned to help in the election; Pasley pointed out that O'Banion prided himself on delivering votes ''as per requirements'' (50), recounted the tag lines: ''Who'll carry the 42nd Ward?'' ''O'Banion in his pistol pockets'' (43), and quoted him as saying ''We're big business without high hats'' (46)
	the Ship	Peterson 125–6 credited *Illinois Policeman* and *Police Journal* for May–June, 1947: he gave names of those at the Ship meeting; McPhaul 198–199 gave a detailed account, specifying that Angelo Genna lost about $6,500 of his own money plus a $5,000 marker, citing Lyle 95—which he had ghostwritten ten years before: but in Lyle he had followed Peterson's 1952 assertion that Genna had lost $30,000: presumably McPhaul had later, more precise, information; it seems better to finesse the problem by not specifying any amount; McPhaul also gave O'Banion's share of the Ship as 15 percent, where others made it 50 percent: 15 percent sounds more reasonable; none of the earlier, more contemporary writers mentioned the incident (and it appeared in no newspaper pieces), but Pasley 49 (followed by Asbury 345) asserted that O'Banion's payment was a beer concession in Cicero, which so blossomed that Torrio wanted a share back, offering O'Banion in return a share of brothels, which the prim Irishman refused; that does not sound like Torrio: either that he'd cede so open-ended a concession or, once given, he'd try to renege

117	Capone again called . . . Yale	Balsamo 188–92
	[Merlo funeral]	DN 11/12,13/24; T 11/14
	Torrio . . . $10,000 . . . Capone . . . $8000	Torrio later told police his order was for $10,000 to show they were friends: DN 11/12/24; Asbury 350 gave Capone's $8,000 order, and Genna's $750 (which Burns 99 gave, without mention of Torrio or Capone); Bennett 11 cited the promise that three men would call as the touch that disarmed O'Banion's suspicion—without mentioning the people involved or the amount of anyone's order
	[O'Banion murder]	Details of murder, funeral and reaction (except as noted separately): DN 11/10–15,17/24; T 11/11–19/24; NYT 11/15/24; Bennett 10–12; Burns 99–103 (esp. the description of the shop): he mistakenly put Mike Genna in Yale's role, with Angelo Genna the driver, Scalise and Anselmi each getting $10,000 and a four-carat, $3,000 diamond ring, Anselmi supposedly sending his home to Italy, also claiming that it was specifically Scalise who fired the sixth shot, something impossible to know; Balsamo 188–193: Yale's summons to Chicago, order to O'Banion, alibi
	That night . . . Yale called	Balsamo int.: the order was in the store's order book
118	*New York Times* marveled	11/12/24
119	Alterie . . . statement	Legend grew that Alterie had made the challenge for "anywhere, anytime," or specifically at the corner of State and Madison, Chicago's busiest intersection, but that was embroidery on Alterie's statement, as given (in much shortened version) in the text
	$10,000 . . . coffin	Browne notes specified $7,500, with a $750 liner, and Browne was meticulous in running down such details
120	picked up Frankie Yale	T 11/19/24; Balsamo 192–3
	Eddie Tancl	T 11/24/24, 5/9,12/25 H&E 4/2/26; n.p. 5/11,12/25; *Lightnin'* vol. 1 #6; Burns 72–3: he gave the correct name of Tancl's place, Hawthorne Park Café (or possibly Inn), which most sources including newspaper accounts got wrong, as "Hawthorne Inn"; *Lightnin'* included a picture of the place with a large, clear sign outside; DN 11/25/24 also gave "Park"
121	coroner's jury	T 4/22/25

Page CHAPTER 11

Page		
122	Four Deuces . . . December	Browne notes
	"A. Brown, MD"	Pasley 68–9; details of the office operation are also in news stories of the April raid, cited below, Chapter 12
	Torrio . . . out of town	T 1/25/25 mentioned the disappearance and that the police "believe" unspecified O'Banion people were trailing the Torrios; "later writers": Bennett 21 claimed that a "police investigation" had shown Torrio on the trip and followed; Pasley 75 added the destinations other than the first three in the *Tribune* story; Lyle 110 followed Pasley, but McPhaul (who, remember, ghosted Lyle) never mentioned any such trip in his own biography of Torrio ten years later: I expect that something like the reasoning in the text persuaded him in the interval that the story—at least the chase—was apocryphal; "sometime by only hours": Kobler 137
123	cut Capone's car off	DN 1/12/25 had the story that day, making it "early today" without a specific time: reference to an "all night party" and the fact that it was too late to make the morning *Tribune* (which never did carry the story) suggests somewhere around dawn; Pasley 75 named Drucci, Weiss and Moran as the gunmen, without saying how he knew (the news story specified that no one had been able to see past the drawn side curtains), and he had Capone just stepping into the restaurant on that inspection tour
	steel-armored body	Pasley 78; the Cadillac people today explain that private customizers built bulletproof bodies on standard chassis, so Cadillac has no record of the cars
	Torrio pleaded guilty	T 1/18/25; the *Tribune* thought O'Donnell was Spike, but see McPhaul 210; A 1/27/25 identified him as East Chicago's former police chief
	[Torrio shooting]	The best, most detailed accounts are found in H&E 1/25/25, which gave Anna Torrio's version, and McPhaul 13–22 and 212–17; however, T 1/25/25 gave some details more correctly; two issues were in dispute: whether there were three or four in the attacking Cadillac, whether it was parked around the corner on Seventieth or across the street from Torrio's apartment, and if around the corner, whether it nosed up parallel to Torrio's car or Moran and Weiss

		ran to the attack; Anna thought she saw four in the car, and remembered that it had pulled up parallel, but all other witnesses agreed on three assailants, and if the account that the Cadillac sped off west along Seventieth rather than down Clyde to Seventy-first is true, it cannot have been parked opposite and cannot have pulled into Clyde before the attack: it makes much more sense for them to have been waiting around the corner, reducing the risk of being spotted by Torrio as he drove up: Clyde is a *very* narrow street; other sources that include the various versions (all 1925): DN 1/25; T 1/26,27; A 1/27; H&E 1/29; J 1/30; also Bennett 20–7; Burns 109–12; Pasley 8, 58, 75–8
125	"Did they get Johnny"	J 1/26/25
	"The gang did it"	T 1/25/25 also "I'll tell you more"
	garlic does not	Froede int. (phone); n.p. 1/28/25 had Dr. Byrne accounting for infection in Torrio's neck wound by speculating that garlic rubbed on the bullet might have caused it
126	"nothing to the theory"	H&E 1/26/25; also "What's the use of telling the police," and " 'No,' said Anna" below
	"Sure, I know who"	H&E 1/25/25
	Peter Veesaert	Later accounts said that he had been shown a picture of Moran as a pallbearer for O'Banion (an argument for gangland's covering-up policy) but the first story (H&E 1/26/25) specified a 1917 mug shot, much more probable for the police to have and show
	"I won't rap them"	n.p. 1/30/25; "rap . . . bum rap": Bennett 21
	"And while I'm there"	J 1/26/25
	"gaining" . . . "recovering"	H&E 1/29/25 . . . DN 1/30/25
127	looked like a dungeon	DN 9/17/25; Bennett 24; a federal agent charged that Torrio was roaming free like Druggan and Lake, but was probably mistaken (DN 9/17/25; J 10/5,6,9/25
	sometime in March	Asbury 54; Burns 112 specified no date (and got the idea that Torrio had been sentenced for ninety days, not nine months); he called it "a transaction that involved several million dollars"; Browne notes made it 25 percent for ten years; as Murray int. says, "When you own something you sell it, you don't give it away"; that seems persuasive

Page	CHAPTER 12	
128	gang killings	Allsop 41: 703 in fourteen years; Pasley 9 set the number at "more than 500" for those killed in the major wars from the taking of Cicero; year totals: Allsop 57 and Pasley 34 gave the same for 1924 and 1925, but differed over 1926: Pasley made it sixty-four, Allsop seventy-six
	Angelo . . . one gallon	Nat. Arch., Chicago: case 9064; James, arrested at the same time, 5/13/20, had sixty-two quarts: case 9516
	Mike Genna . . . skipped	Ibid. case 8992;
129	Paul Notti	T 6/22/22 gave his name as Paul Knotti; DN 6/22/22 as John Notti; Burns 113 as Paul Natti; "Paul" wins by majority vote and "Notti" as the most typical Italian spelling
	Genevieve Court	Nat. Arch., Chicago: case 10300; T 8/11/22
	powerful as . . . Esposito	Pasley 97; Burns 113
	Unione Siciliana presidency	Pasley 101–2
	Angelo Genna married	T 5/27/25; 1/12/26; Bennett 40; Pasley 98–100; Burns 114–15
130	Dever . . . raid	T 4/8/25 ("luxuriously furnished suite"), 4/11/25; 4/11/25; DN 4/11/25; n.p. 4/11/25
	Edward Birmingham . . . $5,000	Kobler 121–2; also Dever's role; date was given as "spring of 1924," surely a typo
	"it would be illegal"	T 4/11/25, quoting the paper, not the officials
	[Angelo Genna murder]	Burns 115–19; T 5/26/25; Pasley 102; today Hudson does not intersect with Ogden; Capone's role is purely speculation; but it does seem odd that the North Siders would have waited so long to revenge themselves on someone who at best figured scantily in O'Banion's killing, whereas Capone's patience had then been exhausted
131	laid Angelo away	Burns 119; T 6/1/25
	[ambush . . . police shoot-out]	T 6/14,15/25, 11/6/25 (also details from trial, below); Pasley 102–8 (he repeated "prominent" Italian story, quoting T 7/10/25); Bennett 35; Burns 119–20; among those who worked over Scalise and Anselmi was Lieutenant Albert Winge, known as Chicago's toughest cop (T 10/26/25), who later became a bootlegger
133	Anthony Genna	T 7/9, 10,12/25, $50 a week: 2/18/26; Pasley 108–10; Bennett 21; Burns 132–5
	Sam and Pete . . . skip	Pasley 111, Burns 131
	[Ammatuna]	Pasley 23 ("silk gloves"), 111; Burns 141–4; Bennett 46; Sullivan 101–2

134	Scalise and Anselmi . . . trial	T 11/1–8,11,12/25; "didn't for a moment suspect that they were battling policemen; they thought, etc.": T 7/9/25; *innocent* citizens: NYT 12/1/24, bank cashier William Perrin shot detectives who were following him because they found his actions suspicious, while he thought them robbers when they trailed him home; yellow Cadillacs: Dornfeld int. went on the plane ride; Bennett 35–43; Allsop 88; Burns 136–141; Sullivan 66
135	Trial . . . Officer Walsh	T 3/19/26; Bennett 42–3 (including Payne quote; Sullivan 65 gave same quote, news story just summarized it)
	"no middle ground"	T 3/16/26
	Ammatuna ended	T 11/11,12/25; also references to Ammatuna above

Page CHAPTER 13

137	padlock the Harlem Inn	DN 1/17/25
	[Hoover raid]	TAX: T 10/18/31
138	judge dismissed . . . charges	Kobler 159; also intimidation of the raiders, including David Morgan's being shot and "left for dead"; it's astounding that Morgan did not mention this at the trial, but Spiering 113 printed a government memo giving details
139	In April 1925 . . . murders	NYT 4/4/25
	by October . . . averaged	NYT 10/7/25
	"keep your nose clean"	Lyle 115; his spelling "ya unnerstand" is changed: Capone did not talk that way
	Henry C. Hassmiller	T 6/8/25
	[killing]	Ibid. and T 6/10/25
	On September 25 . . . McErlane	The various shootings: Allsop 56 ("10 attempts to kill Spike"); Helmer 82 (9/25/25 shooting of O'Donnell) DN 10/5,20/25 12/2/25, 2/5/26; A 1/26/26; T 10/4,17/25
	Brooks . . . Harmening	T 12/23, 24/25
140	Philip D'Andrea . . . drill	Kobler 146; no kin to Anthony D'Andrea, the would-be alderman
	New York, gunmen . . . practiced	NYT 10/21/23
	first crack at Spike	DN 10/20/25
	Another reporter laid it	H&E 9/26/25 quoted in Helmer 82
	John T. Thompson . . . retired	All submachine gun data: Helmer Chapter 1 passim, plus pp 25, 32, 36, 70, 74–81, except 1,000 rounds/minute and WW II use: Kobler 98; short bursts doctrine: author's experience
141	all started innocently	Balsamo 200 (including Dr. Lloyd); Kobler 167 for Capone quotes
142	late 1923 . . . Bill Lovett	Balsamo 170–3

142	a boyhood accident	Ibid. 70–1; Brooklyn *Eagle* 12/26/25 gave it as the right leg, and his age as ten or eleven at the time; Balsamo was sure it was the left leg and Lonergan's age was twelve
	declared war on Yale	Balsamo 181–8, 195–8
	ambushed Yale	New York Police Museum exhibit
	Eddie Lynch . . . meeting	Balsamo 202–4
	known as Jimmy Files	No one seems to know the origin of the nickname or know of anyone else who bears it
143	The Adonis occupied	Details about the Adonis: Balsamo 21–2; it was from the Adonis that Yale stole away the popular *maître* for his new café; action an amalgam of Brooklyn *Eagle* 12/25–29, 31/25; Balsamo 22 (description of club); Balsamo int., amending Balsamo 206–9
	"Stick-em up"	Balsamo int.
	May Wilson	The *Eagle* of 12/26 gave her first name as "Christine" but "May" the next day, presumably correcting its error
	"white men"	Kobler 168; but Wilson was not "chased away"; she was at a table when the shooting started
144	"She's My Baby"	*Eagle* 12/26
	"Alphonso Capone, 'bouncer' "	Ibid. 12/28
	"You can bet"	Kobler 169
	Hart claimed he	Ibid. (also charges dismissed)
	Robert L. McCullough	A 1/26/26: the story called him "a McErlane man," but he was one of Capone's more trusted gunmen—and for the reasons stated below would have been with Sheldon, not McErlane
	machine-gunned . . . Costello's	T 2/10/26, 3/30/26; Helmer 83
145	[Stege, Capone . . . guns]	Helmer 84
	De Laurentis . . . Tuccello	T 4/20,23/26; n.p. 4/20/26; Allsop 127

Page **CHAPTER 14**

146	Orazio Tropea	Burns 145; details of the Little Italy killings: Burns 145–157; T 1/11–14/26 (Spignola), T 2/16–18/26 (Tropea), 2/22/26 (Bascone), 3/8/26
147	a remark by Klondike	E.g., Burns 165; Kobler 176; at least the secret panel version came from a purported eyewitness
	a Cicero card dealer	T 5/2/26 (also Burns 164 & Pasley 131); in a typed copy of a grand jury report in the *Sun-Times* files, he was identified as "one Green" by an otherwise unidentified "Clerk," who had testified that Green had told him the story; as we'll see, there were some crippling though not necessarily fatal

Page	Reference	Note
		problems with it; looking for confirmation, the police asked colleagues in Kansas City to locate traveling salesman W. M. Owens, who supposedly had been eating in the restaurant that night (H&E 5/5/26)
147	Pony Inn . . . 5613 Roosevelt	The address was often given as 5615 (cf., Pasley 128, who also gave Madigan's statement to Captain Stege and H&E 5/5/26) the building, which stood alone on a vacant lot, had two street numbers: 5615 led to the upstairs apartment
	"A few months ago"	T 4/29/26
	Michael J. Windle	His name was regularly spelled a variety of ways: (e.g., H&E 4/30/26 gave it as Wendell; by 5/5/26 H&E made it "Wendle"); the text follows consensus and later spelling (e.g., J 6/2/26) and that in a government indictment, criminal case 14843 of the 21 U.S. District Court
	ancillaries of Klondike's	Wickersham 305; LeVell 41
	Fur Sammons	His nickname was never given with an explanation; perhaps he was especially hairy; he had been convicted of a 1903 murder (T 3/19/26); O'Donnell prized him but thought him strange (Pasley 24)
	"flaunting the collections"	H&E 1/23/27; "strutting" was also the reporter's characterization
	cutting . . . Capone's gambling	T 4/29/26
	three . . . gambling joints *netted*	H&E 4/30/26
	at least three million	H&E 5/3/26
148	only person Capone . . . feared	Pasley 124
	Heeney . . . Moore	Landesco 22; Heeney later became Capone's appointment secretary (Murray int.)
	Ryan . . . shot . . . Quinlan	DN 4/6/26
	William ["Rags"] McCue	H&E 1/21/27
	April 13 . . . primary	H&E 4/30/26, 5/3/16, 6/4/26, 8/31/26; J 5/8/26
	after . . . McSwiggin visited	Date from Capone statement (T 7/28/26) that it was "ten days" after the killing
	"If I told"	Burns 171
149	William Harold McSwiggin	Background and family details, H&E 4/28/26; boxing: J 12/3/25
	sixty-nine assistant state's attorneys	Bennett 74; Burns 159 made the count 70
	seven . . . first-degree . . . verdicts	J 12/3/25
	"Little Mac"	Burns 159; "Specks": DN 7/3/43; "Harold": T 4/29/26
	hold down jobs	T 4/8/26; Bennett 74
	hair . . . eyes . . . joke	DN 4/28/26; "jocund and witty" was byliner Ferdinand Lundberg's description
	good lawyer	Marovitz int.; he did not join Crowe's staff until after McSwiggin's death but heard from other staffers of his reputation
	"The average jury"	DN 4/28/26

149	courted newsmen	Ibid. and J 4/28/26
150	ninety-two slugs	T 4/29/26
	Sammons . . . Doherty	H&E 4/29/26, 5/5/26
	Klondike . . . three companions	T 4/30/26: made it the O'Donnells, Doherty and Duffy in the car; H&E 4/30/26 claimed only it was ''believed'' to be those four, but that only Klondike had been ''identified definitely''; most later sources put only five in the car at the time of the killing (e.g., Pasley 128–9) and contemporary news accounts (below) suggested only five; if so, it makes most sense that Klondike was at home: otherwise why would Myles and Hanley have taken the bodies there? Kobler 176–7 put Klondike in the car for six without saying why he thought so, also had them switch cars *after* they picked up McSwiggin, and made Klondike the Lincoln owner, but it was traced to Doherty by its license plate (H&E 4/29/26), and all contemporary accounts on those points agree with the text version; pollwatching: Burns 163
	Thomas Duffy	E.g., T 12/11/49
	Bill . . . on the phone	T 4/29/26; H&E 4/28/26, without specifying time, wrote that he ''left the table''; Burns 162 had him leaving in the middle of dinner and set the time at six P.M.; most contemporary sources put it near seven-thirty, the earliest being the *Tribune*'s ''shortly after 7''; that came directly from Anthony McSwiggin, whereas the first *Herald and Examiner* story specified no source
	Bill announced . . . play cards	T 4/29/26: ''see a friend'' is the most probable (although he may have said all three to various family members), again since it is the only one attributed to a specific source, Anthony; he named the friend, ''Eddie Moore''—the son of a police lieutenant—obviously *not* the one who worked for Capone
	Willie Heeney . . . spotted	Most likely him, from Anthony McSwiggin's later statements (cf. DN 7/3/43, retrospective); Browne int. agrees
	issued orders. A lead car	Kobler 178; Pasley 131 mentioned the five cars, quoting what he called ''a formal announcement''; however, no witness mentioned seeing more than the attack car, the party just having arrived; timing the attack to coincide with the party's leaving the inn would have been nearly impossible to contrive; the ''official announcement'' Pasley

Page	Phrase	Source
		131 quoted mentioned ''hours of patient trailing''
150	Mrs. Bach	Landesco 13; H&E 5/5/26: Captain Stege quoted her to a grand jury evidently without mentioning her full name; the ''telephone receiver'' was the old upright kind with a long neck
	pocked . . . splintered	T 4/28/26
	Duffy had been riddled	H&E 4/28/26
151	five bullets in him	H&E 5/5/26; also sixteen in Doherty: both coroner's report
	Hanley, the driver	Accounts that mentioned him agreed except J 5/5/26 which claimed that the police no longer thought him the driver
	Frank J. Misek	J 5/4/26; H&E 5/5/26; Misek said two got up from the pavement and pulled the other two into the car, but no other sources believed Myles and Hanley were exposed to the fire in the open
	died six hours later	H&E 4/28/26
	transfer . . . police sergeant	Pasley 130
	memo book	H&E 4/28–9/26; Landesco 13
	''The shock has almost''	T 4/28/26
	Crowe . . . blithely assured	Landesco 10–12; Bennett 74; Allsop 107
	''research into criminology''	H&E 4/29/26
	''it was learned''	T 4/30/26
	why was McSwiggin killed?	Landesco 10–12 for good summary
	voiced by Joseph Klenha	Landesco 10–11
	''Those shots were never''	J 4/28/26
152	''And sometimes quicker''	Bennett 70
	''This machine gun slaughter''	H&E 4/29/26
	nomination . . . seconded	T 5/2/26
	shoot him on sight	T 7/28/26
	cost Capone . . . million	Burns 176: plainly a guess, but likely accurate considering the size, even that early, of Capone's empire and uninhibited take
	protected from . . . raids	Pasley 132
	Shoemaker . . . Harlem Inn	T 5/2/26
	Collins made . . . tour	H&E 5/5/26
	three carloads of vigilantes	T 5/31/26
	rage from . . . Kelly	H&E 4/29/26; T 4/30/26; Allsop 108; Bennett 74; Landesco 12; the entire exchange, while a civic disgrace, was very funny
153	Attorney General Oscar Carlstrom	H&E 5/3/26; Allsop 108–9 pointed out that the *Tribune* thought the move had ''checkmated'' Crowe's critics, precluding appointment of another, truly independent jury
	''If I tell, I'll die''	T 5/1/26
	Korecek *pleaded* to stay	T 5/2/26
	O'Donnell[s] . . . surrender	Landesco 16: some questioned later whether they had been captured or had sur-

		rendered; T 5/27/26 reported at the time that it was the latter, "by arrangement," to a Lieutenant Ryan at six A.M. near Cicero
153	They added . . . details	T 5/29/26; H&E 5/31
	Carlstrom's . . . jury . . . report	H&E 6/5/26; Landesco 17–8 (full text)
154	second grand jury . . . McDonald	Landesco 19
	Capone . . . surrender	T 7/28/26; H&E 7/28/26 stressed that he would surrender only to the feds, fearing that he'd be held incommunicado by Chicago police or Crowe; his long quote is an amalgam of those two stories and J 7/28/26
155	Converse . . . Cowan . . . Shoemaker	T 7/29/26
	Next day . . . Lynch's court	T 7/29,30/26; H&E 7/30/26, including all quotes and Stege's raid; details of grand jury testimony were in a typed report in the *Sun-Times* files
	bullets . . . killed his boy	Associated Press copy prepared 12/1/30 for use at Capone's death in *Sun-Times* file included the story, below, about Capone handing Sergeant McSwiggin a gun; it was repeated by Pasley, who clearly didn't believe it
	"I thought my life work"	Burns 172–3
	"If you think I did it"	Pasley 131–2
	Roger Touhy claimed	A 3/8/50
156	"I believe I know"	J 9/9/26
	"I paid McSwiggin"	DN 5/21/28

Page **CHAPTER 15**

157	Jewish Chicken Killers	Sullivan 120
	Louis Alterie's challenge	Pasley 73–4
	gun at the Friar's	H&E 1/29/24; DN 1/30/24
	Hymie Weiss radiated	Asbury 352 called him "more prudent and far-seeing" than O'Banion
	"You take a picture . . . kill"	Bennett 54; also "shot me, six years ago"
	U.S. deputy marshall	H&E 6/25/26; the friend was Eddie Vogel, Cicero slot machine king when Torrio moved in; lawsuit: Bennett 54
158	"I'd trade places with you"	Bennett 53
	brothers . . . watering . . . horses	T 8/4/26; Lyle 116
	Anthony Curinglone	First name and spelling: Wickersham 307; Burns 185–6 gave the spelling as "Cuiringione" without "Anthony"; he specified the torture (from marks on the body) as being cigar and match burns; Lyle 116 asserted that the boys did not find the body, only told county policeman Nicholas Van Hanxleden about their horses refusing to drink and that *he* investigated; Lyle also wrote—his information coming presum-

		ably from Hanxleden, who much later became manager of Henrici's, where Lyle lunched regularly—that Ross's body had been doubled over a slab of concrete, hands and feet then wired to hold it in place; the news story specified the boys' discovery and the weighting with bricks and stones
158	"They call me heartless"	Burns 186; dialogue Burns gave is quoted sparingly and with trepidation, since he never scrupled to make it up; but this seems a very likely line for Capone
	[Standard Oil shooting]	T 8/11,13/26; Pasley 120–124; Allsop 114–5; the Sanitary District's corruption was of legendary proportions, e.g., Dobyns 16; Eller said, "Why drag me in? Just because some hoodlums want to shoot in front of our offices?"; Allsop 115 gave an account of a *second* attack a week later, with a sedan smashing Weiss's car, shooting, then Weiss and Drucci taking shelter in the Standard Oil building; no newspaper account seems to have appeared; perhaps Allsop construed the account in T 8/18/26, which obliquely mentioned the 8/10 attack, as a new one
159	[Hawthorne shooting]	T 9/21/26; H&E 9/21/26; Pasley 113–20; Allsop 116–8; Pasley gave Twenty-second Street's width as eighty feet, but Klenha proudly insisted on one hundred feet; Pasley also put eight cars in the caravan, but all other sources agreed on ten plus the lead car; Pasley had the father and son's grazings reversed, but the news stories were clear on the point
160	"shown the authorities"	J 10/13/26; Pasley put the figure for Anna Freeman's treatment at $10,000, but all others made it $5,000
	joints . . . slammed shut again	T 9/22/26
	Torrio had decamped	McPhaul 226, quoting Ed Reid *Mafia* (Signet, NY, 1954) p. 37
161	huddled . . . with Torrio	Supposition: DN 9/26/26 reported Capone in Miami talking to a "retired young beer baron" about peace
	"Right after Torrio was shot"	Burns 195; again, this may be his invention, but it rings true for Capone's style and certainly represents his demonstrated attitude; the rest of the speech Burns "quoted" is a pastiche of thoughts and phrases from different accounts of Capone's remarks to reporters after Weiss's death (cf. NYT 10/13/26, DN 10/13/26, Pasley 126–7, Allsop 122–3)

161	"ready for peace . . ."	DN 10/13/26
	[first peace meeting]	Every source differed about the details, hence the repeated "probably"; e.g.: NYT 10/12/26 claimed that "last week reports were current" of a peace meeting, putting it in the Sherman Hotel and making the death of "Al Brown" the deal-breaker; Asbury 362 also put the meeting in the Sherman, but merely suggested Capone's absence, writing that Capone "was told of [Weiss's] ultimatum"; Burns 188–9 also put it in the Sherman but specified Capone's absence: Lombardo had to telephone Weiss's ultimatum; DN 10/13/26 quoted Capone as saying he'd sent Lombardo to Schofield's for the meeting; Allsop 115 put it in the Morrison, following Landesco 102 fn, who wrote that a police official was there to referee; Kobler 192 put it at the Morrison and set the date at precisely October 4; T 10/12/26 claimed the price of peace was the two who had shot at Drucci outside the Standard Oil building
	Capt. John Stege sneered	Asbury 362
	Chief Collins doubted	T 10/13/26
	Torrio came to Chicago	DN 10/13/26
	"Oscar Lundin" took	T 10/12/26; Burns 190 gave the name as Langdon; but the news stories were clear and unanimous, except for T 10/14/27, which gave "Landon"; Harry Keeler: Bennett 55; crabbed, dingy room: Burns 193; flower shop jutted out: photographs in Bennett 53 and J 10/12/26
162	"Mrs. Thomas Schultz"	T 10/20/26; Burns 190 gave her name as Mrs. Theodore Schultz, her address as Mitchell, SD
	Rotariu remembered seeing	Burns 190; there may have been more: Pasley 125 put three chairs at the window of each room; most others (cf. Sullivan 54 and Burns) made it two; no contemporary news story specified the number of chairs
	shoes soiling the spread	Burns 193
	[Weiss murder]	T 10/12,13,20/26, 10/20/26; J 10/12/26; supplementary details as noted
	William W. O'Brien	Kobler 194 for "William"; all other accounts gave "W. W."
	Murray . . . Weiss's bodyguard	Newspapers and Bennett 56 just described him as a minor bootlegger (and brother of a holdup man) "allied with Weiss"; Burns 191 called him bodyguard—reasoning, presumably, that no one that minor would otherwise have been there

162	O'Brien . . . eluded disbarment	Kobler 194
	parked on Superior	J 10/12/26 had Weiss parking on State, but other sources agree on Superior
163	EVERY KNEE SHOULD	Bennett 52 for exactly what was left; T 10/12/26 only quoted the entire inscription and noted it had been defaced; it is at Chapter 2, verse 10
	"I don't want to encourage"	T 10/12/25
	"I'm sorry Weiss was killed"	T 10/13/26
	"He knows why"	Pasley 127
	"It's getting to be a joke"	DN 10/13/26
164	"Hymie Weiss is dead"	Sullivan 55; except for next line, the rest of the quote is an amalgam of two news reports of the press conference—which Sullivan may well have attended; the text amalgamates different accounts to avoid obvious overlaps, yet give a complete picture of what Capone said; source of each segment is noted, DN and NYT both 10/13/26
	phone kept ringing	DN
	spartanly furnished room	NYT
	"butchery"	NYT
	"But don't get me wrong"	DN; emphasis added on "dead"
	"I've got a boy"	NYT
	"And it's pretty terrible"	DN
	"I don't want to die"	NYT
	"Competition needn't be"	Sullivan 55; "But they don't see it," and what follows resumes NYT to ". . . stay out of the shooting stuff."
	"in the game to stay"	DN; and what follows
165	Chief Collins explained	T 10/13/26
	[Weiss funeral]	Allsop 121 except two hundred mourners and honorary pallbearers: NYT 10/16/26
	WEISS MURDER LIFTS VEIL	T 10/12/26
	[McErlane kills Fancher]	T 4/23,25/26, 9/11/26, 3/4/27; Kobler 97 for marksmanship challenge; NYT 7/24/26 for drunk in court; 3/4/27 gave the bludgeoned witness's first name as "Frank," but 9/11/26 had, otherwise, more precise details so seems more reliable
166	ride for Jules Portuguese	DN 7/14/26; NYT 7/15/26; supposition that it was Saltis: while Jules was not formally associated with Sheldon (Wickersham 304 listed him in the "20th Ward" group), his brother Alex was (DN 5/14/27)
	[Foley murder]	T 8/7–9/26, 10/12/26, 11/5/26
	rifled [McDonald's] files	T 10/11/26
	[Peace conference]	T 10/21/26; *Lightnin'* Vol. 1, #9 (quoting a version by John Stege); Pasley 138–46; Allsop 121–5; the *Tribune* snidely called the preliminary talks "*pourparlers*," the

		term borrowed by later writers (how amusing to apply the language of international diplomacy to hoodlums!), but these were as deadly serious as any held in Geneva; word circulated beforehand that a peace conference was to be held, but only because politicians demanded a truce until after the elections (NYT 10/17/26)
166	"give each other a break"	Allsop 124; rest of quote: Asbury 263
167	probably . . . at . . . Sherman	The majority opinion, though Asbury 62 and Landesco 102 put it at the Morrison; so did Murray 129 (though his version was seriously flawed); Murray int. explains that "the manager of the Sherman, Ernie Byfield, didn't like gangsters, didn't want them around his place"
	Christian P. Bertsche	Pasley 142 gave his full name, all other contemporaries calling him Barney Bertsche
	"making a shooting gallery"	Sullivan 57 quoted Capone as having said later that he told the conference what the text gives as a direct quote
	"Remember that night"	Bennett 59–60
168	Two days later a car	H&E 10/23/26
	Druggan and Lake	T 12/18/26
	"different verdict"	T 11/10/26
	Theodore Anton	Gambler: T 10/23/23; missing November 28: T 12/5/25
	"sobbing like a child"	Pasley 68
	Later . . . story circulated	DN 7/9/43, 1/21/47
	a piano player . . . charged	Jerome Nachtman, of Brookfield, Illinois, who played piano to work his way through dental school: *Suburban Life Citizen* 9/13/89 (courtesy of Mike Graham), confirmed and amplified by int. (phone)
	the police thought	LeVell int.; a friend on the force found the records "browsing where he shouldn't have been"
	Hillary Clements	T 12/31/26
169	Bill Thompson announced	NYT 12/12/26
	nomination on December 28	T 12/29/26

Page **CHAPTER 16**

170	"Chase me out of Cook"	H&E 1/22/27; he also said, "I'm getting tired of fellows like Hughes using me to attract glory to themselves"
	[Metropole set up]	Pasley 68–9, 165; T 10/13/31
	O'Donnell . . . Sammons	T 3/6/27; 3/13/27
	Koncil . . . Hubacek (p. 171)	"Hubacek" (Pasley 147; Allsop 127–8) was also given as "Hrubeck" (Bennett 60; Burns 61–3); first news stories gave "Hrubeck" (e.g., T 3/12/27; NYT 3/13/27 [after

		"Ruebeck" on 3/12]), later (T 3/13–14; NYT 3/14) corrected to "Hubacek"
171	farm, summer resort and estate	Burns and many newspaper references, e.g., T 2/4/31; DN 6/22/43
	"most notorious wowser"	*Chicago Tribune* (no date) in LD 11/5/27
	Big Bill	Stuart 206–8; W&K 216–8 quoted Thompson as saying he'd bring back movies of the fish, though he did talk of aquariums; Stuart, his supporter, had him promising the fish itself; Elmer Davis in *Harpers* July 1927 pointed out that both movies and mounted specimens already existed in Chicago
	"The Rat Show"	DN 4/6/26; T 4/4,7/26; H&E 4/7/26
	Dever did not want to run	Bright 250
	"In some mysterious way"	LD 11/5/27
172	"I shouldn't be surprised"	Bright 231
	"punch him in the snoot"	Ibid. 256
	"I try to get my opponent"	Stuart 297 (also "Thompson supporter claimed")
	Peoria *Star* put it	LD 4/16/27; people especially resented scenes of the police exultantly pouring out real whiskey seized from small-time operators when they knew the big shots' rotgut flowed freely: a Back of the Yards priest said, "They give the good stuff to the sewers and the bad stuff to the people" (Slayton 104); one pundit charged that Dever's policy "showered gold" on bootleggers by eliminating their two-bit competition (*Century* November 1927)
	"adequate schools"	Bright 252
	"Who the hell"	*Harpers* July 1927; even the decency seemed problematic at times: ceding the black vote, the Democrats ran a blatantly racist campaign (T 3/14/27; Stuart 302–6; W&K 253–7), and when Thompson protested, Dever said, "I don't know what ails the man" (NYT 3/8/27)
	"anything wetter than that"	W&K 248; he called his opponents dry, insisting that Robertson was so dry "he never even takes a bath"; Pasley 159 quoted him as saying "I'm wetter than the middle of the Atlantic ocean"
	"open ten thousand new ones"	Stuart 298; W&K 244; supporter Stuart claimed he meant factories and such; but everyone knew what he meant; Dever hadn't closed any factories
	"I will discharge"	Stuart 291
	"no copper will invade"	Bright 255
	[Capone's] men collected	W&K 249–50
	Serritella . . . Loesch . . . Zuta	Ibid. 268–9; Pasley 153 called $250,000 an "excessive" estimate, quoted the head of the Illinois Association for Criminal Jus-

		tice as saying only that Capone ''contributed substantially''
172	''fellow hoodlums''	Stuart 311
	''The Dever decent element''	W&K 268
173	Dever demanded police	NYT 2/19/27
	[Drucci shooting]	T 4/5/27; Pasley 159–162; Stuart 315–6; W&K 269–70
	''Capone men . . . enlisted''	Stuart 315
	''trying . . . to beat Bill''	*Harpers* July, 1927
174	A contemporary academic	Bright xxi; introduction by historian Harry Elmer Barnes
	''Tell 'em, cowboys''	Pasley 155; W&K 271–2
	Elmer Davis asked himself	*Harpers* July, 1927
	Fish Fans clubhouse	Stuart 235–43; W&K 272–3
	''observation convinced us''	H&E 8/6/27
	''Al Brown owns the place''	J 8/6/27; T 8/6/27
	a tiny tragedy occurred	NYT 4/7/27
175	On hand to greet him	Pasley 82 for Capone's role, which was not mentioned in the following news accounts; T 5/15–17, 22,25/27
	yacht . . . Zenith radio	Inference: Pasley specified only a radio magnate; Berardi int. saw Capone on the Zenith owner's yacht on another occasion
	''a fervent handshaker''	Pasley 11
	Christmas . . . $100,000	Asbury 364
	belt buckle . . . $275	TAX: Samuel J. Steinberg of B. Weinstein, jeweler; T 10/13/31
	a whitefish dinner . . . coffee	T 7/10/27
	''If he gives you his word''	Bennett 85
176	''hand it to Napoleon''	Burns 33; the quote as given is much longer; this much of it *sounds* like Capone, who read Emil Ludwig's famous biography of Napoleon when he was in Philadelphia in 1929
	''Sure,'' he'd say when asked	T 3/22/30
	''Well,'' Capone . . . declare	Bennett 83
	''Yes, it's bootleg while''	Ibid.
	''People respect nothing''	*Liberty* magazine 10/17/31, Vanderbilt interview; also: deplored birth control
	loathed homosexuals	Meyer int.
	execrated the flapper's	Browne int.
	''The trouble with women''	Allsop 309
	''Reform . . . end prostitution''	Murray 97
	''Snorky'' . . . ''stylish''	T 10/18/31; DT 3/14/39 (definition, quoting *The Dictionary of American Slang*); MDN 4/20/59; Meyer was the gang member
	Busloads would drive by	Pasley 62
177	a flivver . . . pilot fish	Pasley 78–9; Bennett 80
	''There goes Al''	Pasley 79
	$135 custom sack suits	Dress details, TAX: Oscar D. de Feo, Ira Gay, Earl A. Corbett, Peter M. Arl, J. Pankan and M. A. Oles of Marshall Field & Co.; T 10/13/31

177	$50,000, 11.5 carat	Pasley 78–9; Burns 32 (who made it only eleven carats) for Jagersfontein
	[bodyguards]	Pasley 10; Bennett 80; Burns 31; "four . . . ahead": *Forum* October, 1927; others made it always at least two ahead and two behind
	One patron . . . behind Capone	Lyle 146
	salaries that ranged	Meyer 9, 19; Bennett 80 wrote that bodyguards received "about $100 a week"; Fox 112 quoted Capone (in an interview with newsman Arthur Brisbane) as saying the minimum was $150 to 200; Meyer insists that he made $500; but he did have skill as a driver and later "got" people (his word) for Capone; "Capone owed me," he says
	Olson estimated . . . gross	Sullivan 33
	net at $3,000,000	H&E 5/3/26
	the Stockade alone	Bennett 84
	Smoke Shop, cleared	Peterson 131, quoting Treasury man Elmer Irey in November 1947 *Coronet*
	receipts . . . $100,000,000	United Press biographical material dated 12/28/42 in *Sun-Times* file
	five thousand cases of booze	T 5/5/32
	$30,000,000 . . . protection	Sullivan 33
	$218,057.04 for 1927	TAX: "Brief Statement of Indictment"
178	"Al," said one observer	Bennett 84
	[gambling]	Pasley 63–4; Bennett 84–5; Asbury 364–5;
	plunging on horses in 1924	T 11/3/24
	"with the bag over his head"	Bennett 85
	"I've lost a million"	H&E 4/19/27
	fooled away $7.5 million	Bennett 84; Sullivan 106 put the gambling loss according to Capone at $4 million, stipulating that Capone was "anything but boastful" and that the figure was likely much higher
	parties . . . Saturnalia	Sullivan 106; Bennett 84
	he and a dozen henchmen	Condon 123–4, 135; Shapiro & Hentoff, *Hear Me Talkin' to Ya* 130
	tipped . . . $100	Burns 32
	"very gay and noisy"	Mezzrow 63
	Banquets at the Metropole	T 10/12/31: testimony at tax trial
179	party with Fats Waller	Fox 85, quoting the Waller-Calabrese book, *Fats Waller* 62–3
	When Scalise and Anselmi	Pasley 63–4
	before Jack Dempsey fought	Allsop 291
	lose on Dempsey	Fox 91, quoting the autobiography *Dempsey*
	Anton liked to tell	Pasley 67
	easy and openhanded touch	References are legion and widespread, cf. MDN 3/9/30
	ex-policeman still remembers	Trilling int.
	Merchants in Cicero	Burns 32; Browne int.; Murray int.: a well-known benefaction; Murray int. for soup kitchens and clouted makings

Page	Text	Source
179	A. J. Liebling visited	Liebling 134
	"say what you want about Al"	Baldwin int.; her maiden name was Russo, and they lived in the North Side's Little Sicily; her son is Detective Sergeant William Baldwin of the Cicero Police Department
180	"never knew what gravy was"	Ibid.; Spirko int. and Staley int. made the same point
	"real gift for organization"	Bennett 81–2
	in shirtsleeves, desk buried	Murray int.; Berardi int.; nine phones: Bennett 85
	"hires nothing but gentlemen"	Brooklyn *Eagle* 7/10/29
	"I thought I told you"	Bennett 85–6; Pasley 81–2 gave the same story, probably following Bennett, in a slightly different version: Capone had ordered the judge to arrange for a henchman to be released on some charge; the judge had written a memo for his bailiff, who had forgotten to give it to the sitting judge
	"He always tipped his hat"	Baldwin int.
181	"I got nothing against"	Murray 126–7
	squad of young overreachers	Pasley 81
	"transferred to Hegeswich"	Murray int.; Spirko int.
	One who was shot	Meyer 15
	wife . . . gang . . . recalled	LeVell int.; he spoke with her
	At the Roamer Inn	Mezzrow 23–4
182	young caddy at . . . Burnham	Fox 87: Sullivan and Kobler in *Sports Illustrated* 11/6/72
	$1,500 . . . Chinese rugs	TAX: Paul H. Miner of Peck and Hills
	reporter . . . spaghetti	Bennett 81; Pasley 90
	sometimes, "Jiggs"	T 10/15/32; S-T 4/15/51
	The Richards School	Pasley 90
	Mafalda . . . later complain	T 3/17/30
	Though . . . 28 . . . Capone looked	Sullivan 106
	fined for disorderly conduct	Kobler 104
	Burnham's Arrowhead Inn	Mezzrow 64; his "Mitzi" is corrected in the text to "Mimi"

Page **CHAPTER 17**

Page	Text	Source
184	Joseph Aiello	Details about the family and their struggle with Capone, in addition to those specified below: Pasley 169–73; Burns 225–6, 232–5; Landesco 80
	mansion . . . built	LeVell int.; Kobler 209 for three stories
	"never left . . . Five Points"	Pasley 60; Lucky Luciano called Chicago "A real damned crazy place. Nobody's safe on the streets" (Feder 61 [Popular Library])
185	Torchio, was found	T 5/26/27, T 7/18/27; Pasley 164–5 wrote that he was expensively dressed, but the news story specified a "laborer's" blue suit

		and old army shirt, with no mention of the $1,200 wad Pasley claimed; also no mention of the nickel (below in the text); indeed the only such story I've ever seen was a squib in MDN 6/30/30 about an obscure Chicago hood, found murdered, supposedly clutching one
185	Jack McGurn	Balsamo 139–164, emended by Balsamo int.; LeVell int. points out that only two letters changed in the name from Gibaldi; Pasley 164–70 thought De Mora his real name, didn't know about Brooklyn; fight manager: Lyle 206; dancing "splits": Nachtman int.
186	William O'Conner	Presumably the former police lieutenant who nabbed O'Banion hijacking after the Miller shooting
	"known criminals" . . . insanity	NYT 8/17/27
	"something wrong with McGurn"	NYT 8/18/27
	Frank McErlane wriggled	NYT 8/26/27
	Dingbat Oberta showed up	NYT 8/28/27
	[Capone] did it to Aiello	DN 11/21–2/27, including "Listen, Billy . . ."
	[Aiello ambush plot]	DN 11/21/27; T 11/21/27; W&K *Lords* 344–5 told how Capone had declared that Coughlin and Kenna could keep their power only if they kept their noses clean. "My God," said the alderman, ashen after the interview, "what could I say?" Capone could have taken over their entire organization. "We're lucky to get as good a break as we did"; that settled, Capone treated both with friendship
187	He sent six cabs full	Pasley 170–1, 174–5
	"You're dead, friend"	Allsop 135; the text dialogue is an amalgam of Pasley and DN 11/22/27
	a nervous breakdown	DN 11/23/27
	Trenton, New Jersey	DN 1/5/28; also T and H&E, same date
	"I'm willing to talk"	Kobler 213
188	"I'm the boss"	DN 11/22/27
	Omaha *World-Herald* noted	LD 11/5/27
	symbol, around the *world*	Pasley 182
	Siege . . . Detective Bureau	Pasley 174; at 182 Pasley wrote that it "presumably" decided Capone's exile; the timing makes it stronger than that
	"leaving for Saint Petersburg"	T 12/6/27
	"He isn't . . . come back"	n.p. 12/13/27
	[press conference]	T 12/6/27; order rearranged for greater coherence
189	"They treated me fine"	T 12/17/27
	"This gang war stuff"	Los Angeles *Examiner* 12/13/27
190	"We have no room here"	Ibid. 12/14/27

190	"I'm just tired of all this"	Los Angeles *Examiner* 12/14/27
	"Why should everybody"	Los Angeles *Times* 12/14/27; movies "a grand racket" and Mary Pickford's house interpolated from T 12/17/27
	"We will have a reception"	n.p. 12/14/27
191	[Joliet]	Accounts from J 12/16,17/27; T 12/17/27; H&E 12/17/27
	back . . . Thursday	J 12/22/27; T 12/23/27
192	"We don't want Capone here"	NYT 12/18/27
	In Italy . . . "forced domicile"	NYT 8/17/27, 10/23/27

Page CHAPTER 18

193	"garden of America"	n.p. 1/11/28
	"Billion-dollar Sandbar"	Redford title
	police moonlighted	Detective Tom Hoolahan int.
	One [operator] complained	Fisher File (also following bootlegger story)
	"Furthermore," he said	n.p. 1/11/28
194	no legal way to bar	H&E 1/9/28
	"If he's just here"	T 1/10/28
	registered as "A. Costa"	TAX: T 10/10/31; includes Capone's use of Henderson and purchase of Palm Island house, below; details about Ponce de Leon: Koivu int.
	"palatial home"	T 10/8/31
	Miami Beach chamber of commerce	H&E 1/21/28
	"We have talked over"	Ibid.
	"Mr. Capone was one"	n.p. 1/21/28
	"If I am not wanted here"	H&E 1/22/28
	St. Petersburg . . . Bahamas	Pasley 180
	brothers Ralph and Albert	H&E 1/30/28
	Back in Chicago	T 1/3–5/28, 2/15/28
	eighteen times . . . Henderson took (195)	In the TAX testimony cited above, "Circella" appears as "Sorello"
195	93 Palm Island	Tampa *Tribune* 1/26/69: a UPI story found in the Kutner file of the Chicago Historical Society; Kobler 222
	Thompson kept denying	He had said back in August 20, 1927, "I deeply regret that the rumor will not down that I am a candidate . . ." (Pasley 186); but his travels continued: as late as February 22, 1928, he was in Washington, D.C., ostensibly talking about flood control (Pasley 185); his supporter, Stuart 337 wrote, "If Chicago's mayor was not a candidate . . . he had all the appearance of a candidate . . ."
	[primary background]	Stuart 337–8; T 2/9/28; Pasley 211–2
	[bombings]	T 2/19,22/28; Pasley 183–4; Landesco 126; Asbury 241–2; Allsop 218
196	Deneen faction . . . routed	Pasley 207–10; Allsop 221

196	[Esposito murder]	T 3/22–24/28; Pasley 198; Esposito background: Bennett 76–9; Pasley 193–7; Burns 216–21
	eight thousand brave a storm	T 3/27/28
	[bombings]	T 3/27,28/28, "got out alive": T 3/23/28; Swanson insurance: T 3/30/28; Bennett 73; Pasley 201; Thompson supporter Stuart 373–4 claimed that Ben Newmark had arranged the bombing, guided by Deneen strategists, though probably without Deneen's knowledge, using Capone men as free-lancers, again not necessarily with Capone's knowledge; he pointed out that Newmark was murdered a month later, suggesting that the routed Thompson-Crowe organization harbored enough hate "to have destroyed a hundred Newmarks"; this seems unpersuasive: the Crowe-Thompson reaction, not the bombings themselves, did the damage, and Deneen strategists could not anticipate that; also, Swanson's point in refutation (below) seems cogent; Thompson embraced: T 3/30/28
	"After having bombed"	T 3/27/27 (also "plain poppycock")
	"drive the crooks out"	T 7/16/27
197	"Bombs!" . . . "Pineapples!"	Pasley 204; "Pineapple Barasa": 214
	rebel Augusto César Sandino	T 3/27/28
	"It seems . . . American"	T 3/28/28; Octavius R. Granady, a black lawyer who dared challenge Morris Eller for committeeman in the Twentieth Ward, was shot and killed just after the polls closed (Pasley 217; Landesco 128 fn); Stuart 375–6 claimed Eller had nothing to do with it; but he was charged with conspiracy, while seven—including four policemen and James Belcastro, Capone's chief bomber—were tried for the murder, all acquitted: T 8/4/28
	[Capone] did his . . . best	Kobler 226–7
	[election results]	Stuart 369–70; Pasley 217–8
	Thompson, crushed, sulked	Pasley 221–2
	"Mayor Thompson was elected"	DN 3/15/28
	Comment across the country	Pasley 219–20; Landesco 127
	refurbishment . . . Palm Island	TAX: transcripts of testimony, passim; MDN 6/11/30, 2/13/52 (retrospective)
	crew of tile layers	Koivu int. who was told it by one of the workmen
198	Daniel J. Brown	Tampa *Tribune* 1/26/69
	"Honestly," Capone . . . ask	n.p. 2/25/29
	[principal lieutenants]	Asbury 369; Kobler 145–7; Meyer 10, 16–7; Humphreys . . . 61 locals: Murray 88
	one . . . driver-gunman hankered	Meyer 19

198	New York Stock Exchange	T 3/22/28
	volume mushroomed	T 3/31/28; 4,890,000 shares was a record day; RCA was up $24.50 to $195 and GM up $7 to $192
	"a racket"	Pasley 64
	lakeside Drake Hotel	T 8/7/28
199	Smokers got two packs	T 3/15/28
	roller skating cost	T 9/20/28
	Model A Fords, replacing	T 12/1/27
	(Clarence Darrow asserted)	Lyle 187 fn
	unions were "owned"	Murray passim; Hostetter passim
	"The members," Capone once	Murray 87 (also following quote)
	The Employers' . . . issued	Asbury 366; Lyle 187
	[Becker]	T 5/27/28; H&E 5/28/28; Hostetter 30–41 printed Becker's 12/13/27 letter to Crowe in full; also gave "not a process server"; Landesco 156–60; Murray 155–9; Pasley 248–50
201	"Capone did not voluntarily"	Pasley *Muscling In* 73
	twenty-five Loop skyscrapers	McPhaul 243
	drugs . . . kidnapping	Drugs: Murray int.; Baldwin int.: it was common knowledge, certainly throughout Italian neighborhoods; kidnapping: Pasley *Muscling In* 93; *Liberty* 10/17/31: "Al Capone does not tolerate some kinds of rackets, and kidnapping is one of them"
	pulled out of Sanitary	n.p. 11/1/29
	"Al Capone was scrupulous"	Murray 159
	owners . . . did not know	*New Yorker* 3/2/29; Sullivan 45–6 and Pasley 85 embroidered it, but the original antedated both, though Pasley said the story came to him from a friend of a friend of the lessor—unnamed even in the original magazine story, which came from a cousin of the lessor
	[Yale murder]	Brooklyn *Eagle* 7/8/27 (DeAmato killing), 7/2,3,5,6,10/28; Balsamo 211–8, much corrected and expanded by Balsamo int.; he also corrected items like the typo in *Eagle* 7/2, that put the speakeasy at *Fourth*, not Fourteenth Avenue, adding the fact that it was Yale's own place; New York Police Department Museum exhibit, drawing on records of the NYPD property clerk; DN 7/3/28; Pasley 242–5; Burns 224–8
202	But Yale soon resented	All sources agree, and it helps account for Yale's dangerous game; but it makes little sense: the national Unione head would have swung only modest weight in Chicago, anyway; and it's not clear why Lombardo would have had to remit anything; Balsamo in his *Clock* followed the same logic but says he is not really sure that Yale was in

		fact national head of the Unione, or that such a post even existed; in any case, he stresses that the Unione was in no way identical with what we today call La Cosa Nostra, in which Yale was merely Brooklyn underlord of Joe Masseria (''the Boss'')
202	Knoxville . . . black Nash	The *Eagle* 7/6/28 had the car sold to ''Cox'' in Memphis, Pasley 243, Knoxville (and costing $1,050); however, Knoxville is on the route from Miami to Chicago, Memphis not; on the other hand, Balsamo is convinced that the *Eagle*'s figure of $2,040 is correct
	Brooklyn, McGurn . . . guide	Balsamo int.; they stayed the night before the murder at the Hotel Bossert; they also showed up at a bar run by the father-in-law of Yale's lieutenant and inheritor, ''Little Augie'' Pisano, who Balsamo thinks was betraying Yale; details of the escape also from Balsamo
203	Dade County . . . Taylor	T 7/8/28, 7/10/28; H&E 7/10/28
	''in the cleaning business''	TAX: T 10/10/31
	July 30 . . . Lexington	T 10/13/31
	Lexington had deteriorated	T 1/29/62 (retrospective)
	''What!'' cried the Lexington's	H&E 8/7/28
	Eventually . . . would occupy	Asbury 367–8; Burns 42; Kobler 234–5; Murray 31
	Capone installed . . . girl	Kobler 235
	Frank J. Loesch	Background: Fox 131, 133–4 (''The real Americans''), 148 (''xenophobic bigot''); Allsop 279–80 (admired Mussolini's methods: quoting Loesch); Capone interview: Dobyns 1–3 gave a version, quoting what Loesch told the Southern California Academy of Criminology, reported in T 3/25/31; text relies on the less colored (and sworn-to) version (also used by Fox) given by Loesch to the U.S. Senate Judiciary Committee (*Nomination of James H. Wilkerson* [Part 2], Mar. 1, 3, 8, 1932)
204	Black Hand . . . in cahoots	The consensus of all sources, although no evidence of exact responsibility transpired; the killing was much more precisely planned and executed than any Aiello or his people had shown they could carry off alone
	[Lombardo murder]	T 9/8/28; Pasley 229–32; Burns 229–32
205	$12 dollars in his pocket	Pasley 226–7: Lombardo's autobiographical sketch
	FRATELLI	Bennett 65 showed a picture of the sign
	''From Al Capone''	T 9/11/28
206	[Lolordo murder]	T 1/9/29; curiously, Pasley 236–8 had the spread laid for the three killers and Burns

		235–9 had the killers slipping the pillow under Lolordo's head, although the news story is emphatic on both points
206	"She didn't identify"	DN 6/17/29; Stege's title was actually deputy police *commissioner*: the chief now called "commissioner," unlike the days of Major Funkhouser (Chapter 3) when the commissioners were civilians; to avoid confusion, the text sticks with "chief"
	Pete and Frank Gusenberg	Browne notes; Pasley gave James Clark, and later news stories agreed it was those three
	Capone . . . pneumonia	T 1/15/29: reported he'd been in bed "for the last ten days"; he was reported as convalescing in T 1/21/29

Page CHAPTER 19

207	Walter Spirko . . .	Spirko int.
208	[St. Valentine's]	Basic events: Pasley 251–263; Sullivan 191–99; Burns 258–276; however, the text relies more heavily on contemporary accounts in T, DN, J and NYT, as noted; specific references only for disputed points and those that have not appeared in the more usual accounts
	fronted . . . twenty-five feet	Pasley had it 40 feet by 150 feet, Burns 60 feet by 120 feet; J 2/14/29 gave the text dimensions with enough circumstance to be more credible than either—especially since contemporary news photos, with people and cars to give scale, clearly showed the front as about 25 feet wide
	terra-cotta brick	J 2/14/29; the others either specified no color or made it, reflexively, red: again the pictures showed it a lighter color; also "for sale" sign
	since October . . . leased	Ibid.
	office . . . raw unused look	Ibid.
	Wigwam, a bar	Wickersham 312; Browne notes changed Wickersham's "Race" to the correct "Grace" Street
	storage area . . . rafters	DN 12/4/30; the next tenant discovered the hiding place, complete with hoists that let the gang haul out of sight (presumably) crates of liquor
209	"the Beast . . . rot-gut"	Lyle 202
	Moran . . . Purple Gang	story in both Pasley 252 and Sullivan 196–7 with more circumstance than in contemporary news accounts: e.g., T 2/15/29, 2/19/29
	Mrs. Frank Arvidson	Sullivan 196 gave it as "Atkinson," and Allsop 142 followed him; the police report,

		traced by Mark LeVell, gave the spelling clearly, as did the news accounts
209	Chicago expected killings	T 2/14/29; all three got stays, Walz and Grecco actually being executed shortly after midnight, February 20: T 2/20/29; Shanks received a reprieve on grounds of insanity: DN 3/7/29
	temperature had "warmed"	T 2/15/29
	Capone . . . appointment . . . twelve-thirty	Brooklyn *Eagle* 2/24/29 quoted the assistant D.A. from New York (see below) who had instanced the meeting; affidavits (by participants in the interview) in connection with Capone's contempt trial (Nat. Arch., DC) confirmed the time
210	John May . . . truck driver	Coroner's test. (brother, James May: also age "about thirty-five"); "ordinary mechanic": NYT 2/15/29; arrest record: T 2/15/29
	Alsatian, Highball	Some sources (e.g., Pasley 254) made him Heyer's dog; Herbert Asbury in a 1947 magazine piece made it definitely May's: supplementary detail by Kobler 249–50 is convincing
	"Frank Snyder" . . . Heyer	T 2/15/29: also arrest record; "son . . . by a first marriage": Coroner's test. (son, Howard Heyer)
	Albert Kachellek	Age, born Germany, alias: Coroner's test. (sister Marie Neubauer); arrest record T 2/15/29
	Gusenberg brothers	Ages: Coroner's test. (brother Henry); arrest records T 2/15; "Henry . . . projectionist" Murray 233–4; Coroner's test., where he called himself a movie "operator"; Browne says their death certificates gave the family name as "Gusenberger"
	Reinhart H. Schwimmer	Coroner's test. (mother, Josephine); sideline in fortune-telling: J 2/15/29; photographed prominently: funeral picture reprinted, T 2/16/29
211	Albert R. Weinshank	Age: Coroner's test. (uncle, Maurice: he spelled it "Weinshenk," as did the Coroner's report for Albert; but Albert evidently used the *a*
	Elmer R. Lewis	Detail from LeVell, who obtained the unpublished police report, which included Lewis's statement
213	Frank's pistol . . . dropped	Most reports either did not identify its owner (e.g. T 2/15/29. J 2/14/29) or attributed it to Weinshank (e.g., Burns 262); but LeVell reports than an associate of his, Ron Kovar, uncovered the police report that made it definitely Gusenberg's; instead of being dropped then, it's possible that the

		pistol was missed by the frisker, and that Gusenberg fumbled for it and dropped it after the raiders left, while he crawled toward the door
213	raiders carried two Thompson	Coroner's test., also shotguns, below; ballistics expert Calvin Goddard (q.v., below) could identify two separate Thompsons and two pump shotguns from marks on the shell casings found; LeVell points out that Lieutenant Loftus (q.v., below) said he found "about 85" machine gun casings, argues that it would have been foolish to go in with only a twenty-round clip on one gun; but Goddard counted exactly seventy, identifiable from markings as coming twenty from one gun and fifty from another, strongly suggesting the clip sizes; and LeVell does note that the police report of Loftus—which he also tracked down—does not specify that the lieutenant actually *counted* the casings he saw; as noted above, the presence of only four guns resolves the dispute by some sources over whether four or five raiders went in the garage
	hands still thrust . . . air	Coroner's test.; pathologists could tell during autopsy from the path of bullets in the bodies
	"long guns . . . pointing"	J 2/15/29; she was quoted in T 2/15/29 as saying the two men in uniform were preceded by "two or three men . . . with their hands up in air," which presumably gave rise to the later confusion about the number that went into the garage
	"took it so easy"	J 2/14/29
214	"I got a little curious"	J 2/15/29
	"Do you know me, Frank?"	Loftus police report, from LeVell; Pasley 256–7 and Sullivan 193 reported a Sergeant Clarence Sweeney as the one questioning Gusenberg, but LeVell found no mention of him in any of the police reports and it's clear from the difference in accounts that Pasley and Burns were making up the dialogue: e.g., "It's getting dark, Sarg, so long"; the dialogue quoted in the text is verbatim from Loftus's report—which of course was still only the lieutenant's recollection
	Heyer . . . so riddled	Coroner's test.
	"rather obscenely"	Allsop 141

Page **CHAPTER 20**

216 limousine . . . pulled up — Brooklyn *Eagle* 2/24/29; also ''Well . . . what can I do''

Capone . . . asked . . . about Yale — The dialogue that follows is an amalgam of accounts in TAX: T, DN and NYT, all 10/10/31; the accounts did not clearly distinguish the source of all quoted material as to whether it came from the 2/14/29 meeting or the one held the summer before; the way DN was written strongly suggests that the dialogue had been from the first meeting, and only T mentioned the 2/14/29 date—but specified that the stenographer who attested to the accuracy of the transcript of the meeting read into evidence saw Capone for the first time on that date; only the NYT seems unequivocal: after reporting Miami D.A. Vernon Hawthorne's testimony about the previous summer's meeting, the NYT correspondent wrote, ''The government then began to put into the record the transcript of another meeting . . .''—but never specified the date of the meeting; T presented the dialogue in a ''Q and A'' format, suggesting that they had a copy of the transcript, but one key difference between T's and that of the other two indicate otherwise: both NYT and DN reported Taylor's warning, following Capone's reluctance to name names, while the T version had a different sequence that eliminated that important warning; Kobler 247–253 cinematically cut between the interview in Miami and the massacre in Chicago as though they had occurred at the same time, having given the starting time of the interview as ''shortly after nine,'' which anyway would have been eight in Chicago; as noted, the interview started at twelve-thirty Eastern Standard Time, an hour after the massacre

''All I did was answer'' — J 2/15/29

217 ''mental capacity of Capone'' — NYT (above); the DN reporter made the same point

first major tax trials — 11/1/29

IRS had been looking — Casey 114, 118

specifically targeted Capone — TAX: T 10/10/29

''Dan Serritella . . . living'' — NYT gave it as ''Sorello'' with no first name mentioned, T as Serritella; I use, however, the NYT response, T's having been ''he's just a friend''; it's possible though, that both reporters misheard; as

Page	Phrase	Note
		noted, Taylor earlier asked about bodyguard Nick Circella, which T rendered as "Sorello," as did NYT, DN as "Serello"
217	"He's a fighter"	The DN answer; T gave the less plausible "He fights"
	"Al Brown"	Only NYT recorded this exchange
218	"money at the Western Union"	T version is confusing, first having Capone answer "I don't remember, I'll try and find out"—then soon answering essentially as in the text version, which is the NYT's
	"Fifty thousand dollars"	So pointless a lie, Capone may have forgotten that he gave Henderson $10,000 and had a $30,000 mortgage on the $40,000 property
	"Do you keep a record"	All three accounts gave slightly different versions whose substance was the same; only NYT gave the "my personal affair" exchange—but it certainly seems a likely follow-up question and inevitable response
	"money from . . . Fischetti"	T and DN versions were essentially the same (NYT did not give the exchange), but DN gave the text "amount of money" while T gave "What has money to do with it," which doesn't make much sense
219	"It's war to the finish"	Most sources; this version NYT 2/15/29
	loaded gun . . . Schwimmer	Police report from LeVell
	they found Myrtle	T 2/15/29
	Frank's hotel room	Police report from LeVell
	The day after. . . . Silloway	T 2/16/29, J 2/15/29 gave his quote in full, as did NYT 2/16/29 in a slightly different version
	"If Major Silloway has"	NYT 2/16/29
	"toss 'em by the throat"	Burns 268
	255 detectives . . . account	NYT 2/19/29
	transferred . . . fired	NYT 2/27/29 . . . 3/13/29
	"Only Capone's gang kills"	Virtually every source, though none gave its provenance; it quickly became a legend fit for embroidery: e.g., an article written by the usually careful Herbert Asbury (in *'47—the Magazine of the Year*, September) embellished it with "What's the use of all this hollerin'? everyone knows who did it. Only Capone's gang kills like that."
	appeared in Miami papers	T 2/19/29; DN 2/17/29 printed Capone's reply from the Hialeah race track: "Read your wire . . . concerning letter to press from one 'Bugs' Moran in which he blames gang massacre on 'Capone gang.' Such a statement is false, malicious and foolish. "Next they'll be blaming the Uruguay-Paraguay trouble on me. "I deny any connection with the 'massacre' either directly or indirectly."

Page	Note	Source
220	"We don't know what brought"	T & NYT, both 2/17/29 and both verbatim
	Caldwell . . . Tacker	NYT 2/18/29; T 2/18,24/29—which gave the amended report of Caldwell's chauffeur in its story on Tacker's abduction
	George Arthur Brichetti	Although alluded to in news reports, his name was given only by Burns 268–9 and *X Marks* 43; both gave his name as "Brichet," but Browne notes added the Italianate ending, and LeVell found in the contemporary phone book that although there were Brichettis then in Chicago, no Brichets were listed
	J. Thomas Heflin	T 2/17/29
221	"How is it?"	T 2/18/29
	Two nights . . . Capone . . . party	T 2/19/29; a column by Jack Kofoed, MH 2/14/63, put the party on the night of the massacre, but the contemporary news story is explicit that it was the following Saturday; furthermore, a later review of the movie *The St. Valentine's Day Massacre*, MH 7/23/67, made it clear that a press party depicted in the film that afternoon was never reported; it's hard to believe such a party that night would not be mentioned in the press
	Sharkey . . . Stribling	MH 4/26/60; Allsop 289; also Westbrook Pegler column in T 2/28/29 describing fight; Kobler 253 identified Cunningham as then currently a sportscaster
	golf . . . Hollywood Country Club	Purdue int.
	Dr. Samuel D. Light	Light affidavit, Nat. Arch., DC
	[Hialeah]	Ibid.; affidavits of officers J. M. Coroneas and M.C. Wood (et al.); Coroneas testimony, TAX: T 2/26/31
	Nirmaier had flown . . . Bimini	Ibid.; Nirmaier affidavit; also Nirmaier testimony, TAX: T 2/26/31; H&E 2/26/31
	expedition by steamer	Ibid.; affidavits of Capt. William P. Tremblay, et al.
	[burned car]	T 2/22,23/29
222	"There is no doubt"	NYT 2/23/29; T of the same date gave a slightly different, somewhat less enthusiastic version, saying suspicion had been "materially weakened"
	Rewards totaled $50,000	T 2/22/29; NYT 2/23/29
	"Why worry"	n.p., n.d.
	leads in St. Louis	T & NYT, both 2/25/29
223	list of seventeen names	T 2/26/29
	they found McGurn	T 2/28/29; NYT 2/28/29 reported that he'd been registered as "Al Rubenstein," but alas not so; tournament golfer: n.p. 8/14/31; fatal car crash: T 3/20/21
	ten uncombative rounds	Westbrook Pegler in T 2/28/29

223	government process server	DN 2/25/31: testimony at Capone's contempt trial
	"I have nothing to fear"	H&E 3/1/29
	Burke and Ray . . . Yale killing	T 3/5,6/29; NYT 3/6/29; DN 3/7/29
	indict . . . Scalise . . . McGurn	NYT 3/16/29
	"the Blonde Alibi"	Copy editor Boe points out that the correct adjectival form is "blond" even when modifying a blonde; newspapers of the era all used the "e," so the text preserves their error
	"too yellow to be natural"	Kate Susman byline, n.p. 2/28/29
224	"When you're with Jack"	T 3/1/29
	"What does he say?"	T 2/27/29
	Phillips . . . affidavit	Nat. Arch., DC
	"talking to . . . wrong man"	Pasley 48
	"If a federal judge"	T 3/10/29
	"It would be dangerous"	Ibid.
	"time to satisfy himself"	J 3/11/29
	"Why bother"	H&E 3/19/29
	Capone *was* in Florida	T 3/20/29
	jury seemed more interested	NYT 3/27/29
	"pay income tax"	n.p. 3/22/29; similar reports in T 3/26,27/29
	government had . . . looked	T 3/20/29
	Capone posted $5,000 bail	T 3/28/29; NYT 3/28/29 had it $6,000; J 3/27/29 made it $50,000, which seems excessive
225	Inquiry . . . coroner's jury	Except as noted, details and quotes are from a transcript of the jury's sessions
	Daily News . . . claimed	An article in the Milwaukee *Journal* 5/11/30 credited juror Massee with the idea, his lawyer having come up with Goddard's name; the NYT story (below) seems more likely
	hiring of . . . Goddard	NYT 3/5/29; Pasley 262 and Burns 273 gave a middle initial of "C."; so did NYT 2/18/29, although neither the text NYT story nor the *Northwestern University Bulletin*, below, that gave an outline of the Crime Lab's teaching schedule, had a middle initial; Goddard evidently did not use it; Kobler 259 gave it as "H."; as a reserve U.S. Army officer, he was called "Major" and "Lt. Colonel" or just plain courtesy "Colonel"
	Massee . . . Olson	Milwaukee *Journal* 5/11/30; *Northwestern University Bulletin*, 4/6/31—both provided by the Chicago Historical Society
226	Frank Thompson	At first seemed something of a mystery figure: see NYT 4/20/29
	James Shupe . . . "Bozo"	Pasley 244; Kobler 257 gave the connection to Scalise and Anselmi; he also identified

		Thompson as a veteran hoodlum, wanted for "attempting to machine-gun his wife and her lover"; no mention was made in the jury transcript of his record
226	guns showed up . . . ten months	NYT 12/16–18/29, 12/24/29
	A laundry mark . . . FRB	Kobler 260
227	unready to try . . . McGurn	Pasley 260–1
	indict McGurn	n.p. 11/2/29
	advice of a newspaper editor	Berardi int.: the *American*'s Harry Read
	McGurn getting two years	T 10/21/31
	Burke . . . captured	T 7/11/49; TAX: T 10/15/31 had Burke doing life in Marquette and Phil D' Andrea visiting him with a lawyer
	"moral certainty"	Kobler 260
	named . . . other participants	Ibid. 261 fn.
	Bolton . . . Weston	NYT 3/31/29
	"dubiously by . . . photo"	Burns 273
	J. Harvey Bailey	Quoted in *Robbing Banks Was My Business* by J. Evetts Haley, Canyon, Texas: Palo Duvo Press, 1973, p. 85; thanks to Bill Helmer for locating the reference
228	"hillbilly gangsters"	Graham int. quoting a colleague
229	"crimes like this"	LD 3/2/29; also *Globe* and *Record* quotes; *New Republic* 3/13/29 wrote about "insolence"

Page **CHAPTER 21**

230	1927 . . . Supreme Court decided	MacCarthy int. and Mulroy int. both explained the importance of the Sullivan decision; Irey xi (especially) and Casey 114 gave details of the case
	IRS agents . . . scrapbooks	Casey 118–9
	1928 . . . Druggan . . . Lake	Ibid. 135
	hook into Ralph Capone	Irey 28–30
231	"specific item" . . . "net worth"	MacCarthy int.; also Mulroy and Marshall ints.
232	not until 1954	Kobler 286; Mulroy int.
	Conditions . . . Chicago Heights	DN 2/24,25/36: retrospective series by Clarence R. Dore
	Cleveland . . . twenty-seven	H&E 3/17/30; Kobler 262–3; Burns 275
233	called by Frank Costello	Peterson 213, following Asbury in *Colliers* 4/19/47
	Joseph Guinta . . . Lolordo	Allsop 145, 147; Lombardo and Guinta had been picked up together on a concealed weapons charge at the beginning of Capone's exile in 1927: H&E 12/13/27
	"I am the most powerful"	Burns 281
	huddled with Joe Aiello	DN 2/7/30; Waukegan restaurant: Browne int.; Pasley 329–32; Burns 276–282; Allsop 147–8; Murray 113–118
	more vivid version . . . Rio	DN 7/8/43 (retrospective); Allsop stipulated "a roadhouse at Hammond, Indi-

		ana,'' which Murray identified as The Plantation; Meyer int. confirms Murray
234	Spooners' Nook	NYT 5/9/29
	Atlantic City	NYT 5/18/29 specified a three-day conference—which indisputably ended by the 16th; Pasley 326 had Capone in Atlantic City ''a week,'' which Murray 137 presumably followed in specifying that the conference started May 9; Allsop 281 simply had it in May, though his account at pp. 306–7 was very detailed; neither M&G 105 nor McPhaul 250, 256 gave dates, though M&G also gave a detailed account; *Colliers* 4/19/47 referred to a picture of Capone with Nucky Johnson and Frank Costello that ran in local papers
	''I told them''	Kobler 265; Allsop gave a shorter version, beginning with ''We agreed to forget the past''—but Allsop suggested the agreement was between Capone and Moran, who wasn't there, rather than among all the gang leaders
235	According to Capone	Amalgam of BUL 5/17,18/29, 6/26/29; REC 5/17,18/29; LED 5/17,18/29
	''You're 'Scarface Al' ''	BUL 5/17/29
	''My name is Al Brown''	T 5/18/29
236	''Hello, Shooey!''	REC 5/17/29; LED 5/17/29 had it, ''Hello, Shooey . . . Yes, we've got guns''—before the detectives had said a word
	Malone quickly denounced	REC 5/19/29
	Lieutenant Creeden confirmed	BUL 5/18/29
	police . . . roust Magistrate	LED 5/17/29
	Schofield . . . Capone	Amalgam: DN 5/17; LED 5/18; T 5/18; NYT 5/18—all 1929; although a stenographer took it verbatim, various accounts summarized different parts or omitted some material; e.g., only T gave full ''seems to be at a point'' and only NYT gave Capone's line to Rio *after* ''I'll do the talking''; some versions of Capone's ''Look, boy'' to Rio gave ''have been a faithful pal,'' but ''bodyguard'' (NYT) makes more sense, since he'd already called him ''friend''
237	held a two-man lineup	Best accounts are BUL and DN, both 5/17/29
238	''dull reddish hue''	DN 5/17/29
	arrest had been a setup	Except noted, quotes were BUL 5/18/29
	Haggarty . . . might be	BUL 5/29/29
	New York Times decided	5/21/29
	seemed unprecedented	LED 5/18/29 wrote ''almost without precedent'' but was just being cautious; T 5/18/29 agreed

238	Philadelphia *Record* stated	5/18/29
239	"He never gets in jail"	DN 5/17/29
	Mae Capone also doubted	BUL 5/18/29; (slightly different version in REC 5/19/29)
	why . . . carrying a gun?	n.p. 5/18/26; BUL 5/18/29; REC 5/18/29 and T 5/20/29 made the point that Capone no longer carried one
	"enough to make anybody"	NYT 5/19/29; slightly different version in T 5/19/29
	"I didn't give myself up"	BUL 6/4/29
	"I'm here because I'm here"	LED 5/18/29
	"You don't get a reputation"	REC 7/21/29
	"known throughout crookdom"	T 5/19/29
	"manbeaters . . . rheumatism"	T 5/20/29
	food . . . sparked riots	BUL 5/20/29
	Capone's hair . . . suit	BUL 6/4/29
	Why, some argued . . . choose	T 5/19/29
	John Stege pointed out	BUL 5/18/29
	Why not Miami	T 5/9/29
240	ninety days or so	BUL 5/18/29
	"seized eagerly upon"	H&E 5/18/29

Page **CHAPTER 22**

241	trounced Smith by	NYT 11/7/28
	toss a medicine ball	H&E 6/15/31; Irey 36 in greater detail and for quotes
	Rumor . . . Hoover's . . . enmity	Irey 25–6
	subsidiary . . . roistering	Kobler 237
242	"Probably no . . . citizen"	NYT 5/26/29
	"trademark . . . ," better known	DT 5/3/31
	Gridiron dinner	H&E 4/30/31
	FBI had refused	Irey 27
	George Johnson . . . requested	Irey 20
	[Irey]	Irey ix–x
	Martin . . . Tessem	Irey 29–31; DN 2/26/36: also account of tracing Ralph's accounts, with more detail in Irey 31–33; DN gave the name as "James Carson," while Irey gave the sum as $3,200; T 10/10/29 gave the aliases as including "James Carson" (but without details about the check, and without "Harry White"); the DN story, part of a series, is so much more circumstantial about the check (e.g., it was dated June 27, 1928), text follows it for the amount, but follows Irey on the sequence of aliases for like reasons: DN mentioned *only* the Carson alias
243	Mae came to visit	BUL 5/29/29
	offered $50,000	LED 6/16/29
	D.A. John Monaghan	T 8/19/29
	Capone's appeal	REC 6/21; BUL 6/26 and LED 6/27: "coerced," etc.; T 6/27,30, all 1929

243	"You see," . . . a reporter	T 6/5/29
	Capone's $1,000 donation	BUL 6/6/29; H&E 6/7/29
	responsible could not agree	See, for example, J and NYT, both 8/9/29; REC 9/7/29
	in service since 1829	Campbell Collection, Philadelphia Historical Society
244	Smith, "Hard Boiled"	H&E 3/18/30
	warden did not . . . grim	REC 8/21/29
	"Park Avenue" block	Ibid. and T 8/20/29
	"Capone was much pleased"	LED 8/10/29; also BUL 8/9/29
	employ fellow prisoners	BUL 3/18/29
	appeal for parole . . . grounds	REC 9/21/29
	minor sinus surgery . . . tonsils	BUL 9/7/29; he had been operated on for a sinus infection: T 3/17/30; Pasley 334 made Goddard a member of the "Pennsylvania State Board of prison inspectors"
	October 8 . . . arrested Ralph	T 10/9,10/29, Irey 33 specified a front-row seat; Kobler 262 named Converse; indicted: T 11/2/29; trial: T 4/9/30
	"cheat, swindle or defraud"	Irey 34
	in time . . . World Series	REC 10/4/29
245	"Every time a boy falls"	H&E 3/17/30
	"They've hung . . . Chicago fire"	Pasley 352
	"Tell them . . . I deny"	Evening LED 11/25/29
	he started to mingle	REC 12/11/29
	CAPONE LOLLS	T 8/20/29
	Francesco Raffele Nitto	Nat. Arch., Chicago, Cert. 191944
	chain smoked . . . sip only	Meyer int.
	started as a barber	DN 2/25/36
	"brains" of the outfit	H&E 3/23/30
246	"until Nitti has given"	DN 2/7/30
	[Nitti investigation]	DN 2/25/36; also Irey 45–6
	[Nitti indictment]	H&E 3/23/30; DN 3/22/30 gave the same income but different figure for what Nitti owed the government, presumably including penalties and interest, although their total differed on that score from H&E's—which the text follows however, because H&E specified the *tax* mentioned in the indictment
247	case against Jack Guzik	Amalgam: Wilson 29–31, 46–49; Irey 46–8; DN 2/26/36; "nothing that walks": Irey 36; "sweats ice water": Wilson 35; "methods . . . brutal": Spiering 60
248	Randolph . . . Secret Six	Pasley *Muscling In* 63; Allsop 312; Lyle 197
	"Nitti and Guzik . . . careless"	Irey 45
	Broadway Limited	BUL 3/15/30
	chartered . . . Ford trimotor	BUL 3/17/30; LED 3/18/30
	Moran remnant . . . Klondike	BUL 3/17/30
	Herbert Goddard, professed	H&E 3/16/30
249	Fisher . . . had to sign	BUL 3/15/30; Chicago papers had Fisher's trip specifically to Bermuda (e.g. H&E 3/

		16/30), but Philadelphia papers, which presumably knew better, gave it as West Indies
249	governor's secretary	H&E 3/17/30
	"in the routine way"	LED 3/17/30
	Several hundred sensation	Ibid.
	A fast car taking Capone	H&E 3/17/30
	bulletproofed sedan?	BUL 3/17/30
	police escort . . . city limits	LED 3/17/30
	governor had signed . . . 11:10	BUL 3/17/30; also "I don't know whether" and three hundred watchers
	topping out at five hundred	LED 3/18/30
	"What the hell's the use"	T 3/18/30; H&E 3/18/30 gave virtually the same quote, minus the "hell's"
	Smith . . . drove . . . Harrisburg	BUL 3/18/30
250	Goddard explained	LED 3/18/30
	autographed photo of Capone	Meyer int.
	"How much did you get"	T 3/18/30: appears only in this account

Page **CHAPTER 23**

251	"Al Capone is back in town"	DT 3/18/30
	"Boo!" wrote another paper	H&E 3/20/30; also Charlotte police
	Ralph junior . . . snowball fight	H&E 3/19
	"Where's grandma?"	T 3/16/30
252	Carlton—whom Capone	DN 3/21/30
	IT IS REPORTED	Copy of injunction in Miami *Herald* files citing text of telegram; it was sent 3/19/30
	Capone . . . driven to Chicago	BUL 3/22/30; H&E 3/22/30: also "Dr. So-and-So"; Capone gave the reason for needing glasses variously as damage done to his eyesight from the "dazzling white walls" of prison (T 3/17/30) and reading in bad light in his cell (DN 3/21/30)
	[Eliot Ness]	Ness 11–21
	"Is Ralph there?"	Ness 129–30; he wrote that this happened on March 18, day after Capone's release; but Capone specified to Stege (H&E 3/22/30) that he had driven, and had reached Chicago on the 20th–or 19th, according to an offhand comment to a bystander (T 3/22/30)—either date more probable than the 18th in those preturnpike days, especially since Capone also said he and Rio had spent the night of the 17th in Philadelphia
	"I'll bring him over"	T 3/22/30; this is the main account; interpolations from others as noted
	sartorial vision	DN 3/21/30; includes "I burned" and Stege quote below
253	"You're not . . . citizen"	The *Tribune* account specified that the dialogue was recreated by Stege and Ditchburne afterwards
	"If I want to see anybody"	DN 3/21/30

253	"All I ever did"	Text of the interview: T 3/22/30; however, the long quote starting, "All I ever did" down to "see the judge in the morning" is Pasley 349–50; Herrick referred to it in her printed account, and gave "Nobody's on the legit," quoted earlier, but did not print the whole speech; Pasley, a colleague, probably had access to her notes or a longer version of her story; the order is changed for greater narrative coherence
254	"By what authority of law"	T 3/23/30
	federal judge . . . wondered	MDN 3/28/30
255	Rapid City, South Dakota	T 3/28,29/30; DN 3/29/30
	mayor of Monticello, Iowa	n.p. 4/2/30
	A later writer professed	Kobler 281–2
	"I didn't pay income taxes"	T 12/21/30
256	"There was still . . . question"	MacCarthy int.
	until Capone returned	Analysis derived from Marshall, MacCarthy and Mulroy
	happen with Jack Guzik	T 11/19,20/30
	Mattingly wrote . . . Herrick	For their mock trial, the ABA's sponsoring Section of Litigation compiled authenticated versions of the documents quoted below; thanks to the ABA for providing a copy
257	blue double-breasted suit	Wilson 43
258	Scarface stuck his big paw	Wilson 43–4
259	Ralph Capone's jury	T 4/16,17,19/30; guilty: T 4/26/30 and Irey 35; Irey gave "I don't understand" but T 4/26 insisted "Verdict speaks" was his only comment
	[Mattingly Letter]	Appears many places in various forms but particular thanks to the *Sun-Times* which furnished a complete, authenticated copy

Page **CHAPTER 24**

261	"decency of the community"	MDN 3/11/30
	Another in the series	MDN 3/20/30; also "a generation or two"
	Carl Fisher	Biographical details in Fisher Collection, Florida Historical Society
262	first brought James Cox	Redford 153
	"Dear Skip"	Fisher Collection, dated 3/20/30
	Fisher sent Carlton telegram	Ibid., telegram dated 3/26/30
	THE EXPRESSED DECENCY	Ibid., telegram dated 4/12/30
	$45 a *case*	MDN 3/8/30
	"by anyone with a bankroll"	Ibid.
	"liberal . . . for the winter"	MDN 3/11/30
	"a case now of gangsters"	MDN 3/13/30
	Fisher was a lush	Fisher Collection; Redford 112, 120–22
	womanized tirelessly	Redford 52, 59
	"anything that moved"	Ibid. 196

262	"Big Irish" . . . Pancho Villa	Sosin int. (phone)
263	"April 1st at which time"	Fisher Collection, letter dated 3/20/30
	"an imperious bastard"	Sosin int.
	Quigg . . . removed	Muir 199; Pine: MH 11/26/57, 4/26/60, 2/26/66 (retrospective)
	Samuel D. McCreary	Hotel mgr.: testimony, padlock case (Chancery Case 30110); 1935 trial: testimony Criminal Case 5027—records in the archives of the County Clerk's Office, Dade County, as are all following case numbers cited in this chapter
	Sledge . . . had overheard	MDN 6/11/30
	"clearly within the law"	MDN 3/15/30
	Joseph P. Widener	Muir 198
	D. W. Shannon might get shot	Muir 134–5
264	[Capone in Miami]	Koivu int.; Sosin int.; "Capone's Tower": mgr., Westside Country Club; Ivan Rodriguez, Dade County Historical Survey; NFA at site of Everglades mansion, now leveled, only a few stones remaining
	"We give a Negro porter"	MDN 3/21/30
	Young Jack dined out	MH 6/23/68 (retrospective)
265	Sewell . . . acted . . . knowledge	MDN 5/23/30
	identified Ray Nugent	MDN 3/5,9,11,15/30, 5/6/30, 6/11/30
	Lehman . . . master stroke	MH 3/21/30
	Lehman . . . lifted twenty-seven . . . badges	MDN 3/21/30
	Leo J. Brennan	MDN 3/23/30
	Frankie Newton . . . heat	MDN 6/10/30; charges dismissed: MDN 8/1/30
	Twelve bottles in John's closet	MDN 4/23/30
266	they picked up McGurn	MDN 4/1/30; convicted: MDN 6/26/30, 7/3/30
	Twelve days later . . . Albert's	MH 4/12/30; MDN 4/12,16/30
	"I am here for a rest"	MH 4/21/30
	"a place frequented"	"BILL FOR INJUNCTION AND TO ABATE NUISANCE," Chancery case 30110
	sampled . . . Havana	MDN 4/30/30
	[arrest, May 8]	Except as noted, source is transcript of testimony in Chancery case 30276, a hearing to determine the merit of Capone's criminal charge against Miami officials and James Cox for conspiracy and false imprisonment
	"Mr. Director," Capone asked	Testimony of Robert Knight, who had no ax to grind, at least about this part
267	jewelry and $1,160	MDN 5/9/30
268	Someone had seen the arrest	Giblin testimony in Criminal case 5027, five years later
	Capone sat in his box	MH 5/14/30
	Peck . . . Spivey	MDN 5/14/30
	"Roddy Burdine for one"	MH 5/15/30
269	a new vagrancy ordinance	Ibid.

269	hearing to padlock Capone's	MDN 5/16/30; case 30110
	''beer from Tom Harbin''	Fisher Collection, letter dated 5/27/30,
270	police stopped Capone a fourth	MH 5/25/30
	Capone charged the mayor	Ibid.
	Mike Glenn suspected	Fisher Collection, letter to Fisher, dated 5/29/30,
	''And when you arrived''	Case 30276
271	no longer arrested on sight	MH 5/30/30
	teacher at Sonny's school	MH 5/18/30 specified 50; testimony by someone at the party (MDN 6/11/30) put the number at 60–75
	fifty adults showed up	MDN 5/27,28/30
	''In my opinion'' . . . Glenn wrote	Fisher Collection, letter dated 5/27/30
	[padlock hearing resumed]	Testimony transcripts and summaries, MDN 6/10–14/30
272	Two hours after . . . verdict	MH 6/15/30
	McCaskill had blown it	MH 6/20/30, including following testimony; McCaskill lost: MH 6/25/30
273	''telephone'' charge . . . to trial	MDN 7/12,13,15/30
	one frequent guest commented	testimony in padlock case
	estate . . . north . . . Broward	MDN 7/13/30
	''Capone's Island''	Rodney Dillon, Broward County Historical Society

Page **CHAPTER 25**

274	[public enemies list]	DN 4/24/30, which had printed a more extensive list the February before
	[Jake Lingle]	biographical data: T 6/10/30, 6/30/30, 7/1/30; Boettiger 25–40
	soon graduated . . . cub	Browne 9; his novel gave the $12 salary—in line with the $15 that Spirko says was standard in the twenties; Lingle's elevation, Browne wrote, was a matter of *Tribune* policy for copy boys—up or out in four months—not mark of any exceptional merit; all data taken from Browne's novel have been checked with him to distinguish the fruits of his research from those of his novelist's imagination
275	only infrequently pop into	Spirko int.
	For years . . . Bill Russell	T 6/10/30 dated the friendship from twenty years before, when Russell still walked a beat; Boettiger 27–8 from when Russell was already a lieutenant; Sullivan *Surrenders* 8–9 claimed Lingle intervened in Russell's appointment to replace Hughes as chief, but those rumors were vociferously denied by the *Tribune* and the city
	''fonder of him . . . son''	T 6/11/30
	suite 2706 . . . $300	Browne 5; T 7/1/30 gave it as $8 a day, but Browne had the Stevens management look up its records; Lingle also had a mis-

		tress who worked in a brokerage: Browne int.
275	"a rather pretentious" . . . home	T 6/21/30; T 6/10/30 put it in Grand Beach, Michigan, but the later story was Swanson's official accounting of Lingle's finances, straight from records; Grand Beach lies just over the state line; in T 6/30/30, Lingle's city editor explained that before he bought the house, he had summered at Holland, Michigan, favored by his buddies on the police force
	"not because it paid him"	T 6/10/30
	"Where's Al?"	Ness 130–1
276	John J. McLaughlin	T 6/10/30, 7/1/30; Boettiger 42–5; Lingle had a colleague listening on an extension
277	Biltmore Athletic Club	T 7/1/30, quoting H&E
	"Well, the racket is through"	Ibid., quoting DN and H&E
	blue serge . . . skimmer	Browne 4
	bank . . . $1,200 deposit	T 6/29/30
	[Lingle killing]	T 6/10/30; Boettiger 16–25; Browne 23–4; Boettiger later claimed the call of "Hy Schneider" was not directed at Lingle, but Browne has established to his satisfaction from the news dealer's testimony that the call was specifically to "Jake" and that Lingle did indeed respond; at the time it was thought that the call meant to finger Lingle, which theory Boettiger, a *Tribune* man writing to affirm the questioned conviction of Leo Brothers (see below), wanted to dispute: the paper's position was that the murder of its legman *had* been entirely avenged—as publisher McCormick had sworn it would be—and if the gunman had accomplices who had escaped, the vengeance was incomplete; Boettiger later broke with the *Tribune* over its virulent anti–New Deal policy, especially when he, as Washington correspondent, had to ask Franklin Roosevelt hostile questions, and most particularly after he fell in love with the president's daughter, Anna, whom he married, switching to Hearst, who made him publisher of the Seattle *Post-Intelligencer* (McPhaul *Deadlines* 275–7)
	Paraphrase won . . . $23.46	*Sic*; T 6/10/30: tracks apparently had not yet discovered the delights of breakage
278	[Ruthy chase]	T 7/31/30, in addition to above
	"Lingle's money operations"	T 6/26/30
	father . . . $500 . . . uncle	T 6/16/30; Boettiger 60–1 (also stock market operations)

279	Russell . . . forced to resign	T 6/18/30; early as DN 6/13/30 aldermen had called for his ouster, along with Stege (who was later reinstated)
	''Alfred Lingle now takes''	T 6/30/30; the quote ''that Alfred Lingle was killed because . . .'' actually reads, ''that Albert Lingle, etc.''
	''threat to the press''	T 6/18/30
	''Now . . . von Frantzius''	Boettiger 46–9; DN 7/5/30
	Foster . . . Ted Newberry	H&E 7/3/30; Newberry had returned to have the serial numbers removed from the murder weapon
	''Foster . . . ''Frost'' . . . Bruna	Browne int.
	''This town . . . too hot''	DN 7/1/30
	Ruthy . . . identified Foster	DN 7/12/30
	Foster had pulled . . . trigger	DN and H&E, both 7/2/30; Sullivan *Surrenders* 17–8 claimed ''It was obvious from the outset that the police were loath to consider Foster the slayer . . . it would upset all the plans and clews announced by them in the heat of public demand for action in the case''
280	[Sheridan Wave]	DN 7/1/30; T 7/1/30; Sullivan *Surrenders* 21–2; Boettiger 94
	''If this joint is opened''	Allsop 154, quoting H&E; this is the only source giving the full name of the Sheridan Wave
	Lingle . . . $50,000	Same sources apply as for Sheridan Wave
	[Zuta]	He had been a junk dealer (in the old sense, not narcotics) who had put his money in cheap hotels and rooming houses which he converted to more profitable brothels: Boettiger 103, who also wrote that Zuta's family name, back in Russia, had been ''Zoota,'' and that Bratz was his cousin, actually named ''Eli Zoota''; T 7/2,3/30; DN 7/14/30; H&E 7/15/30
	''Lieutenant,'' he pleaded	DT 7/2/30 differs slightly from the T 7/2/30 account; text amalgamates the two; Pontiac sedan: Pasley 302–305 also explained that a bullet had hit the gas tank; the T account merely stated that he had run out of gas, while the DT account had him losing the killers' car in the smoke; Barker Marine record: T 7/3/30
282	John T. Rogers	Allsop 161; tip from Wilson: Kobler 297–8
	series (gleefully reprinted)	Quotes are from T 7/2,3/30; Brundidge offered to name names before a grand jury, but did not print them; they appeared, inter alia (e.g., T 7/20, 25/30; Sullivan *Surrenders* 84, 95–6) in Boettiger 91
	Brundidge caught a train	T 7/19,20/30; H&E 7/19/30; the story carried single, interior quotation marks around

		"you're right; because," etc., indicating as with "dun't esk" an allusion to some well-known catch phrase of the day: "dun't esk" has survived but the text eliminates the other interior quote as probably confusing
283	other papers . . . four-hour	E.g., T 7/13/20
	Harry Reutlinger . . . swore	T 7/25/30
	Brundidge's . . . "hearsay"	DT 7/30/30; T 8/1/30
	"Here's what I want to tell"	Boettiger 179–81; no records were kept of the meeting, which was reconstructed as taking place probably 10/21/30
284	"Al wants to know"	Ibid. 189
	"set a gangster"	Ibid. 202
	"Buster" . . . Bader . . . Brothers	Ibid. 209, 221, 251
	Brothers went on trial	Ibid. 282
285	Perfectly lucid at Foster's	T 7/31/30
	"So they got to him"	Browne 228–29
	"the biggest frame-up"	H&E 4/4/31; Allsop 173 stated that Capone had "handed over to justice" Brothers, when Roche wouldn't take a dead body, but no *proof* exists that Brothers was framed or that Capone did it
	Ruthy . . . shot and killed	Boettiger 298–9
	Brothers emerged from prison	Browne 263
	[Zuta] . . . Lingle's murder	E.g., H&E 8/3/30 headline: GAMING CZAR HIRED SLAYER OF REPORTER SAY POLICE
	[Zuta murder]	DT 8/2/30; H&E 8/3,4,9/30; T 8/2/30; DN 8/4/30; men . . . cottage: the news reports put the number at 15, Boettiger 108 made it nine or ten; nickelodeon: the report circulated that it was a player piano (viz.: the DT and T accounts), but the picture in DN shows clearly it was a primordial jukebox, or nickelodeon, which H&E 8/3 referred to as a "phonograph"; the tune is generally given merely as "Good for You, Bad for Me," but Allsop 170, a music buff, gave the full title, adding that it was from the show *Flying High*; "Turn around": the DT account; sixteen bullets: Pasley 305
286	gun . . . Danny Stanton	DN 10/24/30; T 12/15/30; Boettiger 109–10 gave a convoluted account also implicating Sam Hunt in the first attack on Zuta, ending in any case with the conclusion that Capone had mounted both operations
	Capone . . . hosted a party	T 8/2/30
	[Zuta papers]	H&E 8/16–20; T 8/16, 17, 19, 20; DT 8/18, 19; DN 8/19—all 1930
	"A crook is a crook"	Sullivan *Surrenders* 141; Pasley 312 gave the quote in a shorter form

Page	CHAPTER 26	
287	Prohibition is an awful flop	Allsop 37; Sinclair 366 gave a version different only in that it transposed lines 5, 6 and 7; he identified the source as FPA
	[sugar carrier]	NFA
288	Edwin F. McNichols	McNichols int.
	Joseph Refke remembers	Refke int.
	Arthur T. Ristig	Ristig int.
289	McErlane . . . five charges	H&E 6/9/31
	[attack on McErlane]	DN 2/25/30
	two or three gunmen	No one saw them, their number reconstructed from the number and angle of shots: Pasley 338 made it two, Burns 254, three
290	Oberta was found	DN 3/6/30
	moved in on labor unions	T 4/20,26/30; DN 4/25/30; NYT 4/19/30 discussed his growing political power
	teamsters . . . building trades	T 9/2/30
	World's Fair for 1933	Murray 150; originally, it had been projected for 1937, anniversary of city's charter, but moved up as a Depression cure, the justification being the incorporation of the town
	Roland V. Libonati	He refused an interview for this book, evidently still mad at a reporter from Texas he felt had betrayed him; Judge Marovitz says that when he was an attorney, himself, he always found Libonati a reasonable man to deal with
	increasing . . . booze business	NYT 7/5/31
	$100,000 load . . . bourbon	H&E 5/13/30
	henceforth took from Capone	T 2/7/31
	Matt Kolb	n.p. 5/16/31
	take his pretzels and soda	T 10/2/31
	''Honest to God,'' he'd tell	Murray 163–7
	Daughters of the Nile	DT 5/28/30
	Five Spanish actors	T 12/6/30
291	producer offered . . . million	n.p. 3/31/31
	In Russia . . . Molotov	DT 6/15/32
	French weekly *Voilà*	n.p. 3/18/32; also England
	Cornel Capovici	Brooklyn *Eagle* 7/22/31
	John Gunther . . . reported	DN 2/26/31
	''Why, Al Pecan''	n.p. 2/7/33
	Boy Scouts . . . brought tickets	DN 6/14/64: reformer Saul Alinsky had been there; Allsop 293 gave the number ten thousand, but called it a ''rally,'' not football game
	Cicero high school game	T 8/5/30
	Rahn . . . admiring . . . ability	T 12/13/31
	Merrill C. Meigs	Murray 124–5; Allsop 293 claimed that Capone had considered engaging PR expert

Page	Reference	Note
		Ivy Lee to polish his image but it had come to nothing—questionable, partly for the reason Allsop gave: that Capone knew himself a master at image-making, partly because he would have realized that traditional PR could do little for him
291	"lot of people in Chicago"	Burns 311–12: with the usual qualms; but this not only so exactly mirrors what is known of Capone's views, it reads like him; the entire quote in Burns ran nearly three pages and, although presented as a single quote, had many echoes of past quotes, suggesting that the whole was a pastiche
292	soup kitchens . . . three thousand a day	Chicago *Post* 11/14/30 put the number at 1,100 a day, but consensus (e.g., Los Angeles *Examiner* 11/16/30; NYT 11/14/30) agreed on 3,000 at a cost of about $2,000 a week
	"The public has one idea"	S-T 4/15/51
	hired . . . Jule Styne	BUL 5/20/79
	a man thought as he killed	Kobler 215 interpolated this into Capone's soliloquy with the press before he left for Los Angeles at the end of 1927; it appears in none of the contemporary news accounts of that day's interview
293	shaking down visiting . . . stars	NYT 12/29/29 except for Vallee and Holtz: Jessel 100
	Richman told a different	Richman 6–13
	George Jessel recounted	Jessel 99–100
	Ethel Barrymore got in on	Barrymore 264
294	Joe E. Lewis was warned	Cohn 3-25, 89–90; there's question as to whether McGurn was the one who sent the attackers: he was not shy about doing his own killing, and he did not leave victims "for" dead; Lewis had a lot of enemies, including cuckolded husbands; Berardi doubts that it was McGurn
	Ben Hecht told of	Hecht 486–7
	Clara Bow . . . embroiled	Los Angeles *Examiner* 11/15/30 for the cheering story, *L.A. Times* 11/16/30 for the refutation; the story was so compelling David Stenn included it, without the refutation, in his biography, *Clara Bow: Runnin' Wild*. New York: Doubleday, 1988, pp. 204–5
295	Edward J. O'Hare	DN 5/20/43 (retrospective series by Clem Lane); George Remus episode: T 11/25/39; "best stool pigeon": DN 11/16/39; "most important . . . factor": Wilson 42; MacCarthy int. makes the point that it's hard to understand why O'Hare thought mid-level Treasury men could do more about securing an Annapolis appointment than Capone

Page	Note	Source
		could have done; a good guess is that O'Hare wanted to keep himself safe from government prosecution while continuing to make his fortune as an outfit ancillary for dog racing
295	"greatest natural undercover"	Wilson 33; for Malone: pp 33–35; Irey 38–44, 58–60;
	"Both . . . claimed credit"	Wilson 33–4, Irey 38
	"Michael Lepito"	Irey 40, "Phila. gangster" Wilson 34; Irey evidently anticipated readers' objections that the outfit would wonder why none of them from Brooklyn or Philadelphia knew him or of him: he allowed that gangsters "are never gregarious": Irey 43
296	"We all knew"	Irey 58
	"was sweating bullets"	Wilson 35; Irey wrote that Wilson suggested Malone go armed, and when told all guests had to check their guns at the door, told him to take two and check one (in fact, Meyer says guests at such dinners were patted down); served pepper steak, Malone went into a coughing fit, which dislodged his gun from his belt into his lap, but he bent over and shoved it back, sweating his own bullets; Wilson mentioned the steak without a hint of the gun
	tip from Ed O'Hare	Wilson 44–5
	three targets . . . Irey	Irey 57
	Boettiger added	Boettiger 191–6
297	Another observer saw him	NFA
	"Chicago . . . drying up fast"	Ness 218; "dry as bone" and "put Capone out": 240
	Capone's "undoing"	Ness 83–4
	$2,800-a-year men	Ness 27
	turned down $1,000 a *week*	MDN 4/20/59: also "everyone here in Chicago"
	Capone "had to crush"	Ness 148; (Frank Wilson story: Wilson 44–5); "Maffia [*sic*] gunman": Ness 168–77; lone terrorist: 159–60, Ness's driver: 189–199; for other instances of evoking The Menace see 141–2, 185–6
298	"We all carried guns"	MDN 4/20/59
	One night, everyone . . . gone	Wilson 36–9, including Shumway story, below; Irey 52–55 covered the same incidents, giving a modestly more colorful version, the key ledger supposedly mislabeled "Barracks, Burnham, Ill.," one of Torrio's original suburban whorehouses: that indeed appeared, but only incidentally, scribbled on an inside page; Spiering 105–16, 127–33 had access to the IRS jacket and printed a number of useful memos from it that appear nowhere else (the IRS now

		refuses to let the public see anyone's tax files, however long dead, after the 1976 law in reaction to their shameful abuses in the past); Spiering also reproduced the ledger page described in the text: he dated it 12/2/24 (evidently following Kobler 324) without examining his own exhibit, because the only sure date is at the top where ''Frank Paid 17500 for Al'' appears: December, 1924, with no day specified because the page is for the entire month
299	February, 1931, Ed O'Hare	Kobler 326–7
300	[vagrancy warrants]	T and NYT, both 9/17/30; Lyle background: Lyle 15–18, 22; brawling: T 1/4/22;
	plotting with Moran	DN, H&E, both 10/25/30
301	Pasquale Prestogiacoma	DN and H&E gave it as ''Prest*i*giacoma,'' but agree on ''Presto''; however T gave it as in the text both immediately after the killing and enough later so they could have corrected an error: T 4/1/31
	[Aiello killing]	Amalgam of T, DN and H&E, all 10/24/30; DN gave the number of bullets in Aiello as sixty; Boettiger 182 and Browne 214 gave the number from the first burst only as ''several,'' from the second, thirty-seven; T 10/25/30 made it fifty-seven wounds from ''shotgun slugs and machine gun bullets''
	the second reason Capone	H&E 2/13/31; T 2/24/31
	Galvin . . . McGoorty	DN 11/3/30; T 11/4/30
302	whispered . . . happy couple	H&E 11/18/30; T 11/19/30
	Mafalda denied it all	Chicago *Post* 11/21/30; see T and DT, both 12/15/30 for lavish detail of wedding one woman reporter described as ''neat but not gaudy''; John later . . . problems: SUN 9/25/45

Page CHAPTER 27

303	''Come in,'' said Capone	H&E 1/18/31, reprinted from the day before; she wrote that she had stopped by ''about ten days ago'' though she referred to Sonny's gifts still under a Christmas tree on cotton snow
	Balaban and Katz	Meyer int.
	Danny Serritella . . . wrote	Kobler 242–3 quoted Serritella's letter to Mayor Thompson for use in a campaign tract; Murray 125
304	Dupont Circle mansion	Barrymore 255
305	Capone's trial . . . contempt	T 2/22/31; dodge warrant: H&E 2/24/31
	''nutty'' . . . ''blubbering''	Dobyns 106; also NYT 2/24/31
	296,242 votes . . . 228,401	DN 2/25/31

305	''Lyle . . . make me an issue''	T 2/26/31
	[trial]	DN 2/25–7, T 2/26–8, H&E 2/26–8, all 1931
	''crowds of the worshipful''	H&E 2/26/31
	Capone gazed . . . impressed	DN 2/25/31; and following quotes
	write . . . autobiography	Kobler 316 counted seven books that featured Capone to some extent; he wrote that Capone bought a hundred copies of *Al Capone on the Spot*; McPhaul 247 wrote that a colleague had been asked by Capone to collaborate on a biography—if he'd agree to skip the killings; Frank Mount, editor of this book, thinks $2 million impossible in 1931: ''As late as 1960,'' he writes, ''advances of $20,000 caused hushes to fall on editorial meetings''
306	''I'm only a bondsman''	T 2/26/31 made it a Lieutenant McCarthy who came to the rescue; DN 2/25/31 made it Lieutenant Edward Birmingham (who had refused a $5,000 Guzik bribe), but did not give the bondsman's profession or quote
	[Dr. Phillips description]	T 2/27/31 pictured him standing between the two nurses
	turned jelly	Everyone pushed him around; when Mae's brother, Danny Coughlin, was hurt in an auto accident, Phillips treated him—presenting a bill for $2,000, including specialists' fees, only to settle for half when Capone objected; his current bill ran from $300 to $400, but no money had changed hands: instead Capone had touted him onto other patients; memo dated 7/24/29, while Capone was in Philadelphia, from George Johnson to the U.S. Attorney General (Nat. Arch., DC) shows that the government had contemplated charges against Phillips, too, but held off, anticipating that he would make a better witness against Capone, unindicted; they never did charge him
307	''If the judge thinks''	T 2/28/31; DN 2/27/31 quoted him as saying, ''We'll get another court to overrule this court,'' but surely he knew better from his lawyers
	''no legitimate occupation''	T 2/28/31, quoting Asst. St. Atty. Harry Ditchburne
	needed a continuance	DT 3/20/31
	another gray medley	DT 4/3/31 for first Ditchburne quote; NYT 4/4/31 for gray medley and ''single copper''; Lyle 262–3 claimed Capone had waited ''until I was on assignment in another court'' to surrender so he could appear ''before a judge who had been a severe

		critic of my high bond policy''; first of all, Capone did not pick the date of his contempt trial, the government did, so he could hardly have timed it to coincide with Lyle's absence; worse, while Padden may have been venal—many municipal judges were (and though it proves nothing, later that year his brother would be convicted of forgery [DN 7/31/31])—he had no choice: the state presented no case; so Lyle's implication that Padden somehow let Capone off a charge he himself might have made stick (which isn't a judge's job anyway—indication of why even his crooked critics were right about Lyle's lack of judicial temperament) was cheap and libelous
307	''the state moves''	DT 4/3/31
	Tony, Tony . . . pushcart at?	MDN 3/1/31; election results: T 4/8/31
308	first tax evasion indictment	Details in stories of indictment, below
	Heitler played cards	T 5/3/31
	''a down and out, dope''	T 9/18/31
	[Emily Mulcher] got a call	T 5/3/31
	Hattie Ganusch saw flames	T 5/2/31
	revolver with all six rounds	Ibid.
	found . . . car's license plates	T 5/4/31
	''when we want Capone''	Ibid.
309	[Heitler letter]	T 9/17/31
	''No such letter ever came''	Ibid.
	''No one . . . believe Heitler''	T 9/18/31
	''I know nothing whatever''	T 9/17/31
	grand jury indicted [Capone]	NYT 6/5/31; T 6/6,7/31, 10/17/31
	income of $1,038,660.84	T 6/6/31; the government deducted $1 a year to reach the tax form's ''adjusted income''
310	*The New York Times* joked	6/8/31
	week later . . . third indictment	Criminal Case 23256, Nat. Arch., Chicago; T 6/13–14/31
	Johnson . . . tremendous pressure	n.p. 6/16/31; also REC 8/3/31 for Hoover
	soon as Capone learned	TAX: T 7/31/31: ''in some way unknown to me,'' Johnson said
	''the courts . . . had followed''	Ibid.
	one of his lawyers insisted	NYT 6/16/31; it was Billy Waugh, who seems not to have represented him afterward
	Johnson declared ''We don't''	n.p. 6/7/31
	Mike Ahern said, ''. . . And''	DT 6/16/31
	''I'm going to throw myself''	Ibid.
	''sulphur . . . banana''	T 6/16/31; H&E 6/16/17; Waugh was not in court, instead Capone had with him Ahern and Leopold B. Melnick
311	President Hoover applauded	T 6/16/31; also maximum sentence and 2–4 year estimate

311	United Press heard	LD 6/27/31
	everyone took it . . . *fact*	*Colliers* 10/10/31
	''He was arrested''	*Outlook* 6/17/31
	''a devastating criticism''	LD 6/27/31; also following press quotes
	''a victory'' for Capone	*New Republic* 7/1/31
	Johnson guessed . . . Capone's	Wilkerson nomination hearings, part 3 (see note on Loesch testimony, Chapter 18); Ross 42
	[Wilkerson]	Busch 201; NYT obituary file
	''conspiracies . . . conspiracies''	Philadelphia *Inquirer* 6/30/31; Busch 199 also gave a good account
312	''were unkind enough''	MDN 6/7/31; details of removal: documents in archives of County Clerk, Dade County
	Giblin . . . ''ex–football player''	Kobler 291
	''I found out right away''	MH 6/23/68
	when Capone ''settled''	T 6/24/31; $10,000 was generous: Capone's Philadelphia lawyers had charged $15,000 for representation that included his trial and six complex appeals; Giblin and Gordon had appeared in court only three times for Capone—plus the bailing-out at the police station (ABA trial documents)
	''a Lonesome Town''	H&E 6/21/31
	''Jail's a bad place''	BUL 3/15/30

Page **CHAPTER 28**

313	Capone spent the day	Text is an amalgam of H&E, DT, NYT and of the United Press report as it appeared in REC, all 7/30/31; the H&E interview took place in the ''afternoon,'' though Ted Tod reported Capone was wearing the pajamas mentioned in NYT, while the UP reporter saw him at night in ''polo shirt'' and ''shapeless brown trousers''
314	''The consensus,'' wrote	DT 7/30/31 by F. E. Blankenship
	''There's nothing to all this''	DN 7/30/31
	''Now let me alone''	H&E 7/31/31
	court's opening statement	T 7/31/31: a transcript of proceedings
	Wilkerson had to hear	DN 7/30/31; transcript of the opening: the judge adjourned immediately after his statement
	Johnson appeared . . . marginally	T 7/31/31
315	Capone's motion to withdraw	DT 7/31/31; T 8/1/31
	indictments . . . not reports	H&E 9/11/31
	[Vanderbilt interview]	Vanderbilt 168–74; a much longer version appeared first in *Liberty* 10/17/31, quoting Capone at length on the need to feed the hungry on threat of red riot; but as Kobler 339 put it, ''Here, one suspects, Vanderbilt substituted his own ideas and vocabulary'';

Page	Phrase	Source
		it certainly doesn't read like Capone, and it's perhaps significant that Vanderbilt left out those quotes in his book four years later
315	Capone would deny	T 10/13/31
	heard himself . . . booed	T and n.p., n.d.; "Yea-a-a, Al": NYT—all 10/4/31; the editor of the student newspaper wrote an anti-Capone editorial, then demanded police protection: n.p., n.d., above; NYT 10/16/31 gave a shorter version of the editorial
316	offered $1.5 million	Irey 57–8
	Capone went after the jury	Wilson 50–1
	[tax trial]	Basic account in DN the same date of events, T and H&E the next day; following notes specify only items that appeared either uniquely or in distinctive form; other sources as noted; all newspaper dates are in 1931
	Wilkerson . . . swapped jury	Assistant prosecutor Dwight Green was first to tell the story in a talk at the Press Club: n.p. 12/8/31; Irey 62 claimed he got the word first, and from Malone
	sixty veniremen. . . . Fifty . . . excused	DN 10/6; the defense could complain only that a high proportion had recently been federal jurors in other cases: H&E 10/7
	dismissed . . . six . . . peremptory	NYT 10/7
	When not in court	H&E 10/12
	"told you I wasn't worried"	H&E 10/7; NYT gave it as "Well, who wouldn't be"
317	not even "Snorky"	DN 10/7
	"a nervous, fussy deacon"	NYT 10/8; Berger did not get a byline until later in the trial, but he was clearly, from style, the NYT "Staff Correspondent" given credit over this story
	everyone . . . afternoon off	Mulroy int.
	"I would . . . let them take it"	T 10/8; H&E had slightly different version
318	Ahern . . . Fink	NYT 10/8,9, including picture of Ahern; picture of both: H&E and DN, both 10/8
	Nash . . . litigator of the team	Busch int., confirmed by Nash's son
	"Oh, my conscience"	NYT 10/9; however, the phrase does not appear in any of his transcripted remarks throughout the trial
	"to come in and settle"	H&E 10/9
	"Boston Tea Party"	T 10/9; H&E gave the judge interrupting with "I suppose this is a B.T.P."
319	"This is the last toe"	T 10/9 asserted that it was "an allusion not quite clear, but which some supposed was a technical phrase in wrestling"; my brother, Ted Schoenberg, was a schoolboy wrestler in the mid-thirties, captain of his college team in 1941, and never heard of it; two contemporary professional wrestling promoters, consulted by phone, never heard of it either
	judge replied "I presume"	H&E 10/10

319	Fotre . . . Small . . . horn-rims	Ross 69
320	eyes darting about the court	T 10/10 and questioning, including "You think it over"
	other witness . . . protection	Ibid.
	Fotre squeaked, "What"	Kobler 344; Wilson's own accounts in his book and *Colliers* piece do not mention it, but it would help explain what happened with Phil D'Andrea next day
	Capone's prodigality	In addition to newspaper accounts, ABA documents contained a painstakingly prepared chart showing the sources of testimony for each item
	jurors . . . telephone . . . $36	H&E 10/11
	contractor . . . Koernitzer	T 10/11; H. F. Ryder complained that he was still owed $125 for materials and had been ill-used by some toughs when he went around to collect, Capone not being at home; Capone bridled, but relaxed when cross-examination got Ryder to say he was sure "Mr. Al will pay" (H&E; T gave it as "I think he will give it to me eventually") and allowed that Capone was "a mighty fine man": DN 10/10; H&E & T 10/11
	[D'Andrea arrest]	Wilson 52 claimed he spotted the bulge of a pistol under D'Andrea's coat and asked him to step out during a recess, agents Tessem and Sullivan helping in the arrest; contemporary news accounts all specified that the arrest took place "at the close of the court session" as T 10/11 put it, *not* during a recess (with D'Andrea lured out of the courtroom by a fake message); D'Andrea dismissed as bailiff: T 10/14
	"rural gentlemen of simple"	NYT 10/13
321	"Tendin' bar on Coney Island"	T 10/13; rest of quotation is from DN 10/12, which quoted the scene in court more fully
	"not even to Al"	T 10/14; the quote went on to say "or to Pete Penovich," the manager
322	Penovich, manager . . . Subway	T 10/15 had him answering that, yes, he was connected with the "Smoke Shop," but that was a confusion of nomenclature: the Smoke Shop was in the Anton next to the Hawthorne; when raided, the operation moved within a few blocks on Twenty-second and was called variously "Subway" and "Radio," although as shorthand it was all referred to as the Smoke Shop; T 10/14 gave Ries's response, for instance, that he had been at the Subway with Penovich
	Edward G. Robinson	H&E 10/15; all the papers ran sidebar columns with the daily color surrounding the

		trial; for a particularly good collection, see Kobler passim 340–9
322	Warden David Moneypenny gave	T 10/16
323	''You have the high privilege''	T 10/16
	''If the defendant's name''	H&E 10/17; the rest of the trial and the sentencing were reported by all papers (as noted above)
	Bea Lillie . . . ''Well''	Philadelphia *Inquirer* 10/17/60 (retrospective)

Page CHAPTER 29

326	''I guess it's all over''	Chicago *Post* 10/24/31
	''You've done all you could''	T 10/25/31
	''Goodbye, Al, old man''	*Post* 10/24/31
	''Mr. Capone . . . want to serve''	Ibid.; DN 10/24/31
	''Get enough, boys''	H&E 10/25
	''a little below the belt''	Ibid., a slightly more complete version than T 10/25 gave
	''Please don't . . . picture here''	*Post* 10/24; T 10/25 gave the quote reversed, ''family,'' first
327	''He knows,'' one guard said	T 10/25; also ''The World Is Waiting''
	stay . . . bail	T 10/27/31; the term for *stay* was *supersedeas*
	Wilkerson . . . Phil D'Andrea	T 10/29/31; the warden explained the hospital block accommodations by saying he wanted to spare other prisoners the demoralizing effect of having to associate with gang chiefs (DT 10/26/31); they had their noon meals sent in (T 10/26/31)
	D'Andrea also stayed	Dornfeld int., quoting an ex-prisoner who saw D'Andrea in his cell
	anonymous telegrams	H&E 12/18/31; other details, DN 12/17/31
	''I'm in jail . . . satisfied?''	T 12/18/31
	pattern for . . . rest . . . stay	DT 12/26/31; T 1/26/32; DT 1/27/31; T 2/26/32
	Luciano . . . Schultz . . . Torrio	Kobler 352
328	Observers assumed . . . Mattingly	H&E 10/19/31
	[statute of limitations]	Ross 118–123; T 2/28/32; n.p. 5/1/32
	Capone out of a card game	T 2/28/32
	''most outrageous thing''	n.p. 3/3/32
	Frankie Rio showed	DN 7/9/43 (retrospective)
	editor Arthur Brisbane	Kobler 353
	founder . . . Schwarzkopf	Thomas De Feo (phone), curator N.J. State Police archives
	Lindberghs called in	Wilson 57–8
329	[offer renewed]	T 4/23/32
	Sam Guzik . . . Nitti	BOP letter dated 6/12/43 detailing sentencing, time served and release dates of prominent outfit members
	both . . . had wangled soft jobs	T 10/27/31
	toughest . . . federal prisons	H&E 5/4/32

329	said good-bye . . . family	H&E 5/3,4/32; T 5/5; there was some confusion over exactly who was there (cf. note on Rose Capone for Chapter 5)
	[Capone] heard it . . . radio	T 5/4/32
	"bodies . . . Alphonse Capone"	Ibid.; Morici details: DT 5/3/32
	central yard . . . "Mussolini"	T 5/4/32; H&E 5/4/32 gave it as "I bet Mussolini never got such a send-off as this"
	"I'll take what they give"	DT 5/4/32; T, H&E, both 5/5/32; "I'll take": T; "spaghetti": T and H&E
	"Who are you?"	T 5/5/32; H&E 5/5/32 gave essentially the same account, with slightly different wording, but H&E had the warden, not an outside gatekeeper, speaking from the first and asking "Are you in charge of whoever is in your party"; H&E gave Laubenheimer's quote about the ten years, whereas T merely said he explained the situation
330	A. C. Aderhold	Kobler 356 gave his first name as "Arthur"; newspapers always made him "A. C." and he was addressed that way in a BOP letter dated 9/21/35 from Warden Johnston of Alcatraz
	handed over the $231	BOP receipt memo dated 5/24/32
	"Get on your way"	H&E 5/5/32; Ann Diestel, BOP
	Laubenheimer . . . heard nothing	T 5/6/32
	followed the routine	DT 7/6/36 (*sic*) series by an ex-prisoner, William S. Bolton
	a model prisoner	A 11/18/32, quoting Dwight Green
	"Sleeping is like escaping"	H&E 5/4/32
	crying "No, no!"	Rudensky 57
	April 11, 1932 . . . Court ruled	T 5/1/32, 9/22/32
331	William Leahy . . . Nick Arnstein	A 9/22/32
	Capone's lawyers argued	H&E 4/27/33; the case had been set for October 19, then for November, with no reason given in the papers for the continuances, which the government asked for
	"overpaid dumb bastards"	Rudensky 58
	Investigators heard	MacCarthy int.
	[Prohibition Repeal]	Sinclair 386–92
	Capone had taken over	H&E 1/24/33, except $2 cigars: Rudensky 56; a BOP letter dated 5/31/32, less than a month after Capone's arrival, denied the rumors already circulating
	warden . . . denied . . . reports	H&E 1/24/33; the warden wrote a letter, read into the Congressional Record: n.p. 1/26/33, an AP dispatch; BOP letters: a screenwriter, Homer Croy, was compiling a book of "humorous tombstone epitaphs," some written pre-mortem, and wanted one from Capone; he offered some samples, including "At least I won't have

		to pay taxes''; prison officials refused to pass the request on, observing that prison was no joking matter
332	Capone was ''being accorded''	BOP letter from its director, Sanford Bates, dated 5/31/32
	''lived like a king''	DT 2/15/38 (*sic*) quoting the current *Saturday Evening Post* article by Bryan Conway as told to T. H. Alexander
	special pair of shoes	BOP memos dated from 1/18/34 through 4/20/34 about repair of the shoes which had been built a year before
	too much underwear	BOP dated 10/18/33
	Another guard found . . . bed	BOP dated 12/15/32
	E. W. Yates complained	BOP dated 10/31/33
	album, two rugs	BOP forms for disposition of belongings dated 8/25 (presumably '34: these were items he was not permitted to keep at Alcatraz) and 10/7/34, he had everything shipped to his mother in Chicago, except following note
	Encyclopaedia Britannica	BOP letter dated 9/28/35; he left them in Atlanta and had them shipped to Mae at Palm Island
	a player would leave	Bennett *Prison* 99–100
	A clue to his power	Ibid. 100
	Rudensky later claimed	Kobler 358; it does not appear in *Goniff*; but Kobler talked with Rudensky, and read an unpublished MS. of his
	Capone would give . . . money	DT 7/6/36 (Bolton series)
333	The FBI investigated	FBI file
	''too big a problem''	Bennett *Prison* 99

Page **CHAPTER 30**

334	''would it not be well''	Cummings 29–30
	one and one-quarter miles . . . six-to-nine-knot	Godwin 35–7
	[Alcatraz history]	Ibid. 15–27, 30, 32–35; Johnston 2–8, 12–13
	Depression budget allowed	Johnston 13–15
	Capone's arch supports	Godwin 79; BES wrote of a humiliating search of Theresa in the guard shack
335	the Army cleared out	Johnston 12; his record: Godwin 76–7
	''It was after chow''	Rudensky 61–2; the text reads ''S.O.B.''
	special train . . . shunted	Johnston 18–21; Godwin 83–4 told of the newly arrived prisoner, below, who ''cried like a baby,'' at first sight of Alcatraz; Johnston's ''FIFTY THREE CRATES'' telegram does not appear in capital letters in his text
336	[mail and visits]	Johnston 31–33; BES; Godwin 80
	[food and tobacco]	Godwin 84–5, except ''milanaise soup,'' etc.: Johnston 245

	Capone . . . shipped home	BOP forms
337	offered to equip . . . band	Johnston 35
	a nine-by-five-foot cell	Godwin 77, amended by Anne Diestel, BOP; "so narrow": picture following p. 120
	4:50 . . . 6:30 A.M. wake-up bell	Schedule is exact, supplied by Diestel from a BOP document; also, New York *Herald-Tribune* 9/6/36: interview with an ex-prisoner; Kobler 370–1; Bennett *Prison* 20; Godwin 80; H&E 12/16/34, quoting John Stadig, a prisoner who had escaped on his way back from a court appearance in Oregon; BES claimed they could shave only twice a week
	relaxed the rule of silence	Godwin 121–2; Audett 198 supplied such details as talking only to a guard at work
	"It was apparent," Johnston	Johnston 31
	"most prominent gangster"	Bennett *Prison* 99
	"name that topped the list"	Johnston 26–7
338	"inaccessible . . . outside"	Cummings 33–4
	Johnston found . . . fascinating	Johnston 33–6
	"accomplishment to assault"	Bennett *Prison* 99
	age thirty-five . . . twenty-five-year sentence	Godwin 121
	state had filed detainers	Johnston 39
	"Life gets so monotonous"	New York *Herald-Tribune* 9/6/36
	most "were in a mood"	Johnston 24
339	Capone . . . laundry	Kobler 374 gave the best account; Johnston 36 barely mentioned it but put Capone's time in the hole at eight days; T and H&E, both 3/15/36, giving different versions of an AP story that cited a "prison official" who didn't want to be named, made it four days; BES gave the name of Capone's opponent as "Collyer" and gave many details about the man, but had *Capone* throwing the wet laundry, which seems less probable given the sequence of events; H&E 2/16/37 spelled it "Collier"
	D block . . . "the Hole"	Godwin 78, 122; BES (only reference to their stay in the dungeon, which was seldom used); Audett 206
	"Who's a son of a bitch?"	DT 2/9/36; Audett 217–8 gave a graphic if tendentious and not necessarily reliable account of Allen's death
340	"Those guys are crazy"	Kobler 374
	Lucas . . . force-fed	DT 2/9/26; Godwin 127
	Another prisoner said	BES
	[Lucas attack]	DT 6/23, 24; T 2/24, 26; H&E 6/25—all 1936; monthly haircut: Kobler 375
	choke . . . Whaley	T 6/26/36; T 6/24/36; Johnston 61; H&E 2/16/37; fight with Gardner: H&E 6/27/38;

		warden's denials passim in stories: Godwin 105 averred that only the Collier (whom he did not name) and Lucas assaults happened; also T 2/26/36; Meyer int. for Cleaver, except ''desperado'': T 5/6/32
	Some . . . prisoners asserted	DT 7/2/36 et seq. (Bolton series; also DT and H&E, both 6/26/36, story about Bolton's charges); DT 2/15/38
	soldier wrote home . . . ''wop''	Johnston 36
341	''The boys . . . love him''	T 2/13/36
	''As a matter of fact''	Johnston 27
	''Capone got along as well''	Godwin 105
	One, chopping . . . tires	Ibid. 123–4
	inmate who calmly scaled	Johnston 203–4
	[attack on Johnston]	Godwin 86–9
	''It's hell''	New York *Herald-Tribune* 9/6/36
	''hell nights''	H&E 6/27/38
	deposition to . . . Klein	DT 3/10/39; Kobler 378–9 gave many more details, including Sullivan's participation and ''I carried a gun''
342	enduring myth had it	E.g., *National Enquirer* 3/22/77 story on Edmund J. Ryan, MD, who claimed that he wanted to give Capone a Wassermann in Chicago but was refused because Capone was ''terrified of a needle going into his body''; Kobler 357 cited Dr. Ossenfort, at Atlanta, as having urged a spinal tap when a serum Wassermann ''which Capone took under protest, proved negative'' because analysis of Wassermanns was still ''somewhat crude''—but Capone refused; as the text shows, his Wassermann was positive and he received treatment for syphilis as well as unrelated prostatitis; a spinal tap was no more accurate—as Roger Simon, MD, puts it, both are ''flocculation reactions'': the specimen either clouds or it does not, though false positives can occur—the test of spinal fluid is to determine whether the disease has progressed to a neurosyphilitic stage
	''2 plus positive''	BOP medical report dated 9/7/32
	Mae wrote . . . Johnston wrote	BOP letter, those dates
	Saturday . . . Capone donned	Amalgam of DN 2/8; A, T and DT; T 2/10; DN 2/16—all 1939; and Johnston 37, which gave the calmer account, including ''What happened to you,'' etc.; T 2/10 noted that the straitjacket reports had been exaggerated
343	A spinal tap showed	BOP medical report dated 11/18/39 reiterating the results of spinal fluid taken 2/12/38
	[telegrams]	BOP those dates

344	[tests that summer]	BOP report dated 6/4/38, including the quote of Dr. Ritchey, who later became the prison's chief medical officer
	prison term would expire	regular good time reduced it to 1/19/39, plus, evidently, additional days credit
	pay the fine and costs	BOP memo dated 11/3/38
	government had pressed . . . taxes	T 3/29/35; H&E 11/8/36 gave the original assessment as $386,406.52, compromised to $220,980, and gave the first tax lien against the estate as $51,298
	government served an assessment	Kobler 377, quoting a letter from Ralph
	"managed to borrow"	Ibid.; also DN 9/10/37, which gave the $17,194 payment; it depended on how and when you counted, the interest mounting; the government made various separate claims, e.g., a tax lien originally assessed 4/19/35 for 1924 taxes of $9,443.55 with a $2,082.67 penalty and $5,719.55 interest—total $22,245.77—by 11/9/39 had grown to $28,314.30 because of additional interest: BOP tax lien notice dated 10/25/39 and BOP letter from the IRS to the warden at Terminal Island dated 11/9/39; T 8/27/40 reported the government's summary judgment for $265,887
345	John . . . cashier's check	T 1/5/39; Teitelbaum handed: A 1/5,8/39
	associate warden ferried	Johnston 38; "six weights": DT 1/9/39; other details, including "*The Lark*": A 1/8/39; vigilante attack: A 11/14/39, announcing Capone's release from Terminal Island
	"During the first few days"	BOP letter dated 1/16/39
	"screwy as a bedbug"	DT 1/9/39
	week later, Mae visited	A 1/17/39
	head of the Secret Service	A 2/28/39
	a visiting minister . . . banned	T 3/30/39
346	bureau had rejected . . . claim	BOP letter dated 10/31/39
	[O'Hare murder]	T 11/9/39, including "Mr. Woltz phoned," 11/11/39; A 11/11/39; DN 5/20/43 (*sic*; series on unsolved murders); DT 11/10/39 ran pictures of an Italian poem (the slightly garbled lyrics to the popular song, "Oi Mari, Oi Mari": correct version courtesy of NFA, the Italy-American Chamber of Commerce West, Inc.) and a note found in O'Hare's pockets; his watch was inscribed *Amor Sempiternus*, Latin for "love eternal," with the names Ed-Sue, the latter being his fiancée, Ursula Suzanne Granata; he also had an imitation key ring whose blades, properly arranged, spelled I LOVE U, and a poem with ruminations on how we must

		seize the moment because the fleeting years are slipping by us
347	"confidential secretary"	DN 5/20/39; A 11/29/39
	she would marry Frank Nitti	DN 5/20/43 (retrospective)
	FBI agents shoved	DT 11/16/39
	crossroads twelve miles east	BOP memo dated 11/9/39

Page **CHAPTER 31**

348	"blood-brain" barrier	Drs. Bayne, Rolfs and Simon
	Mae . . . drove Capone	H-A 11/16/39
	two-room suite . . . $30	MH 11/17/39
	Dr. Moore had some question	BOP letter dated 12/21/39
	[Capone] planted . . . tree	Carol Ristau on Union Memorial staff
	a house in Mount Washington	DN 1/23/40; MH 1/24/40
349	only to eat . . . Thirty hours	MH 3/22–3/40
	Daniel Coughlin	DN 4/3/40: business agent; Capone driver: Purdue int.
	"did a tremendous business"	Purdue int.; Waffle Shop: Kobler 382
	Muriel . . . Louis Clark . . . [dog]	Kobler 382
	Brownie . . . Rose, lived out	Ibid.
	fishing . . . golf . . . nightclubs	T 8/1/40; MH 8/3/40
	Capone threw . . . parties	Saxe int.
	Capone owed $265,877.71	H-A 8/26/40; DT 4/25/40
	government wanted Capone to appear	DT 2/6,12,17/41; T 2/13/41; H-A 2/18/41
	Capone . . . cost about $40,000	DT 1/27/47 (*sic*; postmortem piece)
350	witnessed . . . Norman Kassoff	Kassoff int. (phone)
	Vincenzo, or Jimmy—surfaced	Kobler 383–4 gave 1940 as the date of contact; news accounts, one of them contemporary, gave the date as 1941: DT 9/11/41, which gave his appearance as being "about a month ago" (though with such incorrect "details" as his living in Omaha as "a respectable businessman"), plus T 9/20/51 (*sic*) and DN 9/21/51 (which corrects the earlier DN 9/19/51 account that gave 1939), both on the occasion of his appearance under subpoena to testify before a federal grand jury about Ralph's taxes
	"firing from either hip"	DN 9/20/51
	"sole law authority"	T 9/20/51
	[James history]	Kobler (especially for the more discreditable details) plus those cited and S-T 9/20/51
351	[eye] lost . . . gangsters	Kobler; DN 10/1/52, in an obituary, made it lost in a fight in Sioux City without mention of gangsters or relatives
351	classmate . . . Desi Arnaz	St. Patrick's records; John Ingraham int. (sexton at the time, still active) remembers them well; present pastor, the Reverend James P. Murphy, confirms that Arnaz on a visit talked of the friendship; Purdue int.

		says they were "very good buddies"; their patron, dictator: Harris 39–42
	Sonny's slightest misdemeanors	MH 4/23/39: speeding; MDN 11/3/36: accident
	using . . . alias, "Brown"	Kobler 383
	bookies . . . softball game	Saxe int.
	Sonny . . . married	A particularly good account in MDN 12/31/41
	Casey's Oasis	Purdue int.; also Jim Casey's rank
	Her brother, Jim	Letter from Casey to the author; arranged by Miami Beach officer Tom Hoolahan, Casey relayed a request for an interview with Sonny, whose address is kept secret; Casey wrote to say that Sonny wanted only to forget the past and that he, Casey, could not discuss his ex-brother-in-law out of affection and loyalty
	set up house . . . Air Depot	MH 6/28/46; Kobler 385
352	enough penicillin	Kobler 385
	owner . . . least troublesome	DT 9/10/41
	"Mercer . . . Mr. Capone"	T 4/13/42
	reports . . . deathly ill	T 8/13/43; DT 8/16/43
353	James M. Ragen, Sr.	DN 6/25,26/46
	"He hasn't sufficient"	MH 6/28/46
	"nervous and excitable"	MH 6/29/46
	"nutty as a cuckoo"	SUN 7/3/46
	batting the ball . . . hours	H-A 6/27/46
	"likes to chew Sen-Sen"	MH 6/28/46
	"like a spoiled brat"	MH 6/29/46
	"Mrs. Capone . . . borne a cross"	MH 6/28/46
	likelihood, so did Capone	Purdue int. says that Monsignor William Barry told him Capone attended the early Mass nearly every day up to a few days before his final illness; would Barry have been as cooperative as appears below in the text for someone not a communicant?
	Andrew Volstead . . . died	NYT 1/21/47; "continued to believe": NYT 1/22/47
354	[Capone death]	MH 1/22–6/47; SUN 1/26/47
	kept him alive on oxygen	DT 1/25/47
	Ralph would emerge	Pictures exist of Ralph coming laden through the gate; Sosin int. for no help or basket and Cox quote
	Monsignor William Barry . . . shush	NFA: a longtime associate of Monsignor Barry's
	press thrown further off	Spirko int.
	[funeral]	SUN 2/5/47; plot 48: Kobler 386
	"Al had no enemies"	Murray 14
355	[headstone]	SUN 2/5/47 gave Capone's inscription as they assumed it would appear, with the "QUI RIPOSA" ("here rests"), "NATO," etc., extrapolating from the others; LeVell has visited the site and taken

pictures; he calls it "the placebo stone" and speculates that the family never intended to leave the remains there—as we'll see they did not—hoping the stone would satisfy the curious, who trampled the plot for years

CHAPTER 32

Page		
356	Red Barker, was gunned down	DN 6/24/43 (retrospective)
	Three-Finger Jack White	Ibid.; by the outfit: T 3/18/43
	"Diamond Lou" Cowan	n.p. 10/27/33
	James Genna . . . died in bed	DN 6/25/43; Sam and Pete: 6/29/43
	Myles O'Donnell . . . Klondike	DN 6/24/43
	McErlane shot . . . Elfrieda	DN 10/8/31
	died . . . pneumonia	DN 6/24/43
	Frank Rio	T/16/36
357	McGurn . . . laziness	Meyer int.
	[McGurn death]	T 2/15,16,18/36; "You've lost your dough": Kobler 376
	Anthony Demory sat	T 3/3,4/36; H&E 3/3/36; DT 3/2/41 (*sic*)
	[Louise Rolfe]	Accident: DN 10/8/36; DT 12/4/36; married twice: SUN 10/24/45; n.p. 11/26/51 (first ended in 1940, second 1949); gun: DT 2/12/40; "the darnedest thing": DT 3/3/43; living in CA: Graham int., found out during research for TV program, but he lost track of her
	1950s saw an end	Kobler 389 for those not noted below
	Charley Fischetti died	MDN 4/11/51; liaison: DN 9/28/50
	Rocco died in 1964	Browne notes; although, according to Meyer int., Joe was perhaps closest to Capone of the three, he remained less conspicuous, his fate unnoted
	Willie Heeney died	T 7/14/51; n.p. gave his age as sixty-three
	When Louis Campagna . . . died	S-T 6/4/55
	Maddox	S-T 6/22/58; he had been questioned in Stanton's killing, below
	Frank Maritote	MH 10/4/54
	Murray Humphreys	"nicest guy": Meyer int.; MDN 8/20/62; collapsed: MH 11/25/65
	Frank Galluccio	Balsamo int. corrects his piece in *Chicago* magazine, March 1990; Walter Winchell reported Gallucio's death: *Philadelphia Inquirer* 8/11/61
358	Danny Stanton lasted	DN 6/24/43; SUN 8/17/46
	George Moran	Ohio prison: SUN 8/28/46; A 2/11/54; Leavenworth and death: S-T 2/14/67 (*sic*); "I hope": S-T 2/26/57
358	Joe Saltis	Wrenching divorce: H&E 4/26/39; paternity: T 5/26/39; S-T 6/11/43; died broke: DN 8/2/47
	Spike O'Donnell	T 8/27/62

	Druggan . . . Lake	Friendship ended: S-T 1/12/47; Druggan—ulcers: T 3/3/54; died: DN 3/5/54; Lake moved to Detroit and died: S-T, DT & T, all 1/12/47
	Frank Nitti	Indictment 68043, Criminal Court, Chicago; Marovitz int. Murray 175–9; position of wounds: T 3/18/43 (*sic*)
	Ted Newberry's body	Murray 177
	Zangara	NYT 2/16–19/33, 3/7, 10,21,31/33, 9/27/33; four others were also wounded, one seriously, but only Cermak died
359	Jack Guzik	S-T 2/22/56
	John Torrio	S-T 5/8/57
	Big Bill Thompson	Fascist: SUN 3/14/47; death and safe deposits: NYT 3/19–21, 4/9, all 1944
	Dwight Green	A 2/19/56; Casey 202–8
	Jacob Grossman	S-T 3/31/75
	Frank Wilson	*Who Was Who*
	George Johnson	NYT obituary file
	Judge Wilkerson	Ibid.
	Pat Roche	S-T 7/12/55
	Eliot Ness	Cleveland: Nickel passim; Messick *Syndicate* 29, 109, 139–40; death: S-T 5/17/57
360	93 Palm Island	"Stole it": MDN 2/13/52; furnishings: MH 2/14/52; "Capone curse": S-T 1/26/69; Miller offered: MH 5/24/55; swapped: MH 8/25/55; other hands: MH 10/13/67 mentioned two new co-owners, Chimerakis and Williams, while Morrison bought it (MH 8/18/71) from another, Fowler; Morrison says the tourist traffic has diminished
	Theresa's death	S-T and NYT, both 11/30/52; DN 4/1/52
	house on Prairie	T 4/14/89; Chicago *Defender* and S-T, both 4/20/89; S-T 10/28/88;
	Capone's . . . limousines	England: Allsop 15–6; $510: S-T 1/23/58; "insult": INQ 2/16/75; burn: T 8/3/57
	Hawthorne Hotel	DN 2/17/70
	Lexington still stands	T 6/18/81, 7/14/89
361	[massacre wall]	"Hardly a day": DN 2/14/64; housing project: NYT 11/30/67; shopping center . . . moose heads: T 2/14/72; Canadian: S-T 2/15/68; INQ 8/5/78
	[Ralph Capone]	"It's tough": DT 8/11/41 (Jimmy . . . "Ralph runs it": S-T 9/20/51; also S-T 9/22/51 and DN 9/26/51; died: DN 10/1/52); Ralph divorce: A 4/26/36; DN 5/7/38
	Ralph junior	"Risky" . . . businesses: S-T 11/10/50; marriage: SUN 9/20/47; Dear Jeanie: DN 11/10/50 and S-T (above) gave the same version except for S-T's "Only you," which DN omitted; father not imagine why:

		DN 11/10/50; after his death, T 11/10/50 identified him first as Al's nephew, only then as Ralph's son
362	Ralph Capone drifted . . . died	NYT 11/24/74
	Matt	Speeding . . . murder: SUN 8/27/46; died: DN 2/1/67; funeral: DN 2/2/67
	Albert	Aliases: T 1/19/69; BUL 6/19/80; Rayola: S-T 10/15/42; gambling stamps: S-T 4/23/53; he had a handbook in an abandoned steakhouse called Rosie's: DN 8/24,25/60; $25 fine: DN 12/15/58; died: BUL 6/19/80
	John	John Martin: MH 3/7/56; MH 6/19/80 notice of Albert's death reported that he was survived by Mafalda and John Martin (*sic*)
	[Sonny's] first postwar job	Kobler 384–5
	no will . . . no money	Abraham Teitelbaum said that if there had been a will, he'd have known of it as Capone's last lawyer, and if there had been money, the government would have located it: H-A 1/27/47
	"Your father broke my heart"	Balsamo int., who was told by relative of man to whom Sonny told the story
	Ted's Grotto	Kobler 390; MH 3/7/56 alone called it "Ted's Grotto," other Miami references (including MH nine years later, 8/7/65) gave only "Grotto," but the MH alone gave the address, and the fact of John's bar, which suggests they knew better
363	working two jobs	Purdue int.
	Sonny phoned	MH 10/27/65; lawsuit: A 12/29/59; dismissed: S-T and T, both 6/17/64
	better . . . shots	Kobler 393; Southern Florida Historical Society has pictures of Sonny and Diana, also a fine shot, with their guns
	Springfield . . . $25,000	Purdue int.; Sonny's detective brother-in-law, Jim Casey, told him about the offer and conditions
	[shoplifting]	MH 8/7/65 gave it as aspirin and transistor radio batteries, MDN 8/7/65 as "headache pills" and flashlight batteries; but subsequent reports, including MH 11/16/65 on occasion of Sonny's sentencing, uniformly made it aspirin and flashlight batteries
	probation	MH 11/16/65; "real good customer": S-T 8/8/65
	change his name	T 5/10/66
	Mafalda felt differently	S-T 3/7/59; charges of assault: DN 6/26/57, 3/8/58; inscription: picture of marker from LeVell
	Mae continued to live	LeVell, Balsamo and Graham ints.: what they heard
364	1940 presidential election	NYT 11/28/40

364	Brazil's . . . hits	Thanks to Sebastião Santos
	"Al Capone Chair"	BUL 1/5/59
	show's rating	Telephone int. with NFA syndicator
	French radio	Letter to author from Robert Faherty
	"Jessica and Al Capone"	*Harpers* June 1931
365	Capone's marker	Pictures by LeVell

Bibliography

MAJOR SOURCES

Allsop, Kenneth. *The Bootleggers and Their Era*. London: Hutchinson, 1961.

Amfitheatrof, Erik. *The Children of Columbus: An Informal History of the Italians in the New World*. Boston: Little, Brown, 1973.

Asbury, Herbert. *Gem of the Prairie: An Informal History of the Chicago Underworld* Garden City, N.Y.: Garden City Pub., 1942 (Original ed.: New York: Knopf, 1940).

Balsamo, William, and George Carpozi, Jr. *Under the Clock: The Inside Story of the Mafia's First Hundred Years*. Far Hills, N.J.: New Horizon Press, 1988.

Bennett, James O'Donnell. *Chicago Gangland*. Chicago: Chicago Tribune, 1929.

Boettiger, John. *Jake Lingle: Chicago on the Spot*. New York: Dutton, 1931.

Bright, John. *Hizzoner Big Bill Thompson: An Idyll of Chicago*. New York: Jonathan Cape, 1930.

Burns, Walter Noble. *The One-Way Ride: The Red Trail of Chicago Gangland from Prohibition to Jake Lingle*. Garden City, N.Y.: Doubleday, 1931.

Dobyns, Fletcher. *The Underworld of American Politics*. New York: priv. prntd., 1932.

Fitzgerald, Frank J. *I Saw a Century Blossom*. New York: Philosophical Library, 1983.

Godwin, John. *Alcatraz: 1868–1963*. Garden City, N.Y.: Doubleday, 1963.

Irey, Elmer L. *The Tax Dodgers: The Inside Story of the T-Men's War with America's Political and Underworld Hoodlums*. New York: Greenberg, 1948.

Johnston, James A. *Alcatraz Island Prison*. New York: Scribner's, 1949.

Kobler, John. *Capone: The Life and World of Al Capone*. New York: Putnam's, 1971.

Landesco, John. *Organized Crime in Chicago: Part III of the Illinois Crime Survey 1929*. Chicago: U. of Chicago, 1929; 1968 edition. (*NOTE*: this edition is bound separately from the other parts, so page references in the notes do not match the usual citations, which start near page 1000.)

Lyle, John H. (Judge). *The Dry and Lawless Years*. Englewood Cliffs, N.J.: Prentice-Hall, 1960. (*NOTE*: Ghostwritten without credit by John [Jack] McPhaul.)

McPhaul, Jack. *Johnny Torrio: First of the Gang Lords*. New Rochelle, N.Y.: Arlington House, 1970.
Moquin, Wayne (ed., with Charles Van Doren). *Documentary History of the Italian Americans*. New York: Praeger, 1974.
Murray, [Jesse] George. *The Legacy of Al Capone: Portraits and Annals of Chicago's Public Enemies*. New York: Putnam's, 1975.
Ness, Eliot, with Oscar Fraley. *The Untouchables*. New York: Julian Messner, 1957.
Pasley, Fred D. *Al Capone: The Biography of a Self-Made Man*. Pub. 1930; Reprint 1971; Salem, N.H.: Ayer, 1987.
Peterson, Virgil W. *Barbarians in Our Midst: A History of Chicago Crime and Politics*. Boston: Atlantic Monthly Press/Little Brown, 1952.
Pilat, Oliver, and Jo Ransom. *Sodom by the Sea: An Affectionate History of Coney Island*. Garden City, N.Y.: Doubleday, 1941.
Sinclair, Andrew. *Prohibition: The Era of Excess*. Boston: Little, Brown, 1962.
Stuart, William H. *The Twenty Incredible Years*. Chicago: M.A. Donohue, 1935.
Sullivan, Edward D. *Rattling the Cup on Chicago Crime*. New York: Vanguard, 1929.
Wendt, Lloyd, and Herman Kogan. *Big Bill of Chicago*. New York: Bobbs-Merrill, 1953.
———. *Lords of the Levee: The Story of Bathhouse John and Hinky Dink*. Garden City, N.Y.: Garden City Pub., 1944 (Original: New York: Bobbs-Merrill, 1943).
Weld, Ralph Foster. *Brooklyn Is America*. New York: Columbia U. Press, 1950.
Wilson, Frank J., and Beth Day. *Special Agent: A Quarter Century with the Treasury Department and the Secret Service*. New York: Holt, Rinehart & Winston, 1965.

MINOR, BACKGROUND and ANECDOTAL SOURCES

Asbury, Herbert. *The Gangs of New York: An Informal History of the Underworld*. Garden City, N.Y.: Garden City Pub., 1927, 1928.
Audett, James Henry (Blackie). *Rap Sheet: My Life Story*. New York: Sloane, 1954.
Barrymore, Ethel. *Memories: An Autobiography*. New York: Harper, 1955.
Bennett, James V. *I Chose Prison*. New York: Knopf, 1970.
Browne, Howard. *Pork City*. New York: St. Martin's Press/Joan Kahn, 1988. (*NOTE:* A novel about the Lingle murder; exhaustively researched.)
Busch, Francis X. *Enemies of the State*. New York: Bobbs-Merrill, 1954.
Casey, Robert J., and W.A.S. Douglas. *The Midwesterner: The Story of Dwight H. Green*. Chicago: Wilcox & Follett, 1948.
Cohn, Art. *The Joker Is Wild: The Story of Joe E. Lewis*. New York: Random House, 1955.
Condon, Eddie. *We Called It Music*. New York: Holt, 1947.
Cummings, Homer. *Selected Papers of Homer Cummings: Attorney General of the United States, 1933–1939*. New York: Scribner, 1939.
Dedmon, Emmett. *Fabulous Chicago* (Enlarged Edition). New York: Atheneum, 1981 (Original ed. 1953).
Drury, John. *Dining Out*. New York: John Day, 1931. (*NOTE*: A guide to Chicago restaurants; it gives much atmosphere.)
Feder, Sid, and Joachim Joesten. *The Luciano Story*. New York: McKay, 1954.
Fox, Stephen. *Blood and Power: Organized Crime in Twentieth-Century America*. New York: Morrow, 1989.
Gosch, Martin A., and Richard Hamner. *The Last Testament of Lucky Luciano*. Boston: Little, Brown, 1974.
Harris, Warren G. *Lucy and Desi*. New York: Simon and Schuster, 1991.
Helmer, William J. *The Gun That Made the Twenties Roar*. New York: Macmillan, 1969.

Hostetter, Gordon L., and Thomas Quinn Beesley. *It's a Racket!* Chicago: Les Quin Books, 1929.
Hynd, Alan. *The Giant Killers*. New York: Robert M. McBride, 1945. (*NOTE*: An early but almost unusably eccentric version of the Treasury's campaign; Malone is called De Angelo for no discernible reason, and another undercover agent, "Graziano," is invented.)
Jessel, George. *So Help Me*. New York: Random House, 1943.
Karpis, Alvin, with Bill Trent. *The Alvin Karpis Story*. New York: Coward, McCann & Geoghegan, 1971.
Kobler, John. *Ardent Spirits*. New York: Putnam's, 1973.
Kofoed, Jack. *Moon over Miami*. New York: Random House, 1955.
Levell, Mark, and Bill Helmer. *The Quotable Capone*. Chicago: priv. prntd., 1990.
Levi, Carlo. *Christ Stopped at Eboli*. New York: Farrar, Straus, 1947.
Lewis, Lloyd, and Henry Justin Smith. *Chicago: The History of Its Reputation*. New York: Harcourt, Brace, 1929. (*NOTE*: Each wrote half, and only Smith's is relevant.)
Liebling, A. J. *Chicago: The Second City*. New York: Knopf, 1952.
Lombardo, Anthony. *The Italians in America*. Chicago: Claretin, 1975.
Lynch, Denis Tilden. *Criminals and Politicians*. New York: Macmillan, 1932.
McCullough, Edo. *Good Old Coney Island: A Sentimental Journey into the Past*. New York: Scribner's, 1957.
McPhaul, John J. *Deadlines and Monkeyshines: The Fabled World of Chicago Journalism*. Englewood Cliffs, N.J.: Prentice-Hall, 1962.
Mariano, John Horace, Ph.D. *The Italian Contribution to American Democracy*. Boston: Christopher, 1921.
Messick, Hank. *The Silent Syndicate*. New York: Macmillan, 1967.
Messick, Hank, and Burt Goldblatt. *The Mobs and the Mafia: The Illustrated History of Organized Crime*. New York: Galahad Books, 1972.
Mezzrow, Milton, and Bernard Wolfe. *Really the Blues*. New York: Random House, 1946.
Meyer, George H., as told to Chaplain Ray and Max Call. *Al Capone's Devil Driver*. Dallas: Acclaimed Books, 1979.
Miller, Rita Seiden, ed. *Brooklyn, U.S.A.: Fourth Largest City in America*. Brooklyn, N.Y.: Brooklyn College Press, 1979.
Muir, Helen. *Miami, U.S.A.* New York: Holt, 1953.
Nelli, Humbert S. *Italians in Chicago, 1880–1930: A Study in Ethnic Mobility*. New York: Oxford U. Press, 1970.
———. *The Business of Crime: Italians and Syndicate Crime in the United States*. New York: Oxford U. Press, 1976.
Nickel, Steven. *Torso: The Story of Eliot Ness and the Search for a Psychopathic Killer*. Winston-Salem, N.C.: J. F. Blair, 1989.
Pasley, Fred D. *Muscling In*. New York: Ives Washburn, 1931.
Reckless, Walter C. *Vice in Chicago*. Montclair, N.J.: Patterson Smith, 1969 (Original, U. of Chicago, 1933).
Redford, Polly. *Billion-Dollar Sandbar: A Biography of Miami Beach*. New York: Dutton, 1970.
Report of the Senate Vice Committee. Springfield, Ill.: State of Illinois, 1916.
Richman, Harry, with Richard Gehman. *A Hell of a Life*. New York: Duell, Sloan and Pearce, 1966.
Riis, Jacob A. *How the Other Half Lives: Studies Among the Tenements of New York*. New York: Hill & Wang, 1957.
Ross, Robert. *The Trial of Al Capone*. Chicago: Robert Ross, 1933.
Rudensky, Morris (Red), and Don Riley. *The Gonif*. Blue Earth, Minn.: Piper, 1970.

Schmidt, John R. *The Mayor Who Cleaned Up Chicago: A Political Biography of William E. Dever*. De Kalb, Ill.: No. Ill. U. Press, 1989.

Slayton, Robert A. *Back of the Yards: The Making of a Local Democracy*. Chicago: U. of Chicago Press, 1986.

Smith, Robert P. *Brooklyn at Play*. New York: Revisionist Press, 1976.

Snow, Richard. *Coney Island: A Postcard Journey to the City of Fire*. New York: Brightwater Press, 1984.

Spiering, Frank. *The Man Who Got Capone*. New York: Bobbs-Merrill, 1976.

St. John, Robert. *This Was My World*. Garden City, N.Y.: Doubleday, 1953.

Stella, Antonio, M.D. *Some Aspects of Italian Immigration to the United States: Statistical Data and General Considerations Based Chiefly Upon the United States Censuses and Other Official Publications*. San Francisco: R and E Research Associates, 1970 (Original: New York: Putnam's, 1924).

Sullivan, Edward Dean. *Chicago Surrenders*. New York: Vanguard Press, 1930.

Touhy, Roger, with Ray Brennan. *The Stolen Years*. Cleveland: Pennington Press, 1959.

Vanderbilt, Cornelius, Jr. *Farewell to Fifth Avenue*. New York: Simon & Schuster, 1935.

Waller, Irle. *Chicago Uncensored: Firsthand Stories About the Al Capone Era*. New York: Exposition Press, 1965. (*NOTE:* Despite the title, almost nothing: visits to Kenna's bar for a 5¢, 14-oz. schooner of beer and to the black and tan over the Pekin Inn where Morton and Miller killed the police and that's all.)

Wickersham Report: Seventy-first Congress; Enforcement of the Prohibition Laws: Official Records of The National Commission on Law Observance and Enforcement; A Prohibition Survey of the State of Illinois; Part II, Vol. 4. Washington, D.C.: U.S. Printing Office, 1931. (*NOTE*: Report of Guy L. Nichols, Bureau of Prohibition, Treasury Department.)

Willensky, Elliott. *When Brooklyn Was the World: 1920–1957*. New York: Harmony Books, 1986.

X Marks the Spot (Anonymous). The Spot Pub. Co., 1930.

Index